Teach Yourself Electricity and Electronics

Teach Yourself Electricity and Electronics

Fourth Edition

Stan Gibilisco

McGraw-Hill

New York   Chicago   San Francisco   Lisbon   London   Madrid
Mexico City   Milan   New Delhi   San Juan   Seoul
Singapore   Sydney   Toronto

Cataloging-in-Publication Data is on file with the Library of Congress.

1 2 3 4 5 6 7 8 9 0   DOC/DOC   0 1 0 9 8 7 6

ISBN 0-07-145933-2

The sponsoring editor for this book was Judy Bass, the editing supervisor was David E. Fogarty, and the production supervisor was Pamela A. Pelton. It was set in Garamond by North Market Street Graphics. The art director for the cover was Anthony Landi.

Printed and bound by RR Donnelley.

This book was printed on acid-free paper.

McGraw-Hill books are available at special quantity discounts to use as premiums and sales promotions, or for use in corporate training programs. For more information, please write to the Director of Special Sales, McGraw-Hill Professional, McGraw-Hill, Two Penn Plaza, New York, NY 10121-2298. Or contact your local bookstore.

To Tony, Samuel, Tim, Roland, Jack, and Sherri

Contents

Preface

This book is for people who want to learn the fundamentals of electricity, electronics, and related fields without taking a formal course. The book can also serve as a classroom text. This edition contains new material on transducers, sensors, antennas, monitoring, security, and navigation. Material from previous editions has been updated where appropriate.

As you take this course, you'll encounter hundreds of quiz, test, and exam questions that can help you measure your progress. They are written like the questions found in standardized tests used by educational institutions.

There is a short multiple-choice quiz at the end of every chapter. The quizzes are "open-book." You may refer to the chapter texts when taking them. When you have finished a chapter, take the quiz, write down your answers, and then give your list of answers to a friend. Have the friend tell you your score, but not which questions you got wrong. Because you're allowed to look at the text when taking the quizzes, some of the questions are rather difficult.

At the end of each section, there is a multiple-choice test. These tests are easier than chapter-ending quizzes. Don't look back at the text when taking the tests. A satisfactory score is at least three-quarters of the answers correct.

You will find a final exam at the end of this course. As with the section-ending tests, the questions are not as difficult as those in the chapter-ending quizzes. Don't refer back to the text while taking the final exam. A satisfactory score is at least three-quarters of the answers correct.

The answers to all of the multiple-choice quiz, test, and exam questions are listed in an appendix at the back of this book.

You don't need a mathematical or scientific background for this course. Middle-school algebra, geometry, and physics will suffice. There's no calculus here! I recommend that you complete one chapter a week. That way, in a few months, you'll finish the course. You can then use this book, with its comprehensive index, as a permanent reference.

Suggestions for future editions are welcome.

Stan Gibilisco

1
PART

Direct Current

1
CHAPTER

Basic Physical Concepts

IT IS IMPORTANT TO UNDERSTAND SOME SIMPLE, GENERAL PHYSICS PRINCIPLES IN ORDER TO HAVE A full grasp of electricity and electronics. It is not necessary to know high-level mathematics. In science, you can talk about *qualitative* things or *quantitative* things, the "what" versus the "how much." For now, we are concerned only about the "what." The "how much" will come later.

Atoms

All matter is made up of countless tiny particles whizzing around. These particles are extremely dense; matter is mostly empty space. Matter seems continuous because the particles are so small, and they move incredibly fast.

Each chemical *element* has its own unique type of particle, known as its *atom*. Atoms of different elements are always different. The slightest change in an atom can make a tremendous difference in its behavior. You can live by breathing pure oxygen, but you can't live off of pure nitrogen. Oxygen will cause metal to corrode, but nitrogen will not. Wood will burn furiously in an atmosphere of pure oxygen, but will not even ignite in pure nitrogen. Yet both are gases at room temperature and pressure; both are colorless, both are odorless, and both are just about of equal weight. These substances are so different because oxygen has eight *protons*, while nitrogen has only seven. There are many other examples in nature where a tiny change in atomic structure makes a major difference in the way a substance behaves.

Protons, Neutrons, and Atomic Numbers

The part of an atom that gives an element its identity is the *nucleus*. It is made up of two kinds of particles, the *proton* and the *neutron*. These are extremely dense. A teaspoonful of either of these particles, packed tightly together, would weigh tons. Protons and neutrons have just about the same mass, but the proton has an electric charge while the neutron does not.

The simplest element, hydrogen, has a nucleus made up of only one proton; there are usually no neutrons. This is the most common element in the universe. Sometimes a nucleus of hydrogen has a neutron or two along with the proton, but this does not occur very often. These "mutant" forms of hydrogen do, nonetheless, play significant roles in atomic physics.

The second most abundant element is helium. Usually, this atom has a nucleus with two protons and two neutrons. Hydrogen is changed into helium inside the sun, and in the process, energy is given off. This makes the sun shine. The process, called *fusion*, is also responsible for the terrific explosive force of a hydrogen bomb.

Every proton in the universe is just like every other. Neutrons are all alike, too. The number of protons in an element's nucleus, the *atomic number*, gives that element its identity. The element with three protons is lithium, a light metal that reacts easily with gases such as oxygen or chlorine. The element with four protons is beryllium, also a metal. In general, as the number of protons in an element's nucleus increases, the number of neutrons also increases. Elements with high atomic numbers, like lead, are therefore much denser than elements with low atomic numbers, like carbon. Perhaps you've compared a lead sinker with a piece of coal of similar size, and noticed this difference.

Isotopes and Atomic Weights

For a given element, such as oxygen, the number of neutrons can vary. But no matter what the number of neutrons, the element keeps its identity, based on the atomic number. Differing numbers of neutrons result in various *isotopes* for a given element.

Each element has one particular isotope that is most often found in nature. But all elements have numerous isotopes. Changing the number of neutrons in an element's nucleus results in a difference in the weight, and also a difference in the density, of the element. Thus, hydrogen containing a neutron or two in the nucleus, along with the proton, is called *heavy hydrogen*.

The *atomic weight* of an element is approximately equal to the sum of the number of protons and the number of neutrons in the nucleus. Common carbon has an atomic weight of about 12, and is called carbon 12 or C12. But sometimes it has an atomic weight of about 14, and is known as carbon 14 or C14.

Electrons

Surrounding the nucleus of an atom are particles having opposite electric charge from the protons. These are the *electrons*. Physicists arbitrarily call the electrons' charge *negative*, and the protons' charge *positive*. An electron has exactly the same charge quantity as a proton, but with opposite polarity. The charge on a single electron or proton is the smallest possible electric charge. All charges, no matter how great, are multiples of this unit charge.

One of the earliest ideas about the atom pictured the electrons embedded in the nucleus, like raisins in a cake. Later, the electrons were seen as orbiting the nucleus, making the atom like a miniature solar system with the electrons as the planets (Fig. 1-1). Still later, this view was modified further. Today, the electrons are seen as so fast-moving, with patterns so complex, that it is not even possible to pinpoint them at any given instant of time. All that can be done is to say that an electron will just as likely be inside a certain sphere as outside. These spheres are known as electron *shells*. Their centers correspond to the position of the atomic nucleus. The farther away from the nucleus the *shell*, the more energy the electron has (Fig. 1-2).

1-1 An early model of the atom, developed around the year 1900, resembled a miniature solar system. The electrons were held in their orbits around the nucleus by electrostatic attraction.

Electrons can move rather easily from one atom to another in some materials. In other substances, it is difficult to get electrons to move. But in any case, it is far easier to move electrons than it is to move protons. Electricity almost always results, in some way, from the motion of electrons in a material. Electrons are much lighter than protons or neutrons. In fact, compared to the nucleus of an atom, the electrons weigh practically nothing.

Generally, the number of electrons in an atom is the same as the number of protons. The negative charges therefore exactly cancel out the positive ones, and the atom is electrically neutral. But

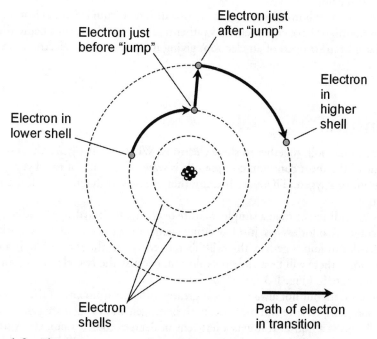

1-2 Electrons move around the nucleus of an atom at defined levels, called *shells*, which correspond to discrete energy states. This is a simplified illustration of an electron gaining energy within an atom.

under some conditions, there can be an excess or shortage of electrons. High levels of radiant energy, extreme heat, or the presence of an electric field (discussed later) can "knock" or "throw" electrons loose from atoms, upsetting the balance.

Ions

If an atom has more or less electrons than protons, that atom acquires an electrical charge. A shortage of electrons results in positive charge; an excess of electrons gives a negative charge. The element's identity remains the same, no matter how great the excess or shortage of electrons. In the extreme case, all the electrons might be removed from an atom, leaving only the nucleus. However, it would still represent the same element as it would if it had all its electrons. A charged atom is called an *ion*. When a substance contains many ions, the material is said to be *ionized*.

A good example of an ionized substance is the atmosphere of the earth at high altitudes. The ultraviolet radiation from the sun, as well as high-speed subatomic particles from space, result in the gases' atoms being stripped of electrons. The ionized gases tend to be found in layers at certain altitudes. These layers are responsible for long-distance radio communications at some frequencies.

Ionized materials generally conduct electricity well, even if the substance is normally not a good conductor. Ionized air makes it possible for a lightning stroke to take place, for example. The ionization, caused by a powerful electric field, occurs along a jagged, narrow channel. After the lightning flash, the nuclei of the atoms quickly attract stray electrons back, and the air becomes electrically neutral again.

An element might be both an ion and an isotope different from the usual isotope. For example, an atom of carbon might have eight neutrons rather than the usual six, thus being the isotope C14, and it might have been stripped of an electron, giving it a positive unit electric charge and making it an ion.

Compounds

Different elements can join together to share electrons. When this happens, the result is a chemical *compound*. One of the most common compounds is water, the result of two hydrogen atoms joining with an atom of oxygen. There are literally thousands of different chemical compounds that occur in nature.

A compound is different than a simple mixture of elements. If hydrogen and oxygen are mixed, the result is a colorless, odorless gas, just like either element is a gas separately. A spark, however, will cause the molecules to join together; this will liberate energy in the form of light and heat. Under the right conditions, there will be a violent explosion, because the two elements join eagerly. Water is chemically illustrated in Fig. 1-3.

Compounds often, but not always, appear greatly different from any of the elements that make them up. At room temperature and pressure, both hydrogen and oxygen are gases. But water under the same conditions is a liquid. If it gets a few tens of degrees colder, water turns solid at standard pressure. If it gets hot enough, water becomes a gas, odorless and colorless, just like hydrogen or oxygen.

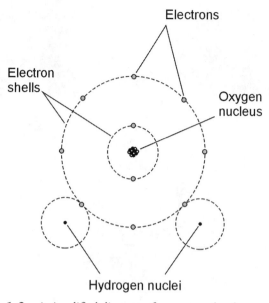

1-3 A simplified diagram of a water molecule.
Note the shared electrons.

Another common example of a compound is rust. This forms when iron joins with oxygen. While iron is a dull gray solid and oxygen is a gas, rust is a maroon-red or brownish powder, completely unlike either of the elements from which it is formed.

Molecules

When atoms of elements join together to form a compound, the resulting particles are *molecules*. Figure 1-3 is an example of a molecule of water, consisting of three atoms put together.

The natural form of an element is also known as its molecule. Oxygen tends to occur in pairs most of the time in the earth's atmosphere. Thus, an oxygen molecule is sometimes denoted by the symbol O_2. The "O" represents oxygen, and the subscript 2 indicates that there are two atoms per molecule. The water molecule is symbolized H_2O, because there are two atoms of hydrogen and one atom of oxygen in each molecule.

Sometimes oxygen atoms exist all by themselves; then we denote the molecule simply as O. Sometimes there are three atoms of oxygen grouped together. This is the gas called *ozone*, which has received much attention lately in environmental news. It is written O_3.

All matter, whether solid, liquid, or gas, is made of molecules. These particles are always moving. The speed with which they move depends on the temperature. The hotter the temperature, the more rapidly the molecules move around. In a solid, the molecules are interlocked in a sort of rigid pattern, although they vibrate continuously (Fig. 1-4A). In a liquid, they slither and slide around (Fig. 1-4B). In a gas, they rush all over the place, bumping into each other and into solids and liquids adjacent to the gas (Fig. 1-4C).

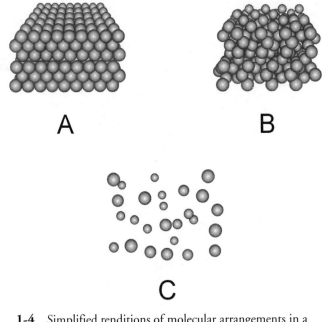

1-4   Simplified renditions of molecular arrangements in a solid (A), a liquid (B), and a gas (C).

Conductors

In some materials, electrons move easily from atom to atom. In others, the electrons move with difficulty. And in some materials, it is almost impossible to get them to move. An electrical *conductor* is a substance in which the electrons are mobile.

The best conductor at room temperature is pure elemental silver. Copper and aluminum are also excellent electrical conductors. Iron, steel, and various other metals are fair to good conductors of electricity. In most electrical circuits and systems, copper or aluminum wire is used. (Silver is impractical because of its high cost.)

Some liquids are good electrical conductors. Mercury is one example. Salt water is a fair conductor. Gases or mixtures of gases, such as air, are generally poor conductors of electricity. This is because the atoms or molecules are usually too far apart to allow a free exchange of electrons. But if a gas becomes ionized, it can be a fair conductor of electricity.

Electrons in a conductor do not move in a steady stream, like molecules of water through a garden hose. Instead, they are passed from one atom to another right next to it (Fig. 1-5). This happens to countless atoms all the time. As a result, literally trillions of electrons pass a given point each second in a typical electrical circuit.

Insulators

An *insulator* prevents electrical currents from flowing, except occasionally in tiny amounts. Most gases are good electrical insulators. Glass, dry wood, paper, and plastics are other examples. Pure water is a

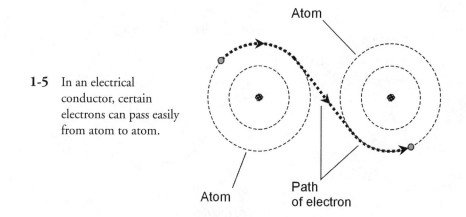

1-5 In an electrical conductor, certain electrons can pass easily from atom to atom.

good electrical insulator, although it conducts some current with even the slightest impurity. Metal oxides can be good insulators, even though the metal in pure form is a good conductor.

Electrical insulators can be forced to carry current. Ionization can take place; when electrons are stripped away from their atoms, they move more or less freely. Sometimes an insulating material gets charred, or melts down, or gets perforated by a spark. Then its insulating properties are lost, and some electrons flow. An insulating material is sometimes called a *dielectric.* This term arises from the fact that it keeps electrical charges apart, preventing the flow of electrons that would equalize a charge difference between two places. Excellent insulating materials can be used to advantage in certain electrical components such as capacitors, where it is important that electrons not flow.

Porcelain or glass can be used in electrical systems to keep short circuits from occurring. These devices, called insulators, come in various shapes and sizes for different applications. You can see them on high-voltage utility poles and towers. They hold the wire up without running the risk of a short circuit with the tower or a slow discharge through a wet wooden pole.

Resistors

Some substances, such as carbon, conduct electricity fairly well but not really well. The conductivity can be changed by adding impurities like clay to a carbon paste, or by winding a thin wire into a coil. Electrical components made in this way are called *resistors.* They are important in electronic circuits because they allow for the control of current flow. The better a resistor conducts, the lower its *resistance;* the worse it conducts, the higher the resistance.

Electrical resistance is measured in units called *ohms.* The higher the value in ohms, the greater the resistance, and the more difficult it becomes for current to flow. For wires, the resistance is sometimes specified in terms of ohms per unit length (foot, meter, kilometer, or mile). In an electrical system, it is usually desirable to have as low a resistance, or ohmic value, as possible. This is because resistance converts electrical energy into heat.

Semiconductors

In a *semiconductor,* electrons flow, but not as well as they do in a conductor. Some semiconductors carry electrons almost as well as good electrical conductors like copper or aluminum; others are almost as bad as insulating materials.

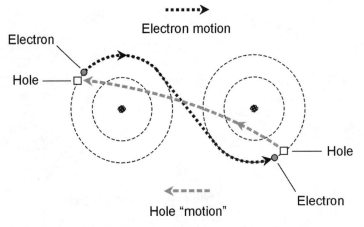

1-6 In a semiconducting material, holes travel in a direction opposite to the direction in which the electrons travel.

Semiconductors are not the same as resistors. In a semiconductor, the material is treated so that it has very special properties.

Semiconductors include certain substances such as silicon, selenium, or gallium, that have been "doped" by the addition of impurities such as indium or antimony. Have you heard of such things as *gallium arsenide, metal oxides,* or *silicon rectifiers?* Electrical conduction in these materials is always a result of the motion of electrons. But this can be a quite peculiar movement, and sometimes engineers speak of the movement of *holes* rather than electrons. A hole is a shortage of an electron—you might think of it as a positive ion—and it moves along in a direction opposite to the flow of electrons (Fig. 1-6).

When most of the *charge carriers* are electrons, the semiconductor is called *N-type,* because electrons are negatively charged. When most of the charge carriers are holes, the semiconductor material is known as *P-type* because holes have a positive electric charge. But P-type material does pass some electrons, and N-type material carries some holes. In a semiconductor, the more abundant type of charge carrier is called the *majority carrier.* The less abundant kind is known as the *minority carrier.* Semiconductors are used in *diodes, transistors,* and *integrated circuits.* These substances are what make it possible for you to have a computer or a television receiver in a package small enough to hold in your hand.

Current

Whenever there is movement of charge carriers in a substance, there is an electric *current.* Current is measured in terms of the number of electrons or holes passing a single point in 1 second.

A great many charge carriers go past any given point in 1 second, even if the current is small. In a household electric circuit, a 100-watt light bulb draws a current of about *six quintillion* (6 followed by 18 zeros) charge carriers per second. Even the smallest bulb carries *quadrillions* (numbers followed by 15 zeros) of charge carriers every second. It is impractical to speak of a current in terms of charge carriers per second, so it is measured in *coulombs per second* instead. A coulomb is equal to approximately 6,240,000,000,000,000,000 electrons or holes. A current of 1 coulomb per second is called an *ampere*, and this is the standard unit of electric current. A 100-watt bulb in your desk lamp draws about 1 ampere of current.

When a current flows through a resistance—and this is always the case because even the best conductors have resistance—heat is generated. Sometimes light and other forms of energy are emitted as well. A light bulb is deliberately designed so that the resistance causes visible light to be generated.

Electric current flows at high speed through any conductor, resistor, or semiconductor. Nevertheless, it is considerably less than the speed of light.

Static Electricity

Charge carriers, particularly electrons, can build up, or become deficient, on things without flowing anywhere. You've experienced this when walking on a carpeted floor during the winter, or in a place where the humidity was low. An excess or shortage of electrons is created on and in your body. You acquire a *charge* of *static electricity.* It's called "static" because it doesn't go anywhere. You don't feel this until you touch some metallic object that is connected to earth ground or to some large fixture; but then there is a *discharge*, accompanied by a spark.

If you were to become much more charged, your hair would stand on end, because every hair would repel every other. Like charges are caused either by an excess or a deficiency of electrons; they repel. The spark might jump an inch, 2 inches, or even 6 inches. Then it would more than startle you; you could get hurt. This doesn't happen with ordinary carpet and shoes, fortunately. But a device called a *Van de Graaff generator*, found in physics labs, can cause a spark this large (Fig. 1-7). Be careful when using this device for physics experiments!

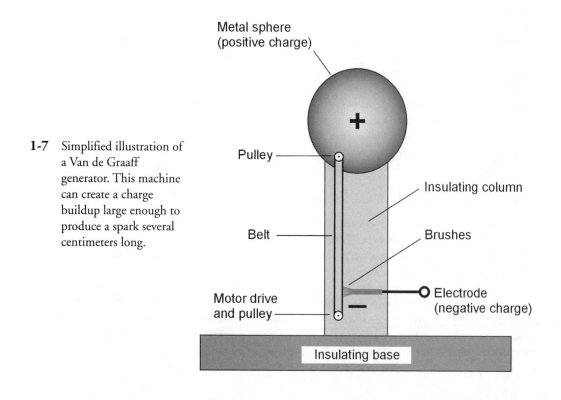

1-7 Simplified illustration of a Van de Graaff generator. This machine can create a charge buildup large enough to produce a spark several centimeters long.

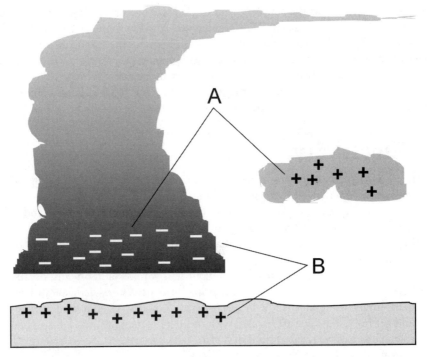

1-8 Electrostatic charges can build up between clouds in a thunderstorm (A), or between a cloud and the surface of the earth (B).

In the extreme, lightning occurs between clouds, and between clouds and ground in the earth's atmosphere. This spark, called a *stroke*, is a magnified version of the spark you get after shuffling around on a carpet. Until the stroke occurs, there is a static charge in the clouds, between different clouds or parts of a cloud, and the ground. In Fig. 1-8, cloud-to-cloud (A) and cloud-to-ground (B) static buildups are shown. In the case at B, the positive charge in the earth follows along beneath the storm cloud. The current in a lightning stroke is usually several tens of thousands, or hundreds of thousands, of amperes. But it takes place only for a fraction of a second. Still, many coulombs of charge are displaced in a single bolt of lightning.

Electromotive Force

Current can only flow if it gets a "push." This can be caused by a buildup of static electric charges, as in the case of a lightning stroke. When the charge builds up, with positive polarity (shortage of electrons) in one place and negative polarity (excess of electrons) in another place, a powerful *electromotive force* (EMF) exists. This force is measured in units called *volts*.

Ordinary household electricity has an effective voltage of between 110 and 130; usually it is about 117. A car battery has an EMF of 12 to 14 volts. The static charge that you acquire when walking on a carpet with hard-soled shoes is often several thousand volts. Before a discharge of lightning, millions of volts exist. An EMF of 1 volt, across a resistance of 1 ohm, will cause a current of 1 ampere to flow. This is a classic relationship in electricity, and is stated generally as *Ohm's Law*. If the EMF is doubled, the current is doubled. If the resistance is doubled, the current is cut in half. This important law of electrical circuit behavior is covered in detail later in this book.

It is possible to have an EMF without having any current. This is the case just before a lightning stroke occurs, and before you touch a metal object after walking on a carpet. It is also true between the two wires of an electric lamp when the switch is turned off. It is true of a dry cell when there is nothing connected to it. There is no current, but a current is possible given a conductive path between the two points. Voltage, or EMF, is sometimes called *potential* or *potential difference* for this reason.

Even a huge EMF does not necessarily drive much current through a conductor or resistance. A good example is your body after walking around on the carpet. Although the voltage seems deadly in terms of numbers (thousands), there are not many coulombs of static-electric charge that can accumulate on an object the size of your body. Therefore, in relative terms, not that many electrons flow through your finger when you touch a radiator. This is why you don't get a severe shock.

If there are plenty of coulombs available, a small voltage, such as 117 volts (or even less) can cause a lethal current. This is why it is dangerous to repair an electrical device with the power on. The power plant will pump an unlimited number of coulombs of charge through your body if you are not careful.

Nonelectrical Energy

In electricity and electronics, there are phenomena that involve other forms of energy besides electrical energy. Visible light is an example. A light bulb converts electricity into radiant energy that you can see. This was one of the major motivations for people like Thomas Edison to work with electricity. Visible light can also be converted into electric current or voltage. A *photovoltaic cell* does this.

Light bulbs always give off some heat, as well as visible light. Incandescent lamps actually give off more energy as heat than as light. You are certainly acquainted with electric heaters, designed for the purpose of changing electricity into heat energy. This heat is a form of radiant energy called *infrared* (IR). It is similar to visible light, except that the waves are longer and you can't see them.

Electricity can be converted into other radiant-energy forms, such as *radio waves*, *ultraviolet* (UV), and *X rays*. This is done by specialized devices such as radio transmitters, sunlamps, and electron tubes. Fast-moving protons, neutrons, electrons, and atomic nuclei are an important form of energy. The energy from these particles is sometimes sufficient to split atoms apart. This effect makes it possible to build an atomic reactor whose energy can be used to generate electricity.

When a conductor moves in a magnetic field, electric current flows in that conductor. In this way, mechanical energy is converted into electricity. This is how an *electric generator* works. Generators can also work backward. Then you have a *motor* that changes electricity into useful mechanical energy.

A magnetic field contains energy of a unique kind. The science of *magnetism* is closely related to electricity. Magnetic phenomena are of great significance in electronics. The oldest and most universal source of magnetism is the *geomagnetic field* surrounding the earth, caused by alignment of iron atoms in the core of the planet.

A changing magnetic field creates a fluctuating electric field, and a fluctuating electric field produces a changing magnetic field. This phenomenon, called *electromagnetism*, makes it possible to send wireless signals over long distances. The electric and magnetic fields keep producing one another over and over again through space.

Chemical energy is converted into electricity in *dry cells*, *wet cells*, and *batteries*. Your car battery is an excellent example. The acid reacts with the metal electrodes to generate an electromotive force. When the two poles of the batteries are connected, current results. The chemical reaction continues, keeping the current going for a while. But the battery can only store a certain amount of chemical energy. Then it "runs out of juice," and the supply of chemical energy must be restored by *charging*. Some cells and batteries, such as lead-acid car batteries, can be recharged by driving current through them, and others, such as most flashlight and transistor-radio batteries, cannot.

Quiz

Refer to the text in this chapter if necessary. A good score is at least 18 correct answers out of these 20 questions. The answers are listed in the back of this book.

1. The atomic number of an element is determined by
   (a) the number of neutrons.
   (b) the number of protons.
   (c) the number of neutrons plus the number of protons.
   (d) the number of electrons.

2. The atomic weight of an element is approximately determined by
   (a) the number of neutrons.
   (b) the number of protons.
   (c) the number of neutrons plus the number of protons.
   (d) the number of electrons.

3. Suppose there is an atom of oxygen, containing eight protons and eight neutrons in the nucleus, and two neutrons are added to the nucleus. What is the resulting atomic weight?
   (a) 8
   (b) 10
   (c) 16
   (d) 18

4. An ion
   (a) is electrically neutral.
   (b) has positive electric charge.
   (c) has negative electric charge.
   (d) can have either a positive or negative charge.

5. An isotope
   (a) is electrically neutral.
   (b) has positive electric charge.
   (c) has negative electric charge.
   (d) can have either a positive or negative charge.

6. A molecule
    (a) can consist of a single atom of an element.
    (b) always contains two or more elements.
    (c) always has two or more atoms.
    (d) is always electrically charged.

7. In a compound,
    (a) there can be a single atom of an element.
    (b) there must always be two or more elements.
    (c) the atoms are mixed in with each other but not joined.
    (d) there is always a shortage of electrons.

8. An electrical insulator can be made a conductor
    (a) by heating it.
    (b) by cooling it.
    (c) by ionizing it.
    (d) by oxidizing it.

9. Of the following substances, the worst conductor is
    (a) air.
    (b) copper.
    (c) iron.
    (d) salt water.

10. Of the following substances, the best conductor is
    (a) air.
    (b) copper.
    (c) iron.
    (d) salt water.

11. Movement of holes in a semiconductor
    (a) is like a flow of electrons in the same direction.
    (b) is possible only if the current is high enough.
    (c) results in a certain amount of electric current.
    (d) causes the material to stop conducting.

12. If a material has low resistance, then
    (a) it is a good conductor.
    (b) it is a poor conductor.
    (c) the current flows mainly in the form of holes.
    (d) current can flow only in one direction.

13. A coulomb
    (a) represents a current of 1 ampere.
    (b) flows through a 100-watt light bulb.
    (c) is equivalent to 1 ampere per second.
    (d) is an extremely large number of charge carriers.

14. A stroke of lightning
    (a) is caused by a movement of holes in an insulator.
    (b) has a very low current.
    (c) is a discharge of static electricity.
    (d) builds up between clouds.

15. The volt is the standard unit of
    (a) current.
    (b) charge.
    (c) electromotive force.
    (d) resistance.

16. If an EMF of 1 volt is placed across a resistance of 2 ohms, then the current is
    (a) half an ampere.
    (b) 1 ampere.
    (c) 2 amperes.
    (d) impossible to determine.

17. A backward-working electric motor, in which mechanical rotation is converted to electricity, is best described as
    (a) an inefficient, energy-wasting device.
    (b) a motor with the voltage connected the wrong way.
    (c) an electric generator.
    (d) a magnetic field.

18. In a battery, chemical energy can sometimes be replenished by
    (a) connecting it to a light bulb.
    (b) charging it.
    (c) discharging it.
    (d) no means known; when a battery is dead, you must throw it away.

19. A fluctuating magnetic field
    (a) produces an electric current in an insulator.
    (b) magnetizes the earth.
    (c) produces a fluctuating electric field.
    (d) results from a steady electric current.

20. Visible light is converted into electricity
    (a) in a dry cell.
    (b) in a wet cell.
    (c) in an incandescent bulb.
    (d) in a photovoltaic cell.

2
CHAPTER

Electrical Units

THIS CHAPTER EXPLAINS, IN MORE DETAIL, STANDARD UNITS THAT DEFINE THE BEHAVIOR OF DIRECT-current (dc) circuits. Many of these rules also apply to utility alternating-current (ac) circuits.

The Volt

In Chap. 1, you learned a little about the volt, the standard unit of electromotive force (EMF) or *potential difference.*

An accumulation of electrostatic charge, such as an excess or shortage of electrons, is always associated with a voltage. There are other situations in which voltages exist. Voltage can be generated at a power plant, produced in an electrochemical reaction, or caused by light rays striking a semiconductor chip. It can be produced when an object is moved in a magnetic field, or is placed in a fluctuating magnetic field.

A potential difference between two points produces an *electric field*, represented by *electric lines of flux* (Fig. 2-1). There is a pole that is relatively positive, with fewer electrons, and one that is relatively negative, with more electrons. The positive pole does not necessarily have a deficiency of electrons compared with neutral objects, and the negative pole does not always have a surplus of electrons relative to neutral objects. But the negative pole always has more electrons than the positive pole.

The abbreviation for volt (or volts) is V. Sometimes, smaller units are used. The *millivolt* (mV) is equal to a thousandth (0.001) of a volt. The *microvolt* (μV) is equal to a millionth (0.000001) of a volt. It is sometimes necessary to use units larger than the volt. One *kilovolt* (kV) is one thousand volts (1000 V). One *megavolt* (MV) is 1 million volts (1,000,000 V) or one thousand kilovolts (1000 kV).

In a dry cell, the voltage is usually between 1.2 and 1.7 V; in a car battery, it is 12 to 14 V. In household utility wiring, it is a low-frequency alternating current of about 117 V for electric lights and most appliances, and 234 V for a washing machine, dryer, oven, or stove. In television sets, transformers convert 117 V to around 450 V for the operation of the picture tube. In some broadcast transmitters, the voltage can be several kilovolts.

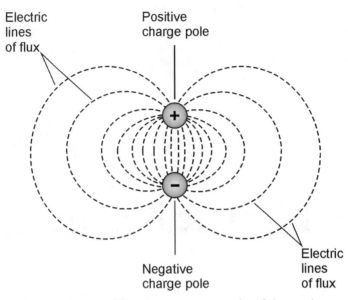

2-1   Electric lines of flux always exist near poles of electric charge.

The largest voltages on our planet occur between clouds, or between clouds and the ground, in thundershowers. This potential difference can build up to several tens of megavolts. The existence of a voltage always means that *charge carriers*, which are electrons in a conventional circuit, flow between two points if a conductive path is provided. Voltage represents the driving force that impels charge carriers to move. If all other factors are held constant, high voltages produce a faster flow of charge carriers, and therefore larger currents, than low voltages. But that's an oversimplification in most real-life scenarios, where other factors are hardly ever constant!

Current Flow

If a conducting or semiconducting path is provided between two poles having a potential difference, charge carriers flow in an attempt to equalize the charge between the poles. This flow of *current* continues as long as the path is provided, and as long as there is a charge difference between the poles.

Sometimes the charge difference is equalized after a short while. This is the case, for example, when you touch a radiator after shuffling around on the carpet while wearing hard-soled shoes. It is also true in a lightning stroke. In these instances, the charge is equalized in a fraction of a second. In other cases, the charge takes longer to be used up. This happens if you short-circuit a dry cell. Within a few minutes, the cell "runs out of juice" if you put a wire between the positive and negative terminals. If you put a bulb across the cell, say with a flashlight, it takes an hour or two for the charge difference to drop to zero.

In household electric circuits, the charge difference is never equalized, unless there's a power failure. Of course, if you short-circuit an outlet (don't!), the fuse or breaker will blow or trip, and the charge difference will immediately drop to zero. But if you put a 100-watt bulb at the outlet, the charge difference will be maintained as the current flows. The power plant can keep a potential difference across a lot of light bulbs indefinitely.

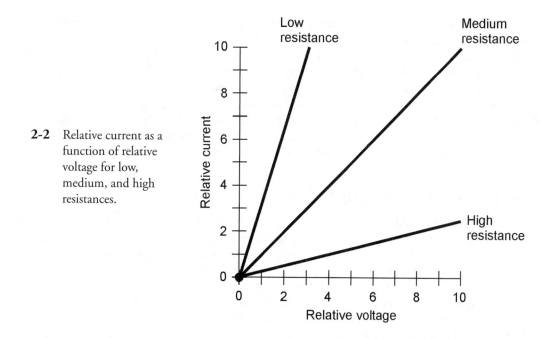

2-2 Relative current as a function of relative voltage for low, medium, and high resistances.

Have you heard that it is current, not voltage, that kills? This is a literal truth, but it plays on semantics. It's like saying "It's the heat, not the fire, that burns you." Naturally! But there can only be a deadly current if there is enough voltage to drive it through your body. You don't have to worry when handling flashlight cells, but you'd better be extremely careful around household utility circuits. A voltage of 1.2 to 1.7 V can't normally pump a dangerous current through you, but a voltage of 117 V almost always can.

In an electric circuit that always conducts equally well, the current is directly proportional to the applied voltage. If you double the voltage, you double the current. If the voltage is cut in half, the current is cut in half too. Figure 2-2 shows this relationship as a graph in general terms. It assumes that the power supply can provide the necessary number of charge carriers.

The Ampere

Current is a measure of the rate at which charge carriers flow. The standard unit is the *ampere*. This represents one coulomb (6,240,000,000,000,000,000) of charge carriers flowing every second past a given point.

An ampere is a comparatively large amount of current. The abbreviation is A. Often, current is specified in terms of *milliamperes*, abbreviated mA, where 1 mA = 0.001 A, or a thousandth of an ampere. You will also sometimes hear of *microamperes* (μA), where 1 μA = 0.000001 A or 0.001 mA, which is a millionth of an ampere. It is increasingly common to hear about *nanoamperes* (nA), where 1 nA = 0.001 μA = 0.000000001 A, which is a thousandth of a millionth of an ampere.

A current of a few milliamperes will give you a startling shock. About 50 mA will jolt you severely, and 100 mA can cause death if it flows through your chest cavity. An ordinary 100-watt light bulb draws about 1 A of current in a household utility circuit. An electric iron draws approximately

10 A; an entire household normally uses between 10 and 50 A, depending on the size of the house and the kinds of appliances it has, and also on the time of day, week, or year.

The amount of current that flows in an electrical circuit depends on the voltage, and also on the *resistance.* There are some circuits in which extremely large currents, say 1000 A, can flow. This will happen through a metal bar placed directly at the output of a massive electric generator. The resistance is extremely low in this case, and the generator is capable of driving huge numbers of charge carriers through the bar every second. In some semiconductor electronic devices, such as microcomputers, a few nanoamperes will suffice for many complicated processes. Some electronic clocks draw so little current that their batteries last as long as they would if left on the shelf without being put to any use.

Resistance and the Ohm

Resistance is a measure of the opposition that a circuit offers to the flow of electric current. You can compare it to the diameter of a hose. In fact, for metal wire, this is an excellent analogy: small-diameter wire has high resistance (a lot of opposition to current), and large-diameter wire has low resistance (not much opposition to current). The type of metal makes a difference too. For example, steel wire has higher resistance for a given diameter than copper wire.

The standard unit of resistance is the *ohm.* This is sometimes symbolized by the uppercase Greek letter omega (Ω). You'll sometimes hear about *kilohms* (symbolized k or kΩ), where 1 kΩ = 1000 Ω, or about *megohms* (symbolized M or MΩ), where 1 MΩ = 1000 kΩ = 1,000,000 Ω.

Electric wire is sometimes rated for *resistivity.* The standard unit for this purpose is the *ohm per foot* (ohm/ft or Ω/ft) or the *ohm per meter* (ohm/m or Ω/m). You might also come across the unit *ohm per kilometer* (ohm/km or Ω/km). Table 2-1 shows the resistivity for various common sizes of solid copper wire at room temperature, as a function of the wire size as defined by a scheme known as the *American Wire Gauge* (AWG).

Table 2-1.   Approximate resistivity at room temperature for solid copper wire as a function of the wire size in American Wire Gauge (AWG).

Wire size, AWG #	Resistivity, ohms/km
2	0.52
4	0.83
6	1.3
8	2.7
10	3.3
12	5.3
14	8.4
16	13
18	21
20	34
22	54
24	86
26	140
28	220
30	350

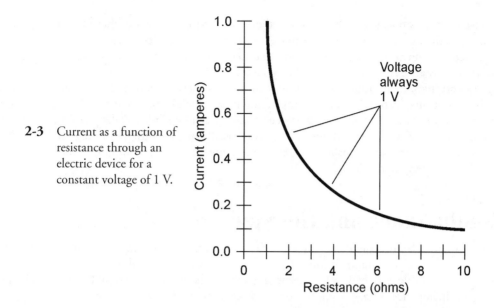

2-3 Current as a function of resistance through an electric device for a constant voltage of 1 V.

When 1 V is placed across 1 Ω of resistance, assuming that the power supply can deliver an unlimited number of charge carriers, there is a current of 1 A. If the resistance is doubled to 2 Ω, the current decreases to 0.5 A. If the resistance is cut by a factor of 5 to 0.2 Ω, the current increases by the same factor, to 5 A. The current flow, for a constant voltage, is said to be *inversely proportional* to the resistance. Figure 2-3 is a graph that shows various currents, through various resistances, given a constant voltage of 1 V across the whole resistance.

Resistance has another property. If there is a current flowing through a resistive material, there is always a potential difference across the resistive component (called a *resistor*). This is shown in Fig. 2-4. In general, this voltage is directly proportional to the current through the resistor. This behavior of resistors is useful in the design of electronic circuits, as you will learn later in this book.

Electrical circuits always have some resistance. There is no such thing as a perfect conductor. When some metals are chilled to temperatures near *absolute zero*, they lose practically all of their resistance, but they never become absolutely perfect, resistance-free conductors. This phenomenon, about which you might have heard, is called *superconductivity.*

2-4 Whenever current passes through a component having resistance, a voltage exists across that component.

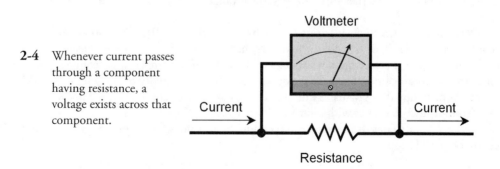

Just as there is no such thing as a perfectly resistance-free substance, there isn't a truly infinite resistance, either. Even air conducts to some extent, although the effect is usually so small that it can be ignored. In some electronic applications, materials are selected on the basis of how "nearly infinite" their resistance is.

In electronics, the resistance of a component often varies, depending on the conditions under which it is operated. A transistor, for example, might have high resistance some of the time, and low resistance at other times. High/low resistance variations can be made to take place thousands, millions, or billions of times each second. In this way, oscillators, amplifiers, and digital devices function in radio receivers and transmitters, telephone networks, digital computers, and satellite links (to name just a few applications).

Conductance and the Siemens

Electricians and electrical engineers sometimes talk about the *conductance* of a material, rather than about its resistance. The standard unit of conductance is the *siemens*, abbreviated S. When a component has a conductance of 1 S, its resistance is 1 Ω. If the resistance is doubled, the conductance is cut in half, and vice versa. Therefore, conductance is the reciprocal of resistance.

If you know the resistance of a component or circuit in ohms, you can get the conductance in siemens: divide 1 by the resistance. If you know the conductance in siemens, you can get the resistance: divide 1 by the conductance. Resistance, as a variable quantity, is denoted by an italicized, uppercase letter R. Conductance, as a variable quantity, is denoted as an italicized, uppercase letter G. If we express R in ohms and G in siemens, then the following two equations describe their relationship:

$$G = 1/R$$
$$R = 1/G$$

Units of conductance much smaller than the siemens are often used. A resistance of 1 kΩ is equal to 1 *millisiemens* (1 mS). If the resistance is 1 MΩ, the conductance is one *microsiemens* (1 μS). You'll sometimes hear about *kilosiemens* (kS) or *megasiemens* (MS), representing resistances of 0.001 Ω and 0.000001 Ω (a thousandth of an ohm and a millionth of an ohm, respectively). Short lengths of heavy wire have conductance values in the range of kilosiemens. Heavy metal rods can have conductances in the megasiemens range.

Suppose a component has a resistance of 50 Ω. Then its conductance, in siemens, is 1/50 S, which is equal to 0.02 S. We can call this 20 mS. Or imagine a piece of wire with a conductance of 20 S. Its resistance is 1/20 Ω, which is equal to 0.05 Ω. You will not often hear the term *milliohm*. But you could say that this wire has a resistance of 50 mΩ, and you would be technically right.

Determining *conductivity* is tricky. If wire has a resistivity of 10 Ω/km, you can't say that it has a conductivity of 1/10, or 0.1, S/km. It is true that a kilometer of such wire has a conductance of 0.1 S, but 2 km of the wire has a resistance of 20 Ω (because there is twice as much wire). That is not twice the conductance, but half. If you say that the conductivity of the wire is 0.1 S/km, then you might be tempted to say that 2 km of the wire has 0.2 S of conductance. That would be a mistake! Conductance decreases with increasing wire length.

Figure 2-5 illustrates the resistance and conductance values for various lengths of wire having a resistivity of 10 Ω/km.

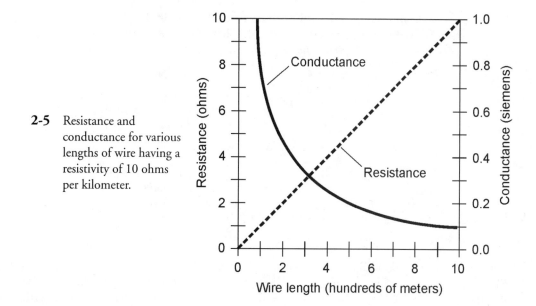

2-5 Resistance and conductance for various lengths of wire having a resistivity of 10 ohms per kilometer.

Power and the Watt

Whenever current flows through a resistance, heat results. The heat can be measured in *watts* (symbolized W) and represents electrical *power*. (As a variable quantity in equations, power is denoted by the uppercase italic letter *P*.) Power can be manifested in many forms, such as mechanical motion, radio waves, visible light, or noise. But heat is always present, in addition to any other form of power, in an electrical or electronic device. This is because no equipment is 100 percent efficient. Some power always goes to waste, and this waste is almost all in the form of heat.

Look again at Fig. 2-4. There is a certain voltage across the resistor, not specifically indicated. There's also a current flowing through the resistance, and it is not quantified in the diagram, either. Suppose we call the voltage *E* and the current *I*, in volts (V) and amperes (A), respectively. Then the power in watts dissipated by the resistance, call it *P*, is the product of the voltage in volts and the current in amperes:

$$P = EI$$

If the voltage *E* across the resistance is caused by two flashlight cells in series, giving 3 V, and if the current *I* through the resistance (a light bulb, perhaps) is 0.1 A, then *E* = 3 V and *I* = 0.1 A, and we can calculate the power *P* in watts as follows:

$$P = EI = 3 \times 0.1 = 0.3 \text{ W}$$

Suppose the voltage is 117 V, and the current is 855 mA. To calculate the power, we must convert the current into amperes: 855 mA = 855/1000 A = 0.855 A. Then:

$$P = EI = 117 \times 0.855 = 100 \text{ W}$$

Table 2-2. **Prefix multipliers from 0.000000000001 (trillionths, or units of 10^{-12}) to 1,000,000,000,000 (trillions, or units of 10^{12}).**

Prefix	Symbol	Multiplier
pico-	p	0.000000000001 (or 10^{-12})
nano-	n	0.000000001 (or 10^{-9})
micro-	μ	0.000001 (or 10^{-6})
milli-	m	0.001 (or 10^{-3})
kilo-	k	1000 (or 10^{3})
mega-	M	1,000,000 (or 10^{6})
giga-	G	1,000,000,000 (or 10^{9})
tera-	T	1,000,000,000,000 (or 10^{12})

You will often hear about *milliwatts* (mW), *microwatts* (μW), *kilowatts* (kW), and *megawatts* (MW). By now, you should be able to tell from the prefixes what these units represent. Otherwise, you can refer to Table 2-2. This table lists the most commonly used *prefix multipliers* in electricity and electronics.

Sometimes you need to use the power equation to find currents or voltages. Then you should use $I = P/E$ to find current, or $E = P/I$ to find voltage. Always remember to convert, if necessary, to the standard units of volts, amperes, and watts before performing the calculations.

A Word about Notation

Have you noticed some strange things about the notation yet? If you're observant, you have! Why, you might ask, are italics sometimes used, and sometimes not used? Something should be said early in this course about notation, because it can get confusing with all the different symbols and abbreviations. Sometimes, symbols and abbreviations appear in italics, and sometimes they do not. You'll see subscripts often, and sometimes even they are italicized! Here are some rules that apply to notation in electricity and electronics:

- Symbols for specific units, such as volts, amperes, and ohms, are not italicized.
- Symbols for objects or components, such as resistors, batteries, and meters, are not italicized.
- Quantifying prefixes, such as "kilo-" or "micro-," are not italicized.
- Labeled points in drawings might or might not be italicized; it doesn't matter as long as a diagram is consistent with itself.
- Symbols for mathematical constants and variables, such as time, are italicized.
- Symbols for electrical quantities, such as voltage, current, resistance, and power, are italicized.
- Symbols and abbreviations for modifiers might or might not be italicized; it doesn't matter as long as a document is consistent with itself.
- Numeric subscripts are not italicized.
- For nonnumeric subscripts, the same rules apply as for general symbols.

Some examples are R (not italicized) for resistor, R (italicized) for resistance, P (italicized) for power, W (not italicized) for watts, V (not italicized) for volts, E or V (italicized) for voltage, A (not italicized) for amperes, I (italicized) for current, f (italicized) for frequency, and t (italicized) for time.

Once in a while you will see the same symbol italicized in one place and not in another—in the same circuit diagram or discussion! We might, for example, talk about "resistor number 3" (symbolized R$_3$), and then later in the same paragraph talk about its value as "resistance number 3" (Symbolized R_3). Still later we might talk about "the nth resistor in a series connection" (R$_n$) and then "the nth resistance in a series combination of resistances" (R_n).

These differences in notation, while subtle (and, some people will say, picayune) are followed in this book, and they are pretty much agreed upon by convention. They are important because they tell the reader exactly what a symbol stands for in a diagram, discussion, or mathematical equation. "Resistor" and "resistance" are vastly different things—as different from each other as a garden hose (the object) and the extent to which it impedes the flow of water (the phenomenon). With this in mind, let us proceed!

Energy and the Watt-Hour

Have you heard the terms "power" and "energy" used interchangeably, as if they mean the same thing? They don't! *Energy* is power dissipated over a length of time. *Power* is the rate at which energy is expended. Physicists measure energy in units called *joules*. One joule (1 J) is the equivalent of a *watt-second*, which is the equivalent of 1 watt of power dissipated for 1 second of time (1 W · s or Ws). In electricity, you'll more often encounter the *watt-hour* (symbolized W · h or Wh) or the *kilowatt-hour* (symbolized kW · h or kWh). As their names imply, a watt-hour is the equivalent of 1 W dissipated for 1 h, and 1 kWh is the equivalent of 1 kW of power dissipated for 1 h.

A watt-hour of energy can be dissipated in an infinite number of different ways. A 60-W bulb consumes 60 Wh in 1 h, the equivalent of a watt-hour per minute (1 Wh/min). A 100-W bulb consumes 1 Wh in 1/100 h, or 36 s. Besides these differences, the rate of power dissipation in real-life circuits often changes with time. This can make the determination of consumed energy complicated, indeed.

Figure 2-6 illustrates two hypothetical devices that consume 1 Wh of energy. Device A uses its power at a constant rate of 60 W, so it consumes 1 Wh in 1 min. The power consumption rate of

2-6 Two devices that burn 1 Wh of energy. Device A dissipates a constant amount of power. Device B dissipates a variable amount of power.

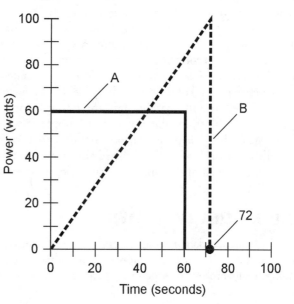

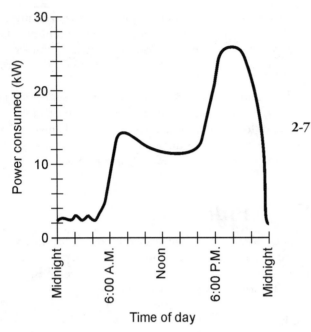

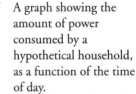

2-7 A graph showing the amount of power consumed by a hypothetical household, as a function of the time of day.

device B varies, starting at zero and ending up at quite a lot more than 60 W. How do you know that this second device really consumes 1 Wh of energy? You must determine the area under the curve in the graph. In this case, figuring out this area is easy, because the enclosed object is a triangle. The area of a triangle is equal to half the product of the base length and the height. Device B is powered up for 72 s, or 1.2 min; this is 1.2/60 = 0.02 h. Then the area under the curve is 1/2 × 100 × 0.02 = 1 Wh.

When calculating energy values, you must always remember the units you're using. In this case the unit is the watt-hour, so you must multiply watts by hours. If you multiply watts by minutes, or watts by seconds, you'll get the wrong kind of units in your answer.

Often, the curves in graphs like these are complicated. Consider the graph of power consumption in your home, versus time, for a day. It might look like the curve in Fig. 2-7. Finding the area under this curve is not easy. But there is another way to determine the total energy burned by your household over a period of time. That is by means of a meter that measures electrical energy in kilowatt-hours. Every month, without fail, the power company sends its representative to read your electric meter. This person takes down the number of kilowatt-hours displayed, subtracts the number from the reading taken the previous month, and a few days later you get a bill. This meter automatically keeps track of total consumed energy, without anybody having to go through high-level mathematical calculations to find the areas under irregular curves such as the graph of Fig. 2-7.

Other Energy Units

The joule, while standard among scientists, is not the only energy unit in existence! Another unit is the *erg*, equivalent to one ten-millionth (0.0000001) of a joule. The erg is used in lab experiments involving small amounts of expended energy.

Table 2-3. **Conversion factors between joules and various other energy units.**

Unit	To convert energy in this unit to energy in joules, multiply by	To convert energy in joules to energy in this unit, multiply by
British thermal units (Btu)	1055	0.000948
Electron volts (eV)	1.6×10^{-19}	6.2×10^{18}
Ergs	0.0000001 (or 10^{-7})	10,000,000 (or 10^7)
Foot-pounds (ft-lb)	1.356	0.738
Watt-hours (Wh)	3600	0.000278
Kilowatt-hours (kWh)	3,600,000 (or 3.6×10^6)	0.000000278 (or 2.78×10^{-7})

Most folks have heard of the *British thermal unit* (Btu), equivalent to 1055 joules. This is the energy unit commonly used to define the cooling or heating capacity of air-conditioning equipment. To cool your room from 85 to 78°F needs a certain amount of energy, perhaps best specified in Btu. If you are getting an air conditioner or furnace installed in your home, an expert will come look at your situation, and determine the size of air-conditioning/heating unit that best suits your needs. That person will likely tell you how powerful the unit should be in terms of its ability to heat or cool in Btu per hour (Btu/h).

Physicists also use, in addition to the joule, a unit of energy called the *electron volt* (eV). This is a tiny unit of energy, equal to just 0.00000000000000000016 joule (there are 18 zeroes after the decimal point and before the 1). The physicists write 1.6×10^{-19} to represent this. It is the energy gained by a single electron in an electric field of 1 V. Machines called *particle accelerators* (or *atom smashers*) are rated by millions of electron volts (MeV), billions of electron volts (GeV), or trillions of electron volts (TeV) of energy capacity.

Another energy unit, employed to denote work, is the *foot-pound* (ft-lb). This is the work needed to raise a weight of one pound by a distance of one foot, not including any friction. It's equal to 1.356 joules.

All of these units, and conversion factors, are given in Table 2-3. Kilowatt-hours and watt-hours are also included in this table. In electricity and electronics, you need to be concerned only with the watt-hour and the kilowatt-hour for most purposes.

Alternating Current and the Hertz

This chapter, and this whole first section, is mostly concerned with *direct current* (dc). That's electric current that always flows in the same direction and that does not change in intensity (at least not too rapidly) with time. But household utility current is not of this kind. It reverses direction periodically, exactly once every 1/120 second. It goes through a complete cycle every 1/60 second. Every repetition is identical to every other. This is *alternating current* (ac).

Figure 2-8 shows the characteristic wave of ac, as a graph of voltage versus time. Notice that the maximum positive and negative voltages are not 117 V, as you've heard about household electricity, but close to 165 V. There is a reason for this difference. The *effective voltage* for an ac wave is never the same as the *instantaneous maximum*, or *peak*, voltage. In fact, for the common waveform shown in Fig. 2-8, the effective value is 0.707 times the peak value. Conversely, the peak value is 1.414 times the effective value.

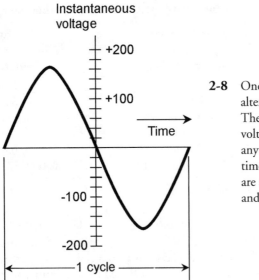

Instantaneous
voltage

2-8   One cycle of utility alternating current (ac). The instantaneous voltage is the voltage at any particular instant in time. The peak voltages are approximately plus and minus 165 V.

Because the whole cycle repeats itself every 1/60 second, the *frequency* of the utility ac wave is said to be 60 *hertz*, abbreviated 60 Hz. The German word *hertz* literally translates to "cycles per second." In the United States, this is the standard frequency for ac. In some places it is 50 Hz.

In wireless communications, higher frequencies are common, and you'll hear about *kilohertz* (kHz), *megahertz* (MHz), and *gigahertz* (GHz). The relationships among these units are as follows:

$$1 \text{ kHz} = 1000 \text{ Hz}$$
$$1 \text{ MHz} = 1000 \text{ kHz} = 1,000,000 \text{ Hz}$$
$$1 \text{ GHz} = 1000 \text{ MHz} = 1,000,000 \text{ kHz} = 1,000,000,000 \text{ Hz}$$

Usually, but not always, the waveshapes are of the type shown in Fig. 2-8. This waveform is known as a *sine wave* or a *sinusoidal* waveform.

Rectification and Pulsating Direct Current

Batteries and other sources of direct current (dc) produce constant voltage. This can be represented by a straight, horizontal line on a graph of voltage versus time (Fig. 2-9). For pure dc, the peak and effective values are identical. But sometimes the value of dc voltage fluctuates rapidly with time. This happens, for example, if the waveform in Fig. 2-8 is passed through a *rectifier* circuit.

Rectification is a process in which ac is changed to dc. The most common method of doing this uses a device called the *diode*. Right now, you need not be concerned with how the rectifier circuit is put together. The point is that part of the ac wave is either cut off, or turned around upside down, so the output is *pulsating dc.* Figure 2-10 illustrates two different waveforms of pulsating dc. In the waveform at A, the negative (bottom) part has been cut off. At B, the negative portion of the wave has been inverted and made positive. The situation at A is known as *half-wave rectification*, because it involves only half the waveform. At B, the ac has been subjected to *full-wave rectification*, because

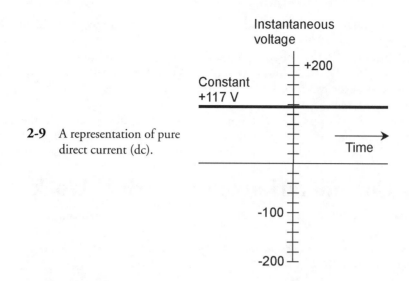

2-9 A representation of pure direct current (dc).

all of the original current still flows, even though the alternating nature has been changed so that the current never reverses.

The effective value, compared with the peak value, for pulsating dc depends on whether half-wave or full-wave rectification is applied to an ac wave. In Fig. 2-10A and B, effective voltage is shown as dashed lines, and the *instantaneous voltage* is shown as solid curves. The instantaneous voltage changes all the time, from instant to instant. (That's how it gets this name!) The peak voltage is

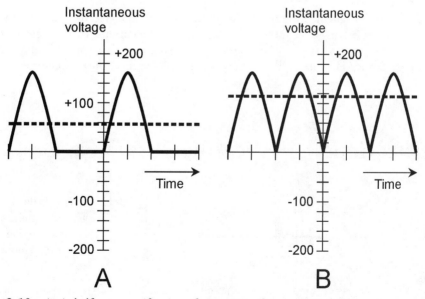

2-10 At A, half-wave rectification of common utility ac. At B, full-wave rectification of common utility ac. Effective voltages are shown by the dashed lines.

the maximum instantaneous voltage. Instantaneous voltage is never any greater than the peak voltage for any wave.

In Fig. 2-10B, the effective voltage is 0.707 times the peak voltage, just as is the case with ordinary ac. The direction of current flow, for many kinds of devices, doesn't make any difference. But in Fig. 2-10A, half of the wave has been lost. This cuts the effective value in half, so that it's only 0.354 times the peak value.

In household ac that appears in wall outlets for conventional appliances in the United States, the peak voltage is about 165 V; the effective value is 117 V. If full-wave rectification is used, the effective value is still 117 V. If half-wave rectification is used, the effective voltage is about 58.5 V.

Safety Considerations in Electrical Work

For our purposes, one rule applies concerning safety around electrical apparatus:

> *If you have any doubt about whether or not something is safe, leave it alone. Let a professional electrician work on it.*

Household voltage, normally about 117 V (but sometimes twice that for large appliances such as electric ranges and laundry machines), is more than sufficient to kill you if it appears across your chest cavity. Certain devices, such as automotive spark coils, can produce lethal currents even from the low voltage (12 to 14 V) in a car battery.

Consult the American Red Cross or your electrician concerning what kinds of circuits, procedures, and devices are safe and which aren't.

Magnetism

Electric currents and magnetic fields are closely related. Whenever an electric current flows—that is, when charge carriers move—a magnetic field accompanies the current. In a straight wire that carries electrical current, *magnetic lines of flux* surround the wire in circles, with the wire at the center, as shown in Fig. 2-11. (The lines of flux aren't physical objects; this is just a convenient way to rep-

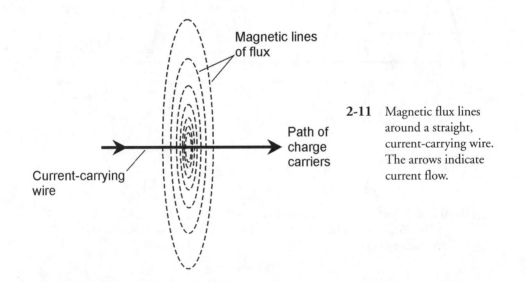

2-11 Magnetic flux lines around a straight, current-carrying wire. The arrows indicate current flow.

resent the magnetic field.) You'll sometimes hear or read about a certain number of flux lines per unit cross-sectional area, such as 100 lines per square centimeter. This is a relative way of talking about the intensity of the magnetic field.

Magnetic fields are produced when the atoms of certain materials align themselves. Iron is the most common metal that has this property. The atoms of iron in the core of the earth have become aligned to some extent; this is a complex interaction caused by the rotation of our planet and its motion with respect to the magnetic field of the sun. The magnetic field surrounding the earth is responsible for various effects, such as the concentration of charged particles that you see as the *aurora borealis* just after a solar eruption.

When a wire is coiled up, the resulting magnetic flux takes a shape similar to the flux field surrounding the earth, or the flux field around a bar magnet. Two well-defined *magnetic poles* develop, as shown in Fig. 2-12.

The intensity of a magnetic field can be greatly increased by placing a special core inside of a coil. The core should be of iron or some other material that can be readily magnetized. Such substances are called *ferromagnetic*. A core of this kind cannot actually increase the total quantity of magnetism in and around a coil, but it will cause the lines of flux to be much closer together inside the material. This is the principle by which an electromagnet works. It also makes possible the operation of electrical transformers for utility current.

Magnetic lines of flux are said to emerge from the magnetic north pole, and to run inward toward the magnetic south pole.

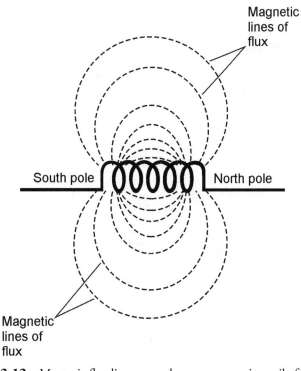

2-12   Magnetic flux lines around a current-carrying coil of wire. The flux lines converge at the magnetic poles.

Magnetic Units

The overall magnitude of a magnetic field is measured in units called *webers*, abbreviated Wb. One weber is mathematically equivalent to one volt-second (1 V · s). For weaker magnetic fields, a smaller unit, called the *maxwell* (Mx), is used. One maxwell is equal to 0.00000001 (one hundred-millionth) of a weber, or 0.01 microvolt-second (0.01 μV · s).

The *flux density* of a magnetic field is given in terms of webers or maxwells per square meter or per square centimeter. A flux density of one weber per square meter (1 Wb/m^2) is called one *tesla* (1 T). One *gauss* (1 G) is equal to 0.0001 T, or one maxwell per square centimeter (1 Mx/cm^2).

In general, as the electric current through a wire increases, so does the flux density near the wire. A coiled wire produces a greater flux density for a given current than a single, straight wire. And the more turns in the coil, the stronger the magnetic field will be.

Sometimes, magnetic field strength is specified in terms of *ampere-turns* (At). This is actually a unit of *magnetomotive force*. A one-turn wire loop, carrying 1 A of current, produces a field of 1 At. Doubling the number of turns, or the current, doubles the number of ampere-turns. Therefore, if you have 10 A flowing in a 10-turn coil, the magnetomotive force is 10 × 10, or 100 At. Or, if you have 100 mA flowing in a 100-turn coil, the magnetomotive force is 0.1 × 100, or 10 At. (Remember that 100 mA = 0.1 A.)

A less common unit of magnetomotive force is the *gilbert* (Gb). This unit is the equivalent of 0.796 At. Conversely, 1 At = 1.26 Gb.

Quiz

Refer to the text in this chapter if necessary. A good score is at least 18 correct answers. The answers are listed in the back of this book.

1. A positive electric pole
   - (a) has a deficiency of electrons.
   - (b) has fewer electrons than the negative pole.
   - (c) has an excess of electrons.
   - (d) has more electrons than the negative pole.

2. An EMF of 1 V
   - (a) cannot drive much current through a circuit.
   - (b) represents a low resistance.
   - (c) can sometimes produce a large current.
   - (d) drops to zero in a short time.

3. A potentially lethal electric current is on the order of
   - (a) 0.01 mA.
   - (b) 0.1 mA.
   - (c) 1 mA.
   - (d) 0.1 A.

4. A current of 25 A is most likely drawn by
   (a) a flashlight bulb.
   (b) a typical household.
   (c) a utility power plant.
   (d) a small radio set.

5. A piece of wire has a conductance of 20 S. Its resistance is
   (a) 20 Ω.
   (b) 0.5 Ω.
   (c) 0.05 Ω.
   (d) 0.02 Ω.

6. A resistor has a value of 300 Ω. Its conductance is
   (a) 3.33 mS.
   (b) 33.3 mS.
   (c) 333 μS.
   (d) 0.333 S.

7. A span of wire 1 km long has a conductance of 0.6 S. What is the conductance of a span of this same wire that is 3 km long?
   (a) 1.8 S
   (b) 0.6 S
   (c) 0.2 S
   (d) More information is necessary to determine this.

8. Approximately how much current can a 2-kW generator reliably deliver at 117 V?
   (a) 17 mA
   (b) 234 mA
   (c) 17 A
   (d) 234 A

9. A circuit breaker is rated for 15 A at 117 V. Approximately how much power does this represent?
   (a) 1.76 kW
   (b) 1760 kW
   (c) 7.8 kW
   (d) 0.0078 kW

10. You are told that an air conditioner has cooled a room by 500 Btu over a certain period of time. What is this amount of energy in kWh?
    (a) 147 kWh
    (b) 14.7 kWh
    (c) 1.47 kWh
    (d) 0.147 kWh

11. Of the following energy units, the one most often used to define electrical energy is
    (a) the Btu.
    (b) the erg.
    (c) the foot-pound.
    (d) the kilowatt-hour.

12. The frequency of common household ac in the United States is
    (a) 60 Hz.
    (b) 120 Hz.
    (c) 50 Hz.
    (d) 100 Hz.

13. Half-wave rectification means that
    (a) half of the ac wave is inverted.
    (b) half of the ac wave is cut off.
    (c) the whole ac wave is inverted.
    (d) the effective voltage is half the peak voltage.

14. In the output of a half-wave rectifier,
    (a) half of the ac input wave is inverted.
    (b) the effective voltage is less than that of the ac input wave.
    (c) the effective voltage is the same as that of the ac input wave.
    (d) the effective voltage is more than that of the ac input wave.

15. In the output of a full-wave rectifier,
    (a) half of the ac input wave is inverted.
    (b) the effective voltage is less than that of the ac input wave.
    (c) the effective voltage is the same as that of the ac input wave.
    (d) the effective voltage is more than that of the ac input wave.

16. A low voltage, such as 12 V,
    (a) is never dangerous.
    (b) is always dangerous.
    (c) is dangerous if it is ac, but not if it is dc.
    (d) can be dangerous under certain conditions.

17. Which of the following units can represent magnetomotive force?
    (a) The volt-turn
    (b) The ampere-turn
    (c) The gauss
    (d) The gauss-turn

18. Which of the following units can represent magnetic flux density?
    (a) The volt-turn
    (b) The ampere-turn
    (c) The gauss
    (d) The gauss-turn

19. A ferromagnetic material
    (a) concentrates magnetic flux lines within itself.
    (b) increases the total magnetomotive force around a current-carrying wire.
    (c) causes an increase in the current in a wire.
    (d) increases the number of ampere-turns in a wire.

20. A coil has 500 turns and carries 75 mA of current. The magnetomotive force is
    (a) 37,500 At.
    (b) 375 At.
    (c) 37.5 At.
    (d) 3.75 At.

3

Measuring Devices

NOW THAT YOU'RE FAMILIAR WITH THE PRIMARY UNITS COMMON IN ELECTRICITY AND ELECTRONICS, let's look at the instruments that are employed to measure these quantities. Many measuring devices work because electric and magnetic fields produce forces proportional to the intensity of the field. Such meters work by means of *electromagnetic deflection* or *electrostatic deflection.* Sometimes, electric current is measured by the extent of heat it produces in a resistance. Such meters work by *thermal heating* principles. Some meters have small motors whose speed depends on the measured quantity. The rotation rate, or the number of rotations in a given time, can be measured or counted. Still other kinds of meters tally up electronic pulses, sometimes in thousands, millions, or billions. These are *electronic counters.*

Electromagnetic Deflection

Early experimenters with electricity and magnetism noticed that an electric current produces a magnetic field. When a magnetic compass is placed near a wire carrying a direct electric current, the compass doesn't point toward magnetic north. The needle is displaced. The extent of the displacement depends on how close the compass is brought to the wire, and also on how much current the wire is carrying.

When this effect was first observed, scientists tried different arrangements to see how much the compass needle could be displaced, and how small a current could be detected. An attempt was made to obtain the greatest possible current-detecting sensitivity. Wrapping the wire in a coil around the compass resulted in a device that could indicate a tiny electric current (Fig. 3-1). This effect is known as *galvanism*, and the meter so devised was called a *galvanometer.* Once this device was made, the scientists saw that the extent of the needle displacement increased with increasing current. Then, the only challenge was to calibrate the galvanometer somehow, and to find a standard so a universal meter could be engineered.

You can make your own galvanometer. Buy a cheap compass, about 2 feet of insulated bell wire, and a 6-volt lantern battery. Set it up as shown in Fig. 3-1. Wrap the wire around the compass four or five times, and align the compass so that the needle points along the wire turns while the wire is disconnected from the battery. Connect one end of the wire to the negative (−) terminal of the bat-

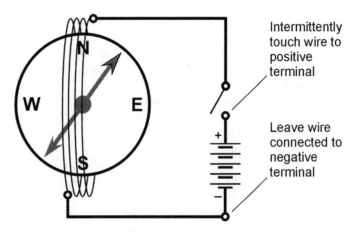

3-1 A simple galvanometer. The compass must lie flat.

tery. Touch the other end to the positive (+) terminal for a second or two, and watch the compass needle. Don't leave the wire connected to the battery for any length of time unless you want to drain the battery in a hurry.

You can buy a *resistor* and a *potentiometer* at a place like RadioShack, and set up an experiment that shows how galvanometers measure current. For a 6-V lantern battery, the fixed resistor should have a value of at least 330 Ω and should be rated for at least ¼ W. The potentiometer should have a maximum value of 10 kΩ. Connect the resistor and potentiometer in series between one end of the bell wire and one terminal of the battery, as shown in Fig. 3-2. The center contact of the potentiometer should be short-circuited to one of the end contacts, and the resulting two terminals used in the circuit.

When you adjust the potentiometer, the compass needle should deflect more or less, depending on the current through the wire. Early experimenters calibrated their meters by referring to the degrees scale around the perimeter of the compass.

3-2 A circuit for demonstrating how a galvanometer indicates relative current.

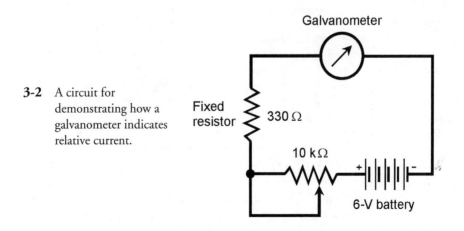

Electrostatic Deflection

Electric fields produce forces, just as magnetic fields do. You have noticed this when your hair feels like it's standing on end in very dry or cold weather. You've heard that people's hair really does stand straight out just before a lightning bolt hits nearby. (This is no myth!)

The most common device for demonstrating electrostatic forces is the *electroscope.* It consists of two foil leaves, attached to a conducting rod, and placed in a sealed container so that air currents cannot move the foil leaves (Fig. 3-3). When a charged object is brought near, or touched to, the contact at the top of the rod, the leaves stand apart from each other. This is because the two leaves become charged with like electric poles—either an excess or a deficiency of electrons—and like poles always repel. The extent to which the leaves stand apart depends on the amount of electric charge. It is difficult to measure this deflection and correlate it with charge quantity; electroscopes do not make very good meters. But variations on this theme can be employed, so that electrostatic forces can operate against tension springs or magnets, and in this way, *electrostatic meters* can be made.

An electrostatic meter can quantify alternating (or ac) electric charges as well as direct (or dc) charges. This gives electrostatic meters an advantage over electromagnetic meters such as the galvanometers. If you connect a source of ac to the coil of the galvanometer device in Fig. 3-1 or Fig. 3-2, the compass needle will not give a clear deflection; current in one direction pulls the meter needle one way, and current in the other direction pushes the needle the opposite way. But if a source of ac is connected to an electrostatic meter, the plates repel whether the charge is positive or negative at any given instant in time.

Most electroscopes aren't sensitive enough to show much deflection with ordinary 117-V utility ac. Don't try connecting 117 V to an electroscope anyway. It can present an electrocution hazard if you bring it out to points where you can easily come into physical contact with it.

An electrostatic meter has another property that is sometimes an advantage in electrical or electronic work. This is the fact that the device does not draw any current, except a tiny initial current needed to put a charge on the plates. Sometimes, an engineer or experimenter doesn't want a measuring device to draw current, because this affects the behavior of the circuit under test. Galvanometers, by contrast, always need some current to produce an indication.

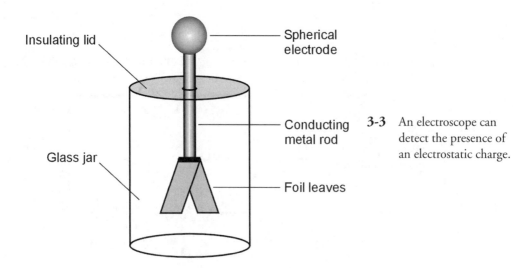

Insulating lid

Spherical electrode

Conducting metal rod

Glass jar

Foil leaves

3-3 An electroscope can detect the presence of an electrostatic charge.

If you have access to a laboratory electroscope, try charging it up with a glass rod that has been rubbed against a cloth. When the rod is pulled away from the electroscope, the foil leaves remain standing apart. The charge just sits there! If the electroscope drew any current, the leaves would fall back together again, just as the galvanometer compass needle returns to magnetic north the instant you take the wire from the battery.

Thermal Heating

Another phenomenon, sometimes useful in the measurement of electric currents, is the fact that whenever current flows through a conductor having any resistance, that conductor is heated. All conductors have some resistance; none are perfect. The extent of this heating is proportional to the amount of current being carried by the wire.

By choosing the right metal or alloy, and by making the wire a certain length and diameter, and by employing a sensitive thermometer, and by putting the entire assembly inside a thermally insulating package, a *hot-wire meter* can be made. The hot-wire meter can measure ac as well as dc, because the current-heating phenomenon does not depend on the direction of current flow.

A variation of the hot-wire principle can be used to advantage by placing two different metals into contact with each other. If the right metals are chosen, the junction heats up when a current flows through it. This is called the *thermocouple principle*. As with the hot-wire meter, a thermometer can be used to measure the extent of the heating. But there is also another effect. A thermocouple, when it gets warm, generates dc. This dc can be measured with a galvanometer. This method is useful when it is necessary to have a fast meter *response time*.

The hot-wire and thermocouple effects are sometimes used to measure ac at high frequencies, in the range of hundreds of kilohertz up to tens of gigahertz.

Ammeters

A magnetic compass doesn't make a very convenient meter. It has to be lying flat, and the coil has to be aligned with the compass needle when there is no current. But of course, electrical and electronic devices aren't all oriented so as to be aligned with the north geomagnetic pole! But the external magnetic field doesn't have to come from the earth. It can be provided by a permanent magnet near or inside the meter. This supplies a stronger magnetic force than does the earth's magnetic field, and therefore makes it possible to make a meter that can detect much weaker currents. Such a meter can be turned in any direction, and its operation is not affected. The coil can be attached directly to the meter pointer, and suspended by means of a spring in the field of the magnet. This type of metering scheme, called the *D'Arsonval movement*, has been around since the earliest days of electricity, but it is still used in some metering devices today. The assembly is shown in Fig. 3-4. This is the basic principle of the *ammeter*.

A variation of the D'Arsonval movement can be obtained by attaching the meter needle to a permanent magnet, and winding the coil in a fixed form around the magnet. Current in the coil produces a magnetic field, and this in turn generates a force if the coil and magnet are aligned correctly with respect to each other. This works all right, but the mass of the permanent magnet causes a slower needle response. This type of meter is also more prone to *overshoot* than the true D'Arsonval movement; the inertia of the magnet's mass, once overcome by the magnetic force, causes the needle to fly past the actual point for the current reading, and then to wag back and forth a couple of times before coming to rest in the right place.

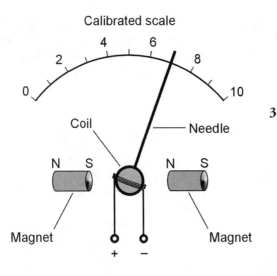

Calibrated scale

3-4 A functional drawing of a D'Arsonval meter movement (spring bearing not shown).

It is possible to use an *electromagnet* in place of the permanent magnet in the meter assembly. This electromagnet can be operated by the same current that flows in the coil attached to the meter needle. This gets rid of the need for a massive, permanent magnet inside the meter. It also eliminates the possibility that the meter sensitivity will change in case the strength of the permanent magnet deteriorates (such as might be caused by heat, or by severe mechanical vibration). The electromagnet can be either in series with, or in parallel with, the meter movement coil.

The sensitivity of the D'Arsonval-type meter, and of similar designs, depends on several factors. First is the strength of the permanent magnet (if the meter uses a permanent magnet). Second is the number of turns in the coil. The stronger the magnet, and the larger the number of turns in the coil, the less current is needed in order to produce a given magnetic force. If the meter is of the electromagnet type, the combined number of coil turns affects the sensitivity. Remember that the strength of a magnetomotive force is given in terms of ampere-turns. For a given current (number of amperes), the force increases in direct proportion to the number of coil turns. The more force in a meter, the greater the needle deflection for a given amount of current, and the smaller the current necessary to cause a certain amount of needle movement. The most sensitive ammeters can detect currents of just a microampere or two. The amount of current for *full-scale deflection* (the needle goes all the way up without banging against the stop pin) can be as little as about 50 µA in commonly available meters.

Sometimes, it is desirable to have an ammeter that will allow for a wide range of current measurements. The full-scale deflection of a meter assembly cannot easily be changed, because that would mean changing the number of coil turns and/or the strength of the magnet. But all ammeters have a certain amount of *internal resistance*. If a resistor, having the same internal resistance as the meter, is connected in parallel with the meter, the resistor will draw half the current. Then it will take twice the current through the assembly to deflect the meter to full scale, as compared with the meter alone. By choosing a resistor of just the right value, the full-scale deflection of an ammeter can be increased by a large factor, such as 10, or 100, or 1000. This resistor must be capable of carrying the current without burning up. It might have to draw practically all of the current flowing through the assembly, leaving the meter to carry only 1/10, or 1/100, or 1/1000 of the current. This is called a *shunt resistance* or *meter shunt* (Fig. 3-5). Meter shunts are used when it is necessary to measure very large currents, such as hundreds of amperes. They also allow microammeters or milliammeters to be used in a versatile *multimeter*, with many current ranges.

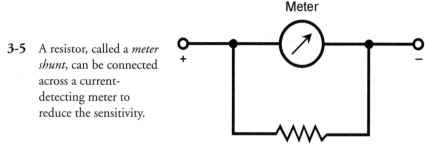

3-5 A resistor, called a *meter shunt*, can be connected across a current-detecting meter to reduce the sensitivity.

Voltmeters

Current, as we have seen, consists of a flow of charge carriers. Voltage, or electromotive force (EMF), or potential difference, is the "pressure" that makes current possible. Given a circuit whose resistance is constant, the current that flows in the circuit is directly proportional to the voltage placed across it. Early electrical experimenters recognized that an ammeter could be used to measure voltage, because an ammeter is a form of constant-resistance circuit. If you connect an ammeter directly across a source of voltage such as a battery, the meter needle deflects. In fact, a milliammeter needle will probably be "pinned" if you do this with it, and a microammeter might well be wrecked by the force of the needle striking the pin at the top of the scale. For this reason, you should never connect milliammeters or microammeters directly across voltage sources. An ammeter, perhaps with a range of 0 to 10 A, might not deflect to full scale if it is placed across a battery, but it's still a bad idea to do this, because it will rapidly drain the battery. Some batteries, such as automotive lead-acid cells, can explode under these conditions.

Ammeters have low internal resistance. They are designed that way deliberately. They are meant to be connected in series with other parts of a circuit, not right across a power supply. But if you place a large resistor in series with an ammeter, and then connect the ammeter across a battery or other type of power supply, you no longer have a short circuit. The ammeter will give an indication that is directly proportional to the voltage of the supply. The smaller the full-scale reading of the ammeter, the larger the resistance that is needed to get a meaningful indication on the meter. Using a microammeter and a very large value of resistance in series, a *voltmeter* can be devised that will draw only a little current from the source.

A voltmeter can be made to have various ranges for the full-scale reading, by switching different values of resistance in series with the microammeter (Fig. 3-6). The internal resistance of the meter is large because the values of the resistors are large. The greater the supply voltage, the larger the internal resistance of the meter, because the necessary series resistance increases as the voltage increases.

A voltmeter should have high internal resistance, and the higher the better! The reason for this is that you don't want the meter to draw much current from the power source. This current should go, as much as possible, toward operating whatever circuit is hooked up to the power supply, and not into getting a reading of the voltage. Also, you might not want, or need, to have the voltmeter constantly connected in the circuit; you might need the voltmeter for testing many different circuits. You don't want the behavior of a circuit to be affected the instant you connect the voltmeter to the supply. The less current a voltmeter draws, the less it affects the behavior of anything that is working from the power supply.

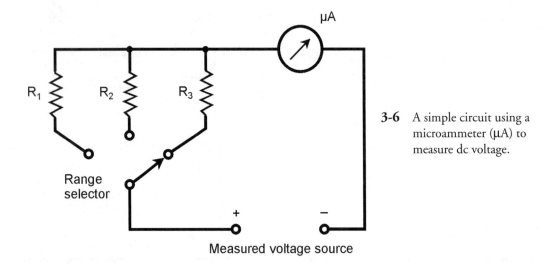

3-6 A simple circuit using a microammeter (µA) to measure dc voltage.

A completely different type of voltmeter uses the effect of electrostatic deflection, rather than electromagnetic deflection. Remember that electric fields produce forces, just as do magnetic fields. Therefore, a pair of plates attract or repel each other if they are charged. The *electrostatic voltmeter* takes advantage of the attractive force between two plates having opposite electric charge, or having a large potential difference. Figure 3-7 is a simplified drawing of the mechanics of an electrostatic voltmeter. It draws almost no current from the power supply. The only thing between the plates is air, and air is a nearly perfect insulator. The electrostatic meter can indicate ac voltage as well as dc voltage. The construction tends to be fragile, however, and mechanical vibration can influence the reading.

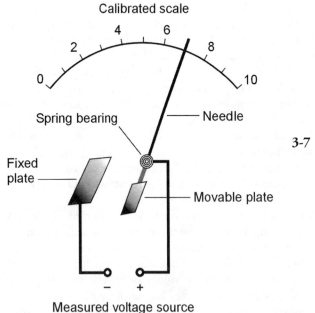

3-7 A functional drawing of an electrostatic voltmeter movement.

Ohmmeters

If all other factors are held constant, the current through a circuit depends on the resistance. This provides us with a means for measuring resistance. An *ohmmeter* can be constructed by placing a milliammeter or microammeter in series with a set of fixed, switchable resistances and a battery that provides a known, constant voltage (Fig. 3-8). By selecting the resistances appropriately, the meter gives indications in ohms over any desired range. The zero point on the milliammeter or microammeter is assigned the value of *infinity ohms*, meaning a perfect insulator. The full-scale value is set at a certain minimum, such as 1 Ω, 100 Ω, 1 kΩ, or 10 kΩ.

An ohmmeter must be calibrated at the factory where it is made, or in an electronics lab. A slight error in the values of the series resistors can cause gigantic errors in measured resistance. Therefore, precise *tolerances* are needed for these resistors. That means their values must actually be what the manufacturer claims they are, to within a fraction of 1 percent if possible. It is also necessary that the battery provide exactly the right voltage.

The scale of an ohmmeter is *nonlinear*. That means the graduations are not of the same width everywhere on the meter scale. The graduations tend to be squashed together toward the infinity end of the scale. Because of this, it is difficult to interpolate for high values of resistance unless the appropriate meter range is selected.

Engineers and technicians usually connect an ohmmeter in a circuit with the meter set for the highest resistance range first. Then they switch the range down until the meter needle is in a part of the scale that is easy to read. Finally, the reading is taken, and is multiplied (or divided) by the appropriate amount as indicated on the range switch. Figure 3-9 shows an ohmmeter reading. The meter itself indicates approximately 4.7, but the range switch says 1 kΩ. This indicates a resistance of about 4.7 kΩ, or 4700 Ω.

Ohmmeters give inaccurate readings if there is a voltage between the points where the meter is connected. This is because such a voltage either adds to, or subtracts from, the ohmmeter's own battery voltage. Sometimes, in this type of situation, an ohmmeter might tell you that a circuit has

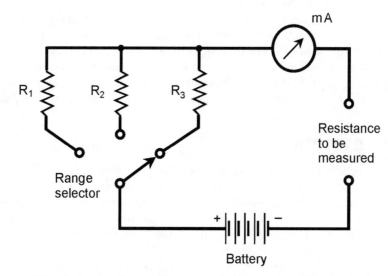

3-8 A circuit using a milliammeter (mA) to measure dc resistance.

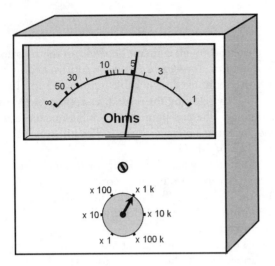

3-9   An example of an ohmmeter reading. This device shows about $4.7 \times 1\ k\Omega = 4.7\ k\Omega = 4700\ \Omega$.

"more than infinity" ohms! The needle will hit the pin at the left end of the scale. Therefore, when using an ohmmeter to measure resistance, you must always be sure that there is no voltage between the points under test. The best way to do this is to switch off the equipment in question.

Multimeters

In the electronics lab, a common piece of test equipment is the *multimeter*, in which different kinds of meters are combined into a single unit. The *volt-ohm-milliammeter* (VOM) is the most often used. As its name implies, it combines voltage, resistance, and current measuring capabilities. You should not have trouble envisioning how a single milliammeter can be used for measuring voltage, current, and resistance. The preceding discussions for measurements of these quantities have all included methods in which a current meter can be used to measure the intended quantity.

Commercially available multimeters have certain limits in the values they can measure. The maximum voltage is around 1000 V. The measurement of larger voltages requires special probes and heavily insulated wires, as well as other safety precautions. The maximum current that a common VOM can measure is about 1 A. The maximum measurable resistance is on the order of several megohms or tens of megohms. The lower limit of resistance indication is around 0.1 to 1 Ω.

FET Voltmeters

A good voltmeter disturbs the circuit under test as little as possible, and this requires that the meter have high internal resistance. Besides the electrostatic-type voltmeter, there is another way to get high internal resistance. This is to sample a tiny current, far too small for any meter to directly indicate, and then amplify this current so a conventional milliammeter or microammeter can display it. When a minuscule current is drawn from a circuit, the equivalent resistance is always extremely high.

The most effective way to accomplish voltage amplification, while making sure that the current drawn is exceedingly small, is to use a *field-effect transistor*, or FET. (Don't worry about how such amplifiers work right now; you'll learn all about that later in this book.) A voltmeter that uses a FET voltage amplifier to minimize the current drain is known as a *FET voltmeter* (FETVM). It has extremely high input resistance, along with good sensitivity and amplification.

Wattmeters

The measurement of electrical power requires that voltage and current both be measured simultaneously. Remember that in a dc circuit, the power (P) in watts is the product of the voltage (E) in volts and the current (I) in amperes. That is, $P = EI$. In fact, watts are sometimes called *volt-amperes* in dc circuits.

Do you think you can connect a voltmeter in parallel with a circuit, thereby getting a reading of the voltage across it, and also hook up an ammeter in series to get a reading of the current through the circuit, and then multiply volts times amperes to get watts consumed by the circuit? Well, you can. For most dc circuits, this is an excellent way to measure power, as shown in Fig. 3-10.

Sometimes, it's simpler yet. In many cases, the voltage from the power supply is constant and predictable. Utility power is a good example. The effective voltage is always very close to 117 V. Although it's ac, and not dc, power in most utility circuits can be measured in the same way as power is measured in dc circuits: by means of an ammeter connected in series with the circuit, and calibrated so that the multiplication (times 117) has already been done. Then, rather than 1 A, the meter will show a reading of 117 W, because $P = EI = 117 \times 1 = 117$ W. If the meter reading is 300 W, the current is $I = P/E = 300/117 = 2.56$ A. An electric iron might consume 1000 W, or a current of $1000/117 = 8.55$ A. A large heating unit might gobble up 2000 W, requiring a current of $2000/117 = 17.1$ A. You should not be surprised if this blows a fuse or trips a circuit breaker, because these devices are often rated for 15 A.

Specialized wattmeters are necessary for the measurement of radio-frequency (RF) power, or for peak audio power in a high-fidelity amplifier, or for certain other specialized applications. But almost all of these meters, whatever the associated circuitry, use simple ammeters, milliammeters, or microammeters as their indicating devices.

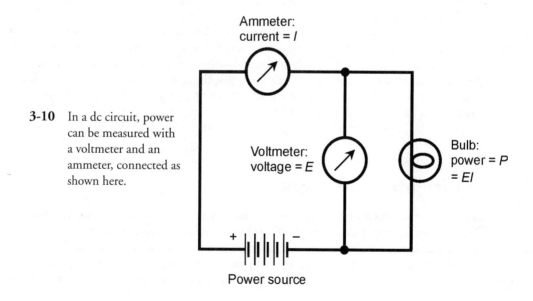

3-10 In a dc circuit, power can be measured with a voltmeter and an ammeter, connected as shown here.

Ammeter: current = I

Voltmeter: voltage = E

Bulb: power = P = EI

Power source

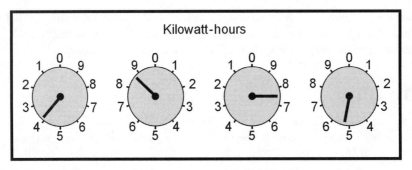

3-11   A utility meter with four rotary analog dials. In this example, the reading is a little more than 3875 kWh.

Watt-Hour Meters

Electrical energy, as you now know, is measured in watt-hours or kilowatt-hours (kWh). Not surprisingly, a metering device that indicates energy in these units is called a *watt-hour meter* or a *kilowatt-hour meter.*

The most often used means of measuring electrical energy is by using a small electric motor, the speed of which depends on the current, and thereby on the power at a constant voltage. The number of turns of the motor shaft, in a given length of time, is directly proportional to the number of kilowatt-hours consumed. The motor is placed at the point where the utility wires enter the building. This is usually at a point where the voltage is 234 V. At this point the circuit is split into some circuits with 234 V (for heavy-duty appliances such as the oven, washer, and dryer) and general household circuits at 117 V (for smaller appliances such as lamps, clock radios, and television sets).

If you've observed a kilowatt-hour meter, you have seen a disk spinning, sometimes fast, other times slowly. Its speed depends on the power being used at any given time. The total number of turns of this little disk, every month, determines the size of the bill you will get, as a function also, of course, of the cost per kilowatt-hour.

Kilowatt-hour meters count the number of disk turns by means of geared rotary drums or pointers. The drum-type meter gives a direct digital readout. The pointer type has several scales calibrated from 0 to 9 in circles, some going clockwise and others going counterclockwise. Reading a pointer-type utility meter is a little tricky, because you must think in whatever direction (clockwise or counterclockwise) the scale goes. An example of a pointer-type utility meter is illustrated in Fig. 3-11. Read from left to right. For each meter scale, take down the number that the pointer has most recently passed. Write down the rest as you go. The meter shown in the figure reads a little more than 3875 kWh.

Digital Readout Meters

Increasingly, metering devices are being designed so that they provide a direct readout. The number on the meter is the indication. It's that simple. Such a meter is called a *digital meter.*

The main advantage of a digital meter is the fact that it's easy for anybody to read, and there is no chance for interpolation errors. This is ideal for utility meters, clocks, and some kinds of ammeters, voltmeters, and wattmeters. It works well when the value of the quantity does not change often or fast.

There are some situations in which a digital meter is a disadvantage. One good example is the signal-strength indicator in a radio receiver. This meter bounces up and down as signals fade, or as you tune the radio, or sometimes even as the signal modulates. A digital meter will show nothing but a constantly changing, meaningless set of numerals. Digital meters require a certain length of time to lock in to the current, voltage, power, or other quantity being measured. If this quantity never settles at any one value for a long enough time, the meter can never lock in.

Meters with a scale and pointer are known as *analog meters.* Their main advantages are that they allow interpolation, they give the operator a sense of the quantity relative to other possible values, and they follow along when a quantity changes. Some engineers and technicians prefer analog metering, even in situations where digital meters would work just as well.

One potential hang-up with digital meters is being certain of where the decimal point goes. If you're off by one decimal place, the error will be by a factor of 10. Also, you need to be sure you know what the units are. For example, a frequency indicator might be reading out in megahertz, and you might forget and think it is giving you a reading in kilohertz. That's a mistake by a factor of 1000! Of course, this latter type of error can happen with analog meters, too.

Frequency Counters

The measurement of energy used by your home is an application to which digital metering is well suited. A digital kilowatt-hour meter is easier to read than the pointer-type meter. When measuring frequencies of radio signals, digital metering is not only more convenient, but far more accurate.

A *frequency counter* measures the frequency of an ac wave by actually counting pulses, in a manner similar to the way the utility meter counts the number of turns of a motor. But the frequency counter works electronically, without any moving parts. It can keep track of thousands, millions, or billions of pulses per second, and it shows the rate on a digital display that is as easy to read as a digital watch.

The accuracy of the frequency counter is a function of the *lock-in time.* Lock-in is usually done in 0.1 second, 1 second, or 10 seconds. Increasing the lock-in time by a factor of 10 will cause the accuracy to increase by one additional digit. Modern frequency counters are good to six, seven, or eight digits; sophisticated lab devices can show frequency to nine or ten digits.

Other Meter Types

Here are a few of the less common types of meters that you will occasionally encounter in electrical and electronics applications.

VU and Decibel Meters

In high-fidelity equipment, especially the more sophisticated amplifiers ("amps"), *loudness meters* are sometimes used. These are calibrated in *decibels,* a unit that you will often have to use, and interpret, in reference to electronic signal levels. A decibel is an increase or decrease in sound or signal level that you can just barely detect, if you are expecting the change.

Audio loudness is given in *volume units* (VU), and the meter that indicates it is called a *VU meter.* The typical VU meter has a zero marker with a red line to the right and a black line to the left, and is calibrated in decibels (dB) below the zero marker and volume units above it (Fig. 3-12). The meter might also be calibrated in *watts rms,* an expression for audio power. As music is played

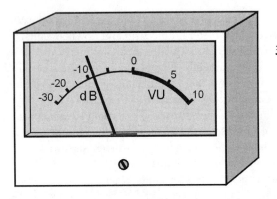

3-12 A VU (volume-unit) meter. The heavy portion of the scale (to the right of 0) is usually red, indicating the risk of audio distortion.

through the system, or as a voice comes over it, the VU meter needle kicks up. The amplifier volume should be kept down so that the meter doesn't go past the zero mark and into the red range. If the meter does kick up into the red scale, it means that distortion is taking place within the amplifier circuit.

Sound level in general can be measured by means of a *sound-level meter*, calibrated in decibels (dB) and connected to the output of a precision amplifier with a microphone of known sensitivity (Fig. 3-13). Have you read that a vacuum cleaner will produce "80 dB" of sound, and a large truck going by will subject your ears to "90 dB"? These figures are determined by a sound-level meter, and are defined with respect to the *threshold of hearing*, which is the faintest sound that a person with good ears can hear.

Light Meters

The intensity of visible light is measured by means of a *light meter* or *illumination meter*. It is tempting to suppose that it's easy to make this kind of meter by connecting a milliammeter to a solar (photovoltaic) cell. As things work out, this is a good way to construct an inexpensive light meter (Fig. 3-14). More sophisticated devices use dc amplifiers, similar to the type found in a FETVM, to enhance sensitivity and to allow for several different ranges of readings.

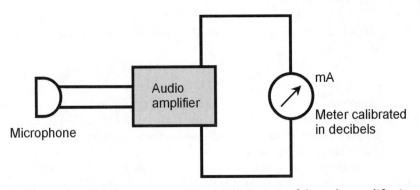

3-13 A meter for measuring sound levels. The output of the audio amplifier is rectified to produce dc that the meter can detect.

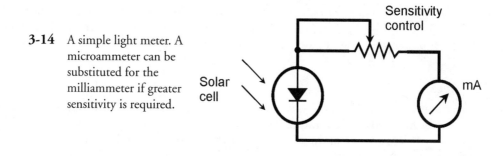

3-14 A simple light meter. A microammeter can be substituted for the milliammeter if greater sensitivity is required.

One problem with this design is that solar cells are not sensitive to light at exactly the same wavelengths as human eyes. This can be overcome by placing a colored filter in front of the solar cell, so that the solar cell becomes sensitive to the same wavelengths, in the same proportions, as human eyes. Another problem is calibrating the meter. This must usually be done at the factory, in standard illumination units such as *lumens* or *candela*.

With appropriate modification, meters such as the one in Fig. 3-14 can be used to measure *infrared* (IR) or *ultraviolet* (UV) intensity. Various specialized photovoltaic cells have peak sensitivity at nonvisible wavelengths, including IR and UV.

Pen Recorders

A meter movement can be equipped with a marking device to keep a graphic record of the level of some quantity with respect to time. Such a device is called a *pen recorder*. The paper, with a calibrated scale, is taped to a rotating drum. The drum, driven by a clock motor, turns at a slow rate, such as one revolution per hour or one revolution in 24 hours. A simplified drawing of a pen recorder is shown in Fig. 3-15.

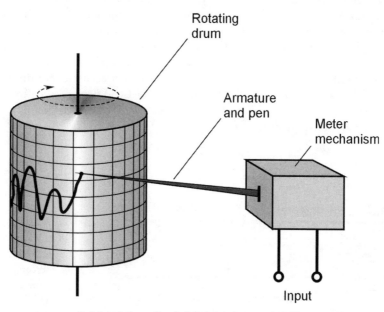

3-15 A functional drawing of a pen recorder.

A device of this kind, along with a wattmeter, can be employed to get a reading of the power consumed by your household at various times during the day. In this way you can find out when you use the most power, and at what particular times you might be using too much.

Oscilloscopes

Another graphic metering device is the *oscilloscope*. This measures and records quantities that vary rapidly, at rates of hundreds, thousands, or millions of times per second. It creates a "graph" by throwing a beam of electrons at a phosphor screen. A *cathode-ray tube*, similar to the kind in a television set, is employed. Some oscilloscopes have electronic conversion circuits that allow for the use of a solid-state *liquid crystal display* (LCD).

Oscilloscopes are useful for observing and analyzing the shapes of signal waveforms, and also for measuring peak signal levels (rather than just the effective levels). An oscilloscope can also be used to approximately measure the frequency of a waveform. The horizontal scale of an oscilloscope shows time, and the vertical scale shows the instantaneous signal voltage. An oscilloscope can indirectly measure power or current, by using a known value of resistance across the input terminals.

Technicians and engineers develop a sense of what a signal waveform should look like, and then they can often tell, by observing the oscilloscope display, whether or not the circuit under test is behaving the way it should. This is a subjective measurement, because it is qualitative as well as quantitative.

Bar-Graph Meters

A cheap, simple kind of meter can be made using a string of light-emitting diodes (LEDs) or an LCD along with a digital scale to indicate approximate levels of current, voltage, or power. This type of meter, like a digital meter, has no moving parts to break. To some extent, it offers the relative-reading feeling you get with an analog meter. Figure 3-16 is an example of a bar-graph meter that is used to show the power output, in kilowatts, for a radio transmitter. This meter can follow along quite well with rapid fluctuations in the reading. In this example, the meter indicates about 0.8 kW, or 800 W.

The chief drawback of the bar-graph meter is that it isn't very accurate. For this reason it is not generally used in laboratory testing. In addition, the LED or LCD devices sometimes flicker when the level is between two values given by the bars. This creates an illusion of circuit instability. With bright LEDs, it can also be quite distracting.

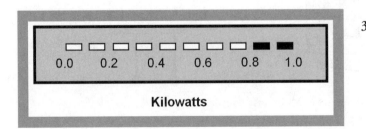

3-16 A bar-graph meter. In this case, the indication is about 80 percent of full-scale, representing 0.8 kW, or 800 W.

Quiz

Refer to the text in this chapter if necessary. A good score is 18 out of 20 correct. Answers are in the back of the book.

1. The attraction or repulsion between two electrically charged objects is called
   - (a) electromagnetic deflection.
   - (b) electrostatic force.
   - (c) magnetic force.
   - (d) electroscopic force.

2. The change in the direction of a compass needle, when a current-carrying wire is brought near, is called
   - (a) electromagnetic deflection.
   - (b) electrostatic force.
   - (c) magnetic force.
   - (d) electroscopic force.

3. Suppose a certain current in a galvanometer causes the compass needle to deflect by 20 degrees, and then this current is doubled while the polarity stays the same. The angle of the needle deflection will
   - (a) decrease.
   - (b) stay the same.
   - (c) increase.
   - (d) reverse direction.

4. One important advantage of an electrostatic meter is the fact that
   - (a) it measures very small currents.
   - (b) it can handle large currents.
   - (c) it can detect and indicate ac voltages as well as dc voltages.
   - (d) it draws a large current from a power supply.

5. A thermocouple
   - (a) gets warm when dc flows through it.
   - (b) is a thin, straight, special wire.
   - (c) generates dc when exposed to visible light.
   - (d) generates ac when heated.

6. An important advantage of an electromagnet-type meter over a permanent-magnet meter is the fact that
   - (a) the electromagnet meter costs much less.
   - (b) the electromagnet meter need not be aligned with the earth's magnetic field.
   - (c) the permanent-magnet meter has a more sluggish coil.
   - (d) the electromagnet meter is more rugged.

7. Ammeter shunts are useful because
   (a) they increase meter sensitivity.
   (b) they make a meter more physically rugged.
   (c) they allow for measurement of large currents.
   (d) they prevent overheating of the meter movement.

8. Voltmeters should generally have
   (a) high internal resistance.
   (b) low internal resistance.
   (c) the greatest possible sensitivity.
   (d) the ability to withstand large currents.

9. In order to measure the power-supply voltage that is applied to an electrical circuit, a voltmeter should be placed
   (a) in series with the circuit that works from the supply.
   (b) between the negative pole of the supply and the circuit working from the supply.
   (c) between the positive pole of the supply and the circuit working from the supply.
   (d) in parallel with the circuit that works from the supply.

10. Which of the following will *not* normally cause a *large* error in an ohmmeter reading?
    (a) A small voltage between points under test
    (b) A slight change in switchable internal resistance
    (c) A small change in the resistance to be measured
    (d) A slight error in the range switch position

11. The ohmmeter in Fig. 3-17 shows a reading of approximately
    (a) 34,000 Ω.
    (b) 3.4 kΩ.
    (c) 340 Ω.
    (d) 34 Ω.

12. The main advantage of a FETVM over a conventional voltmeter is the fact that the FETVM
    (a) can measure lower voltages.
    (b) draws less current from the circuit under test.
    (c) can withstand higher voltages safely.
    (d) is sensitive to ac voltage as well as to dc voltage.

13. Which of the following is *not* a function of a fuse?
    (a) To ensure there is enough current available for an appliance to work right
    (b) To make it impossible to use appliances that are too large for a given circuit
    (c) To limit the amount of power that a device can draw from the electrical circuit
    (d) To make sure the current drawn by an appliance cannot exceed a certain limit

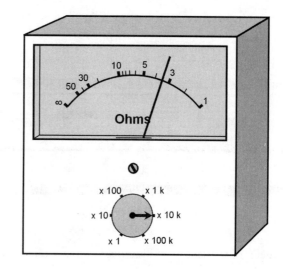

3-17 Illustration for Quiz Question 11.

14. A utility meter's motor speed depends directly on
    (a) the number of ampere-hours being used at the time.
    (b) the number of watt-hours being used at the time.
    (c) the number of watts being used at the time.
    (d) the number of kilowatt-hours being used at the time.

15. A utility meter's readout indicates
    (a) voltage.
    (b) power.
    (c) current.
    (d) energy.

16. A typical frequency counter
    (a) has an analog readout.
    (b) is accurate to six digits or more.
    (c) works by indirectly measuring current.
    (d) works by indirectly measuring voltage.

17. A VU meter is *never* used to get a general indication of
    (a) sound intensity.
    (b) decibels.
    (c) power in an audio amplifier.
    (d) visible light intensity.

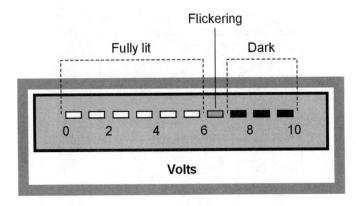

3-18 Illustration for Quiz Question 20.

18. The meter movement in an illumination meter directly measures
    (a) current.
    (b) voltage.
    (c) power.
    (d) energy.

19. An oscilloscope *cannot* be used to indicate
    (a) frequency.
    (b) wave shape.
    (c) energy.
    (d) peak signal voltage.

20. What voltage would be expected to produce the reading on the bar-graph meter shown in Fig. 3-18?
    (a) 6.0 V
    (b) 6.5 V
    (c) 7.0 V
    (d) There is no way to tell because the meter, as shown, is malfunctioning.

4
CHAPTER

Direct-Current Circuit Basics

YOU'VE ALREADY SEEN SOME SIMPLE ELECTRICAL CIRCUIT DIAGRAMS. IN THIS CHAPTER, YOU'LL GET more acquainted with this type of diagram. You'll also learn more about how current, voltage, resistance, and power are related in dc and low-frequency ac circuits.

Schematic Symbols

In this course, the idea is to familiarize you with schematic symbols by getting you to read and use them in action. But right now, why not check out Appendix B, which is a comprehensive table of symbols? Then refer to it frequently in the future, especially when you see a symbol you don't remember or recognize.

The simplest schematic symbol is the one representing a wire or *electrical conductor*: a straight, solid line. Sometimes, dashed lines are used to represent conductors, but usually, dashed lines are drawn to partition diagrams into constituent circuits, or to indicate that certain components interact with each other or operate in step with each other. Conductor lines are almost always drawn either horizontally across or vertically up and down the page. This keeps the diagram neat and easy to read.

When two conductor lines cross, they aren't connected at the crossing point unless a heavy black dot is placed where the two lines meet. The dot should always be clearly visible wherever conductors are to be connected, no matter how many of them meet at the junction. A *resistor* is indicated by a zigzag. A variable resistor, or *potentiometer*, is indicated by a zigzag with an arrow through it, or by a zigzag with an arrow pointing at it. These symbols are shown in Fig. 4-1.

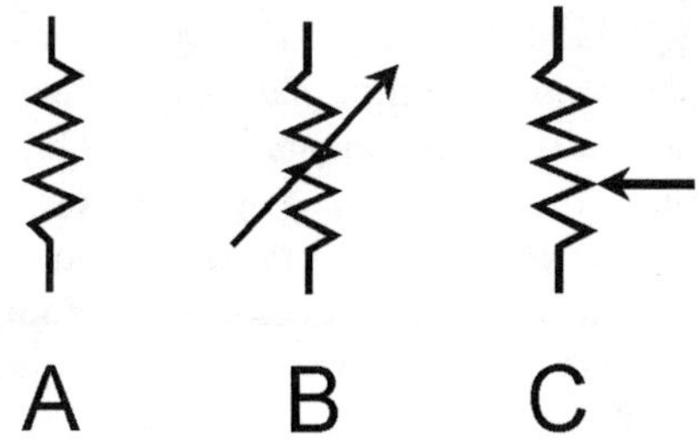

4-1 Schematic symbols for a fixed resistor (A), a two-terminal variable resistor (B), and a three-terminal potentiometer (C).

A   B   C

55

An *electrochemical cell* (such as a common dime-store battery) is shown by two parallel lines, one longer than the other. The longer line represents the plus terminal. A true *battery*, which is a combination of two or more cells in series, is indicated by several parallel lines, alternately long and short. It's not necessary to use more than four lines to represent a battery, although you'll often see 6, 8, 10, or even 12 lines. Symbols for a cell and a battery are shown in Fig. 4-2.

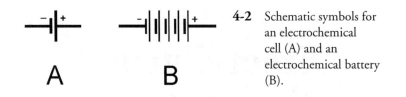

4-2   Schematic symbols for an electrochemical cell (A) and an electrochemical battery (B).

Meters are portrayed as circles. Sometimes the circle has an arrow inside it, and the meter type, such as mA (milliammeter) or V (voltmeter) is written alongside the circle, as shown in Fig. 4-3A. Sometimes the meter type is indicated inside the circle, and there is no arrow (Fig. 4-3B). It doesn't matter which way you draw them, as long as you're consistent throughout a schematic diagram.

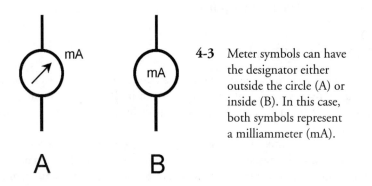

4-3   Meter symbols can have the designator either outside the circle (A) or inside (B). In this case, both symbols represent a milliammeter (mA).

Some other common symbols include the *incandescent lamp*, the *capacitor*, the *air-core coil*, the *iron-core coil*, the *chassis ground*, the *earth ground*, the *ac source*, the set of *terminals*, and the *black box* (general component or device), a rectangle with the designator written inside. These are shown in Fig. 4-4.

Schematic and Wiring Diagrams

Look back through the earlier chapters of this book and observe the electrical diagrams. These are all simple examples of how professionals would draw *schematic diagrams*. In a schematic diagram, the interconnection of the components is shown, but the actual values of the components are not necessarily indicated. You might see a diagram of a two-transistor audio amplifier, for example, with resistors and capacitors and coils and transistors, but without any data concerning the values or ratings of the components. This is a schematic diagram, but not a true *wiring diagram*. It gives the *scheme* for the circuit, but you can't *wire* the circuit and make it work, because there isn't enough information.

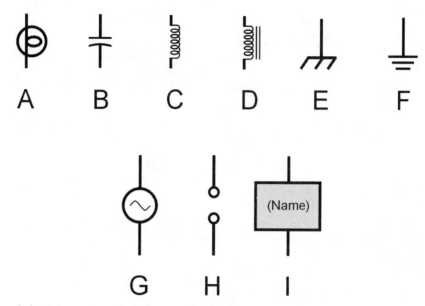

4-4   Schematic symbols for incandescent lamp (A), fixed capacitor (B), fixed inductor with air core (C), fixed inductor with laminated-iron core (D), chassis ground (E), earth ground (F), signal generator or source of alternating current (G), pair of terminals (H), and specialized component or device (I).

Suppose you want to build the circuit. You go to an electronics store to get the parts. What values of resistors should you buy? How about capacitors? What type of transistor will work best? Do you need to wind the coils yourself, or can you get ready-made coils? Are there test points or other special terminals that should be installed for the benefit of the technicians who might have to repair the amplifier? How many watts should the potentiometers be able to handle? All these things are indicated in a wiring diagram. You might have seen this kind of diagram in the back of the instruction manual for a hi-fi amplifier, a stereo tuner, or a television set. Wiring diagrams are especially useful when you want to build, modify, or repair an electronic device.

Voltage/Current/Resistance Circuits

Most dc circuits can be boiled down to three major components: a voltage source, a set of conductors, and a resistance. This is shown in Fig. 4-5. The voltage or EMF source is E; the current in the conductor is I; the resistance is R.

You already know that there is a relationship among these three quantities. If one of them changes, then one or both of the others will change. If you make the resistance smaller, the current will get larger. If you reduce the applied voltage, the current will also decrease. If the current in the circuit increases, the voltage across the resistor will increase. There is a simple arithmetic relationship among these three quantities.

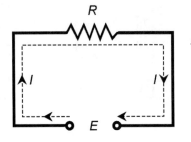

4-5   The basic elements of a dc circuit. The voltage is *E*, the current is *I*, and the resistance is *R*.

Ohm's Law

The interdependence among current, voltage, and resistance in dc circuits is called *Ohm's Law*, named after the scientist who supposedly first quantified it. Three formulas denote this law:

$$E = IR$$
$$I = E/R$$
$$R = E/I$$

You need only remember the first of these formulas in order to derive the others. The easiest way to remember it is to learn the abbreviations *E* for voltage, *I* for current, and *R* for resistance, and then remember that they appear in alphabetical order with the equal sign after the *E*. Sometimes the three symbols are arranged in the so-called Ohm's Law triangle, shown in Fig. 4-6. To find the value of a quantity, cover it up and read the positions of the others.

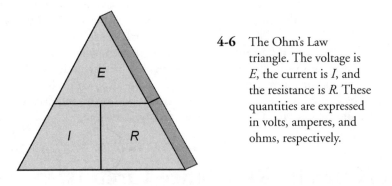

4-6   The Ohm's Law triangle. The voltage is *E*, the current is *I*, and the resistance is *R*. These quantities are expressed in volts, amperes, and ohms, respectively.

Remember that you must use units of volts, amperes, and ohms for the Ohm's Law formulas to yield a meaningful result! If you use, say, volts and microamperes to calculate a resistance, you cannot be sure of the units you'll end up with when you derive the final result. If the initial quantities are given in units other than volts, amperes, and ohms, convert to these units, and then calculate. After that, you can convert the calculated current, voltage, or resistance value to whatever size unit you want. For example, if you get 13,500,000 Ω as a calculated resistance, you might prefer to say that it's 13.5 MΩ.

Current Calculations

The first way to use Ohm's Law is to determine current in dc circuits. In order to find the current, you must know the voltage and the resistance, or be able to deduce them. Refer to the schematic diagram of Fig. 4-7. It consists of a dc voltage source, a voltmeter, some wire, an ammeter, and a calibrated, wide-range potentiometer.

Problem 4-1

Suppose that the dc generator in Fig. 4-7 produces 10 V and the potentiometer is set to a value of 10 Ω. What is the current?

This is solved by the formula $I = E/R$. Plug in the values for E and R; they are both 10, because the units are given in volts and ohms. Then $I = 10/10 = 1.0$ A.

Problem 4-2

Imagine that dc generator in Fig. 4-7 produces 100 V and the potentiometer is set to 10 kΩ. What is the current?

First, convert the resistance to ohms: 10 kΩ = 10,000 Ω. Then plug the values in: $I = 100/10,000 = 0.01$ A. You might prefer to express this as 10 mA.

Problem 4-3

Suppose that dc generator in Fig. 4-7 is set to provide 88.5 V, and the potentiometer is set to 477 MΩ. What is the current?

This problem involves numbers that aren't exactly round, and one of them is huge. But you can use a calculator. First, change the resistance value to ohms, so you get 477,000,000 Ω. Then plug into the Ohm's Law formula: $I = E/R = 88.5 / 477,000,000 = 0.000000186$ A. It is more reasonable to express this as 0.186 μA or 186 nA.

4-7　A circuit for working Ohm's Law problems.

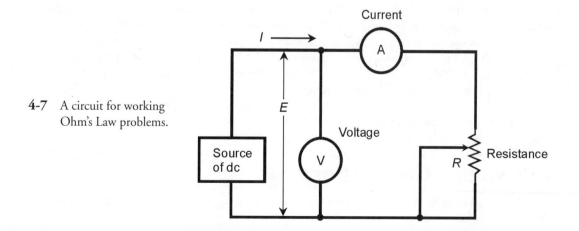

Voltage Calculations

The second application of Ohm's Law is to find unknown dc voltages when the current and the resistance are known. Let's work out some problems of this kind.

Problem 4-4

Suppose the potentiometer in Fig. 4-7 is set to 100 Ω, and the measured current is 10 mA. What is the dc voltage?

Use the formula $E = IR$. First, convert the current to amperes: 10 mA = 0.01 A. Then multiply: $E = 0.01 \times 100 = 1.0$ V. That's a little less than the voltage produced by a flashlight cell.

Problem 4-5

Adjust the potentiometer in Fig. 4-7 to a value of 157 kΩ, and suppose the current reading is 17.0 mA. What is the voltage of the source?

You must convert both the resistance and the current values to their proper units. A resistance of 157 kΩ is 157,000 Ω, and a current of 17.0 mA is 0.0170 A. Then $E = IR = 0.017 \times 157,000 = 2669$ V = 2.669 kV. You should round this off to 2.67 kV. This is a dangerously high voltage.

Problem 4-6

Suppose you set the potentiometer in Fig. 4-7 so that the meter reads 1.445 A, and you observe that the potentiometer scale shows 99 Ω. What is the voltage?

These units are both in their proper form. Therefore, you can plug them right in and use your calculator: $E = IR = 1.445 \times 99 = 143.055$ V. This can and should be rounded off—but to what extent? This is a good time to state an important rule that should be followed in all technical calculations.

The Rule of Significant Figures

Competent engineers and scientists go by the *rule of significant figures*, also called the *rule of significant digits*. After completing a calculation, you should always round the answer off to the *least* number of digits given in the input data numbers.

If you follow this rule in Problem 4-6, you must round off the answer to two significant digits, getting 140 V, because the resistance (99 Ω) is only specified to that level of accuracy. If the resistance were given as 99.0 Ω, then you would round off the answer to 143 V. If the resistance were given as 99.00 Ω, then you could state the answer as 143.1 V. However, any further precision in the resistance value would not entitle you to go to any more digits in your answer, unless the current were specified to more than four significant figures.

This rule takes some getting used to if you haven't known about it or practiced it before. But after a while, it will become a habit.

Resistance Calculations

Ohms' Law can be used to find a resistance between two points in a dc circuit when the voltage and the current are known.

Problem 4-7

If the voltmeter in Fig. 4-7 reads 24 V and the ammeter shows 3.0 A, what is the resistance of the potentiometer?

Use the formula $R = E/I$, and plug in the values directly, because they are expressed in volts and amperes: $R = 24/3.0 = 8.0\ \Omega$. Note that you can specify this value to two significant figures, the 8 and the 0, rather than saying simply 8 Ω. This is because you are given both the voltage and the current to two significant figures. If the ammeter reading had been given as 3 A, you would only be entitled to express the answer as 8 Ω, to one significant digit. The digit 0 can be, and often is, just as important in calculations as any of the other digits 1 through 9.

Problem 4-8

What is the value of the resistance in Fig. 4-7 if the current is 18 mA and the voltage is 229 mV?

First, convert these values to amperes and volts. This gives $I = 0.018$ A and $E = 0.229$ V. Then plug into the equation: $R = E/I = 0.229/0.018 = 13\ \Omega$.

Problem 4-9

Suppose the ammeter in Fig. 4-7 reads 52 μA and the voltmeter indicates 2.33 kV. What is the resistance?

Convert to amperes and volts, getting $I = 0.000052$ A and $E = 2330$ V. Then plug into the formula: $R = E/I = 2330/0.000052 = 45,000,000\ \Omega = 45\ M\Omega$.

Power Calculations

You can calculate the power P, in watts, in a dc circuit such as that shown in Fig. 4-7, by using the formula $P = EI$. This formula tells us that the power in watts is the product of the voltage in volts and the current in amperes. If you are not given the voltage directly, you can calculate it if you know the current and the resistance.

Recall the Ohm's Law formula for obtaining voltage: $E = IR$. If you know I and R but you don't know E, you can get the power P this way:

$$P = EI = (IR)I = I^2R$$

Suppose you're given only the voltage and the resistance. Remember the Ohm's Law formula for obtaining current: $I = E/R$. Therefore:

$$P = EI = E(E/R) = E^2/R$$

Problem 4-10

Suppose that the voltmeter in Fig. 4-7 reads 12 V and the ammeter shows 50 mA. What is the power dissipated by the potentiometer?

Use the formula $P = EI$. First, convert the current to amperes, getting $I = 0.050$ A. (Note that the last 0 counts as a significant digit.) Then multiply by 12 V, getting $P = EI = 12 \times 0.050 = 0.60$ W.

Problem 4-11

If the resistance in the circuit of Fig. 4-7 is 999 Ω and the voltage source delivers 3 V, what is the power dissipated by the potentiometer?

Use the formula $P = E^2/R = 3 \times 3/999 = 9/999 = 0.009$ W $= 9$ mW. You are justified in going to only one significant figure here.

Problem 4-12

Suppose the resistance in Fig. 4-7 is 47 kΩ and the current is 680 mA. What is the power dissipated by the potentiometer?

Use the formula $P = I^2R$, after converting to ohms and amperes. Then $P = 0.680 \times 0.680 \times 47,000 = 22,000$ W $= 22$ kW. (This is an unrealistic state of affairs: an ordinary potentiometer, such as the type you would use as the volume control in a radio, dissipating 22 kW, several times more than a typical household!)

Problem 4-13

How much voltage would be necessary to drive 680 mA through a resistance of 47 kΩ, as is described in the previous problem?

Use Ohm's Law to find the voltage: $E = IR = 0.680 \times 47,000 = 32,000$ V $= 32$ kV. That's the level of voltage you'd expect to find on a major utility power line, or in a high-power tube-type radio broadcast transmitter.

Resistances in Series

When you place resistances in series, their ohmic values add together to get the total resistance. This is easy to imagine, and it's easy to remember!

Problem 4-14

Suppose resistors with the following values are connected in series, as shown in Fig. 4-8: 112 Ω, 470 Ω, and 680 Ω. What is the total resistance of the series combination?

Simply add up the values, getting a total of $112 + 470 + 680 = 1262$ Ω. You might round this off to 1260 Ω. It depends on the *tolerances* of the resistors—how precise their actual values are to the ones specified by the manufacturer.

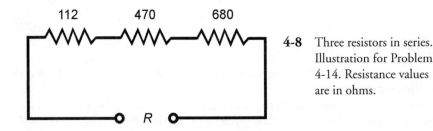

4-8   Three resistors in series. Illustration for Problem 4-14. Resistance values are in ohms.

Resistances in Parallel

When resistances are placed in parallel, they behave differently than they do in series. One way to look at resistances in parallel is to consider them as conductances instead. In parallel, conductances add up directly, just as resistances add up in series. If you change all the ohmic values to siemens, you can add these figures up and convert the final answer back to ohms.

The symbol for conductance is *G*. This figure, in siemens, is related to the resistance *R*, in ohms, by these formulas, which you learned in Chap. 2:

$$G = 1/R$$
$$R = 1/G$$

Problem 4-15

Consider five resistors in parallel. Call them R_1 through R_5, and call the total resistance *R* as shown in Fig. 4-9. Let the resistance values be as follows: $R_1 = 100\ \Omega$, $R_2 = 200\ \Omega$, $R_3 = 300\ \Omega$, $R_4 = 400\ \Omega$, and $R_5 = 500\ \Omega$. What is the total resistance, *R*, of this parallel combination?

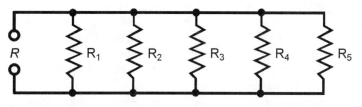

4-9   Five resistors of values R_1 through R_5, connected in parallel, produce a net resistance *R*. Illustration for Problems 4-15 and 4-16.

Converting the resistances to conductance values, you get: $G_1 = 1/100 = 0.01$ S, $G_2 = 1/200 = 0.005$ S, $G_3 = 1/300 = 0.00333$ S, $G_4 = 1/400 = 0.0025$ S, and $G_5 = 1/500 = 0.002$ S. Adding these gives $G = 0.01 + 0.005 + 0.00333 + 0.0025 + 0.002 = 0.0228$ S. The total resistance is therefore $R = 1/G = 1/0.0228 = 43.8\ \Omega$.

Problem 4-16

Suppose you have five resistors, called R_1 through R_5, connected in parallel as shown in Fig. 4-9. Suppose all the resistances, R_1 through R_5, are 4.70 kΩ. What is the total resistance, *R*, of this combination?

When you have two or more resistors connected in parallel and their resistances are all the same, the total resistance is equal to the resistance of any one component divided by the number of components. In this example, convert the resistance of any single resistor to 4700 Ω, and then divide this by 5. Thus, you can see that the total resistance is $4700/5 = 940\ \Omega$.

In a situation like this, where you have a bunch of resistors connected together to operate as a single unit, the total resistance is sometimes called the *net resistance*. Take note, too, that R is not italicized when it means *resistor*, but *R* is italicized when it means *resistance!*

Division of Power

When combinations of resistances are connected to a source of voltage, they draw current. You can figure out how much current they draw by calculating the total resistance of the combination, and then considering the network as a single resistor.

If the resistors in the network all have the same ohmic value, the power from the source is evenly distributed among them, whether they are hooked up in series or in parallel. For example, if there are eight identical resistors in series with a battery, the network consumes a certain amount of power, each resistor bearing $1/8$ of the load. If you rearrange the circuit so that the resistors are in parallel, the circuit will dissipate a certain amount of power (a lot more than when the resistors were in series), but again, each resistor will handle $1/8$ of the total power load.

If the resistances in the network do not all have identical ohmic values, they divide up the power unevenly. Situations like this are discussed in the next chapter.

Resistances in Series-Parallel

Sets of resistors, all having identical ohmic values, can be connected together in parallel sets of series networks, or in series sets of parallel networks. By doing this, the total power-handling capacity of the resistance can be greatly increased over that of a single resistor.

Sometimes, the total resistance of a *series-parallel network* is the same as the value of any one of the resistors. This is always true if the components are identical, and are in a network called an *n-by-n matrix*. That means, when n is a whole number, there are n parallel sets of n resistors in series (Fig. 4-10A), or else there are n series sets of n resistors in parallel (Fig. 4-10B). Either arrangement gives the same practical result.

Engineers and technicians sometimes use series-parallel networks to obtain resistances with large power-handling capacity. A series-parallel array of n by n resistors will have n^2 times that of a single resistor. Thus, a 3×3 series-parallel matrix of 2 W resistors can handle up to $3^2 \times 2 = 9 \times$

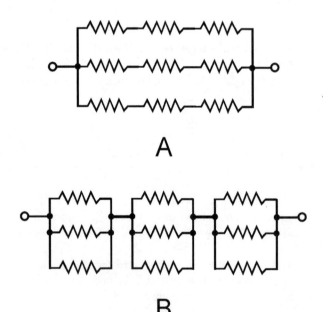

A

B

4-10   Series-parallel resistances. At A, sets of series resistors are connected in parallel. At B, sets of parallel resistances are connected in series. These examples show symmetrical *n-by-n* matrices with $n = 3$.

2 = 18 W, for example. A 10 × 10 array of 1-W resistors can dissipate up to 100 W. The total power-handling capacity is multiplied by the total number of resistors in the matrix. But this is true only if all the resistors have the same ohmic values, and the same power-dissipation ratings.

It is unwise to build series-parallel arrays from resistors with different ohmic values or power ratings. If the resistors have values and/or ratings that are even a little nonuniform, one of them might be subjected to more current than it can withstand, and it will burn out. Then the current distribution in the network can change so a second component fails, and then a third. It's hard to predict the current and power distribution in an array when its resistor values are all different.

If you need a resistance with a certain power-handling capacity, you must be sure the network can handle at least that much power. If a 50-W rating is required, and a certain combination will handle 75 W, that's fine. But it isn't good enough to build a circuit that will handle only 48 W. Some extra tolerance, say 10 percent over the minimum rating needed, is good, but it's silly to make a 500-W network using far more resistors than necessary, unless that's the only convenient combination given the parts available.

Nonsymmetrical series-parallel networks, made up from identical resistors, can increase the power-handling capability over that of a single resistor. But in these cases, the total resistance is not the same as the value of the single resistors. The overall power-handling capacity is always multiplied by the total number of resistors, whether the network is symmetrical or not, provided all the ohmic values are identical. In engineering work, cases sometimes arise where nonsymmetrical networks fit the need.

Quiz

Refer to the text in this chapter if necessary. A good score is at least 18 correct answers. The answers are in the back of the book.

1. Suppose you double the voltage in a simple dc circuit, and cut the resistance in half. The current will
    (a) become four times as great.
    (b) become twice as great.
    (c) stay the same as it was before.
    (d) become half as great.

2. You can expect to find a wiring diagram
    (a) on a sticker on the back of a television receiver.
    (b) in an advertisement for an electric oven.
    (c) in the service/repair manual for a two-way radio.
    (d) in the photograph of the front panel of a stereo hi-fi tuner.

For questions 3 through 11, please refer to Fig. 4-7. Remember to take significant figures into account when completing your calculations!

3. Given a dc voltage source delivering 24 V and a resistance of 3.3 kΩ, what is the current?
    (a) 0.73 A
    (b) 138 A
    (c) 138 mA
    (d) 7.3 mA

4. Suppose the resistance is 472 Ω, and the current is 875 mA. The source voltage must therefore be

    (a) 413 V.

    (b) 0.539 V.

    (c) 1.85 V.

    (d) none of the above.

5. Suppose the dc voltage is 550 mV and the current is 7.2 mA. Then the resistance is

    (a) 0.76 Ω.

    (b) 76 Ω.

    (c) 0.0040 Ω.

    (d) none of the above.

6. Given a dc voltage source of 3.5 kV and a resistance of 220 Ω, what is the current?

    (a) 16 mA

    (b) 6.3 mA

    (c) 6.3 A

    (d) None of the above

7. Suppose the resistance is 473,332 Ω, and the current flowing through it is 4.4 mA. The best expression for the voltage of the source is

    (a) 2082 V.

    (b) 110 kV.

    (c) 2.1 kV.

    (d) 2.08266 kV.

8. A source delivers 12 V and the current is 777 mA. The best expression for the resistance is

    (a) 15 Ω.

    (b) 15.4 Ω.

    (c) 9.3 Ω.

    (d) 9.32 Ω.

9. Suppose the voltage is 250 V and the current is 8.0 mA. The power dissipated by the potentiometer is

    (a) 31 mW.

    (b) 31 W.

    (c) 2.0 W.

    (d) 2.0 mW.

10. Suppose the voltage from the source is 12 V and the potentiometer is set for 470 Ω. The power dissipated in the resistance is approximately

    (a) 310 mW.

    (b) 25.5 mW.

    (c) 39.2 W.

    (d) 3.26 W.

11. If the current through the potentiometer is 17 mA and its resistance is set to 1.22 kΩ, what is the power dissipated by it?

    (a) 0.24 µW

    (b) 20.7 W

    (c) 20.7 mW

    (d) 350 mW

12. Suppose six resistors are hooked up in series, and each of them has a value of 540 Ω. What is the resistance across the entire combination?

    (a) 90 Ω

    (b) 3.24 kΩ

    (c) 540 Ω

    (c) None of the above

13. If four resistors are connected in series, each with a value of 4.0 kΩ, the total resistance is

    (a) 1 kΩ.

    (b) 4 kΩ.

    (c) 8 kΩ.

    (d) 16 kΩ.

14. Suppose you have three resistors in parallel, each with a value of 0.069 MΩ. Then the total resistance is

    (a) 23 Ω.

    (b) 23 kΩ.

    (c) 204 Ω.

    (d) 0.2 MΩ.

15. Imagine three resistors in parallel, with values of 22 Ω, 27 Ω, and 33 Ω. If a 12-V battery is connected across this combination, as shown in Fig. 4-11, what is the current drawn from the battery?

    (a) 1.4 A

    (b) 15 mA

    (c) 150 mA

    (d) 1.5 A

4-11 Illustration for Quiz Question 15. Resistance values are in ohms.

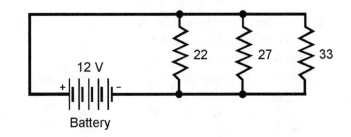

16. Imagine three resistors, with values of 47 Ω, 68 Ω, and 82 Ω, connected in series with a 50-V dc generator, as shown in Fig. 4-12. The total power consumed by this network of resistors is

   (a) 250 mW.
   (b) 13 mW.
   (c) 13 W.
   (d) impossible to determine from the data given.

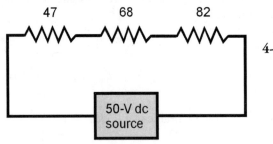

4-12 Illustration for Quiz Question 16. Resistance values are in ohms.

17. Suppose you have an unlimited supply of 1-W, 100-Ω resistors. You need to get a 100-Ω, 10-W resistor. This can be done most cheaply by means of a series-parallel matrix of

   (a) 3 × 3 resistors.
   (b) 4 × 3 resistors.
   (c) 4 × 4 resistors.
   (d) 2 × 5 resistors.

18. Suppose you have an unlimited supply of 1-W, 1000-Ω resistors, and you need a 500-Ω resistance rated at 7 W or more. This can be done by assembling

   (a) four sets of two resistors in series, and connecting these four sets in parallel.
   (b) four sets of two resistors in parallel, and connecting these four sets in series.
   (c) a 3 × 3 series-parallel matrix of resistors.
   (d) a series-parallel matrix, but something different than those described above.

19. Suppose you have an unlimited supply of 1-W, 1000-Ω resistors, and you need to get a 3000-Ω, 5-W resistance. The best way is to

   (a) make a 2 × 2 series-parallel matrix.
   (b) connect three of the resistors in parallel.
   (c) make a 3 × 3 series-parallel matrix.
   (d) do something other than any of the above.

20. Good engineering practice usually requires that a series-parallel resistive network be assembled

   (a) from resistors that are all different.
   (b) from resistors that are all identical.
   (c) from a series combination of resistors in parallel but not from a parallel combination of resistors in series.
   (d) from a parallel combination of resistors in series, but not from a series combination of resistors in parallel.

5
CHAPTER

Direct-Current Circuit Analysis

IN THIS CHAPTER, YOU'LL LEARN MORE ABOUT DC CIRCUITS AND HOW THEY BEHAVE UNDER VARIOUS conditions. These principles apply to most ac utility circuits as well.

Current through Series Resistances

Have you ever used those tiny holiday lights that come in strings? If one bulb burns out, the whole set of bulbs goes dark. Then you have to find out which bulb is bad, and replace it to get the lights working again. Each bulb works with something like 10 V; there are about a dozen bulbs in the string. You plug in the whole bunch and the 120-V utility mains drive just the right amount of current through each bulb.

In a series circuit, such as a string of light bulbs (Fig. 5-1), the current at any given point is the same as the current at any other point. The ammeter, A, is shown in the line between two of the bulbs. If it were moved anywhere else along the current path, it would indicate the same current.

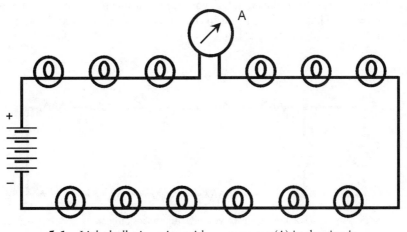

5-1   Light bulbs in series, with an ammeter (A) in the circuit.

This is true in any series dc circuit, no matter what the components actually are, and regardless of whether or not they all have the same resistance.

If the bulbs in Fig. 5-1 had different resistances, some of them would consume more power than others. In case one of the bulbs in Fig. 5-1 burns out, and its socket is then shorted out instead of filled with a replacement bulb, the current through the whole chain will increase, because the overall resistance of the string will go down. This will force each of the remaining bulbs to carry more current, and pretty soon another bulb would burn out because of the excessive current. If it, too, were replaced with a short circuit, the current would be increased still further. A third bulb would blow out almost right away thereafter.

Voltages across Series Resistances

The bulbs in the string of Fig. 5-1, being all the same, each get the same amount of voltage from the source. If there are a dozen bulbs in a 120-V circuit, each bulb has a potential difference of 10 V across it. This will remain true even if the bulbs are replaced with brighter or dimmer ones, as long as all the bulbs in the string are identical.

Look at the schematic diagram of Fig. 5-2. Each resistor carries the same current. Each resistance R_n has a potential difference E_n across it equal to the product of the current and the resistance of that particular resistor. The voltages E_n are in series, like cells in a battery, so they add together. What if the voltages across all the resistors added up to something more or less than the supply voltage, E? Then there would have to be a "phantom EMF" someplace, adding or taking away voltage. But that's impossible. Voltage cannot come out of nowhere!

Look at this another way. The voltmeter V in Fig. 5-2 shows the voltage E of the battery, because the meter is hooked up across the battery. The voltmeter V also shows the sum of the voltages E_n across the set of resistances, because it's connected across the whole combination. The meter says the same thing whether you think of it as measuring the battery voltage E or as measuring the sum of the voltages E_n across the series combination of resistances. Therefore, E is equal to the sum of the voltages E_n.

How do you find the voltage across any particular resistance R_n in a circuit like the one in Fig. 5-2? Remember Ohm's Law for finding voltage: $E = IR$. Remember, too, that you must use volts, ohms, and amperes when making calculations.

In order to find the current in the circuit, I, you need to know the total resistance and the supply voltage; then $I = E/R$. First find the current in the whole circuit; then find the voltage across any particular resistor.

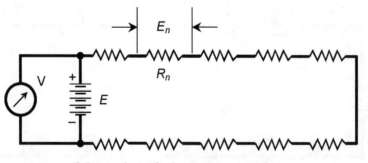

5-2   Analysis of voltages in a series circuit.

Problem 5-1

In Fig. 5-2, there are 10 resistors. Five of them have values of 10 Ω, and the other five have values of 20 Ω. The power source is 15-V dc. What is the voltage across any one of the 10-Ω resistors? Across any one of the 20-Ω resistors?

First, find the total resistance: $R = (10 \times 5) + (20 \times 5) = 50 + 100 = 150\ \Omega$. Then find the current: $I = E/R = 15/150 = 0.10$ A. This is the current through each of the resistances in the circuit.

- If $R_n = 10\ \Omega$, then $E_n = IR_n = 0.1 \times 10 = 1.0$ V.
- If $R_n = 20\ \Omega$, then $E_n = IR_n = 0.1 \times 20 = 2.0$ V.

Let's check to be sure all of these voltages add up to the supply voltage. There are five resistors with 1.0 V across each, for a total of 5.0 V; there are also five resistors with 2.0 V across each, for a total of 10 V. So the sum of the voltages across the resistors is $5.0 + 10 = 15$ V.

Problem 5-2

In the circuit of Fig. 5-2, what will happen to the voltages across the resistances if one of the 20-Ω resistances is replaced with a short circuit?

In this case the total resistance becomes $R = (10 \times 5) + (20 \times 4) = 50 + 80 = 130\ \Omega$. The current is therefore $I = E/R = 15/130 = 0.12$ A. This is the current at any point in the circuit, rounded off to two significant figures.

The voltage E_n across any of the 10-Ω resistances R_n is equal to IR_n, which is $0.12 \times 10 = 1.2$ V. The voltage E_n across any of the 20-Ω resistances R_n is equal to IR_n, which is $0.12 \times 20 = 2.4$ V. Checking the total voltage, add $(5 \times 1.2) + (4 \times 2.4) = 6.0 + 9.6 = 15.6$ V. This rounds off to 16 V when we cut it down to two significant figures.

A "Rounding-Off Bug"

Compare the result for total voltage in Problem 5-2 with the result for total voltage in Problem 5-1. What is going on here? Where does the extra volt come from in the second calculation? Certainly, shorting out one of the resistances cannot cause the battery voltage to change!

This is an example of what can happen when you round off to a certain number of significant figures after calculating the value of some parameter X in a circuit, then change a different parameter Y in the circuit, and finally calculate the value of X again, rounding off to the same number of significant digits as you did the first time. The discrepancy is the result of a "rounding-off bug."

If this bug bothers you (and it should), keep all the digits your calculator will hold while you go through the solution process for Problem 5-2. The current in the circuit, as obtained by means of a calculator that can show 10 digits, should come out as 0.115384615 A. When you find the voltages across all the resistances R_n, accurate to all these extra digits, and then add them up, you'll get a final rounded-off voltage of 15 V.

This example shows why it is a good idea to wait until you get the final answer in a calculation, or set of calculations, involving a particular circuit before you round off to the allowed number of significant digits. Rounding-off bugs of the sort we have just seen can be more than mere annoyances. They are easy to overlook, but they can generate large errors in *iterative processes* involving calculations that are done over and over.

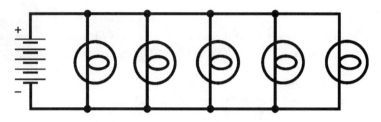

5-3   Light bulbs in parallel.

Voltage across Parallel Resistances

Imagine a set of ornamental light bulbs connected in parallel (Fig. 5-3). This is the method used for outdoor holiday lighting or for bright indoor lighting. It's easier to repair a parallel-wired string of such lights if one bulb should burn out than it is to fix a series-wired string. And in the parallel configuration, the failure of one bulb does not cause total system failure.

In a parallel circuit, the voltage across each component is equal to the supply or battery voltage. The current drawn by each component depends only on the resistance of that particular device. In this sense, the components in a parallel-wired circuit operate independently, as opposed to the series-wired circuit in which they all interact.

If any one branch of a parallel circuit opens up, is disconnected, or is removed, the conditions in the other branches do not change. If new branches are added, assuming the power supply can handle the load, conditions in previously existing branches are not affected.

Currents through Parallel Resistances

Refer to the schematic diagram of Fig. 5-4. The resistances are called R_n. The total parallel resistance in the circuit is R. The battery voltage is E. The current in any particular branch n, containing resistance R_n, is measured by ammeter A and is called I_n. The sum of all the currents I_n is equal to the total current, I, drawn from the battery. The current is divided up in the parallel circuit in a manner similar to the way that voltage is divided up in a series circuit.

Conventional Current

Have you noticed that the direction of current flow in Fig. 5-4 is portrayed as outward from the positive battery terminal? Don't electrons, which are the actual charge carriers in a wire, flow out of the minus terminal of a battery? Yes, that's true; but scientists consider *theoretical current*, more often called *conventional current* (because it is defined by convention), to flow from positive to negative voltage points, rather than from negative to positive.

Problem 5-3

Suppose that the battery in Fig. 5-4 delivers 12 V. Further suppose that there are 12 resistors, each with a value of 120 Ω in the parallel circuit. What is the total current, I, drawn from the battery?

First, find the total resistance. This is easy, because all the resistors have the same value. Just divide $R_n = 120$ by 12 to get $R = 10\ \Omega$. Then the current can be found by Ohm's Law: $I = E/R = 12/10 = 1.2$ A.

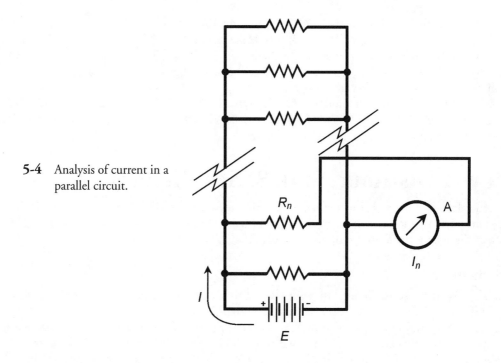

5-4 Analysis of current in a parallel circuit.

Problem 5-4

In the circuit of Fig. 5-4, what does the ammeter say?

This involves finding the current in any given branch. The voltage is 12 V across every branch, and $R_n = 120\ \Omega$. Therefore I_n, the ammeter reading, is found by Ohm's Law: $I_n = E/R_n = 12/120 = 0.10$ A.

Because this is a parallel circuit, all of the branch currents I_n should add up to get the total current, I. There are 12 identical branches, each carrying 0.10 A; therefore the total current is $0.10 \times 12 = 1.2$ A. It checks out.

Problem 5-5

Suppose three resistors are in parallel across a battery that supplies $E = 12$ V. The resistances are $R_1 = 22\ \Omega$, $R_2 = 47\ \Omega$, and $R_3 = 68\ \Omega$. These resistances carry currents I_1, I_2, and I_3, respectively. What is the current, I_3, through R_3?

This problem is solved by means of Ohm's Law as if R_3 is the only resistance in the circuit. There's no need to worry about the parallel combination. The other branches do not affect I_3. Thus $I_3 = E/R_3 = 12/68 = 0.18$ A.

Problem 5-6

What is the total current drawn by the circuit described in Problem 5-5?

There are two ways to go at this. One method involves finding the total resistance, R, of R_1, R_2, and R_3 in parallel, and then calculating I based on R. Another way is to find the currents through R_1, R_2, and R_3 individually, and then add them up.

Using the first method, first change the resistances R_n into conductances G_n. This gives $G_1 = 1/R_1 = 1/22 = 0.04545$ S, $G_2 = 1/R_2 = 1/47 = 0.02128$ S, and $G_3 = 1/R_3 = 1/68 = 0.01471$ S. Adding these gives $G = 0.08144$ S. The resistance is therefore $R = 1/G = 1/0.08144 = 12.279$ Ω. Use Ohm's Law to find $I = E/R = 12/12.279 = 0.98$ A. Note that extra digits are used throughout the calculation, rounding off only at the end.

Now let's try the other method. Find $I_1 = E/R_1 = 12/22 = 0.5455$ A, $I_2 = E/R_2 = 12/47 = 0.2553$ A, and $I_3 = E/R_3 = 12/68 = 0.1765$ A. Adding these gives $I = I_1 + I_2 + I_3 = 0.5455 + 0.2553 + 0.1765 = 0.9773$ A, which rounds off to 0.98 A.

Power Distribution in Series Circuits

When calculating the power in a circuit containing resistors in series, all you need to do is find out the current, I, that the circuit is carrying. Then it's easy to calculate the power P_n dissipated by any one of the resistances R_n, based on the formula $P_n = I^2 R_n$.

Problem 5-7

Suppose we have a series circuit with a supply of 150 V and three resistances: $R_1 = 330$ Ω, $R_2 = 680$ Ω, and $R_3 = 910$ Ω. What is the power dissipated by R_2?

First, find the current that flows through the circuit. Calculate the total resistance first. Because the resistors are in series, the total is $R = 330 + 680 + 910 = 1920$ Ω. The current is $I = 150/1920 = 0.07813$ A. The power dissipated by R_2 is therefore $P_2 = I^2 R_2 = 0.07813 \times 0.07813 \times 680 = 4.151$ W. Round this off to three significant digits, because that's all we have in the data, to obtain 4.15 W.

The total wattage dissipated in a series circuit is equal to the sum of the wattages dissipated in each resistance.

Problem 5-8

Calculate the total dissipated power P in the circuit of Problem 5-7 by two different methods.

First, let's figure out the power dissipated by each of the three resistances separately, and then add the figures up. The power P_2 is already known. Let's use all the significant digits we have while we calculate. Thus, as found in Problem 5-7, $P_2 = 4.151$ W. Recall that the current is $I = 0.07813$ A. Then $P_1 = 0.07813 \times 0.07813 \times 330 = 2.014$ W, and $P_3 = 0.07813 \times 0.07813 \times 910 = 5.555$ W. Adding the three power figures gives us $P = P_1 + P_2 + P_3 = 2.014 + 4.151 + 5.555 = 11.720$ W. We should round this off to 11.7 W.

The second method is to find the total series resistance and then calculate the power. The series resistance is $R = 1920$ Ω, as found in Problem 5-7. Then $P = I^2 R = 0.07813 \times 0.07813 \times 1920 = 11.72$ W. Again, we should round this to 11.7 W.

Power Distribution in Parallel Circuits

When resistances are wired in parallel, they each consume power according to the same formula, $P = I^2 R$. But the current is not the same in each resistance. An easier method to find the power P_n dissipated by each of the various resistances R_n is to use the formula $P_n = E^2/R_n$, where E is the voltage of the supply or battery. This voltage is the same across every branch resistance in a parallel circuit.

Problem 5-9

Suppose a dc circuit contains three resistances $R_1 = 22\ \Omega$, $R_2 = 47\ \Omega$, and $R_3 = 68\ \Omega$ across a battery that supplies a voltage of $E = 3.0$ V. Find the power dissipated by each resistance.

Let's find the square of the supply voltage, E^2, first. We'll be needing this figure often: $E^2 = 3.0 \times 3.0 = 9.0$. Then the wattages dissipated by resistances R_1, R_2, and R_3 respectively are $P_1 = 9.0/22 = 0.4091$ W, $P_2 = 9.0/47 = 0.1915$ W, and $P_3 = 9.0/68 = 0.1324$ W. These should be rounded off to $P_1 = 0.41$ W, $P_2 = 0.19$ W, and $P_3 = 0.13$ W. (But let's remember the values to four significant figures for the next problem!)

In a parallel circuit, the total dissipated wattage is equal to the sum of the wattages dissipated by the individual resistances.

Problem 5-10

Find the total consumed power of the resistor circuit in Problem 5-9 using two different methods.

The first method involves adding P_1, P_2, and P_3. Let's use the four-significant-digit values to avoid the possibility of encountering the rounding-off bug. The total power thus calculated is $P = 0.4091 + 0.1915 + 0.1324 = 0.7330$ W. Now that we've finished the calculation, we should round it off to 0.73 W.

The second method involves finding the net resistance R of the parallel combination. You can do this calculation yourself. Determining it to four significant digits, you should get a net resistance of $R = 12.28\ \Omega$. Then $P = E^2/R = 9.0/12.28 = 0.7329$ W. Now that the calculation is done, this can be rounded to 0.73 W.

It's the Law!

In electricity and electronics, dc circuit analysis can be made easier if you are acquainted with certain axioms, or *laws.* Here they are:

- The current in a series circuit is the same at every point along the way.
- The voltage across any resistance in a parallel combination of resistances is the same as the voltage across any other resistance, or across the whole set of resistances.
- The voltages across resistances in a series circuit always add up to the supply voltage.
- The currents through resistances in a parallel circuit always add up to the total current drawn from the supply.
- The total wattage consumed in a series or parallel circuit is always equal to the sum of the wattages dissipated in each of the resistances.

Now, let's get acquainted with two of the most famous laws that govern dc circuits. These rules are broad and sweeping, and they make it possible to analyze complicated series-parallel dc networks.

Kirchhoff's First Law

The physicist Gustav Robert Kirchhoff (1824–1887) was a researcher and experimentalist in a time when little was understood about how electric currents flow. Nevertheless, he used certain common-sense notions to deduce two important properties of dc circuits.

Kirchhoff reasoned that dc ought to behave something like water in a network of pipes, and that the current going into any point ought to be the same as the current going out of that point. This, Kirchhoff thought, must be true for any point in a circuit, no matter how many branches lead into or out of the point.

Two examples of this principles are shown in Fig. 5-5. Examine illustration A. At point X, I, the current going in, equals $I_1 + I_2$, the current going out. At point Y, $I_2 + I_1$, the current going in, equals I, the current going out. Now look at illustration B. In this case, at point Z, the current $I_1 + I_2$ going in is equal to the current $I_3 + I_4 + I_5$ going out. These are examples of *Kirchhoff's First Law*. We can also call it *Kirchhoff's Current Law* or the principle of *conservation of current*.

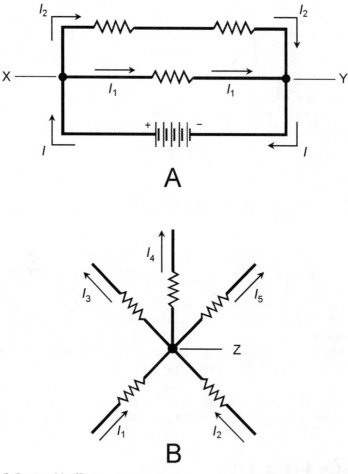

5-5   Kirchhoff's First Law. At A, the current into point X or point Y is the same as the current out of that point. That is, $I = I_1 + I_2$. At B, the current into point Z equals the current flowing out of point Z. That is, $I_1 + I_2 = I_3 + I_4 + I_5$. Illustration for Quiz Questions 13 and 14.

Problem 5-11

Refer to Fig. 5-5A. Suppose all three resistors have values of 100 Ω, and that $I_1 = 2.0$ A and $I_2 = 1.0$ A. What is the battery voltage?

First, find the current I drawn from the battery: $I = I_1 + I_2 = 2.0 + 1.0 = 3.0$ A. Next, find the resistance of the entire network. The two 100-Ω resistances in series give a value of 200 Ω, and this is in parallel with 100 Ω. You can do the calculations and find that the total resistance, R, connected across the battery is 66.67 Ω. Then $E = IR = 66.67 \times 3.0 = 200$ V.

Problem 5-12

In Fig. 5-5B, suppose each of the two resistors below point Z has a value of 100 Ω, and all three resistors above point Z have values of 10.0 Ω. Suppose the current through each 100-Ω resistor is 500 mA. What is the current through any one of the 10.0-Ω resistors, assuming that the current through all three 10.0-Ω resistors is the same? What is the voltage across any one of the three 10.0-Ω resistors?

The total current into point Z is 500 mA + 500 mA = 1.00 A. This is divided equally among the three 10-Ω resistors. Therefore, the current through any one of them is 1.00/3 A = 0.333 A. The voltage across any one of the 10.0-Ω resistors can thus found by Ohm's Law: $E = IR = 0.333 \times 10.0 = 3.33$ V.

Kirchhoff's Second Law

The sum of all the voltages, as you go around a circuit from some fixed point and return there from the opposite direction, and taking polarity into account, is always zero. Does this seem counterintuitive? Let's think about it a little more carefully.

What Kirchhoff was expressing, when he wrote his second law, is the principle that voltage cannot appear out of nowhere, nor can it vanish. All the potential differences must ultimately cancel each other out in any closed dc circuit, no matter how complicated that circuit happens to be. This is *Kirchhoff's Second Law*. We can also call it *Kirchhoff's Voltage Law* or the principle of *conservation of voltage*.

Remember the rule you've already learned about series dc circuits: The sum of the voltages across all the individual resistances adds up to the supply voltage. This statement is true as far as it goes, but it is an oversimplification, because it ignores polarity. The *polarity* of the potential difference across each resistance is *opposite* to the polarity of the potential difference across the battery. So when you add up the potential differences all the way around the circuit, taking polarity into account for every single component, you always get a net voltage of zero.

An example of Kirchhoff's Second Law is shown in Fig. 5-6. The voltage of the battery, E, has polarity opposite to the sum of the potential differences across the resistors, $E_1 + E_2 + E_3 + E_4$. Therefore, $E + E_1 + E_2 + E_3 + E_4 = 0$.

5-6 Kirchhoff's Second Law. The sum of the voltages across the resistances is equal to, but has opposite polarity from, the supply voltage. Therefore, $E + E_1 + E_2 + E_3 + E_4 = 0$. Illustration for Quiz Questions 15 and 16.

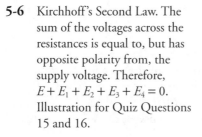
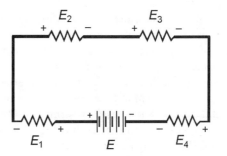

Problem 5-13

Refer to the diagram of Fig. 5-6. Suppose the four resistors have values of 50 Ω, 60 Ω, 70Ω, and 80 Ω, and that the current through each of them is 500 mA. What is the battery voltage, E?

Find the voltages E_1, E_2, E_3, and E_4 across each of the resistors. This can be done using Ohm's Law. For E_1, say with the 50-Ω resistor, calculate $E_1 = 0.500 \times 50 = 25$ V. In the same way, you can calculate $E_2 = 30$ V, $E_3 = 35$ V, and $E_4 = 40$ V. The supply voltage is the sum $E_1 + E_2 + E_3 + E_4 = 25 + 30 + 35 + 40 = 130$ V. Kirchhoff's Second Law tells us that the polarities of the voltages across the resistors are in the opposite direction from that of the battery.

Problem 5-14

In the situation shown by Fig. 5-6, suppose the battery provides 20 V. Suppose the resistors labeled with voltages E_1, E_2, E_3, and E_4 have ohmic values in the ratio 1:2:3:4 respectively. What is the voltage E_3?

This problem does not provide any information about current in the circuit, nor does it give you the exact resistances. But you don't need to know these things to solve for E_3. Regardless of what the actual ohmic values are, the ratio $E_1{:}E_2{:}E_3{:}E_4$ will be the same as long as the resistances are in the ratio 1:2:3:4. We can plug in any ohmic values we want for the values of the resistors, as long as they are in that ratio.

Let R_n be the resistance across which the voltage is E_n, where n can range from 1 to 4. Now that we have given the resistances specific names, suppose $R_1 = 1.0$ Ω, $R_2 = 2.0$ Ω, $R_3 = 3.0$ Ω, and $R_4 = 4.0$ Ω. These are in the proper ratio. The total resistance is $R = R_1 + R_2 + R_3 + R_4 = 1.0 + 2.0 + 3.0 + 4.0 = 10$ Ω. You can calculate the current as $I = E/R = 20/10 = 2.0$ A. Then the voltage E_3, across the resistance R_3, is given by Ohm's Law as $E_3 = IR_3 = 2.0 \times 3.0 = 6.0$ V.

Voltage Divider Networks

Resistances in series produce ratios of voltages, and these ratios can be tailored to meet certain needs by means of *voltage divider networks.*

When a voltage divider network is designed and assembled, the resistance values should be as small as possible without causing too much current drain on the battery or power supply. (In practice, the optimum values depend on the nature of the circuit being designed. This is a matter for engineers, and specific details are beyond the scope of this course.) The reason for choosing the smallest possible resistances is that, when the divider is used with a circuit, you do not want that circuit to upset the operation of the divider. The voltage divider "fixes" the intermediate voltages best when the resistance values are as small as the current-delivering capability of the power supply will allow.

Figure 5-7 illustrates the principle of voltage division. The individual resistances are R_1, R_2, R_3, . . . , R_n. The total resistance is $R = R_1 + R_2 + R_3 + \ldots + R_n$. The supply voltage is E, and the current in the circuit is therefore $I = E/R$. At the various points P_1, P_2, P_3, . . . , P_n, the potential differences relative to the negative battery terminal are E_1, E_2, E_3, . . . , E_n, respectively. The last voltage, E_n, is the same as the battery voltage, E. All the other voltages are less than E, and ascend in succession, so that $E_1 < E_2 < E_3 < \ldots < E_n$. (The mathematical symbol < means "is less than.")

The voltages at the various points increase according to the sum total of the resistances up to each point, in proportion to the total resistance, multiplied by the supply voltage. Thus, the voltage E_1 is equal to ER_1/R. The voltage E_2 is equal to $E(R_1 + R_2)/R$. The voltage E_3 is equal to $E(R_1 + R_2 + R_3)/R$. This process goes on for each of the voltages at points all the way up to $E_n = E(R_1 + R_2 + R_3 + \ldots + R_n)/R = ER/R = E$.

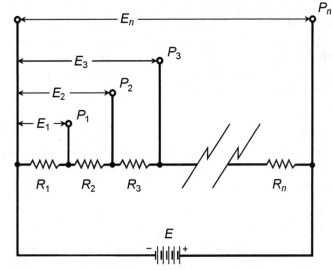

5-7 General arrangement for a voltage-divider circuit. Illustration for Quiz Questions 19 and 20.

Problem 5-15

Suppose you are building an electronic circuit, and the battery supplies 9.0 V. The minus terminal is at common (chassis) ground. You need to provide a circuit point where the dc voltage is +2.5 V. Give an example of a pair of resistors that can be connected in a voltage divider configuration, such that +2.5 V appears at some point.

Examine the schematic diagram of Fig. 5-8. There are infinitely many different combinations of resistances that will work here! Pick some total value, say $R = R_1 + R_2 = 1000 \ \Omega$. Keep in mind that the ratio $R_1{:}R$ will always be the same as the ratio $E_1{:}E$. In this case, $E_1 = 2.5$ V, so $E_1{:}E = 2.5/9.0 = 0.28$. This means that you want the ratio $R_1{:}R$ to be equal to 0.28. You have chosen to

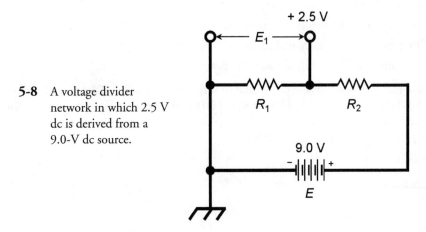

5-8 A voltage divider network in which 2.5 V dc is derived from a 9.0-V dc source.

make R equal to 1000 Ω. This means R_1 must be 280 Ω in order to get the ratio R_1:R = 0.28. The value of R_2 is the difference between R and R_1. That is 1000 − 280 = 720 Ω.

In a practical circuit, you would want to choose the smallest possible value for R. This might be less than 1000 Ω, or it might be more, depending on the nature of the circuit and the current-delivering capability of the battery. It's not the actual values of R_1 and R_2 that determine the voltage you get at the intermediate point, but their ratio.

Problem 5-16

What is the current I, in milliamperes, drawn by the entire network of series resistances in the situation described in Problem 5-15 and its solution?

Use Ohm's Law to get $I = E/R = 9.0/1000 = 0.0090$ A = 9.0 mA.

Problem 5-17

Suppose that it is all right for the voltage divider network to draw up to 100 mA of current in the situation shown by Fig. 5-8 and posed by Problem 5-15. You want to design the network to draw this amount of current, because that will offer the best voltage regulation for the circuit to be operated from the network. What values of resistances R_1 and R_2 should you use?

Calculate the total resistance first, using Ohm's Law. Remember to convert 100 mA to amperes! That means you use the figure $I = 0.100$ A in your calculations. Then $R = E/I = 9.0/0.100 = 90$ Ω. The ratio of resistances that you need is R_1:R_2 = 2.5/9.0 = 0.28. You should use R_1 = 0.28 × 90 = 25 Ω. The value of R_2 is the difference between R and R_1. That is, $R_2 = R - R_1 = 90 − 25 = 65$ Ω.

Quiz

Refer to the text in this chapter if necessary. A good score is at least 18 correct answers. The answers are in the back of the book.

1. In a series-connected string of ornament bulbs, if one bulb gets shorted out, which of the following will occur?

(a) All the other bulbs will go out.

(b) The current in the string will go up. ,

(c) The current in the string will go down.

(d) The current in the string will stay the same.

2. Imagine that four resistors are connected in series across a 6.0-V battery, and the ohmic values are R_1 = 10 Ω, R_2 = 20 Ω, R_3 = 50 Ω, and R_4 = 100 Ω, as shown in Fig. 5-9. What is the voltage across the resistance R_2?

(a) 0.18 V

(b) 33 mV

(c) 5.6 mV

(d) 0.67 V

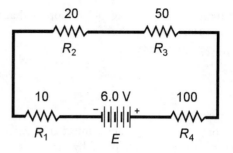

5-9 Illustration for Quiz Questions 2, 3, 8, and 9. Resistance values are in ohms.

3. In the scenario shown by Fig. 5-9, what is the voltage across the combination of R_3 and R_4?
   (a) 0.22 V
   (b) 0.22 mV
   (c) 5.0 V
   (d) 3.3 V

4. Suppose three resistors are connected in parallel across a battery that delivers 15 V, and the ohmic values are $R_1 = 470\ \Omega$, $R_2 = 2.2\ k\Omega$, and $R_3 = 3.3\ k\Omega$, as shown in Fig. 5-10. The voltage across the resistance R_2 is
   (a) 4.4 V.
   (b) 5.0 V.
   (c) 15 V.
   (d) not determinable from the data given.

5-10 Illustration for Quiz Questions 4, 5, 6, 7, 10, and 11. Resistances are in ohms, where k indicates multiplication by 1000.

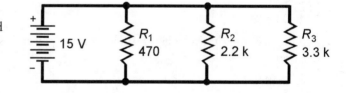

5. In the situation shown by Fig. 5-10, what is the current through R_2?
   (a) 6.8 mA
   (b) 43 mA
   (c) 0.15 A
   (d) 6.8 A

6. In the situation shown by Fig. 5-10, what is the total current drawn from the source?
   (a) 6.8 mA
   (b) 43 mA
   (c) 0.15 A
   (d) 6.8 A

7. In the situation shown by Fig. 5-10, suppose that resistor R_2 opens up. The current through the other two resistors will

(a) increase.

(b) decrease.

(c) drop to zero.

(d) not change.

8. Suppose that four resistors are connected in series with a 6.0-V supply, with values shown in Fig. 5-9. What is the power dissipated by the whole combination?

(a) 0.2 W

(b) 6.5 mW

(c) 200 W

(d) 6.5 W

9. In the situation shown by Fig. 5-9, what is the power dissipated by R_4?

(a) 11 mW

(b) 0.11 W

(c) 0.2 W

(d) 6.5 mW

10. Suppose that three resistors are in parallel as shown in Fig. 5-10. What is the power dissipated by the whole set of resistors?

(a) 5.4 W

(b) 5.4 µW

(c) 650 W

(d) 0.65 W

11. In the situation shown by Fig. 5-10, what is the power dissipated in resistance R_1?

(a) 32 mW

(b) 0.48 W

(c) 2.1 W

(d) 31 W

12. Fill in the blank in the following sentence to make it true: "In a series or parallel dc circuit, the sum of the _____s in each component is equal to the total _____ provided by the power supply."

(a) current

(b) voltage

(c) wattage

(d) resistance

13. Look at Fig. 5-5A. Suppose the resistors each have values of 33 Ω and the battery supplies 24 V. What is the current I_1?

(a) 1.1 A

(b) 0.73 A

(c) 0.36 A

(d) Not determinable from the information given

14. Look at Fig. 5-5B. Let each resistor have a value of 820 Ω. Suppose the top three resistors all lead to identical light bulbs. If $I_1 = 50$ mA and $I_2 = 70$ mA, what is the power dissipated in the resistor carrying current I_4?

    (a) 33 W

    (b) 40 mW

    (c) 1.3 W

    (d) It can't be found using the information given.

15. Refer to Fig. 5-6. Suppose the resistances R_1, R_2, R_3, and R_4 are in exactly the ratio 1:2:4:8 from left to right, and the battery supplies 30 V. What is the voltage E_2?

    (a) 4.0 V

    (b) 8.0 V

    (c) 16 V

    (d) It is not determinable from the data given.

16. Refer to Fig. 5-6. Suppose the resistances are each 3.3 kΩ, and the battery supplies 12 V. If the plus terminal of a dc voltmeter is placed between resistances R_1 and R_2 (with voltages E_1 and E_2 across them, respectively), and the minus terminal of the voltmeter is placed between resistances R_3 and R_4 (with voltages E_3 and E_4 across them, respectively), what will the meter register?

    (a) 0.0 V

    (b) 3.0 V

    (c) 6.0 V

    (d) 12 V

17. In a voltage divider network, the total resistance

    (a) should be large to minimize current drain.

    (b) should be as small as the power supply will allow.

    (c) is not important.

    (d) should be such that the current is kept to 100 mA.

18. The maximum voltage output from a voltage divider

    (a) is a fraction of the power supply voltage.

    (b) depends on the total resistance.

    (c) is equal to the supply voltage.

    (d) depends on the ratio of resistances.

19. Refer to Fig. 5-7. Suppose the battery voltage E is 18.0 V, and there are four resistances in the network such that $R_1 = 100$ Ω, $R_2 = 22.0$ Ω, $R_3 = 33.0$ Ω, and $R_4 = 47.0$ Ω. What is the voltage E_3 at P_3?

    (a) 4.19 V

    (b) 13.8 V

(c) 1.61 V

(d) 2.94 V

20. Refer to Fig. 5-7. Suppose the battery voltage is 12 V, and you want to obtain intermediate voltages of 3.0 V, 6.0 V, and 9.0 V. Suppose that a maximum of 200 mA is allowed to be drawn from the battery. What should the resistances, R_1, R_2, R_3, and R_4 be, respectively?

(a) 15 Ω, 30 Ω, 45 Ω, and 60 Ω

(b) 60 Ω, 45 Ω, 30 Ω, and 15 Ω

(c) 15 Ω, 15 Ω, 15 Ω, and 15 Ω

(d) There isn't enough information given here to design the circuit.

CHAPTER

Resistors

ALL ELECTRICAL COMPONENTS, DEVICES, AND SYSTEMS HAVE SOME RESISTANCE. IN EVERYDAY PRACTICE, there is no such thing as a perfect electrical conductor. You've seen some examples of circuits containing components that are deliberately designed to oppose the flow of current. These components are *resistors*. In this chapter, you'll learn all about them.

Purpose of the Resistor

Resistors play diverse roles in electrical and electronic equipment. Here are a few of the more common ways they are used.

Voltage Division

You've learned how voltage dividers can be designed using resistors. The resistors dissipate some power in doing this job, but the resulting voltages can provide the proper *biasing* of electronic circuits. This ensures, for example, that an amplifier or oscillator will function in the most efficient, reliable way possible.

Bias

The term *bias* means, in the case of a *bipolar transistor*, a *field-effect transistor*, or a *vacuum tube*, that the control electrode—the *base*, *gate*, or *grid*—is provided with a certain voltage, or made to carry a certain current, relative to the *emitter*, *source*, or *cathode*. Networks of resistors can accomplish this.

A radio transmitting amplifier is biased differently than an oscillator or a low-level receiving amplifier. Sometimes voltage division is required for biasing. Other times it isn't necessary. Figure 6-1 shows a bipolar transistor whose base is biased using a pair of resistors in a voltage divider configuration.

Current Limiting

Resistors interfere with the flow of electrons in a circuit. Sometimes this is essential to prevent damage to a component or circuit. A good example is a receiving amplifier. A resistor can keep the

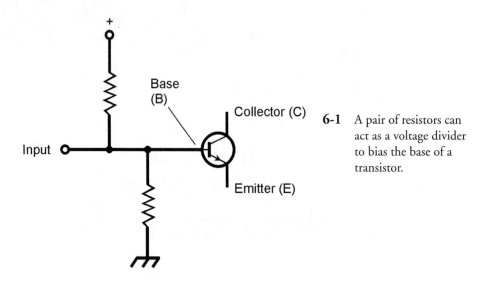

6-1 A pair of resistors can act as a voltage divider to bias the base of a transistor.

transistor from using up a lot of power just getting hot. Without resistors to limit or control the current, the transistor can be overstressed carrying direct current that doesn't contribute to the signal. Figure 6-2 shows a current-limiting resistor between the emitter of a bipolar transistor and electrical ground.

Power Dissipation

The dissipation of power in the form of heat is not always a bad thing. Sometimes a resistor can be used as a dummy component, so a circuit sees the resistor as if it were something more complicated. When testing a radio transmitter, for example, a resistor can be used to take the place of an antenna. This keeps the transmitter from interfering with communications on the airwaves. The transmitter output heats the resistor without radiating any signal. But as far as the transmitter knows, it's connected to a real antenna (Fig. 6-3)—and a perfect one, too, if the resistor has just the right ohmic value!

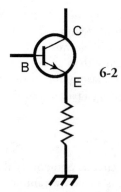

6-2 A resistor can limit the current that passes through the emitter of a transistor.

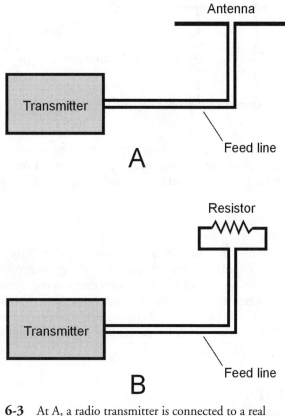

6-3 At A, a radio transmitter is connected to a real antenna. At B, the same transmitter is connected to a resistive dummy antenna.

Another situation in which power dissipation is useful is at the input of a power amplifier, such as the sort used in high-fidelity audio equipment. Sometimes the circuit *driving* the amplifier (supplying its input signal) has too much power. A resistor, or network of resistors, can dissipate this excess so that the amplifier doesn't get too much *drive*. In any type of amplifier, *overdrive* (an excessively strong input signal) can cause distortion, inefficiency, and other problems.

Bleeding Off Charge

In a high-voltage, dc power supply, capacitors are used to smooth out the fluctuations in the output. These capacitors acquire an electric charge, and they store it for a while. In some power supplies, these *filter capacitors* hold the full output voltage of the supply, say something like 750 V, even after the supply has been turned off, and even after it is unplugged from the wall outlet. If you attempt to repair such a power supply, you can be electrocuted by this voltage. *Bleeder resistors*, connected across the filter capacitors, drain their stored charge so that servicing the supply is not dangerous. In Fig. 6-4, the bleeder resistor, R, should have a value high enough so that it doesn't interfere with the operation of the power supply, but low enough so it will discharge the capacitor, C, in a short time after the power supply has been shut down.

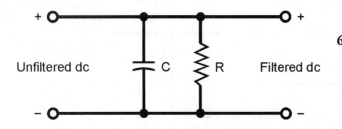

6-4 A bleeder resistor (R) is connected across the filter capacitor (C) in a power supply.

It's a good idea to short out all filter capacitors, using a screwdriver with an insulated handle and wearing heavy, insulated gloves, before working on a dc power supply. Even if the supply has bleeder resistors, they might take a while to get rid of the residual charge. In addition, bleeder resistors can, and sometimes do, fail.

Impedance Matching

A more sophisticated application for resistors is in the *coupling* in a chain of amplifiers, or in the input and output circuits of amplifiers. In order to produce the greatest possible amplification, the *impedances* must agree between the output of a given amplifier and the input of the next. The same is true between a source of signal and the input of an amplifier. Also, this applies between the output of an amplifier and a *load*, whether that load is a speaker, a headset, or whatever.

Impedance is the ac "big brother" of dc resistance. You will learn about impedance in Part 2 of this book.

Fixed Resistors

There are several ways in which *fixed resistors* (units whose resistance does not change, or cannot be adjusted) are manufactured. Here are the most common types.

Carbon-Composition Resistors

The cheapest method of making a resistor is to mix up powdered carbon (a fair electrical conductor) with some nonconductive substance, press the resulting claylike stuff into a cylindrical shape, and insert wire leads in the ends (Fig. 6-5). The resistance of the final product depends on the ratio

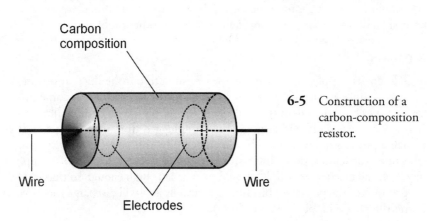

6-5 Construction of a carbon-composition resistor.

of carbon to the nonconducting material, and also on the physical distance between the wire leads. This results in a *carbon-composition resistor.*

Carbon-composition resistors can be manufactured in a wide range of resistance values. This kind of resistor also has the advantage of being *nonreactive,* meaning that it introduces almost pure resistance into the circuit, and not much capacitance or inductance. This makes carbon-composition resistors useful in radio receivers and transmitters.

Carbon-composition resistors dissipate power according to how big, physically, they are. Most of the carbon-composition resistors you see in electronics stores can handle $1/4$ W or $1/2$ W. There are $1/8$-W units available for miniaturized, low-power circuitry, and 1- or 2-W units for circuits where some electrical ruggedness is needed. Occasionally you'll see a carbon-composition resistor with a much higher power rating, but these are rare.

Wirewound Resistors

Another way to get resistance is to use a length of wire that isn't a good conductor. The wire can be wound around a cylindrical form as a coil (Fig. 6-6). The resistance is determined by how well the wire metal conducts, by its diameter or *gauge,* and by its stretched-out length. This type of component is called a *wirewound resistor.*

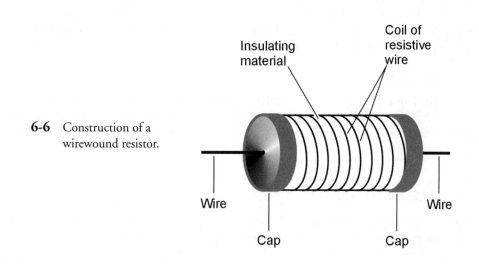

6-6 Construction of a wirewound resistor.

Wirewound resistors can be manufactured to have values within a very close range. They are *precision components.* Also, wirewound resistors can be made to handle large amounts of power. A disadvantage of wirewound resistors, in some applications, is that they act like inductors. This makes them unsuitable for use in most radio-frequency circuits. Wirewound resistors usually have low to moderate values of resistance.

Film-Type Resistors

Carbon, resistive wire, or some mixture of ceramic and metal can be applied to a cylindrical form as a film, or thin layer, in order to obtain a specific resistance. This type of component is called a *carbon-film resistor* or *metal-film resistor.* Superficially, it looks like a carbon-composition resistor, but the construction is different (Fig. 6-7).

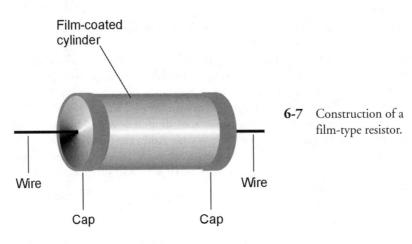

Film-coated
cylinder

Wire

Cap

Wire

Cap

6-7 Construction of a film-type resistor.

The cylindrical form is made of an insulating substance, such as porcelain. The film is deposited on this form by various methods, and the value tailored as desired. Metal-film resistors can be made to have nearly exact values. Film-type resistors usually have low to medium-high resistance.

A major advantage of film-type resistors is that they, like carbon-composition resistors, do not have much inductance or capacitance. A disadvantage, in some applications, is that they can't handle as much power as carbon-composition or wirewound types.

Integrated-Circuit (IC) Resistors

Resistors can be fabricated on a semiconductor wafer known as an *integrated circuit* (IC), also called a *chip*. The thickness, and the types and concentrations of impurities added, control the resistance of the component. Integrated-circuit resistors can handle only a tiny amount of power because of their small size.

The Potentiometer

Figure 6-8 is a simplified drawing of the construction of a *potentiometer*, or variable resistor. A resistive strip, similar to that found on film-type fixed resistors, is bent into a nearly complete circle, and terminals are connected to either end. This forms a fixed resistance. To obtain the variable resistance, a sliding contact is attached to a rotatable shaft and bearing, and is connected to a third terminal. The resistance between the middle terminal and either of the end terminals can vary from zero up to the resistance of the whole strip.

Some potentiometers use a straight strip of resistive material, and the control moves up and down or from side to side. This type of variable resistor, called a *slide potentiometer*, is used in hi-fi audio *graphic equalizers*, as the volume controls in some hi-fi audio amplifiers, and in other applications when a linear scale is preferable to a circular scale. Potentiometers are manufactured to handle low levels of current, at low voltage.

Linear-Taper Potentiometer

One type of potentiometer uses a strip of resistive material whose density is constant all the way around. This results in a *linear taper*. The resistance between the center terminal and either end terminal changes at a steady rate as the control shaft is turned.

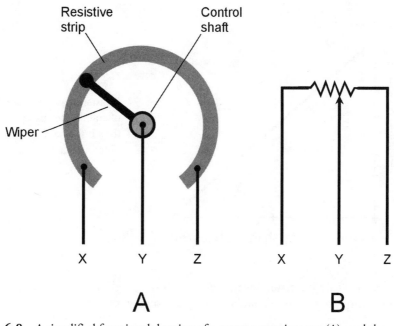

6-8 A simplified functional drawing of a rotary potentiometer (A), and the schematic symbol (B).

Suppose a *linear-taper potentiometer* has a value of zero to 280 Ω. In most units the shaft can be rotated through about 280°, or a little more than three-quarters of a circle. The resistance between the center and one end terminal will increase right along with the number of angular degrees that the shaft is turned. The resistance between the center and the other end terminal will be equal to 280 minus the number of degrees the shaft is turned. The resistance is a *linear function* of the angular shaft position.

Linear-taper potentiometers are commonly used in electronic test instruments and in various consumer electronic devices. Figure 6-9 is a graph of relative resistance versus relative angular shaft displacement for a linear-taper potentiometer.

Audio-Taper Potentiometer

In some applications, linear taper potentiometers don't work well. The volume control of a radio receiver or hi-fi audio amplifier is a good example. Humans perceive sound intensity according to the *logarithm* of the actual sound power. If you use a linear-taper potentiometer as the volume control for a radio or other sound system, the sound volume will vary too slowly in some parts of the control range, and too fast in other parts of the control range.

To compensate for the way in which people perceive sound level, an *audio-taper potentiometer* is used. In this device, the resistance between the center and end terminal increases as a nonlinear function of the angular shaft position. The device is sometimes called a *logarithmic-taper potentiometer* or *log-taper potentiometer* because the nonlinear function is logarithmic. This precisely compensates for the way the human ear-and-brain "machine" responds to sounds of variable intensity. Audio-taper potentiometers are manufactured so that as you turn the shaft, the sound intensity

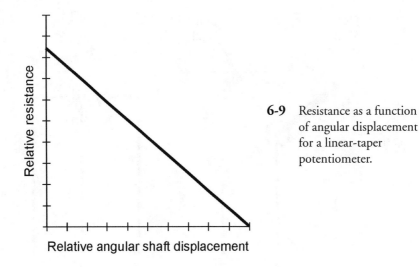

6-9 Resistance as a function of angular displacement for a linear-taper potentiometer.

seems to increase in a smooth, natural way. Figure 6-10 is a graph of relative resistance versus relative angular shaft displacement for an audio-taper potentiometer.

The Rheostat

A variable resistor can be made from a wirewound element, rather than a solid strip of material. This is called a *rheostat*. It can have either a rotary control or a sliding control. This depends on whether the resistive wire is wound around a donut-shaped form (*toroid*) or a cylindrical form (*solenoid*). Rheostats have inductance as well as resistance. They share the advantages and disadvantages of fixed wirewound resistors.

A rheostat is not continuously adjustable, as is a potentiometer. This is because the movable contact slides along from turn to turn of the wire coil. The smallest possible increment is the resistance in one turn of the coil.

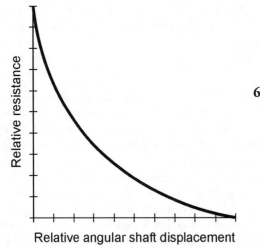

6-10 Resistance as a function of angular displacement for an audio-taper potentiometer.

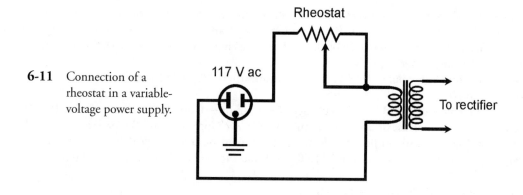

6-11 Connection of a rheostat in a variable-voltage power supply.

Rheostats are used in high-voltage, high-power applications. A good example is in a variable-voltage power supply. This kind of supply uses a transformer that steps up the voltage from the 117-V utility mains, and diodes to change the ac to dc. The rheostat can be placed between the utility outlet and the transformer (Fig. 6-11). This results in a variable voltage at the power-supply output.

The Decibel

As stated in the preceding paragraphs, perceived levels of sound change according to the logarithm of the actual sound power level. The same is true for various other phenomena, too, such as visible-light intensity and radio-frequency signal strength. Specialized units have been defined to take this into account.

The fundamental unit of sound-level change is called the *decibel*, symbolized as dB. A change of +1 dB is the minimum increase in sound level that you can detect if you are expecting it. A change of −1 dB is the minimum detectable decrease in sound volume, when you are anticipating the change. Increases in volume are given positive decibel values, and decreases in volume are given negative decibel values.

If you aren't expecting the level of sound to change, then it takes about +3 dB or −3 dB to make a noticeable difference.

Changes in intensity, when expressed in decibels, are sometimes called *gain* and *loss*. Positive decibel changes represent gain, and negative decibel changes represent loss. The sign (plus or minus) is usually absent when speaking of changes in terms of decibel gain or decibel loss. If you say that a certain system causes 5 dB of loss, you are saying that the gain of that circuit is −5 dB.

Calculating Decibel Values

Decibel values are calculated according to the logarithm of the ratio of change. Suppose a sound produces a power of P watts on your eardrums, and then it changes (either getting louder or softer) to a level of Q watts. The change in decibels is obtained by dividing out the ratio Q/P, taking its base-10 logarithm (symbolized as $\log_{10}$ or simply as $\log$), and then multiplying the result by 10. Mathematically:

$$dB = 10 \log (Q/P)$$

As an example, suppose a speaker emits 1 W of sound, and then you turn up the volume so that it emits 2 W of sound power. Then $P = 1$ and $Q = 2$, and dB = 10 log (2/1) = 10 log 2 = 10 × 0.3 = 3 dB. This is the minimum detectable level of volume change if you aren't expecting it: doubling of the actual sound power!

If you turn the volume level back down again, then $P/Q = 1/2 = 0.5$, and you can calculate dB = 10 log 0.5 = 10 × −0.3 = −3 dB.

A gain or loss of 10 dB (that is, a change of +10 dB or −10 dB, often shortened to ±10 dB) represents a 10-fold increase or decrease in sound power. A change of ±20 dB represents a 100-fold increase or decrease in sound power. It is not unusual to encounter sounds that vary in intensity over ranges of ±60 dB, which represents a 1,000,000-fold increase or decrease in sound power!

Sound Power in Terms of Decibels

The preceding formula can be worked inside out, so that you can determine the final sound power, given the initial sound power and the decibel change. To do this, you use the *inverse of the logarithmic function*, symbolized as $\log^{-1}$ or antilog. This function, like the logarithmic function, can be performed by any good scientific calculator, or by the calculator program in a personal computer when set to scientific mode.

Suppose the initial sound power is P, and the change in decibels is dB. Let Q be the final sound power. Then:

$$Q = P \text{ antilog (dB/10)}$$

As an example, suppose the initial power, P, is 10 W, and the perceived volume change is −3 dB. Then the final power, Q, is equal to 10 antilog (−3/10) = 10 × 0.5 = 5 W.

Decibels in the Real World

Sound levels are sometimes specified in decibels relative to the *threshold of hearing*, defined as the faintest possible sound that a person can detect in a quiet room, assuming his or her hearing is normal. This threshold is assigned the value 0 dB. Other sound levels can then be quantified as figures such as 30 dB or 75 dB.

If a certain noise has a loudness of 30 dB, that means it's 30 dB above the threshold of hearing, or 1000 times as loud as the quietest detectable noise. A noise at 60 dB is 1,000,000 (or 10^6) times as powerful as a sound at the threshold of hearing. *Sound-level meters* are used to determine the decibel levels of various noises and acoustic environments.

A typical conversation occurs at a level of about 70 dB. This is 10,000,000 (or 10^7) times the threshold of hearing, in terms of actual sound power. The roar of the crowd at a rock concert might be 90 dB, or 1,000,000,000 (10^9) times the threshold of hearing. A sound at 100 dB, typical of the music at a large rock concert if you are sitting in the front row, is 10,000,000,000 (10^{10}) times as loud, in terms of power, as a sound at the threshold of hearing.

Resistor Specifications

When choosing a resistor for a particular application in an electrical or electronic device, it's important to get a unit that has the correct properties, or *specifications*. Here are some of the most important specifications to watch for.

Ohmic Value

In theory, a resistor can have any ohmic value from the lowest possible (such as a shaft of solid silver) to the highest (dry air). In practice, it is unusual to find resistors with values less than about 0.1 Ω or more than about 100 MΩ.

Resistors are manufactured with ohmic values in power-of-10 multiples of 1.0, 1.2, 1.5, 1.8, 2.2, 2.7, 3.3, 3.9, 4.7, 5.6, 6.8, and 8.2. Thus, you will often see resistors with values of 47 Ω, 180 Ω, 6.8 kΩ, or 18 MΩ, but hardly ever with values such as 384 Ω, 4.54 kΩ, or 7.297 MΩ.

In addition to these standard values, there are others that are used for resistors made with greater precision, or tighter *tolerance*. These are power-of-10 multiples of 1.1, 1.3, 1.6, 2.0, 2.4, 3.0, 3.6, 4.3, 5.1, 6.2, 7.5, and 9.1.

Tolerance

The first set of numbers above represents standard resistance values available in tolerances of plus or minus 10 percent (±10%). This means that the resistance might be as much as 10 percent more or 10 percent less than the indicated amount. In the case of a 470-Ω resistor, for example, the value can be larger or smaller than the rated value by as much as 47 Ω, and still be within tolerance. That's a range of 423 to 517 Ω.

Tolerance is calculated according to the specified value of the resistor, not the actual value. You might measure the value of a 470-Ω resistor and find it to be 427 Ω, and it would be within ±10% of the specified value. But if it measures 420 Ω, it's outside the rated range, and is therefore a reject. The second set, along with the first set, of numbers represents standard resistance values available in tolerances of plus or minus 5 percent (±5%). A 470-Ω, 5 percent resistor will have an actual value of 470 Ω plus or minus 24 Ω, or a range of 446 to 494 Ω.

Some resistors are available in tolerances tighter than ±5%. These precision units are employed in circuits where a little error can make a big difference. In most audio and radio-frequency oscillators and amplifiers, the ±10% or ±5% tolerance is good enough. In many cases, even a ±20% tolerance is satisfactory.

Power Rating

All resistors are given a specification that determines how much power they can safely dissipate. Typical values are ¼ W, ½ W, and 1 W. Units also exist with ratings of ⅛ W or 2 W. These dissipation ratings are for *continuous duty*, meaning they can dissipate this amount of power constantly and indefinitely.

You can figure out how much current a given resistor can handle by using the formula for power (P) in terms of current (I) and resistance (R). That formula, you should recall, is $P = I^2R$. Work this formula backward, plugging in the power rating in watts for P and the resistance in ohms for R, and solve for the current I in amperes. Alternatively, you can find the square root of P/R.

The power rating for a given resistor can, in effect, be increased by using a network of 2×2, 3×3, 4×4, or more units in series-parallel. If you need a 47-Ω, 45-W resistor, but all you have is a bunch of 47-Ω, 1-W resistors, you can make a 7×7 network in series-parallel, and this will handle 49 W.

Resistor power dissipation ratings are specified with a margin for error. A good engineer never tries to take advantage of this and use, say, a ¼-W unit in a situation that needs to draw 0.27 W. In fact, good engineers usually include their own safety margin. Allowing 10 percent, a ¼-W resistor should not be called upon to handle more than about 0.225 W.

Temperature Compensation

All resistors change value when the temperature changes dramatically. And because resistors dissipate power, they can get hot just because of the current they carry. Often, this current is so tiny that it doesn't appreciably heat the resistor. But in some cases it does, and the resistance will change. Then a circuit might behave differently than it did when the resistor was still cool.

There are various ways to approach problems of resistors changing value when they get hot. One method is to use specially manufactured resistors that do not appreciably change value when they get hot. Such units are called *temperature-compensated*. But one of these can cost several times as much as an ordinary resistor. Another approach is to use a power rating that is much higher than the actual dissipated power in the resistor. This will keep the resistor from getting very hot. Still another scheme is to use a series-parallel network of identical resistors to increase the power dissipation rating. Alternatively, you can take several resistors, say three of them, each with about three times the intended resistance, and connect them all in parallel. Or you can take several resistors, say four of them, each with about one-fourth the intended resistance, and connect them in series.

It is unwise to combine resistors with different values. This can result in one of them taking most of the load while the others "loaf," and the combination will be no better than the single hot resistor you started with.

How about using two resistors with half (or twice) the value you need, but with *opposite* resistance-versus-temperature characteristics, and connecting them in series or parallel? It is tempting to suppose that if you do this, the component whose resistance decreases with heat (*negative temperature coefficient*) will have a canceling-out effect on the component whose resistance goes up (*positive temperature coefficient*). This can sometimes work, but in practice it's difficult to find a pair of resistances that will do this job just right.

The Color Code for Resistors

Some resistors have *color bands* that indicate their values and tolerances. You'll see three, four, or five bands around carbon-composition resistors and film resistors. Other units are large enough so that the values can be printed on them in ordinary numerals.

On resistors with *axial leads* (wires that come straight out of both ends), the first, second, third, fourth, and fifth bands are arranged as shown in Fig. 6-12A. On resistors with *radial leads* (wires that come off the ends at right angles to the axis of the component body), the colored regions are arranged as shown in Fig. 6-12B. The first two regions represent numbers 0 through 9, and the third region represents a multiplier of 10 to some power. (For the moment, don't worry about the fourth and fifth regions.) Refer to Table 6-1.

Suppose you find a resistor whose first three bands are yellow, violet, and red, in that order. Then the resistance is 4700 Ω. Read yellow = 4, violet = 7, red = ×100. As another example, suppose you find a resistor with bands of blue, gray, orange. Refer to Table 6-1 and determine blue = 6, gray = 8, orange = ×1000. Therefore, the value is 68,000 Ω = 68 kΩ.

The fourth band, if there is one, indicates tolerance. If it's silver, it means the resistor is rated at ±10%. If it's gold, the resistor is rated at ±5%. If there is no fourth band, the resistor is rated at ±20%.

The fifth band, if there is one, indicates the maximum percentage that the resistance can be expected to change after 1000 hours of use. A brown band indicates a maximum change of ±1% of the rated value. A red band indicates ±0.1%. An orange band indicates ±0.01%. A yellow band indicates ±0.001%. If there is no fifth band, it means that the resistor might deviate by more than ±1% of the rated value after 1000 hours of use.

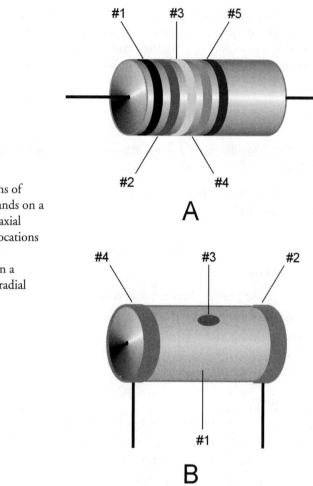

6-12 At A, locations of color-code bands on a resistor with axial leads. At B, locations of color code designators on a resistor with radial leads.

Table 6-1. The color code for the first three bands that appear on fixed resistors. See text for discussion of the fourth and fifth bands.

Color of band	Numeral (first and second bands)	Multiplier (third band)
Black	0	1
Brown	1	10
Red	2	100
Orange	3	1000 (1 k)
Yellow	4	10^4 (10 k)
Green	5	10^5 (100 k)
Blue	6	10^6 (1 M)
Violet	7	10^7 (10 M)
Gray	8	10^8 (100 M)
White	9	10^9 (1000 M or 1 G)

A competent engineer or technician always tests a resistor with an ohmmeter before installing it in a circuit. If the component happens to be labeled wrong, or if it is defective, it's easy to catch this problem while assembling or servicing a circuit. But once the circuit is all together, and it won't work because some resistor is labeled wrong or is bad, it's difficult to troubleshoot.

Quiz

Refer to the text in this chapter if necessary. A good score is at least 18 correct. Answers are in the back of the book.

1. Proper biasing in an amplifier circuit
   (a) causes it to oscillate.
   (b) prevents an impedance match.
   (c) can be obtained using a voltage divider network.
   (d) maximizes current flow.

2. A transistor can be protected from needless overheating by
   (a) a current-limiting resistor.
   (b) bleeder resistors.
   (c) maximizing the drive.
   (d) shorting out the power supply when the circuit is off.

3. A bleeder resistor
   (a) is connected across the capacitor in a power supply.
   (b) keeps a transistor from drawing too much current.
   (c) prevents an amplifier from being overdriven.
   (d) optimizes the efficiency of an amplifier.

4. Carbon-composition resistors
   (a) can handle gigantic levels of power.
   (b) have capacitance or inductance along with resistance.
   (c) have essentially no capacitance or inductance.
   (d) work better for ac than for dc.

5. A logical place for a wirewound resistor is
   (a) in a radio-frequency amplifier.
   (b) in a circuit where a noninductive resistor is called for.
   (c) in a low-power radio-frequency circuit.
   (d) in a high-power dc circuit.

6. A metal-film resistor
   (a) is made using a carbon-based paste.
   (b) does not have much inductance.

(c) can dissipate large amounts of power.

(d) has considerable inductance.

7. What type of resistor, or combination of resistors, would you use as the meter-sensitivity control in a test instrument, when continuous adjustment is desired?

(a) A set of switchable, fixed resistors

(b) A linear-taper potentiometer

(c) An audio-taper potentiometer

(d) A wirewound resistor

8. What type of resistor, or combination of resistors, would you use as the volume control in a stereo compact-disc (CD) player?

(a) A set of switchable, fixed resistors

(b) A linear-taper potentiometer

(c) An audio-taper potentiometer

(d) A wirewound resistor

9. If a sound triples in actual power level, approximately what is this, expressed in decibels?

(a) +3 dB

(b) +5 dB

(c) +6 dB

(d) +9 dB

10. Suppose a sound changes in volume by −13 dB. If the original sound power is 1.0 W, what is the final sound power?

(a) 13 W

(b) 77 mW

(c) 50 mW

(d) There is not enough information given here to answer this question.

11. The sound from a portable radio is at a level of 50 dB. How many times the threshold of hearing is this, in terms of actual sound power?

(a) 50

(b) 169

(c) 5000

(d) 100,000

12. An advantage of a rheostat over a potentiometer is the fact that

(a) a rheostat can handle higher frequencies.

(b) a rheostat is more precise.

(c) a rheostat can handle more current.

(d) a rheostat works better with dc.

13. A resistor is specified as having a value of 68 Ω, but is measured with an ohmmeter as 63 Ω. The value is off by which of the following percentages?

(a) 7.4%

(b) 7.9%

(c) 5%

(d) 10%

14. Suppose a resistor is rated at 3.3 kΩ ±5%. This means it can be expected to have a value between

(a) 2970 Ω and 3630 Ω.

(b) 3295 Ω and 3305 Ω.

(c) 3135 Ω and 3465 Ω.

(d) 2.8 kΩ and 3.8 kΩ.

15. A package of resistors is rated at 56 Ω ±10%. You test them with an ohmmeter. Which of the following values indicates a reject?

(a) 50.0 Ω

(b) 53.0 Ω

(c) 59.7 Ω

(d) 61.1 Ω

16. A resistor has a value of 680 Ω, and you expect that it will have to draw 1 mA maximum continuous current in a circuit you're building. What power rating is good for this application, but not needlessly high?

(a) ¼ W

(c) ½ W

(c) 1 W

(d) 2 W

17. Suppose a 1-kΩ resistor will dissipate 1.05 W, and you have a good supply of 1-W resistors of various ohmic values. If there's room for 20 percent resistance error, the cheapest solution is to use

(a) four 1-kΩ, 1-W resistors in series-parallel.

(b) a pair of 2.2-kΩ, 1-W resistors in parallel.

(c) a set of three 3.3-kΩ, 1-W resistors in parallel.

(d) a single 1-kΩ, 1-W resistor, because all manufacturers allow for a 10 percent margin of safety when rating resistors for their power-handling capability.

18. Suppose a carbon-composition resistor has the following colored bands on it: red, red, red, gold. This indicates a resistance of

(a) 22 Ω.

(b) 220 Ω.

(c) 2.2 kΩ.

(d) 22 kΩ.

19. The actual resistance of the component described in the previous question can be expected to vary above or below the specified ohmic value by up to what amount?

(a) 11 Ω

(b) 110 Ω

(c) 22 Ω

(d) 220 Ω

20. Suppose a carbon-composition resistor has the following colored bands on it: gray, red, yellow. This unit can be expected to have a value within approximately what range?

(a) 660 kΩ to 980 kΩ

(b) 740 kΩ to 900 kΩ

(c) 7.4 kΩ to 9.0 kΩ

(d) The manufacturer does not make any claim.

7
CHAPTER

Cells and Batteries

IN ELECTRICITY AND ELECTRONICS, A *CELL* IS A UNIT SOURCE OF DC ENERGY. WHEN TWO OR MORE cells are connected in series, the result is known as a *battery.* There are many types of cells and batteries, and new types are constantly being invented.

Electrochemical Energy

Early in the history of electrical science, laboratory physicists found that when metals came into contact with certain chemical solutions, voltages appeared between the pieces of metal. These were the first *electrochemical cells.*

A piece of lead and a piece of lead dioxide immersed in an acid solution (Fig. 7-1) acquire a persistent potential difference. This can be detected by connecting a galvanometer between the pieces of metal. A resistor of about 1000 Ω must be used in series with the galvanometer in experiments of this kind, because connecting the galvanometer directly will cause too much current to flow, possibly damaging the galvanometer and causing the acid to boil.

The chemicals and the metal have an inherent ability to produce a constant exchange of charge carriers. If the galvanometer and resistor are left hooked up between the two pieces of metal for a long time, the current will gradually decrease, and the electrodes will become coated. All the *chemical energy* in the acid will have been turned into electrical energy as current in the wire and galvanometer. In turn, this current will have heated the resistor (another form of kinetic energy), and escaped into the air and into space.

Primary and Secondary Cells

Some electrical cells, once their chemical energy has all been changed to electricity and used up, must be thrown away. These are called *primary cells.* Other kinds of cells, such as the lead-and-acid type, can get their chemical energy back again by means of *recharging.* Such a cell is a *secondary cell.*

Primary cells include the ones you usually put in a flashlight, in a transistor radio, and in various other consumer devices. They use dry electrolyte pastes along with metal electrodes. They go by names such as *dry cell, zinc-carbon cell,* or *alkaline cell.* Go into a department store and find a rack of batteries, and you'll see various sizes and types of primary cells, such as *AAA batteries, D batteries,*

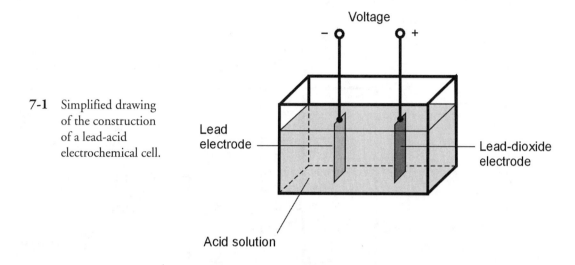

7-1 Simplified drawing of the construction of a lead-acid electrochemical cell.

camera batteries, and *watch batteries*. (These are actually cells, not true batteries.) You'll also see 9-V *transistor batteries* and large 6-V *lantern batteries*.

Secondary cells can also be found in consumer stores. *Nickel-based cells* are common. The most common sizes are AA, C, and D. These cost several times as much as ordinary dry cells, and a charging unit also costs a few dollars. But if you take care of them, these rechargeable cells can be used hundreds of times and will pay for themselves several times over if you use a lot of batteries in everyday life.

The battery in your car is made from secondary cells connected in series. These cells recharge from the alternator or from an outside charging unit. This battery has cells like the one in Fig. 7-1. It is dangerous to short-circuit the terminals of such a battery, because the acid (sulfuric acid) can bubble up and erupt out of the battery casing. Serious skin and eye injuries can result. In fact, it's a bad idea to short-circuit any cell or battery, because it can get extremely hot and cause a fire, or rupture and damage surrounding materials, wiring, and components.

The Weston Standard Cell

Most electrochemical cells produce 1.2 to 1.8 V. Different types vary slightly. A mercury cell has a voltage that is a little less than that of a zinc-carbon or alkaline cell. The voltage of a cell can also be affected by variables in the manufacturing process. Most consumer-type dry cells can be assumed to produce 1.5 V.

There are certain cells whose voltages are predictable and exact. These are called *standard cells*. A good example is the *Weston cell*, which produces 1.018 V at room temperature. It has a solution of cadmium sulfate, a positive electrode made from mercury sulfate, and a negative electrode made from mercury and cadmium. The device is set up in a container, as shown in Fig. 7-2.

Storage Capacity

Recall that the common electrical units of energy are the watt-hour (Wh) and the kilowatt-hour (kWh). Any electrochemical cell or battery has a certain amount of electrical energy that can be obtained from it, and this can be specified in watt-hours or kilowatt-hours. More often, though, it's given in *ampere-hours* (Ah).

A battery with a rating of 2 Ah can provide 2 A for 1 h, or 1 A for 2 h, or 100 mA for 20 h.

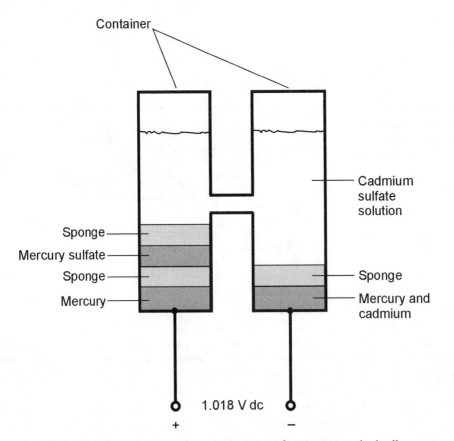

7-2   Simplified drawing of the construction of a Weston standard cell.

There are infinitely many possibilities here, as long as the product of the current in amperes and the use time in hours is equal to 2. The limitations are the *shelf life* at one extreme, and the *maximum deliverable current* at the other. Shelf life is the length of time the battery will last if it is never used; this can be years. The maximum deliverable current is the highest amount of current that the battery can provide before its voltage drops because of its own *internal resistance*.

Small cells have storage capacity of a few milliampere-hours (mAh) up to 100 or 200 mAh. Medium-sized cells can supply 500 mAh to 1 Ah. Large automotive or truck batteries can provide upward of 50 Ah. The energy capacity in watt-hours is the ampere-hour capacity multiplied by the battery voltage.

An *ideal cell* or *ideal battery* (a theoretically perfect cell or battery) delivers a constant current for a while, and then the current starts to drop (Fig. 7-3). Some types of cells and batteries approach this level of perfection, which is represented by a *flat discharge curve*. But many cells and batteries are far from perfect; they deliver current that declines gradually, almost right from the start. When the current that a battery can provide has tailed off to about half of its initial value, the cell or battery is said to be *weak*. At this time, it should be replaced. If it's allowed to run all the way out, until the current actually goes to zero, the cell or battery is *dead*. The area under the curve in Fig. 7-3 is a graphical representation the total capacity of the cell or battery in ampere-hours.

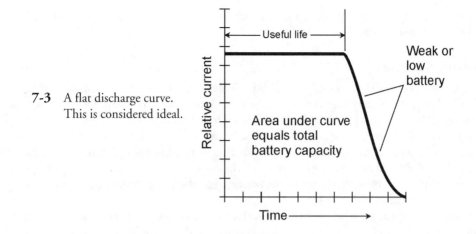

7-3 A flat discharge curve. This is considered ideal.

Grocery Store Cells and Batteries

The cells you see in grocery stores, department stores, drugstores, and hardware stores provide 1.5 V, and are available in sizes known as AAA (very small), AA (small), C (medium large), and D (large). Batteries are widely available that deliver 6 or 9 V.

Zinc-Carbon Cells

Figure 7-4 is a translucent drawing of a *zinc-carbon cell*. The zinc forms the case and is the negative electrode. A carbon rod serves as the positive electrode. The electrolyte is a paste of manganese dioxide and carbon. Zinc-carbon cells are inexpensive and are good at moderate temperatures and in applications where the current drain is moderate to high. They are not very good in extreme cold.

Alkaline Cells

The *alkaline cell* has granular zinc as the negative electrode, potassium hydroxide as the electrolyte, and a device called a *polarizer* as the positive electrode. The construction is similar to that of the zinc-carbon cell. An alkaline cell can work at lower temperatures than a zinc-carbon cell. It lasts longer in most electronic devices, and is therefore preferred for use in transistor radios, calculators,

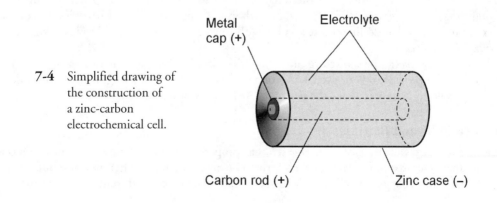

7-4 Simplified drawing of the construction of a zinc-carbon electrochemical cell.

and portable cassette players. Its shelf life is much longer than that of a zinc-carbon cell. As you might expect, it costs more.

Transistor Batteries

A *transistor battery* consists of six tiny zinc-carbon or alkaline cells in series. Each of the six cells supplies 1.5 V. Thus, the battery supplies 9 V. Even though these batteries have more voltage than individual cells, the total energy available from them is less than that from a C cell or D cell. This is because the electrical energy that can be obtained from a cell or battery is directly proportional to the amount of chemical energy stored in it, and this, in turn, is a direct function of the *volume* (physical size) of the cell or the *mass* (quantity of chemical matter) of the cell. Cells of size C or D have more volume and mass than a transistor battery, and therefore contain more stored energy for the same chemical composition.

Transistor batteries are used in low-current electronic devices such as remote-control garage-door openers, television (TV) and hi-fi remote controls, and electronic calculators.

Lantern Batteries

The *lantern battery* has much greater mass than a common dry cell or transistor battery, and consequently it lasts much longer and can deliver more current. Lantern batteries are usually rated at 6 V, and consist of four good-size zinc-carbon or alkaline cells. Two lantern batteries connected in series make a 12-V battery that can power a 5-W citizens band (CB) or ham radio transceiver for a while. They're also good for scanner radio receivers in portable locations, for camping lamps, and for other medium-power needs.

Miniature Cells and Batteries

In recent years, cells and batteries—especially cells—have become available in many different sizes and shapes besides the old cylindrical cells, transistor batteries, and lantern batteries. These are used in wristwatches, small cameras, and various microminiature electronic devices.

Silver-Oxide Cells and Batteries

A *silver-oxide cell* is usually found in a buttonlike shape, and can fit inside a small wristwatch. These types of cells come in various sizes and thicknesses, all with similar appearances. They supply 1.5 V, and offer excellent energy storage for the weight. They also have a nearly flat discharge curve, like the one shown in the graph of Fig. 7-3. Zinc-carbon and alkaline cells and batteries, in contrast, have current output that declines more steadily with time, as shown in Fig. 7-5. This is known as a *declining discharge curve.*

Silver-oxide cells can be stacked to make batteries. Several of these miniature cells, one on top of the other, can provide 6, 9, or even 12 V for a transistor radio or other light-duty electronic device. The resulting battery is about the size of an AAA cylindrical cell.

Mercury Cells and Batteries

A *mercury cell*, also called a *mercuric-oxide cell*, has properties similar to those of silver-oxide cells. They are manufactured in the same general form. The main difference, often not of significance, is a somewhat lower voltage per cell: 1.35 V. If six of these cells are stacked to make a battery, the re-

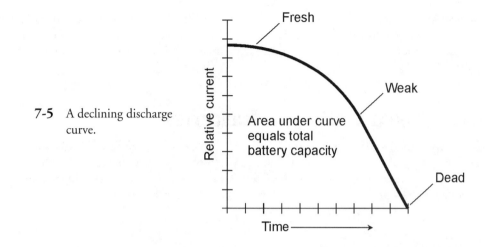

7-5 A declining discharge
curve.

sulting voltage will be about 8.1 V rather than 9 V. One additional cell can be added to the stack, yielding about 9.45 V.

There has been a decline in the popularity of mercury cells and batteries in recent years, because of the fact that mercury is toxic to humans and animals, even in trace amounts. When mercury cells and batteries are dead, they must be discarded. Eventually the mercury or mercuric oxide leaks into the soil and groundwater. Mercury pollution has become a significant concern throughout the world.

Lithium Cells and Batteries

Lithium cells gained popularity in the early 1980s. There are several variations in the chemical makeup of these cells; they all contain lithium, a light, highly reactive metal. Lithium cells can be made to supply 1.5 to 3.5 V, depending on the particular chemistry used. These cells, like silver-oxide and mercury cells, can be stacked to make batteries.

The first application of lithium batteries was in memory backup for electronic microcomputers. Lithium cells and batteries have superior shelf life, and they can last for years in very-low-current applications such as memory backup or the powering of a digital liquid crystal display (LCD) watch or clock. These cells also provide high energy capacity per unit volume or mass.

Lead-Acid Batteries

You've seen the basic configuration for a lead-acid cell. This has a solution of sulfuric acid, along with a lead electrode (negative) and a lead-dioxide electrode (positive). These cells are rechargeable.

Automotive batteries are made from sets of lead-acid cells having a free-flowing liquid acid. You cannot tip such a battery on its side, or turn it upside-down, without running the risk of having some of the acid electrolyte spill out. Lead-acid batteries are also available in a construction that uses a semisolid electrolyte. These batteries are sometimes used in consumer electronic devices that require a moderate amount of current. The most common example is an uninterruptible power supply (UPS) that can keep a desktop personal computer running for a few minutes if the utility power fails.

A large lead-acid battery, such as the kind in your car or truck, can store several tens of ampere-hours. The smaller ones, like those in a UPS, have less capacity but more versatility. Their main attributes are that they can be charged and recharged many times, and they are not particularly expensive.

Nickel-Based Cells and Batteries

Nickel-based cells include the *nickel-cadmium* (NICAD or NiCd) type and the *nickel-metal-hydride* (NiMH) type. *Nickel-based batteries* are available in packs of cells. These packs can be plugged into equipment, and sometimes form part of the case for a device such as a portable radio transceiver. All nickel-based cells are rechargeable, and can be put through hundreds or even thousands of *charge/discharge cycles* if they are properly cared for.

Configurations and Applications

Nickel-based cells are found in various sizes and shapes. *Cylindrical cells* look like ordinary dry cells. *Button cells* are those little things you find in cameras, watches, memory backup applications, and other places where miniaturization is important. *Flooded cells* are used in heavy-duty applications, and can have storage capacity in excess of 1000 Ah. *Spacecraft cells* are made in packages that can withstand the rigors of a deep-space environment.

Most orbiting satellites are in darkness half the time and in sunlight half the time. *Solar panels* can be used while the satellite is in sunlight, but during the times that the earth eclipses the sun, batteries are needed to power the electronic equipment on board the satellite. The solar panels can charge a nickel-based battery, in addition to powering the satellite, for the daylight half of each orbit. The nickel-based battery can provide the power during the dark half of each orbit.

Cautions

Never discharge nickel-based cells all the way until they totally die. This can cause the polarity of a cell, or of one or more cells in a battery, to reverse. Once this happens, the cell or battery is ruined.

A phenomenon peculiar to nickel-based cells and batteries is known as *memory* or *memory drain.* If a nickel-based unit is used over and over, and is discharged to the same extent every time, it might begin to die at that point in its discharge cycle. Memory problems can usually be solved. Use the cell or battery almost all the way up, and then fully recharge it. Repeat the process several times.

Nickel-based cells and batteries work best if used with charging units that take several hours to fully replenish the charge. So-called high-rate or quick chargers are available, but these can sometimes force too much current through a cell or battery. It's best if the charger is made especially for the cell or battery type being charged. An electronics dealer, such as the manager at a RadioShack store, should be able to tell you which chargers are best for which cells and batteries.

In recent years, concern has grown about the toxic environmental effects of discarded heavy metals, including cadmium. For this reason, NiMH cells and batteries have replaced NICAD types in many applications. In most practical scenarios, a NICAD battery can be directly replaced with a NiMH battery of the same voltage and current-delivering capacity, and the powered-up device will work satisfactorily.

Some vendors and dealers will call a nickel-based cell or battery a NICAD, even when it is actually a NiMH cell or battery.

Photovoltaic Cells and Batteries

The *photovoltaic* (PV) *cell* is different from any electrochemical cell. It's also known as a *solar cell.* This device converts visible light, infrared (IR), and/or ultraviolet (UV) directly into electric current.

Solar Panels

Several, or many, photovoltaic cells can be combined in series-parallel to make a *solar panel.* An example is shown in Fig. 7-6. Although this shows a 3 × 3 series-parallel array, the matrix does not have to be symmetrical. And it's often very large. It might consist of, say, 50 parallel sets of 20 series-connected cells. The series scheme boosts the voltage to the desired level, and the parallel scheme increases the current-delivering ability of the panel. It's not unusual to see hundreds of solar cells combined in this way to make a large panel.

Construction and Performance

The construction of a photovoltaic cell is shown in Fig. 7-7. The device is a flat semiconductor *P-N junction*, and the assembly is made transparent so that light can fall directly on the *P-type* silicon. The metal ribbing, forming the positive electrode, is interconnected by means of tiny wires. The negative electrode is a metal backing or *substrate*, placed in contact with the *N-type* silicon.

Most solar cells provide about 0.5 V. If there is very low current demand, dim light will result in the full-output voltage from a solar cell. As the current demand increases, brighter light is needed to produce the full-output voltage. There is a maximum limit to the current that can be provided from a solar cell, no matter how bright the light. This limit is increased by connecting solar cells in parallel.

Practical Applications

Solar cells have become cheaper and more efficient in recent years, as researchers have looked to them as an alternative energy source. Solar panels are used in satellites. They can be used in conjunction with rechargeable batteries, such as the lead-acid or nickel-cadmium types, to provide power independent of the commercial utilities.

A completely independent solar/battery power system is called a *stand-alone* system. It uses large solar panels, large-capacity lead-acid batteries, power converters to convert the dc into ac, and a sophisticated charging circuit. These systems are best suited to environments where there is sunshine a high percentage of the time.

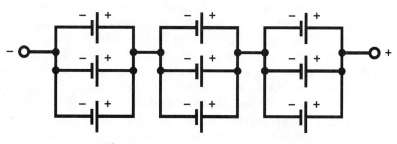

7-6   Connection of cells in series-parallel.

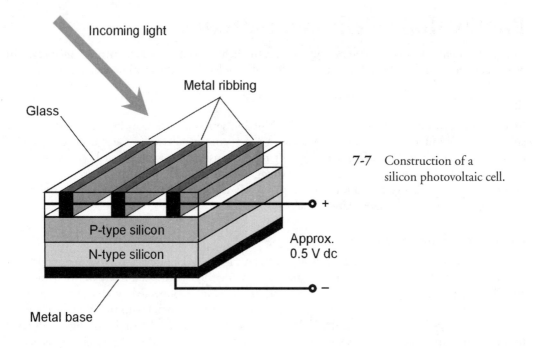

7-7 Construction of a silicon photovoltaic cell.

Solar cells, either alone or supplemented with rechargeable batteries, can be connected into a home electric system in an *interactive* arrangement with the electric utilities. When the solar power system can't provide for the needs of the household all by itself, the utility company can take up the slack. Conversely, when the solar power system supplies more than enough for the needs of the home, the utility company can buy the excess.

Fuel Cells

In the late 1900s, a new type of electrochemical power device emerged that is believed by some scientists and engineers to hold promise as an alternative energy source: the *fuel cell.*

Hydrogen Fuel

The most talked-about fuel cell during the early years of research and development became known as the *hydrogen fuel cell.* As its name implies, it derives electricity from hydrogen. The hydrogen combines with oxygen (that is, it *oxidizes*) to form energy and water. There is no pollution, and there are no toxic by-products. When a hydrogen fuel cell "runs out of juice," all that is needed is a new supply of hydrogen, because its oxygen is derived from the atmosphere.

Instead of combusting, the hydrogen in a fuel cell oxidizes in a more controlled fashion, and at a much lower temperature. There are several schemes for making this happen. The *proton exchange membrane* (PEM) *fuel cell* is one of the most widely used. A PEM hydrogen fuel cell generates approximately 0.7 V of dc. In order to obtain higher voltages, individual cells are connected in series. A series-connected set of fuel cells is technically a battery, but the term used more often is *stack*.

Fuel-cell stacks are available in various sizes. A stack about the size and weight of an airline suitcase filled with books can power a subcompact electric car. Smaller cells, called *micro fuel cells*, can provide dc to run devices that have historically operated from conventional cells and batteries. These include portable radios, lanterns, and notebook computers.

Other Fuels

Hydrogen is not the only chemical that can be used to make a fuel cell. Almost anything that will combine with oxygen to form energy has been considered.

Methanol, a form of alcohol, has the advantage of being easier to transport and store than hydrogen, because it exists as a liquid at room temperature. *Propane* is another chemical that has been used for powering fuel cells. This is the substance that is stored in liquid form in tanks for barbecue grills and some rural home heating systems. *Methane*, also known as natural gas, has been used as well.

Some scientists and engineers object to the use of these fuels because they, especially propane and methane, closely resemble fuels that are already commonplace, and on which society has developed the sort of dependence that purists would like to get away from. In addition, they are derived from so-called *fossil fuel* sources, the supplies of which, however great they might be today, are nevertheless finite.

A Promising Technology

As of this writing (2006), fuel cells have not yet replaced conventional electrochemical cells and batteries. Cost is the main reason. Hydrogen is the most abundant and simplest chemical element in the universe, and it does not produce any toxic by-products. This would at first seem to make it the ideal choice for use in fuel cells. But storage and transport of hydrogen has proven to be difficult and expensive. This is especially true for fuel cells and stacks intended for systems that aren't fixed to permanent pipelines.

An interesting scenario, suggested by one of my physics teachers all the way back in the 1970s, is the piping of hydrogen gas through the lines designed to carry methane. Some modification of existing lines would be required in order to safely handle hydrogen, which escapes through small cracks and openings more easily than methane. But hydrogen, if obtained at reasonable cost and in abundance, could be used to power large fuel-cell stacks in common households and businesses. The dc from such a stack could be converted to utility ac by power inverters similar to those used with PV energy systems. The entire home power system would be about the size of a gas furnace.

Quiz

Refer to the text in this chapter if necessary. A good score is 18 correct. Answers are in the back of the book.

1. The chemical energy in a battery or cell
   (a) is a form of kinetic energy.
   (b) cannot be replenished once it is gone.
   (c) changes to electrical energy when the cell is used.
   (d) is caused by electric current.

2. A cell that cannot be recharged is known as
   (a) a dry cell.
   (b) a wet cell.
   (c) a primary cell.
   (d) secondary cell.

3. A Weston cell is generally used
   (a) as a current reference source.
   (b) as a voltage reference source.
   (c) as a power reference source.
   (d) as a fuel cell.

4. The voltage produced by a battery of multiple cells connected in series is
   (a) less than the voltage produced by a cell of the same composition.
   (b) the same as the voltage produced by a cell of the same composition.
   (c) more than the voltage produced by a cell of the same composition.
   (d) always a whole-number multiple of 1.018 V.

5. A direct short-circuit of a large battery can cause
   (a) an increase in its voltage.
   (b) no harm other than a rapid discharge of its energy.
   (c) the current to drop to zero.
   (d) a physical rupture or explosion.

6. Suppose a cell of 1.5 V delivers 100 mA for 7 hours and 20 minutes, and then it is replaced. How much energy is supplied during this time?
   (a) 0.49 Wh
   (b) 1.1 Wh
   (c) 7.33 Wh
   (d) 733 mWh

7. Suppose a 12-V automotive battery is rated at 36 Ah. If a 100-W, 12-V bulb is connected across this battery, approximately how long will the bulb stay aglow, assuming the battery has been fully charged?
   (a) 4 hours and 20 minutes
   (b) 432 hours
   (c) 3.6 hours
   (d) 21.6 minutes

8. Alkaline cells
   (a) are cheaper than zinc-carbon cells.
   (b) generally work better in radios than zinc-carbon cells.
   (c) have higher voltages than zinc-carbon cells.
   (d) have shorter shelf lives than zinc-carbon cells.

9. The energy in a cell or battery depends mainly on
    (a) its physical size.
    (b) the current drawn from it.
    (c) its voltage.
    (d) all of the above.

10. In which of the following devices would a lantern battery most likely be found?
    (a) A heart pacemaker
    (b) An electronic calculator
    (c) An LCD wall clock
    (d) A two-way portable radio

11. In which of the following devices would a transistor battery be the best power choice?
    (a) A heart pacemaker
    (b) An electronic calculator
    (c) An LCD wall clock
    (d) A two-way portable radio

12. For which of the following applications would you choose a lithium battery?
    (a) A microcomputer memory backup
    (b) A two-way portable radio
    (c) A stand-alone solar-electric system
    (d) A rechargeable lantern

13. Where would you most likely find a lead-acid battery?
    (a) In a portable audio CD player
    (b) In an uninterruptible power supply
    (c) In an LCD wall clock
    (d) In a flashlight

14. A cell or battery that maintains a constant current-delivering capability almost until it dies is said to have
    (a) a large ampere-hour rating.
    (b) excellent energy capacity.
    (c) a flat discharge curve.
    (d) good energy storage capacity per unit volume.

15. Where might you find a nickel-based battery?
    (a) In a satellite
    (b) In a portable cassette player
    (c) In a handheld radio transceiver
    (d) More than one of the above

16. A disadvantage of mercury cells and batteries is the fact that
    (a) they don't last as long as other types.
    (b) they have a flat discharge curve.
    (c) mercury is destructive to the environment.
    (d) they need to be recharged often.

17. Which kind of battery should never be used until it dies?
    (a) Silver-oxide
    (b) Lead-acid
    (c) Nickel-based
    (d) Mercury

18. The useful current that is delivered by a solar panel can be increased by
    (a) connecting capacitors in parallel with the solar cells.
    (b) connecting resistors in series with the solar cells.
    (c) connecting two or more groups of solar cells in parallel.
    (d) connecting resistors in parallel with the solar cells.

19. An interactive solar power system
    (a) allows a homeowner to sell power to the electric company.
    (b) lets the batteries recharge at night.
    (c) powers lights, but not electronic devices.
    (d) is totally independent from the electric company.

20. An advantage of methanol over hydrogen for use in fuel cells is the fact that
    (a) methanol is the most abundant element in the universe.
    (b) methanol is not flammable.
    (c) methanol is a solid at room temperature.
    (d) methanol is easier to transport and store.

8
CHAPTER

Magnetism

ELECTRIC AND MAGNETIC PHENOMENA INTERACT. MAGNETISM WAS MENTIONED BRIEFLY NEAR THE end of Chap. 2. Here, we'll look at it more closely.

The Geomagnetic Field

The earth has a core made up largely of iron, heated to the extent that some of it is liquid. As the earth rotates, the iron flows in complex ways. It is thought that this flow is responsible for the magnetic field that surrounds the earth. Some other planets, notably Jupiter, have magnetic fields as well. Even the sun has one.

The Poles and Axis

The *geomagnetic field*, as it is called, has poles, just as a bar magnet does. The *geomagnetic poles* are near, but not at, the *geographic poles*. The *north geomagnetic pole* is located in far northern Canada. The *south geomagnetic pole* is near Antarctica. The *geomagnetic axis* is therefore tilted relative to the axis on which the earth rotates.

The Solar Wind

Charged subatomic particles from the sun, streaming outward through the solar system, distort the geomagnetic *lines of flux* (Fig. 8-1). This stream of particles is called the *solar wind*. That's a good name for it, because the fast-moving particles produce measurable forces on sensitive instruments in space. This force has actually been suggested as a possible means to drive space ships, equipped with *solar sails*, out of the solar system!

At and near the earth's surface, the geomagnetic field is not affected very much by the solar wind, so the geomagnetic field is nearly symmetrical. As the distance from the earth increases, the distortion of the field also increases, particularly on the side of the earth away from the sun.

The Magnetic Compass

The presence of the geomagnetic field was first noticed in ancient times. Some rocks, called *lodestones*, when hung by strings, would always orient themselves a certain way. This was correctly at-

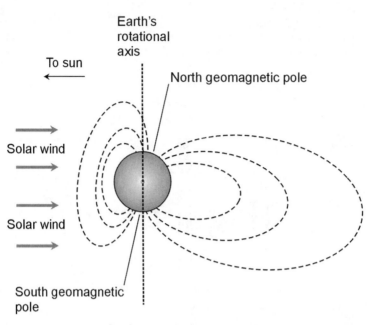

8-1 Geomagnetic flux lines (dashed curves) are distorted by the solar wind, so the geomagnetic field is not symmetrical with respect to the earth.

tributed to the presence of a "force" in the air. This effect was put to use by early seafarers and land explorers. Today, a *magnetic compass* can still be a valuable navigation aid, used by mariners, backpackers, and others who travel far from familiar landmarks.

The geomagnetic field interacts with the magnetic field around a compass needle, and a force is thus exerted on the needle. This force works not only in a horizontal plane (parallel to the earth's surface), but vertically at most latitudes. The vertical component is zero only at the *geomagnetic equator*, a line running around the globe equidistant from both geomagnetic poles.

As the geomagnetic latitude increases, toward either the north or the south geomagnetic pole, the magnetic force pulls up and down on the compass needle more and more. One end of the needle seems to insist on touching the compass face, while the other end tilts up toward the glass. The needle tries to align itself parallel to the geomagnetic *lines of flux*. The vertical angle, in degrees, at which the geomagnetic lines of flux intersect the earth's surface at any given location is called the *geomagnetic inclination*.

Because geomagnetic north is not the same as geographic north in most places on the earth's surface, there is an angular difference between the two. This horizontal angle, in degrees, is called *geomagnetic declination*. It, like inclination, varies with location.

Causes and Effects

Magnets are attracted to some, but not all, metals. Iron, nickel, and alloys containing either or both of these elements are known as *ferromagnetic* materials. They "stick" to magnets. They can also be made into *permanent magnets.* When a magnet is brought near a piece of ferromagnetic material, the atoms in the material become lined up, so that the material is temporarily magnetized. This produces a *magnetic force* between the atoms of the ferromagnetic substance and those in the magnet.

Attraction and Repulsion

If a magnet is brought near another magnet, the force can be repulsive or attractive, depending on the way the magnets are oriented. The force gets stronger as the magnets are brought near each other. Some magnets are so strong that no human being can pull them apart if they get stuck together, and no person can bring them all the way together against their mutual repulsive force. This is especially true of *electromagnets*, discussed later in this chapter.

The tremendous forces produced by electromagnets are of use in industry. A large electromagnet can be used to carry heavy pieces of scrap iron from place to place. Other electromagnets can provide sufficient repulsion to suspend one object above another. This phenomenon is called *magnetic levitation.* It is the basis for low-friction, high-speed commuter trains now in use in some metropolitan areas.

Charge in Motion

Whenever the atoms in a ferromagnetic material are aligned, a *magnetic field* exists. A magnetic field can also be caused by the motion of electric charge carriers, either in a wire or in free space.

The magnetic field around a permanent magnet arises from the same cause as the field around a wire that carries an electric current. The responsible factor in either case is the motion of electrically charged particles. In a wire, electrons move along the conductor, being passed from atom to atom. In a permanent magnet, the movement of orbiting electrons occurs in such a manner that an effective electrical current is produced.

Magnetic fields are also generated by the motion of charged particles through space. The sun is constantly ejecting protons and helium nuclei. These particles carry a positive electric charge. Because of this, and the fact that they are in motion, they are surrounded by tiny magnetic fields. When the particles approach the earth and their magnetic fields interact with the geomagnetic field, the particles are accelerated toward the geomagnetic poles.

When there is a *solar flare*, the sun ejects far more charged particles than normal. When these approach the geomagnetic poles, the result is considerable disruption of the geomagnetic field. This type of event is called a *geomagnetic storm.* It causes changes in the earth's ionosphere, affecting long-distance radio communications at certain frequencies. If the fluctuations are intense enough, even wire communications and electric power transmission can be interfered with. Aurora (northern or southern lights) are frequently observed at night during these events.

Flux Lines

Have you seen the well-known experiment in which iron filings are placed on a horizontal sheet of paper, and then a magnet is placed underneath the paper? The filings arrange themselves in a pattern that shows, roughly, the shape of the magnetic field in the vicinity of the magnet. A bar magnet has a field with a characteristic form (Fig. 8-2). Another popular experiment involves passing a current-carrying wire through a horizontal sheet of paper at a right angle, as shown in Fig. 8-3. The iron filings become grouped along circles centered at the point where the wire passes through the paper.

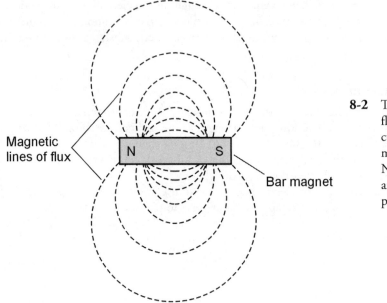

Magnetic
lines of flux

Bar magnet

8-2 The pattern of magnetic flux lines (dashed curves) around a bar magnet (rectangle). The N and S represent north and south magnetic poles, respectively.

The intensity of a magnetic field is determined according to the number of flux lines passing through a certain cross section, such as a square centimeter or a square meter. The lines don't exist as real objects, but it is intuitively appealing to imagine them that way. The iron filings on the paper really do bunch themselves into lines (curves, actually) when there is a magnetic field of sufficient strength to make them move. Sometimes lines of flux are called *lines of force*. But technically, this is a misnomer.

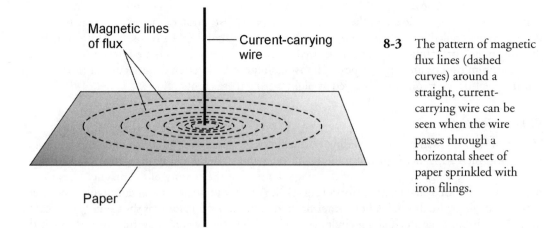

Magnetic lines
of flux

Current-carrying
wire

Paper

8-3 The pattern of magnetic flux lines (dashed curves) around a straight, current-carrying wire can be seen when the wire passes through a horizontal sheet of paper sprinkled with iron filings.

Poles

A magnetic field has a specific direction, as well as a specific intensity, at any given point in space near a current-carrying wire or a permanent magnet. The flux lines run parallel with the direction of the field. A magnetic field is considered to begin at the north magnetic pole, and to terminate at the south magnetic pole. In the case of a permanent magnet, it is obvious where the magnetic poles are. In the case of a current-carrying wire, the magnetic field goes in endless circles around the wire.

A charged electric particle, such as a proton or electron, hovering all by itself in space, constitutes an *electric monopole*. The *electric lines of flux* around an isolated, charged particle in free space are straight, and they "run off to infinity" (Fig. 8-4). A positive electric charge does not have to be mated with a negative electric charge.

A magnetic field is different. All magnetic flux lines, at least in ordinary real-world situations, are *closed loops*. With permanent magnets, there is a starting point (the north pole) and an ending point (the south pole). Around a straight, current-carrying wire, the loops are closed circles, even though the starting and ending points are not obvious. A pair of magnetic poles is called a *magnetic dipole*.

At first you might think that the magnetic field around a current-carrying wire is caused by a monopole, or that there aren't any poles at all, because the concentric circles don't actually converge anywhere. But you can envision a half plane, with the edge along the line of the wire, as a magnetic dipole. Then the lines of flux go around once in a 360° circle from the "north face" of the half plane to the "south face."

The greatest flux density, or field strength, around a bar magnet is near the poles, where the lines converge. Around a current-carrying wire, the greatest field strength is near the wire.

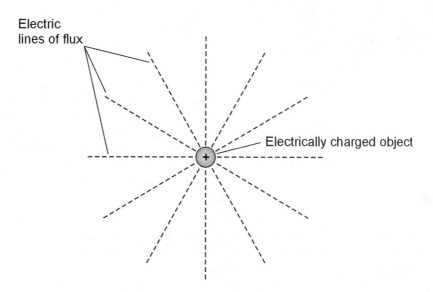

8-4   Electric flux lines (dashed lines) around an electrically charged object. This example shows a positive charge. The pattern of flux lines for a negative charge is identical.

Magnetic Field Strength

The overall magnitude of a magnetic field is measured in units called *webers* (Wb). A smaller unit, the *maxwell* (Mx), is sometimes used if a magnetic field is weak. One weber is equivalent to 100,000,000 (10^8) maxwells. Conversely, 1 Mx = 0.00000001 Wb = 10^{-8} Wb.

The Tesla and the Gauss

If you have access to a permanent magnet or electromagnet, you might see its strength expressed in terms of webers or maxwells. But usually you'll hear units called *teslas* (T) or *gauss* (G). These units are expressions of the concentration, or intensity, of the magnetic field within a certain cross section. The *flux density*, or number of lines per square meter or per square centimeter, is a more useful expression for magnetic effects than the overall quantity of magnetism. A flux density of 1 tesla (1 T) is equal to 1 weber per square meter (1 Wb/m^2). A flux density of 1 gauss (1 G) is equal to 1 maxwell per square centimeter (1 Mx/cm^2). It turns out that the gauss is equal to 0.0001 tesla (10^{-4} T). Conversely, the tesla is equivalent to 10,000 gauss (10^4 G).

The Ampere-Turn and the Gilbert

With electromagnets, another unit is employed: the *ampere-turn* (At). This is technically a unit of *magnetomotive force*, which is the magnetic counterpart of electromotive force. A wire, bent into a circle and carrying 1 A of current, produces 1 At of magnetomotive force. If the wire is bent into a loop having 50 turns, and the current stays the same, the resulting magnetomotive force is 50 At. If the current is then reduced to 1/50 A or 20 mA, the magnetomotive force will go back down to 1 At.

The *gilbert* (Gb) is also used to express magnetomotive force, but it is less common than the ampere-turn. One gilbert (1 Gb) is equal to 0.796 At. Conversely, 1 At = 1.26 Gb.

Electromagnets

Any electric current, or movement of charge carriers, produces a magnetic field. This field can become intense in a tightly coiled wire that has many turns and carries a large current. When a ferromagnetic core is placed inside the coil, the magnetic lines of flux are concentrated in the core, and the field strength in and near the core can become tremendous. This is the principle of an *electromagnet* (Fig. 8-5). Electromagnets are almost always cylindrical in shape. Sometimes the cylinder is long and thin; in other cases it is short and fat. But whatever the ratio of diameter to length for the core, the principle is the same: the magnetic field produced by the current results in magnetization of the core.

Direct-Current Types

You can build a dc electromagnet by taking a large bolt, such as a stove bolt, and wrapping a few dozen or a few hundred turns of wire around it. These items are available in any good hardware store. Be sure the bolt is made of ferromagnetic material. (If a permanent magnet sticks to the bolt, the bolt is ferromagnetic.) Ideally, the bolt should be at least 1 cm (approximately ⅜ in) in diameter and several inches long. You must use insulated wire, preferably made of solid, soft copper. "Bell wire" works well. Be sure all the wire turns go in the same direction. A large 6-V lantern battery can provide plenty of current to work the electromagnet. Never leave the coil connected to the battery for more than a few seconds at a time. And never use a car battery for this experiment! The acid can boil out of this type of battery, because the electromagnet places a heavy load on it.

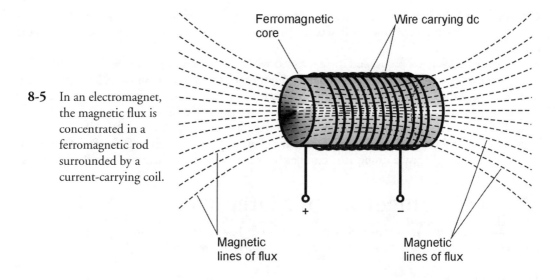

8-5 In an electromagnet, the magnetic flux is concentrated in a ferromagnetic rod surrounded by a current-carrying coil.

Direct-current electromagnets have defined north and south poles, just like permanent magnets. The main difference is that an electromagnet can get much stronger than any permanent magnet. You will see evidence of this if you do the preceding experiment with a large enough bolt and enough turns of wire.

Alternating-Current Types

Do you get the idea that an electromagnet can be made far stronger if, rather than using a lantern battery for the current source, you plug the wires into a wall outlet? In theory, this is true. In prac-

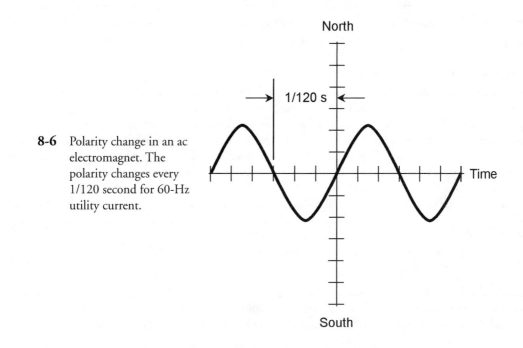

8-6 Polarity change in an ac electromagnet. The polarity changes every 1/120 second for 60-Hz utility current.

tice, you'll blow the fuse or circuit breaker. *Do not try this!* The electrical circuits in some buildings are not adequately protected and it can create a fire hazard. Also, you can get a lethal shock from the utility mains.

Some electromagnets use ac, and these magnets will stick to ferromagnetic objects. But the polarity of the magnetic field reverses every time the direction of the current reverses. With conventional household ac in the United States, there are 120 fluctuations, or 60 complete north-to-south-to-north polarity changes (Fig. 8-6), per second. If a permanent magnet, or a dc electromagnet, is brought near either "pole" of an ac electromagnet, there is no net force because the poles are alike half the time and opposite half the time, producing an equal amount of attractive and repulsive force. But if a piece of iron or steel is brought near a strong ac electromagnet, watch out! The attractive force will be powerful.

Magnetic Properties of Materials

There are four important properties that materials can have with respect to magnetic flux. These properties are *ferromagnetism*, *diamagnetism*, *permeability*, and *retentivity*.

Ferromagnetism

Some substances cause magnetic lines of flux to bunch closer together than they would in the medium of air or a vacuum. This property is called *ferromagnetism*, and materials that exhibit it are called *ferromagnetic*. You've already learned something about this!

Diamagnetism

Another property is known as *diamagnetism*, and materials that exhibit it are called *diamagnetic*. This type of substance decreases the magnetic flux density by causing the magnetic flux lines to diverge. Wax, dry wood, bismuth, and silver are examples. No diamagnetic material reduces the strength of a magnetic field by anywhere near the factor that ferromagnetic substances can increase it. Diamagnetic materials are generally used to keep magnetic objects apart, while minimizing the interaction between them. In recent years, they have also found some application in *magnetic levitation* devices.

Permeability

Permeability is a quantitative indicator of the extent to which a ferromagnetic material concentrates magnetic lines of flux. It is measured on a scale relative to a vacuum, or free space. Free space is assigned permeability 1. If you have a coil of wire with an air core, and a current is forced through the wire, then the flux in the coil core is at a certain density, just about the same as it would be in a vacuum. Therefore, the permeability of pure air is about equal to 1. If you place an iron core in the coil, the flux density increases by a large factor. The permeability of iron can range from 60 (impure) to as much as 8000 (highly refined).

If you use certain ferromagnetic alloys as the core material in electromagnets, you can increase the flux density, and therefore the local strength of the field, by as much as a million times. Such substances thus have permeability as great as 1,000,000 (10^6).

Table 8-1 gives permeability values for some common materials.

Retentivity

When a substance, such as iron, is subjected to a magnetic field as intense as it can handle, say by enclosing it in a wire coil carrying a massive current, there will be some *residual magnetism* left

Table 8-1.   Permeability values for some common materials.

Substance	Permeability (approx.)
Air, dry, at sea level	1
Alloys, ferromagnetic	3000–1,000,000
Aluminum	Slightly more than 1
Bismuth	Slightly less than 1
Cobalt	60–70
Iron, powdered and pressed	100–3000
Iron, solid, refined	3000–8000
Iron, solid, unrefined	60–100
Nickel	50–60
Silver	Slightly less than 1
Steel	300–600
Vacuum	1
Wax	Slightly less than 1
Wood, dry	Slightly less than 1

when the current stops flowing in the coil. *Retentivity*, also sometimes called *remanence*, is a measure of how well the substance "memorizes" the magnetism and thereby becomes a permanent magnet.

Retentivity is expressed as a percentage, and is symbolized B_r. If the flux density in the material is x tesla or gauss when it is subjected to the greatest possible magnetomotive force, and then goes down to y tesla or gauss when the current is removed, the retentivity is equal to $100(y/x)\%$.

Suppose that a metal rod can be magnetized to 135 G when it is enclosed by a coil carrying an electric current. Imagine that this is the maximum possible flux density that the rod can be forced to have. (For any substance, there is always such a maximum.) Now suppose that the current is shut off, and 19 G remain in the rod. Then the retentivity, B_r, is calculated as follows:

$$B_r = 100(19/135)\% = (100 \times 0.14)\% = 14\%$$

Some ferromagnetic substances have high retentivity. These materials are excellent for making permanent magnets. Other substances have low retentivity. They work well as electromagnets, but not as permanent magnets.

If a ferromagnetic substance has poor retentivity, it is especially well-suited for use as the core material for an ac electromagnet, because the polarity of the magnetic flux can reverse within the material at a rapid rate. Materials with high retentivity do not work well for ac electromagnets, because they resist the polarity reversal that takes place with ac.

Practical Magnetism

Magnetism has numerous applications in common consumer devices and systems. Here are some of the more common ways in which magnetic phenomena can be put to use.

Permanent Magnets

Permanent magnets are manufactured by using a high-retentivity ferromagnetic material as the core of an electromagnet for an extended period of time. The coil of the electromagnet carries a large direct current, causing intense magnetic flux of constant polarity within the material. (Don't try to do this at home. The high current can heat the coil and overload a battery or power supply, which produces a fire hazard and/or the risk of battery explosion.)

If you want to magnetize a screwdriver a little bit so that it will hold onto screws, just stroke the shaft of the screwdriver with the end of a bar magnet several dozen times. Once you have magnetized a tool in this way, however, it is nearly impossible to demagnetize it.

A Ringer Device

Figure 8-7 is a simplified diagram of a *bell ringer*, also called a *chime*. The main functional component is called a *solenoid*, and it is an electromagnet. The core has a hole going along its axis. The coil has several layers, but the wire is always wound in the same direction, so that the electromagnet is powerful. A movable steel rod runs through the hole in the electromagnet core.

When there is no current flowing in the coil, the steel rod is held down by the force of gravity. When a pulse of current passes through the coil, the rod is pulled forcibly upward so that it strikes the ringer plate. This plate is like one of the plates in a xylophone. The current pulse is short, so the steel rod falls back down again to its resting position, allowing the plate to reverberate.

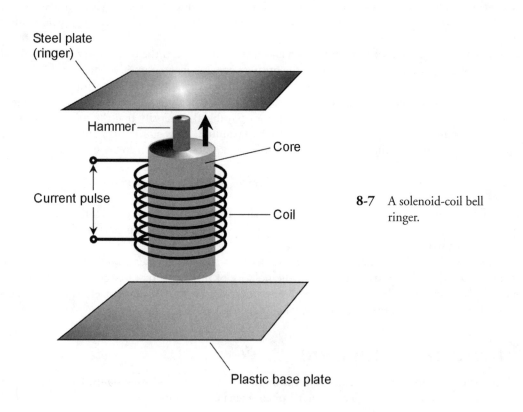

8-7 A solenoid-coil bell ringer.

The Relay

A *relay* makes use of a solenoid to allow remote-control switching of high-current circuits. A diagram of a relay is shown in Fig. 8-8. The movable lever, called the *armature*, is held to one side by a spring when there is no current flowing through the electromagnet. Under these conditions, terminal X is connected to Y, but not to Z. When a sufficient current is applied, the armature is pulled over to the other side. This disconnects terminal X from terminal Y, and connects X to Z.

There are numerous types of relays. Some are meant for use with dc, and others are for ac; a few will work with either dc or ac. A *normally closed relay* completes the circuit when there is no current flowing in its electromagnet coil, and breaks the circuit when current flows through the coil. A *normally open relay* is just the opposite, completing the circuit when current flows through the electromagnet coil, and opening the circuit when current ceases to flow through the coil. *Normal*, in this context, refers to the condition of no current applied to the electromagnet.

The relay shown in Fig. 8-8 can be used as either a normally open or normally closed relay, depending on which contacts are selected. It can also be used to switch a line between two different circuits.

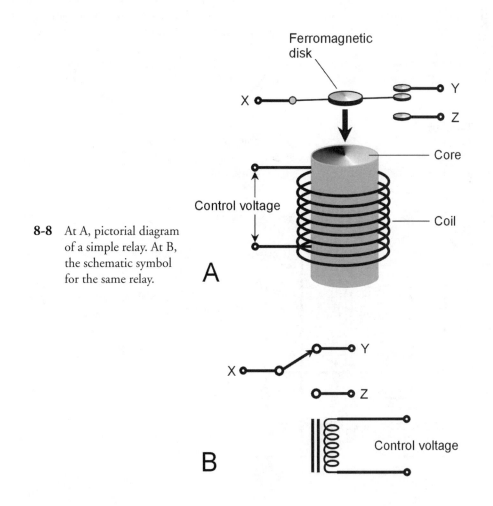

8-8 At A, pictorial diagram of a simple relay. At B, the schematic symbol for the same relay.

Some relays have several sets of contacts. Some relays are meant to remain in one state (either with current or without) for a long time, while others are meant to switch several times per second. The fastest relays can operate several dozen times per second. In recent years, relays have been largely supplanted by switching transistors and diodes, except in applications where extremely high current or high voltage is involved.

The DC Motor

Magnetic forces can be harnessed to do work. One common device that converts direct-current energy into rotating mechanical energy is a *dc motor*. In a dc motor, the source of electricity is connected to a set of coils, producing magnetic fields. The attraction of opposite poles, and the repulsion of like poles, is switched in such a way that a constant *torque*, or rotational force, results. As the current in the coils increases, the torque that the motor can provide also increases.

Figure 8-9 is a simplified, cutaway drawing of a dc motor. One set of coils, called the *armature coil*, rotates along with the motor shaft. The other set of coils, called the *field coil*, is stationary. The current direction is periodically reversed during each rotation by means of the *commutator*. This keeps the rotational force going in the same angular direction, so the motor continues to rotate rather than oscillating back and forth. The shaft is carried along by its own inertia, so that it doesn't come to a stop during those instants when the current is being switched in polarity.

Some dc motors can also be used to generate dc. These motors contain permanent magnets in place of one of the sets of coils. When the shaft is rotated, a pulsating dc flows in the coil.

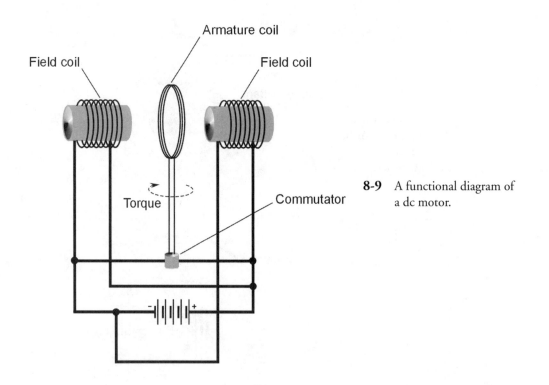

8-9   A functional diagram of a dc motor.

Magnetic Tape

Magnetic tape, also called *recording tape*, consists of millions of ferromagnetic particles attached to a flexible, thin plastic strip. In the *tape recorder*, a fluctuating magnetic field, produced by the *recording head*, polarizes these particles. As the field changes in strength next to the recording head, the tape passes by at a constant speed. This produces regions in which the ferromagnetic particles are polarized in either direction (Fig. 8-10).

When the tape is run at the same speed through the recorder in the playback mode, the magnetic fields around the individual particles cause a fluctuating field that is detected by the *pickup head.* This field has the same pattern of variations as the original field from the recording head.

Magnetic tape is available in various widths and thicknesses. Thicker tapes result in cassettes that don't play as long, but the tape is more resistant to stretching. The speed of the tape determines the fidelity of the recording. Higher speeds are preferred for music and video, and lower speeds for voice and data.

The impulses on a magnetic tape can be distorted or erased by external magnetic fields. Therefore, tapes should be protected from such fields. Keep the tape away from magnets. Extreme heat can also result in loss of data, and can cause permanent physical damage to the tape.

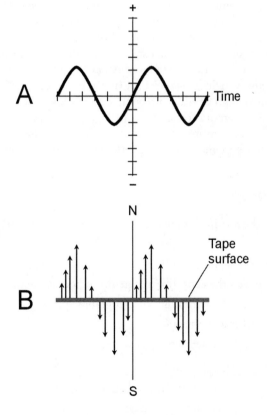

8-10 On recording tape, particles are magnetized in a pattern that follows the input waveform. Graph A shows an example of an audio input waveform. Graph B shows relative polarity and intensity of magnetization for selected particles on the tape surface.

Magnetic Disk

Since the advent of the personal computer, ever-more compact data-storage systems have evolved. One of the most versatile is the *magnetic disk.*

Hard disks, also called *hard drives*, store the most data, and are generally found inside of computer units. *Diskettes* are 8.9 cm (3.5 in) across, and can be inserted and removed from recording/playback machines called *diskette drives.* In recent years, magnetic diskettes have been largely supplanted by non-magnetic *compact disc recordable* (CD-R) and *compact disc rewritable* (CD-RW) media.

The principle of the magnetic disk, on the microscale, is the same as that of magnetic tape. The information is stored in *binary digital* form; that is, there are only two different ways that the particles are magnetized. This results in almost perfect, error-free storage. On a larger scale, the disk works differently than the tape because of the difference in geometry. On a tape, the information is spread out over a long span, and some bits of data are far away from others as measured along the medium itself. But on a disk, no two bits are ever farther apart than the diameter of the disk. This means that data can be stored to, and retrieved from, a disk much faster than is possible with tape.

The same precautions should be observed when handling and storing magnetic disks as are necessary with magnetic tape.

Bubble Memory

Bubble memory is a sophisticated method of storing data that gets rid of the need for moving parts such as are required in tape machines and disk drives. Data is stored as tiny magnetic fields, in a medium that is made from magnetic film and semiconductor materials.

Bubble memory makes use of all the advantages of magnetic data storage, as well as the favorable aspects of electronic data storage. Advantages of electronic memory include rapid storage and recovery, and high density (a lot of data can be put in a tiny volume of space). Advantages of magnetic memory include *nonvolatility* (it can be stored for a long time without needing a constant current source), high density, and comparatively low cost.

Bubble memory seems to go through phases. Just as it is declared obsolete, someone comes up with a new and improved way to make it work. Check the Internet to find out its current status; enter "bubble memory" or "magnetic bubble memory" into a search engine.

Quiz

Refer to the text in this chapter if necessary. A good score is at least 18 correct. Answers are in the back of the book.

1. The geomagnetic field
    (a) makes the earth like a huge horseshoe magnet.
    (b) runs exactly through the geographic poles.
    (c) makes a compass work.
    (d) makes an electromagnet work.

2. Geomagnetic lines of flux
    (a) are horizontal at the geomagnetic equator.
    (b) are vertical at the geomagnetic equator.

(c) are never horizontal, no matter where you go.

(d) are perfectly symmetrical around the earth, even far out in space.

3. A material that can be permanently magnetized is generally said to be

   (a) ultramagnetic.

   (b) electromagnetic.

   (c) diamagnetic.

   (d) ferromagnetic.

4. The force between a magnet and a piece of ferromagnetic metal that has not been magnetized

   (a) can be either repulsive or attractive.

   (b) is never repulsive.

   (c) gets smaller as the magnet gets closer to the metal.

   (d) depends on the geomagnetic field.

5. The presence of a magnetic field can always be attributed to

   (a) ferromagnetic materials.

   (b) diamagnetic materials.

   (c) motion of electric charge carriers.

   (d) the north geomagnetic pole.

6. Lines of magnetic flux are said to originate

   (a) in atoms of ferromagnetic materials.

   (b) at a north magnetic pole.

   (c) at points where the lines are straight.

   (d) in electric charge carriers.

7. The magnetic flux around a straight, current-carrying wire

   (a) gets stronger with increasing distance from the wire.

   (b) is strongest near the wire.

   (c) does not vary in strength with distance from the wire.

   (d) consists of straight lines parallel to the wire.

8. The gauss is a unit of

   (a) overall magnetic field strength.

   (b) ampere-turns.

   (c) magnetic flux density.

   (d) magnetic power.

9. A unit of overall magnetic field quantity is the

   (a) maxwell.

   (b) gauss.

   (c) tesla.

   (d) ampere-turn.

10. If a wire coil has 10 turns and carries 500 mA of current, what is the magnetomotive force?

    (a) 5000 At

    (b) 50 At

    (c) 5.0 At

    (d) 0.02 At

11. If a wire coil has 100 turns and carries 1.30 A of current, what is the magnetomotive force?

    (a) 130 Gb

    (b) 76.9 Gb

    (c) 164 Gb

    (d) 61.0 Gb

12. Which of the following can occur during a geomagnetic storm?

    (a) Charged particles stream out from the sun.

    (b) The earth's magnetic field is affected.

    (c) Electrical power transmission is disrupted.

    (d) More than one of the above can occur.

13. An ac electromagnet

    (a) attracts only permanent magnets.

    (b) attracts pure, unmagnetized iron.

    (c) repels all permanent magnets.

    (d) either attracts or repels permanent magnets, depending on the polarity.

14. An advantage of an electromagnet over a permanent magnet is the fact that

    (a) an electromagnet can be switched on and off.

    (b) an electromagnet does not have specific polarity.

    (c) an electromagnet requires no power source.

    (d) permanent magnets must always be cylindrical, but electromagnets can have any shape.

15. A substance with high retentivity

    (a) can make a good ac electromagnet.

    (b) repels both north and south magnetic poles.

    (c) is always a diamagnetic material.

    (d) is well suited to making a permanent magnet.

16. Suppose a relay is connected into a circuit so that a device gets a signal only when the relay coil carries current. The relay is

    (a) an ac relay.

    (b) a dc relay.

    (c) normally closed.

    (d) normally open.

17. A device that repeatedly reverses the polarity of a magnetic field in order to keep a dc motor rotating is known as

    (a) a solenoid.

    (b) an armature coil.

    (c) a commutator.

    (d) a field coil.

18. A high tape-recorder motor speed is generally used for

    (a) voice recording and playback.

    (b) video recording and playback.

    (c) digital data storage and retrieval.

    (d) all of the above.

19. An advantage of a magnetic disk, compared with magnetic tape, for data storage and retrieval is that

    (a) a disk lasts longer.

    (b) data can be stored and retrieved more quickly with disks than with tapes.

    (c) disks look better.

    (d) disks are less susceptible to magnetic fields.

20. A magnetic hard disk is usually part of

    (a) a computer.

    (b) a dc motor.

    (c) a tape recorder.

    (d) an electromagnet.

Test: Part 1

Do not refer to the text when taking this test. A good score is at least 37 correct. Answers are in the back of the book. It's best to have a friend check your score the first time, so you won't memorize the answers if you want to take the test again.

1. An application in which an analog meter would almost always be preferred over a digital meter is
    (a) the signal-strength indicator in a radio receiver.
    (b) a meter that shows power-supply voltage.
    (c) a utility watt-hour meter.
    (d) a clock.
    (e) a device in which a direct numeric display is wanted.

2. Which of the following statements is false?
    (a) The current in a series dc circuit is divided up among the resistances.
    (b) In a parallel dc circuit, the voltage is the same across each component.
    (c) In a series dc circuit, the sum of the voltages across all the components, going once around a complete circle and taking polarity into account, is zero.
    (d) The net resistance of a parallel set of resistors is less than the value of the smallest resistor.
    (e) The total wattage consumed in a series circuit is the sum of the wattages consumed by each of the components.

3. The ohm is a unit of
    (a) electrical charge quantity.
    (b) the rate at which charge carriers flow.
    (c) opposition to electrical current.
    (d) electrical conductance.
    (e) potential difference.

4. A wiring diagram differs from a schematic diagram in that
    (a) a wiring diagram is less detailed than a schematic diagram.
    (b) a wiring diagram always shows the component values, but a schematic diagram might not.
    (c) a schematic does not show all the interconnections between the components, but a wiring diagram does.
    (d) a schematic diagram shows pictures of components, while a wiring diagram shows the electronic symbols.
    (e) a schematic diagram shows the electronic symbols, while a wiring diagram shows pictures of the components.

5. In which of the following places would you be most likely to find a wirewound resistor?
    (a) A dc circuit location where a large amount of power must be dissipated
    (b) The input circuit of a radio-frequency amplifier
    (c) The output circuit of a radio-frequency amplifier
    (d) In an antenna system, to limit the transmitter power
    (e) Between ground and the chassis of a power supply

6. The number of protons in the nucleus of an element is known as the
    (a) electron number.
    (b) atomic number.
    (c) valence number.
    (d) charge number.
    (e) proton number.

7. A hot-wire ammeter
    (a) can measure ac as well as dc.
    (b) registers current changes very fast.
    (c) can indicate very low voltages.
    (d) measures electrical energy.
    (e) works only when current flows in one direction.

8. Which of the following units indicates the rate at which energy is expended?
    (a) The volt
    (b) The ampere
    (c) The coulomb
    (d) The ampere-hour
    (e) The watt

9. Which of the following correctly states Ohm's Law?
    (a) Volts equal amperes divided by ohms.
    (b) Ohms equal amperes divided by volts.
    (c) Amperes equal ohms divided by volts.

(d) Amperes equal ohms times volts.

(e) Ohms equal volts divided by amperes.

10. The current flowing into a point in a dc circuit is always equal to the current

   (a) delivered by the power supply.

   (b) through any one of the resistances.

   (c) flowing out of that point.

   (d) at any other point.

   (e) in any single branch of the circuit.

11. A loudness meter in a hi-fi system is generally calibrated in

   (a) volts.

   (b) amperes.

   (c) decibels.

   (d) watt-hours.

   (e) ohms.

12. An electrically charged atom (either positive or negative) is known as

   (a) a molecule.

   (b) an isotope.

   (c) an ion.

   (d) an electron.

   (e) a fundamental particle.

13. Suppose a battery delivers 12.0 V to a bulb, and current flowing through the bulb is 3.00 A. The resistance of the bulb is which of the following?

   (a) 36.0 Ω

   (b) 4.00 Ω

   (c) 0.250 Ω

   (d) 108 Ω

   (e) 0.750 Ω

14. The peak voltage in an ac wave is always

   (a) greater than the average voltage.

   (b) less than the average voltage.

   (c) greater than or equal to the average voltage.

   (d) less than or equal to the average voltage.

   (e) fluctuating.

15. Suppose a resistor is specified a having a value of 680 Ω, and a tolerance of $\pm 5\%$. You measure the actual resistance with a precision digital ohmmeter. Which of the following meter readings indicates a reject?

   (a) 648 Ω

   (b) 712 Ω

(c) 699 Ω

(d) 636 Ω

(e) 707 Ω

16. A primitive device for indicating the presence of an electric current is
    (a) an electrometer.
    (b) a galvanometer.
    (c) a voltmeter.
    (d) a coulometer.
    (e) a wattmeter.

17. A disadvantage of mercury cells is the fact that they
    (a) can adversely affect the environment when discarded.
    (b) supply dangerously high voltage.
    (c) can reverse polarity unexpectedly.
    (d) must be physically larger than other types of cells that have the same current-delivering capacity.
    (e) must be kept right-side up to keep the mercury from spilling out.

18. Suppose a battery supplies 6.0 V to a bulb rated at 12 W. The bulb draws how much current?
    (a) 2.0 A
    (b) 0.5 A
    (c) 72 A
    (d) 40 mA
    (e) 72 mA

19. Which of the following is not a common use for a resistor or set of resistors?
    (a) Biasing for a transistor
    (b) Voltage division
    (c) Current limiting
    (d) As a dummy antenna
    (e) Helping a capacitor to hold its charge for a long time

20. When an electrical charge exists but there is no flow of current, the charge is said to be
    (a) ionizing.
    (b) atomic.
    (c) molecular.
    (d) electronic.
    (e) static.

21. The sum of the voltages, going around a dc circuit, but not including the power supply, has
    (a) an equal value and the same polarity as the supply.
    (b) a value that depends on the ratio of the resistances.
    (c) a different value from, but the same polarity as, the supply.

(d) an equal value as, but the opposite polarity from, the supply.

(e) a different value from, and the opposite polarity from, the supply.

22. A watt-hour meter measures

(a) voltage.

(b) current.

(c) power.

(d) energy.

(e) charge.

23. Every chemical element has its own unique type of particle, which is known as its

(a) neutron.

(b) electron.

(c) proton.

(d) atom.

(e) isotope.

24. An advantage of a magnetic disk over magnetic tape for data storage is the fact that

(a) data is too closely packed on the tape.

(b) the disk is immune to the effects of magnetic fields.

(c) data storage and retrieval is faster on disk.

(d) disks store computer data in analog form.

(e) tapes cannot be used to store digital data.

25. Suppose a 6-V battery is connected across a series combination of resistors. The resistance values are 1.0 Ω, 2.0 Ω, and 3.0 Ω. What is the current through the 2.0-Ω resistor?

(a) 1.0 A

(b) 3.0 A

(c) 12 A

(d) 24 A

(e) 72 A

26. A sample of material with resistance so high that it can be considered infinite for most practical purposes is known as

(a) a semiconductor.

(b) a paraconductor.

(c) an insulator.

(d) a resistor.

(e) a diamagnetic substance.

27. Primary cells

(a) can be used over and over.

(b) have higher voltage than other types of cells.

(c) all supply exactly 1.500 V.

(d) cannot be recharged.

(e) are made of zinc and carbon.

28. A rheostat

    (a) can be used in high-voltage and/or high-power dc circuits.

    (b) is ideal for tuning a radio receiver.

    (c) is often used as a bleeder resistor.

    (d) is better than a potentiometer for low-power audio.

    (e) offers the advantage of having no inductance.

29. How much dc voltage does a typical dry cell provide?

    (a) 12 V

    (b) 6 V

    (c) 1.5 V

    (d) 117 V

    (e) Any of the above

30. A geomagnetic storm

    (a) causes solar wind.

    (b) causes the earth's magnetic field to disappear.

    (c) can disturb the earth's magnetic field.

    (d) can pollute the earth's atmosphere.

    (e) stabilizes the ac utility grid.

31. An advantage of an alkaline cell over a zinc-carbon cell is the fact that

    (a) the alkaline cell provides more voltage.

    (b) the alkaline cell can be recharged.

    (c) the alkaline cell can deliver useful current at lower temperatures.

    (d) the alkaline cell is far less bulky for the same amount of energy capacity.

    (e) the alkaline cell can produce ac as well as dc.

32. Suppose a battery delivers 12 V across a set of six 4.0-Ω resistors in a series voltage dividing combination. This provides six different voltages, differing by equal increments of which of the following?

    (a) 0.25 V

    (b) 0.33 V

    (c) 1.0 V

    (d) 2.0 V

    (e) 3.0 V

33. A unit of electrical charge quantity is the

    (a) volt.

    (b) ampere.

    (c) watt.

(d) tesla.

(e) coulomb.

34. A unit of conductance is the

    (a) volt per meter.

    (b) ampere per meter.

    (c) anti-ohm.

    (d) siemens.

    (e) ohm per meter.

35. Suppose a 24-V battery is connected across a set of four resistors in parallel. Each resistor has a value of 32 Ω. What is the total power dissipated by the set of resistors?

    (a) 0.19 W

    (b) 3.0 W

    (c) 0.19 kW

    (d) 0.33 W

    (e) 72 W

36. The main difference between a lantern battery and a transistor battery is the fact that

    (a) a lantern battery has higher voltage than a transistor battery.

    (b) a fresh lantern battery has more energy stored in it than a fresh transistor battery.

    (c) a lantern battery cannot be used with electronic devices such as transistor radios, but a transistor battery can.

    (d) a lantern battery can be recharged, but a transistor battery cannot.

    (e) a lantern battery is more compact than a transistor battery.

37. Nickel-based batteries would most likely be found

    (a) in disposable flashlights.

    (b) in large lanterns.

    (c) as car and truck batteries.

    (d) in handheld radio transceivers.

    (e) in electromagnets.

38. A voltmeter should have

    (a) low internal resistance.

    (b) electrostatic plates.

    (c) a sensitive amplifier.

    (d) high internal resistance.

    (e) the highest possible full-scale value.

39. The purpose of a bleeder resistor is to

    (a) provide bias for a transistor.

    (b) serve as a voltage divider.

(c) protect people against the danger of electric shock.

(d) reduce the current in a power supply.

(e) smooth out the ac ripple in a power supply.

40. A dc electromagnet

(a) has constant polarity.

(b) requires an air core.

(c) does not attract or repel a permanent magnet.

(d) has polarity that periodically reverses.

(e) cannot be used to permanently magnetize anything.

41. The rate at which charge carriers flow is measured in

(a) amperes.

(b) coulombs.

(c) volts.

(d) watts.

(e) watt-hours.

42. Suppose a 12-V battery is connected to a set of three resistors in series. The resistance values are $1.0\ \Omega$, $2.0\ \Omega$, and $3.0\ \Omega$. What is the voltage across the $3.0\text{-}\Omega$ resistor?

(a) 1.0 V

(b) 2.0 V

(c) 4.0 V

(d) 6.0 V

(e) 12 V

43. Suppose nine $90\text{-}\Omega$ resistors are connected in a 3×3 series-parallel network. What is the total (net) resistance of the network?

(a) $10\ \Omega$

(b) $30\ \Omega$

(c) $90\ \Omega$

(d) $270\ \Omega$

(e) $810\ \Omega$

44. A device commonly used for remote switching of high-current circuits is

(a) a solenoid.

(b) an electromagnet.

(c) a potentiometer.

(d) a photovoltaic cell.

(e) a relay.

45. Memory in a nickel-based cell or battery

(a) occurs whenever the battery is discharged.

(b) indicates that the cell or battery is dead.

(c) can usually be remedied by repeated discharging and recharging.

(d) can cause an explosion.

(e) causes a reversal in polarity.

46. Suppose a 100-W bulb burns for 100 hours. It has consumed how many units of energy?

    (a) 0.10 kWh

    (b) 1.00 kWh

    (c) 10.0 kWh

    (d) 100 kWh

    (e) 1000 kWh

47. A material with high permeability

    (a) increases magnetic field quantity.

    (b) is necessary if a coil is to produce a magnetic field.

    (c) always has high retentivity.

    (d) concentrates magnetic lines of flux.

    (e) reduces flux density.

48. A chemical compound

    (a) consists of two or more atoms.

    (b) contains an unusual number of neutrons.

    (c) is technically the same as an ion.

    (d) has a shortage of electrons.

    (e) has an excess of electrons.

49. Suppose a 6.00-V battery is connected to a parallel combination of two resistors whose values are 8.00 Ω and 12.0 Ω. What is the power dissipated in the 8-Ω resistor?

    (a) 0.300 W

    (b) 0.750 W

    (c) 1.25 W

    (d) 1.80 W

    (e) 4.50 W

50. The main problem with bar-graph meters is the fact that

    (a) they are not very sensitive.

    (b) they are unstable.

    (c) they cannot give very precise readings.

    (d) you need special training to read them.

    (e) they can display only peak values.

2
PART

Alternating Current

9

CHAPTER

Alternating-Current Basics

DIRECT CURRENT CAN BE EXPRESSED IN TERMS OF TWO VARIABLES: DIRECTION (POLARITY) AND intensity (amplitude). Alternating current (ac) is a little more complicated. This chapter will acquaint you with some common forms of ac.

Definition of Alternating Current

You have learned that dc has polarity that stays constant over time. Although the amplitude (the number of amperes, volts, or watts) can fluctuate from moment to moment, the charge carriers always flow in the same direction at any point in the circuit.

In ac, the polarity reverses at regular intervals. The *instantaneous amplitude* (that is, the amplitude at any given instant in time) of ac usually varies because of the repeated reversal of polarity. But there are certain cases where the amplitude remains constant, even though the polarity keeps reversing.

The rate of change of polarity is the variable that makes ac so much different from dc. The behavior of an ac wave depends largely on this rate: the *frequency*.

Period and Frequency

In a *periodic ac wave*, the kind that is discussed in this chapter (and throughout the rest of this book), the function of *instantaneous amplitude versus time* repeats itself over and over, so that the same pattern recurs indefinitely. The length of time between one repetition of the pattern, or one *cycle*, and the next is called the *period* of the wave. This is illustrated in Fig. 9-1 for a simple ac wave. The period of a wave can, in theory, be anywhere from a minuscule fraction of a second to many centuries. Period, when measured in seconds, is denoted by *T*.

Originally, ac frequency was specified in *cycles per second* (cps). High frequencies were sometimes given in *kilocycles*, *megacycles*, or *gigacycles*, representing thousands, millions, or billions (thousand-millions) of cycles per second. But nowadays, the unit is known as the *hertz* (Hz). Thus, 1 Hz = 1 cps, 10 Hz = 10 cps, and so on. Higher frequencies are given in *kilohertz* (kHz), *megahertz* (MHz), or *gigahertz* (GHz). The relationships are as follows:

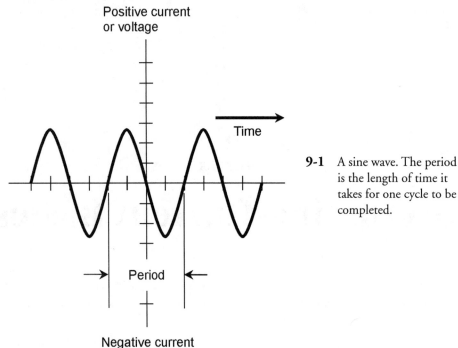

9-1 A sine wave. The period is the length of time it takes for one cycle to be completed.

$$1 \text{ kHz} = 1000 \text{ Hz}$$
$$1 \text{ MHz} = 1000 \text{ kHz} = 1,000,000 \text{ Hz} = 10^6 \text{ Hz}$$
$$1 \text{ GHz} = 1000 \text{ MHz} = 1,000,000,000 \text{ Hz} = 10^9 \text{ Hz}$$

Sometimes an even bigger unit, the *terahertz* (THz), is used to specify ac frequency. This is a trillion (1,000,000,000,000, or 10^{12}) hertz. Electrical currents generally do not attain such frequencies, although some forms of *electromagnetic radiation* do.

The frequency of an ac wave, denoted *f*, in hertz is the reciprocal of the period in seconds. Mathematically, these two equations express the relationship:

$$f = 1/T \quad \text{and} \quad T = 1/f$$

Some ac waves have only one frequency. These waves are called *pure*. But often, there are components at multiples of the main, or *fundamental*, frequency. There can also be components at odd frequencies. Some ac waves have hundreds, thousands, or even infinitely many different component frequencies.

The Sine Wave

Sometimes, alternating current has a *sine-wave*, or *sinusoidal*, nature. This means that the direction of the current reverses at regular intervals, and that the current-versus-time curve is shaped like the trigonometric *sine function*. The waveform in Fig. 9-1 is a sine wave.

Any ac wave that consists of a single frequency has a perfectly sinusoidal shape. Any perfect sinusoidal ac source has only one component frequency. In practice, a wave might be so close to a sine wave that it looks exactly like the sine function on an oscilloscope, when in reality there are traces of other frequencies present. Imperfections are often too small to see. But pure, single-frequency ac not only looks perfect, but actually is a perfect replication of the trigonometric sine function.

The current at the wall outlets in your house is an almost perfect ac sine wave with a frequency of 60 Hz.

Square Waves

Earlier in this chapter, it was said that there can be an ac wave whose instantaneous amplitude remains constant, even though the polarity reverses. Does this seem counterintuitive? Think some more! A *square wave* is such a wave.

On an oscilloscope, a square wave looks like a pair of parallel, dashed lines, one with positive polarity and the other with negative polarity (Fig. 9-2A). The oscilloscope shows a graph of voltage on the vertical scale and time on the horizontal scale. The transitions between negative and positive for a theoretically perfect square wave would not show up on the oscilloscope, because they would be instantaneous. But in practice, the transitions can often be seen as vertical lines (Fig. 9-2B).

True square waves have equal negative and positive peaks. Thus, the absolute amplitude of the wave is constant. Half of the time it's $+x$, and the other half of the time it's $-x$ (where x can be expressed in volts, amperes, or watts).

Some squared-off waves are lopsided; the negative and positive amplitudes are not the same. Still others remain at positive polarity longer than they remain at negative polarity (or vice versa). These are examples of *asymmetrical square waves*, more properly called *rectangular waves*.

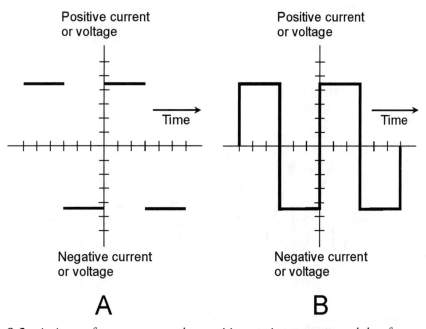

9-2   At A, a perfect square wave; the transitions are instantaneous and therefore do not show up on the graph. At B, the more common rendition of a square wave, showing the transitions as vertical lines.

Sawtooth Waves

Some ac waves rise and/or fall in straight, sloping lines as seen on an oscilloscope screen. The slope of the line indicates how fast the magnitude is changing. Such waves are called *sawtooth waves* because of their appearance. Sawtooth waves are generated by certain electronic test devices. They can also be generated by electronic sound synthesizers.

Fast Rise, Slow Decay

Figure 9-3 shows a sawtooth wave in which the positive-going slope (called the *rise*) is extremely steep, as with a square wave, but the negative-going slope (called the *decay*) is not so steep. The period of the wave is the time between points at identical positions on two successive pulses.

Slow Rise, Fast Decay

Another form of sawtooth wave is just the opposite, with a defined, finite rise and an instantaneous decay. This type of wave is often called a *ramp* because it looks like an incline going upward (Fig. 9-4). This waveshape is useful for scanning in television sets and oscilloscopes. It tells the electron beam to move, or *trace*, at constant speed from left to right across the screen during the rise. Then it retraces, or brings the electron beam back, instantaneously during the decay so the beam can trace across the screen again.

Variable Rise and Decay

Sawtooth waves can have rise and decay slopes in an infinite number of different combinations. One common example is shown in Fig. 9-5. In this case, the rise and the decay are both finite and equal. This is known as a *triangular wave*.

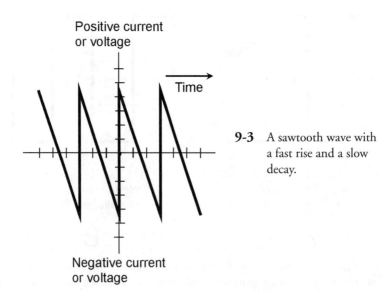

9-3 A sawtooth wave with a fast rise and a slow decay.

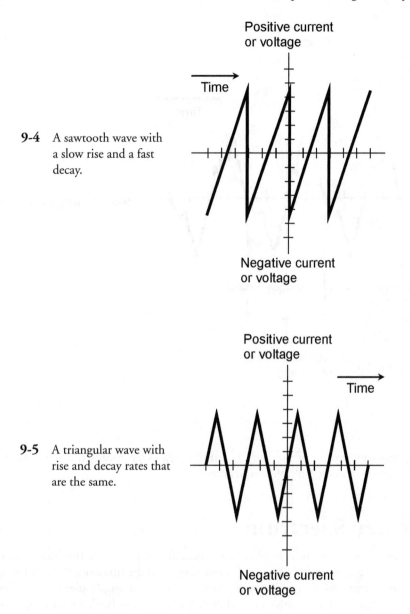

9-4 A sawtooth wave with a slow rise and a fast decay.

9-5 A triangular wave with rise and decay rates that are the same.

Complex and Irregular Waveforms

As long as a wave has a definite period, and as long as the polarity keeps switching back and forth between positive and negative, it is ac, no matter how complicated the actual shape of the waveform. Figure 9-6 shows an example of a complex ac wave. There is a definable period, and therefore a definable frequency. The period is the time between two points on succeeding wave repetitions.

With some waves, it can be difficult or almost impossible to ascertain the period. This is because the wave has two or more components that are of nearly the same amplitude. When this hap-

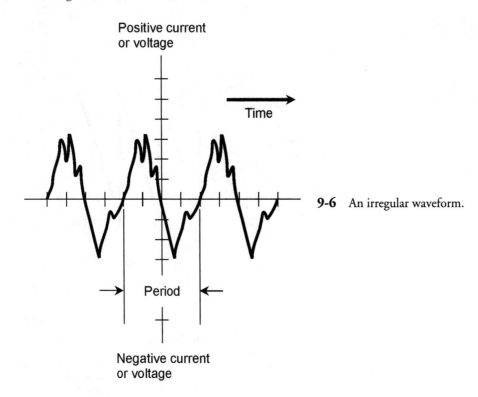

9-6   An irregular waveform.

pens, the *frequency spectrum* of the wave is multifaceted. That means the wave energy is split up more or less equally among multiple frequencies.

Frequency Spectrum

An oscilloscope shows a graph of amplitude as a function of time. Because time is on the horizontal axis and represents the *independent variable* or *domain* of the function, the oscilloscope is said to be a *time-domain* instrument. But suppose you want to see the amplitude of a complex signal as a function of frequency, rather than as a function of time? This can be done with a *spectrum analyzer.* It is a *frequency-domain* instrument. Its horizontal axis shows frequency as the independent variable, ranging from some adjustable minimum frequency (at the extreme left) to some adjustable maximum frequency (at the extreme right).

An ac sine wave, as displayed on a spectrum analyzer, appears as a single *pip*, or vertical line (Fig. 9-7A). This means that all of the energy in the wave is concentrated at one frequency. But many, if not most, ac waves contain *harmonic* energy along with energy at the fundamental frequency. A harmonic frequency is a whole-number multiple of the fundamental frequency. For example, if 60 Hz is the fundamental frequency, then harmonics can exist at 120 Hz, 180 Hz, 240 Hz, and so on. The 120-Hz wave is the *second harmonic;* the 180-Hz wave is the *third harmonic;* the 240-Hz wave is the *fourth harmonic;* and so on.

In general, if a wave has a frequency equal to *n* times the fundamental (where *n* is some whole number), then that wave is called the n*th harmonic*. In Fig. 9-7B, a wave is shown along with several harmonics, as it would look on the display screen of a spectrum analyzer.

Square waves and sawtooth waves contain harmonic energy in addition to energy at the fundamental frequency. Other waves can get more complicated. The exact shape of a wave depends on the amount of energy in the harmonics, and the way in which this energy is distributed among them.

Irregular waves can have any imaginable frequency distribution. Figure 9-8 shows an example. This is a spectral (frequency-domain) display of an *amplitude-modulated* (AM) voice radio signal. Much of the energy is concentrated at the center of the pattern, at the frequency shown by the vertical line. That is the *carrier frequency.* There is also plenty of energy near, but not exactly at, the carrier frequency. That's the part of the signal that contains the voice.

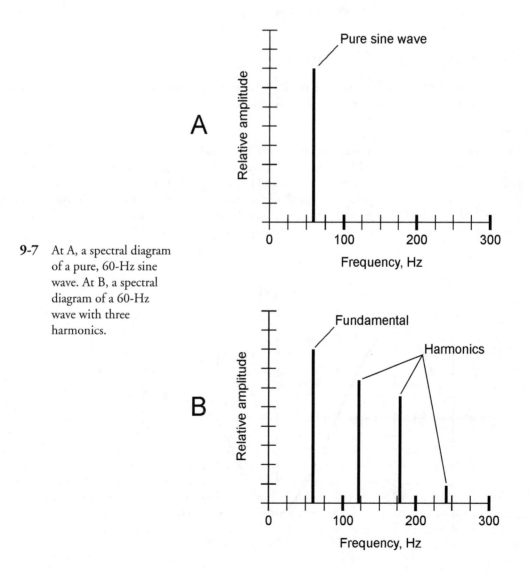

9-7 At A, a spectral diagram of a pure, 60-Hz sine wave. At B, a spectral diagram of a 60-Hz wave with three harmonics.

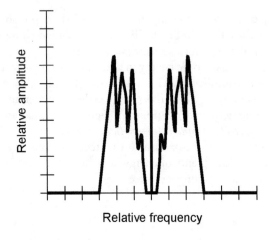

9-8 A spectral diagram of a modulated radio signal.

Fractions of a Cycle

Engineers break the ac cycle down into small parts for analysis and reference. One complete cycle can be compared to a single revolution around a circle.

Degrees

One method of specifying the phase of an ac cycle is to divide it into 360 equal parts, called *degrees* or *degrees of phase*, symbolized by a superscript, lowercase letter *o* (°). The value 0° is assigned to the point in the cycle where the magnitude is zero and positive-going. The same point on the next cycle is given the value 360°. The point one-fourth of the way through the cycle is 90°; the point halfway through the cycle is 180°; the point three-fourths of the way through the cycle is 270°. This is illustrated in Fig. 9-9. Degrees of phase are used mainly by engineers and technicians.

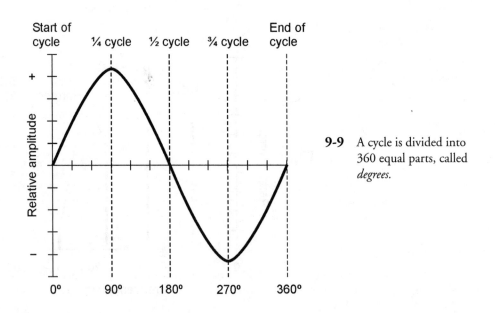

9-9 A cycle is divided into 360 equal parts, called *degrees*.

Radians

The other method of specifying phase is to divide the cycle into 2π equal parts, where π (pi) is a geometric constant equal to the number of diameters of any circle that can be laid end to end around the circumference of that circle. This constant is approximately equal to 3.14159. A *radian* (rad) of phase is thus equal to about 57.3°. Sometimes, the frequency of an ac wave is measured in *radians per second* (rad/s) rather than in hertz. Because there are about 6.28 radians in a complete cycle of 360°, the *angular frequency* of a wave, in radians per second, is equal to about 6.28 times the frequency in hertz. Radians of phase are used mainly by physicists.

Phase Difference

Even if two ac waves have exactly the same frequency, they can have different effects because they are out of sync with each other. This is especially true when ac waves are added together to produce a third, or *composite*, wave.

If two pure ac sine waves have identical frequencies and identical amplitudes but differ in phase by 180° (a half cycle), they cancel each other out, and the composite wave is zero; it ceases to exist! If the two waves are exactly in phase, the composite wave has the same frequency, but twice the amplitude, of either signal alone.

If two pure ac sine waves have the same frequency but different amplitudes, and if they differ in phase by 180°, the composite signal has the same frequency as the originals, and an amplitude equal to the difference between the two. If two such waves are exactly in phase, the composite has the same frequency as the originals, and an amplitude equal to the sum of the two.

If two pure ac sine waves have the same frequency but differ in phase by some odd amount such as 75° or 110°, the resulting signal has the same frequency, but does not have the same waveshape as either of the original signals. The variety of such cases is infinite.

Household electricity from 117-V wall outlets consists of a 60-Hz sine wave with only one phase component. But the energy is transmitted over long distances in three phases, each differing by 120° or one-third of a cycle. This is what is meant by *three-phase ac*. Each of the three ac waves carries one-third of the total power in a utility transmission line.

Expressions of Amplitude

Amplitude is also called *magnitude, level, strength,* or *intensity*. Depending on the quantity being measured, the amplitude of an ac wave can be specified in amperes (for current), volts (for voltage), or watts (for power). In addition to this, there are several different ways in which amplitude can be expressed.

Instantaneous Amplitude

The *instantaneous amplitude* of an ac wave is the amplitude at some precise moment, or instant, in time. This constantly changes. The manner in which it varies depends on the waveform. Instantaneous amplitudes are represented by individual points on the wave curves.

Peak Amplitude

The *peak* (pk) *amplitude* of an ac wave is the maximum extent, either positive or negative, that the instantaneous amplitude attains. In many situations, the positive and negative peak amplitudes of

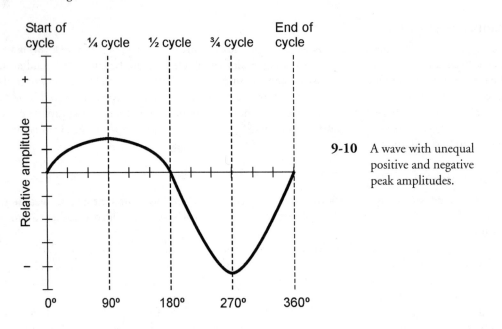

9-10 A wave with unequal positive and negative peak amplitudes.

an ac wave are the same. But sometimes they differ. Figure 9-9 is an example of a wave in which the positive peak amplitude is the same as the negative peak amplitude. Figure 9-10 is an illustration of a wave that has different positive and negative peak amplitudes.

Peak-to-Peak Amplitude

The *peak-to-peak* (pk-pk) *amplitude* of a wave is the net difference between the positive peak amplitude and the negative peak amplitude (Fig. 9-11). The peak-to-peak amplitude is equal to the positive peak amplitude plus the negative peak amplitude. When the positive and negative peak amplitudes of an ac wave are equal, the peak-to-peak amplitude is exactly twice the peak amplitude.

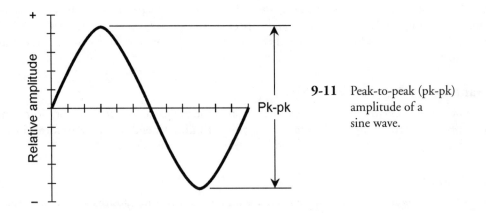

9-11 Peak-to-peak (pk-pk) amplitude of a sine wave.

Root-Mean-Square Amplitude

Often, it is necessary to express the *effective amplitude* of an ac wave. This is the voltage, current, or power that a dc source would have to produce in order to have the same general effect as a given ac wave. When you say a wall outlet provides 117 V, you mean 117 effective volts. This is not the same as the peak or peak-to-peak voltage.

The most common expression for effective ac intensity is called the *root-mean-square* (rms) *amplitude.* The terminology reflects the fact that the ac wave is mathematically operated on by taking the square root of the mean (average) of the square of all its instantaneous amplitudes.

In the case of a perfect ac sine wave, the rms value is equal to 0.707 times the peak value, or 0.354 times the peak-to-peak value. Conversely, the peak value is 1.414 times the rms value, and the peak-to-peak value is 2.828 times the rms value. The rms amplitude is often specified when talking about utility ac, radio-frequency (RF) ac, and audio-frequency (AF) ac.

For a perfect square wave, the rms value is the same as the peak value, and half the peak-to-peak value. For sawtooth and irregular waves, the relationship between the rms value and the peak value depends on the exact shape of the wave. But the rms value is never greater than the peak value for any type of ac wave.

Superimposed DC

Sometimes a wave has components of both ac and dc. The simplest example of an ac/dc combination is illustrated by the connection of a dc voltage source, such as a battery, in series with an ac voltage source, like the utility mains. An example is shown in the schematic diagram of Fig. 9-12. Imagine connecting a 12-V automotive battery in series with the wall outlet. (Do not try this experiment in real life!) When this is done, the ac wave is displaced either positively or negatively by 12 V, depending on the polarity of the battery. This results in a sine wave at the output, but one peak is 24 V (twice the battery voltage) more than the other.

Any ac wave can have dc components along with it. If the dc component exceeds the peak value of the ac wave, then fluctuating, or pulsating, dc will result. This would happen, for example, if a 200-V dc source were connected in series with the output of a common utility ac outlet, which has peak voltages of approximately ±165 V. Pulsating dc would appear, with an average value of 200 V but with instantaneous values much higher and lower. The waveshape in this case is shown in Fig. 9-13.

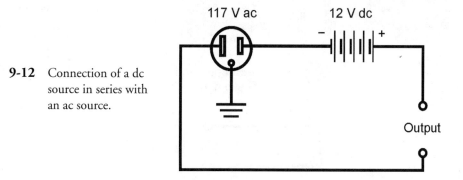

9-12   Connection of a dc source in series with an ac source.

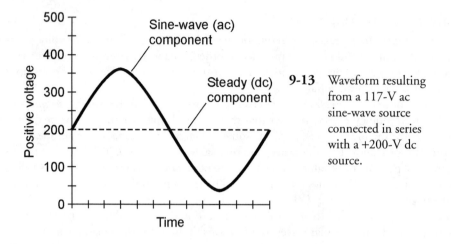

9-13 Waveform resulting from a 117-V ac sine-wave source connected in series with a +200-V dc source.

The Generator

Alternating current can be generated by a rotating coil of wire inside a powerful magnet, as shown in Fig. 9-14. An ac voltage appears between the ends of the wire coil. The ac voltage that a generator can produce depends on the strength of the magnet, the number of turns in the wire coil, and the speed at which the magnet or coil rotates. The ac frequency depends only on the speed of rota-

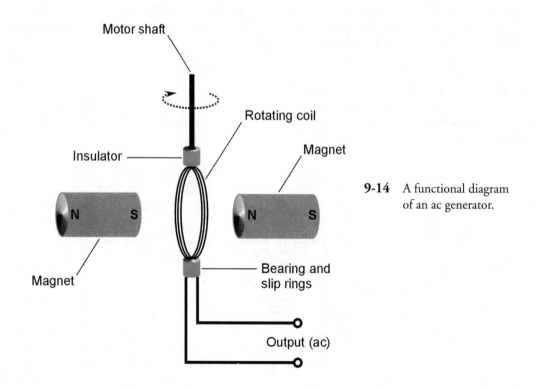

9-14 A functional diagram of an ac generator.

tion. Normally, for utility ac, this speed is 3600 revolutions per minute (rpm), or 60 complete revolutions per second (rps), so the ac output frequency is 60 Hz.

When a load, such as a light bulb or heater, is connected to an ac generator, it becomes more difficult, mechanically, to turn the generator shaft, compared to when there is nothing connected to the output. As the amount of electrical power demanded from a generator increases, so does the mechanical power required to drive it. This is why it is impossible to connect a generator to a stationary bicycle and pedal an entire city into electrification. There's no way to get something for nothing. The electrical power that comes out of a generator can never be more than the mechanical power driving it. In fact, there is always some energy lost, mainly as heat in the generator. Your legs might generate enough power to run a small radio or television set, but nowhere near enough to provide electricity for a household.

The *efficiency* of a generator is the ratio of the electrical power output to the mechanical driving power, both measured in the same units (such as watts or kilowatts), multiplied by 100 to get a percentage. No generator is 100 percent efficient, but a good one can come fairly close.

At power plants, generators are driven by massive turbines. The turbines are turned by various natural sources of energy such as moving water, steam heated by combustion of fossil fuels, or steam taken directly from deep inside the earth. These energy sources can provide tremendous mechanical power, and this is why power plants can produce megawatts of electrical power.

Why Alternating and Not Direct?

Do you wonder why ac is used at all? Isn't it a lot more complicated than dc? Well, ac may be more complicated in theory, but in practice it is a lot simpler to use when it is necessary to provide electricity to a large number of people.

Alternating current lends itself well to being transformed to lower or higher voltages, according to the needs of electrical apparatus. It is not so easy to change dc voltages. Electrochemical cells produce dc directly, but they are impractical for the needs of large populations. Serving millions of consumers requires the immense power of falling or flowing water, the ocean tides, wind, fossil fuels, controlled nuclear reactions, or geothermal heat. All of these energy sources can be used to drive turbines that turn ac generators.

Technology is advancing in the realm of solar-electric energy; someday a significant part of our electricity might come from photovoltaic power plants. These would generate dc. High voltages could be attained by connecting giant arrays of solar panels in series. But there would be a problem transforming this voltage down to manageable levels for consumer use.

Thomas Edison is said to have favored dc over ac for electrical power transmission in the early days, as the electric utilities were first being devised and constructed. His colleagues argued that ac would work better. But perhaps Edison knew something that his contemporaries did not. There is one advantage to dc in utility applications, and it involves the transmission of energy over great distances using wires. Direct currents, at extremely high voltages, are transported more efficiently than alternating currents. The wire has less effective resistance with dc than with ac, and there is less energy lost in the magnetic fields around the wires. Direct-current *high-tension* transmission lines are being considered for future use. Right now, the main problem is expense. Sophisticated power-conversion equipment is needed. If the cost can be brought within reason, Edison will be vindicated.

Quiz

Refer to the text in this chapter if necessary. A good score is at least 18 correct. Answers are in the back of the book.

1. Which of the following can vary with ac, but never with dc?
    (a) Power
    (b) Voltage
    (c) Frequency
    (d) Amplitude

2. The length of time between a point in one cycle and the same point in the next cycle of an ac wave is the
    (a) frequency.
    (b) magnitude.
    (c) period.
    (d) polarity.

3. On a spectrum analyzer, an ac signal having only one frequency component looks like
    (a) a single pip.
    (b) a sine wave.
    (c) a square wave.
    (d) a sawtooth wave.

4. The period of an ac wave, in seconds, is
    (a) the same as the frequency in hertz.
    (b) not related to the frequency in any way.
    (c) equal to 1 divided by the frequency in hertz.
    (d) equal to the peak amplitude in volts divided by the frequency in hertz.

5. The sixth harmonic of an ac wave whose period is 1.000 millisecond (1.000 ms) has a frequency of
    (a) 0.006 Hz.
    (b) 167.0 Hz.
    (c) 7.000 kHz.
    (d) 6.000 kHz.

6. A degree of phase represents
    (a) 6.28 cycles.
    (b) 57.3 cycles.
    (c) $\frac{1}{60}$ of a cycle.
    (d) $\frac{1}{360}$ of a cycle.

7. Suppose that two ac waves have the same frequency but differ in phase by exactly $^1/_{20}$ of a cycle. What is the phase difference between these two waves?

(a) 18°

(b) 20°

(c) 36°

(d) 5.73°

8. Suppose an ac signal has a frequency of 1770 Hz. What is its angular frequency?

(a) 1770 rad/s

(b) 11,120 rad/s

(c) 282 rad/s

(d) Impossible to determine from the data given

9. A triangular wave exhibits

(a) an instantaneous rise and a defined decay.

(b) a defined rise and an instantaneous decay.

(c) a defined rise and a defined decay, and the two are equal.

(d) an instantaneous rise and an instantaneous decay.

10. Three-phase ac

(a) has sawtooth waves that add together in phase.

(b) consists of three sine waves in different phases.

(c) is a sine wave with exactly three harmonics.

(d) is of interest only to physicists.

11. If two perfect sine waves have the same frequency and the same amplitude, but are in opposite phase, the composite wave

(a) has twice the amplitude of either input wave alone.

(b) has half the amplitude of either input wave alone.

(c) is complex, but has the same frequency as the originals.

(d) has zero amplitude (that is, it does not exist), because the two input waves cancel each other out.

12. If two perfect sine waves have the same frequency and the same phase, the composite wave

(a) is a sine wave with an amplitude equal to the difference between the amplitudes of the two input waves.

(b) is a sine wave with an amplitude equal to the sum of the amplitudes of the two original waves.

(c) is not a sine wave, but has the same frequency as the two input waves.

(d) has zero amplitude (that is, it does not exist), because the two input waves cancel each other out.

13. In a 117-V rms utility circuit, the positive peak voltage is approximately
    (a) +82.7 V.
    (b) +165 V.
    (c) +234 V.
    (d) +331 V.

14. In a 117-V rms utility circuit, the peak-to-peak voltage is approximately
    (a) 82.7 V.
    (b) 165 V.
    (c) 234 V.
    (d) 331 V.

15. In a perfect sine wave, the peak-to-peak amplitude is equal to
    (a) half the peak amplitude.
    (b) the peak amplitude.
    (c) 1.414 times the peak amplitude.
    (d) twice the peak amplitude.

16. If a 45-V dc battery is connected in series with the 117-V rms utility mains as shown in Fig. 9-15, the peak voltages will be approximately
    (a) +210 V and −120 V.
    (b) +162 V and −72 V.
    (c) +396 V and −286 V.
    (d) +117 V and −117V.

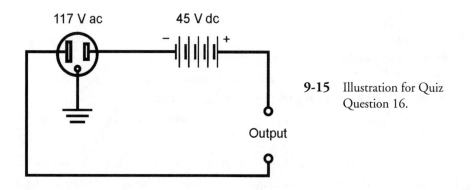

9-15   Illustration for Quiz Question 16.

17. In the situation described in question 16 and illustrated in Fig. 9-15, the peak-to-peak voltage will be approximately
    (a) 117 V.
    (b) 210 V.
    (c) 331 V.
    (d) 396 V.

18. Which one of the following does *not* affect the power output available from a particular ac generator?

    (a) The strength of the magnet

    (b) The number of turns in the coil

    (c) The type of natural energy source used

    (d) The speed of rotation of the coil or magnet

19. If a 175-V dc source were connected in series with the utility mains from a standard wall outlet, the result would be

    (a) smooth dc at a constant voltage.

    (b) pure ac with equal peak voltages.

    (c) ac with one peak voltage greater than the other.

    (d) fluctuating dc.

20. An advantage of ac over dc in utility applications is the fact that

    (a) ac is easier to transform from one voltage to another.

    (b) ac is transmitted with lower loss in wires.

    (c) ac can be easily obtained from dc generators.

    (d) ac can be generated with less-dangerous by-products.

10
CHAPTER

Inductance

IN THIS CHAPTER, YOU'LL LEARN ABOUT ELECTRICAL COMPONENTS THAT OPPOSE THE FLOW OF AC BY temporarily storing energy as magnetic fields. These devices are called *inductors*, and their action is known as *inductance*. Inductors often, but not always, consist of wire coils. Sometimes a length of wire, or a pair of wires, is used as an inductor.

The Property of Inductance

Suppose you have a wire 1 million miles long (about 1.6 million kilometers). Imagine that you make this wire into a huge loop, and connect its ends to the terminals of a battery (Fig. 10-1). An electrical current will flow through the loop of wire, but this is only part of the picture.

If the wire was short, the current would begin to flow immediately, and it would attain a level limited by the resistance in the wire and in the battery. But because the wire is extremely long, it takes a while for the electrons from the negative terminal to work their way around the loop to the positive terminal. It will take a little time for the current to build up to its maximum level.

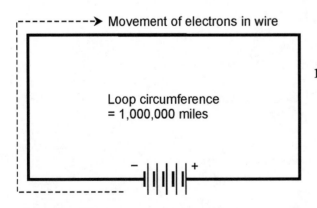

10-1 A huge, imaginary loop of wire can be used to illustrate the principle of inductance.

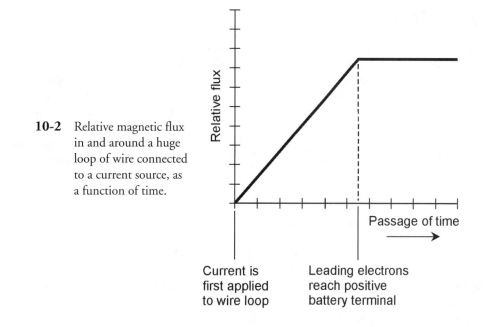

10-2 Relative magnetic flux in and around a huge loop of wire connected to a current source, as a function of time.

The magnetic field produced by the loop will be small during the first few moments when current flows in only part of the loop. The magnetic field will build up as the electrons get around the loop. Once a steady current is flowing around the entire loop, the magnetic field will have reached its maximum quantity and will level off (see Fig. 10-2). A certain amount of energy is stored in this magnetic field. The amount of stored energy depends on the *inductance* of the loop, which is a function of its overall size. Inductance, as a property or as a mathematical variable, is symbolized by an italicized, uppercase letter L. The loop constitutes an *inductor*, the symbol for which is an uppercase, nonitalicized letter L.

Practical Inductors

It is impractical to make wire loops 1 million miles in circumference. But lengths of wire can be coiled up. When this is done, the magnetic flux is increased for a given length of wire compared with the flux produced by a single-turn loop.

The magnetic flux density inside a coil is multiplied when a ferromagnetic core is placed within it. The increase in flux density has the effect of increasing the inductance, too, so L is many times greater with a ferromagnetic core than with an air core or a nonmagnetic core such as plastic or wood. The current that an inductor can handle depends on the diameter (gauge) of the wire. But the value of L is a function of the number of turns in the coil, the diameter of the coil itself, and the overall shape of the coil.

In general, the inductance of a coil is directly proportional to the number of turns of wire. Inductance is directly proportional to the diameter of the coil. The length of a coil, given a certain number of turns and a certain diameter, has an effect as well. If a coil having a certain number of turns and a certain diameter is "stretched out," its inductance decreases. Conversely, if it is "squashed up," its inductance increases.

The Unit of Inductance

When a battery is first connected across an inductor, the current builds up at a rate that depends on the inductance. The greater the inductance, the slower the rate of current buildup for a given battery voltage. The unit of inductance is an expression of the ratio between the rate of current buildup and the voltage across an inductor. An inductance of 1 *henry* (1 H) represents a potential difference of 1 volt (1 V) across an inductor within which the current is changing at the rate of 1 ampere per second (1 A/s).

The henry is a huge unit of inductance. You won't often see an inductor this large, although some power-supply filter chokes have inductances up to several henrys. Usually, inductances are expressed in *millihenrys* (mH), *microhenrys* (μH), or *nanohenrys* (nH). You should know your prefix multipliers by now, but in case you've forgotten:

$$1 \text{ mH} = 0.001 \text{ H} = 10^{-3} \text{ H}$$
$$1 \text{ μH} = 0.001 \text{ mH} = 10^{-6} \text{ H}$$
$$1 \text{ nH} = 0.001 \text{ μH} = 10^{-9} \text{ H}$$

Small coils with few turns of wire produce small inductances, in which the current changes quickly and the induced voltages are small. Large coils with ferromagnetic cores, and having many turns of wire, have high inductances in which the current changes slowly and the induced voltages are large. The current from a battery, building up or dying down through a high-*L* coil, can give rise to a deadly potential difference between the end terminals of the coil—many times the voltage of the battery itself. This is how spark coils work in internal combustion engines. Be careful around them!

Inductors in Series

When the magnetic fields around inductors do not interact, inductances in series add like resistances in series. The total value is the sum of the individual values. It's important to be sure that you are using the same size units for all the inductors when you add their values. After that, you can convert the result to any inductance unit you want.

Problem 10-1

Suppose three 40.0-μH inductors are connected in series, and there is no interaction, or *mutual inductance*, among them (Fig. 10-3). What is the total inductance?

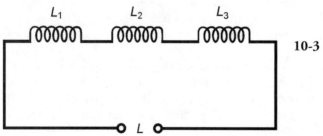

10-3 Inductances in series simply add up, as long as the inductors do not interact.

Add up the values. Call the inductances of the individual components L_1, L_2, and L_3, and the total inductance L. Then $L = L_1 + L_2 + L_3 = 40.0 + 40.0 + 40.0 = 120$ μH.

Problem 10-2

Imagine three inductors, with no mutual inductance, with values of 20.0 mH, 55.0 μH, and 400 nH. What is the total inductance, in millihenrys, of these components if they are connected in series as shown in Fig. 10-3?

First, convert all the inductances to the same units. Microhenrys are a good choice because that unit makes the calculation process the least messy. Call $L_1 = 20.0$ mH = 20,000 μH, $L_2 = 55.0$ μH, and $L_3 = 400$ nH = 0.400 μH. The total inductance is therefore $L = 20,000 + 55.0 + 0.400 = 20,055.4$ μH. This is 20.1 mH after converting and rounding off.

Inductors in Parallel

If there is no mutual inductance among two or more parallel-connected inductors, their values add up like the values of resistors in parallel. Suppose you have inductances L_1, L_2, L_3, ... , L_n all connected in parallel. Then you can find the reciprocal of the total inductance, $1/L$, using the following formula:

$$1/L = 1/L_1 + 1/L_2 + 1/L_3 + \ldots + 1/L_n$$

The total inductance, L, is found by taking the reciprocal of the number you get for $1/L$. Again, as with inductances in series, it's important to remember that all the units have to agree during the calculation process. Once you have completed the calculation, you can convert the result to any inductance unit.

Problem 10-3

Suppose there are three inductors, each with a value of 40 μH, connected in parallel with no mutual inductance, as shown in Fig. 10-4. What is the net inductance of the combination?

Let's call the inductances $L_1 = 40$ μH, $L_2 = 40$ μH, and $L_3 = 40$ μH. Use the preceding formula to obtain $1/L = 1/40 + 1/40 + 1/40 = 3/40 = 0.075$. Then $L = 1/0.075 = 13.333$ μH. This should be rounded off to 13 μH, because the original inductances are specified to only two significant digits.

Problem 10-4

Imagine four inductors in parallel, with no mutual inductance and values of $L_1 = 75.0$ mH, $L_2 = 40.0$ mH, $L_3 = 333$ μH, and $L_4 = 7.00$ H. What is the net inductance of this combination?

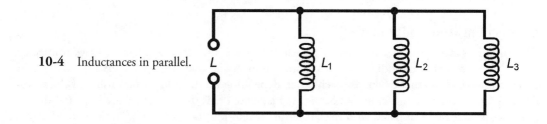

10-4   Inductances in parallel.

You can use henrys, millihenrys, or microhenrys as the standard units in this problem. Suppose you decide to use henrys. Then $L_1 = 0.0750$ H, $L_2 = 0.0400$ H, $L_3 = 0.000333$ H, and $L_4 = 7.00$ H. Use the preceding formula to obtain $1/L = 13.33 + 25.0 + 3003 + 0.143 = 3041.473$. The reciprocal of this is the inductance $L = 0.00032879$ H $= 328.79$ μH. This should be rounded off to 329 μH. This is only a little less than the value of the 333 μH inductor alone.

If there are several inductors in parallel, and one of them has a value that is much smaller than the values of all the others, then the total inductance is a little smaller than the value of the smallest inductor.

Interaction among Inductors

In real-world circuits, there is almost always some mutual inductance between or among solenoidal coils. The magnetic fields extend significantly outside such coils, and mutual effects are difficult to avoid or eliminate. The same is true between and among lengths of wire, especially at high ac frequencies. Sometimes, mutual inductance has no detrimental effect, but in some situations it is not wanted. Mutual inductance can be minimized by using *shielded* wires and *toroidal* inductors. The most common shielded wire is *coaxial cable*. Toroidal inductors are discussed later in this chapter.

Coefficient of Coupling

The *coefficient of coupling*, symbolized k, is an expression of the extent to which two inductors interact. It is specified as a number ranging from 0 (no interaction) to 1 (the maximum possible interaction). Two coils separated by a sheet of solid iron, or by a great distance, have a coefficient of coupling of zero ($k = 0$); two coils wound on the same form, one right over the other, have the maximum possible coefficient of coupling ($k = 1$). Sometimes, the coefficient of coupling is multiplied by 100 and expressed as a percentage from 0 to 100 percent.

Mutual Inductance

The *mutual inductance* between two inductors is symbolized M, and is expressed in the same units as inductance: henrys, millihenrys, microhenrys, or nanohenrys. The value of M is a function of the values of the inductors, and also of the coefficient of coupling.

In the case of two inductors having values of L_1 and L_2 (both expressed in the same size units), and with a coefficient of coupling equal to k, the mutual inductance M is found by multiplying the inductance values, taking the square root of the result, and then multiplying by k. Mathematically:

$$M = k(L_1 L_2)^{1/2}$$

where the $^1\!/_2$ power represents the square root. The value of M thus obtained will be in the same size unit as the values of the inductance you input to the equation.

Effects of Mutual Inductance

Mutual inductance can either increase or decrease the net inductance of a pair of series-connected coils, compared with the condition of zero mutual inductance. The magnetic fields around the coils either reinforce each other or oppose each other, depending on the phase relationship of the ac applied to them. If the two ac waves (and thus the magnetic fields they produce) are in phase, the inductance is increased compared with the condition of zero mutual inductance. If the two waves are in opposing phase, the net inductance is decreased relative to the condition of zero mutual inductance.

When two inductors are connected in series and there is *reinforcing* mutual inductance between them, the total inductance L is given by the following formula:

$$L = L_1 + L_2 + 2M$$

where L_1 and L_2 are the inductances, and M is the mutual inductance. All inductances must be expressed in the same size units.

When two inductors are connected in series and the mutual inductance is *opposing*, the total inductance L is given by this formula:

$$L = L_1 + L_2 - 2M$$

where, again, L_1 and L_2 are the values of the individual inductors.

It is possible for mutual inductance to increase the total series inductance of a pair of coils by as much as a factor of 2, if the coupling is total and if the flux reinforces. Conversely, it is possible for the inductances of two coils to completely cancel each other. If two equal-valued inductors are connected in series so their fluxes oppose (or *buck* each other) and $k = 1$, the result is theoretically zero inductance.

Problem 10-5

Suppose two coils, having inductances of 30 µH and 50 µH, are connected in series so that their fields reinforce, as shown in Fig. 10-5. Suppose that the coefficient of coupling is 0.500. What is the total inductance of the combination?

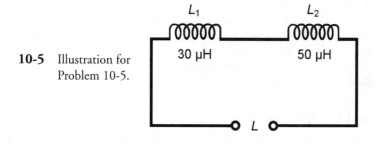

10-5 Illustration for Problem 10-5.

First, calculate M from k. According to the formula for this, given previously, $M = 0.500(50 \times 30)^{1/2} = 19.4$ µH. Then figure the total inductance. It is equal to $L = L_1 + L_2 + 2M = 30 + 50 + 38.8 = 118.8$ µH, rounded to 120 µH because only two significant digits are justified.

Problem 10-6

Imagine two coils with inductances of $L_1 = 835$ µH and $L_2 = 2.44$ mH. Suppose they are connected in series so that their coefficient of coupling is 0.922, acting so that the coils oppose each other, as shown in Fig. 10-6. What is the net inductance of the pair?

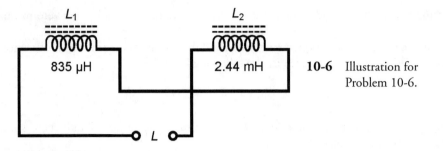

10-6 Illustration for Problem 10-6.

First, calculate M from k. The coil inductances are specified in different units. Let's use microhenrys for our calculations, so $L_2 = 2440$ μH. Then $M = 0.922(835 \times 2440)^{1/2} = 1316$ μH. Then figure the total inductance. It is $L = L_1 + L_2 - 2M = 835 + 2440 - 2632 = 643$ μH.

Air-Core Coils

The simplest inductors (besides plain, straight lengths of wire) are coils. A coil can be wound on a hollow cylinder of plastic or other nonferromagnetic material, forming an *air-core coil*. In practice, the maximum attainable inductance for such coils is about 1 mH.

Air-core coils are used mostly in radio-frequency transmitters, receivers, and antenna networks. In general, the higher the frequency of ac, the less inductance is needed to produce significant effects. Air-core coils can be made to have almost unlimited current-carrying capacity, simply by using heavy-gauge wire and making the radius of the coil large. Air does not dissipate much energy in the form of heat. It's efficient, even though it has low permeability.

Ferromagnetic Cores

Ferromagnetic substances can be crushed into dust and then bound into various shapes, providing core materials that greatly increase the inductance of a coil having a given number of turns. Depending on the mixture used, the increase in flux density can range from a factor of a few times, up through many thousands of times. A small coil can thus be made to have a large inductance. There are two main types of ferromagnetic material in common use as coil cores. These substances are known as *powdered iron* and *ferrite*.

Advantages and Limitations

Powdered-iron cores are common at high and very high radio frequencies. Ferrite is a special form of powdered iron that has exceptionally high permeability, causing a great concentration of magnetic flux lines within the coil. Ferrite is used at audio frequencies, as well as at low, medium, and high radio frequencies. Coils using these materials can be made much smaller, physically, than can air-core coils having the same inductance.

The main trouble with ferromagnetic cores is that, if the coil carries more than a certain amount of current, the core will *saturate*. This means that the ferromagnetic material is holding as much flux as it possibly can. When a core becomes saturated, any further increase in coil current will not produce a corresponding increase in the magnetic flux in the core. The result is that the inductance changes, decreasing with coil currents that are more than the critical value. In extreme cases, ferromagnetic cores can also waste considerable power as heat. This makes a coil *lossy*.

Permeability Tuning

Solenoidal coils can be made to have variable inductance by sliding ferromagnetic cores in and out of them. The frequency of a radio circuit can be adjusted in this way, as you'll learn later in this book.

Because moving the core in and out of a coil changes the effective permeability within the coil, this method of tuning is called *permeability tuning*. The in/out motion can be precisely controlled by attaching the core to a screw shaft, and anchoring a nut at one end of the coil (Fig. 10-7). As the screw shaft is rotated clockwise, the core enters the coil, and the inductance increases. As the screw shaft is rotated counterclockwise, the core moves out of the coil, and the inductance decreases.

Toroids

Inductor coils do not have to be wound on cylindrical forms, or on cylindrical ferromagnetic cores. There's another coil geometry, called the *toroid*. It gets its name from the shape of the ferromagnetic core. The coil is wound over a core having this shape (Fig. 10-8), which resembles a donut or bagel.

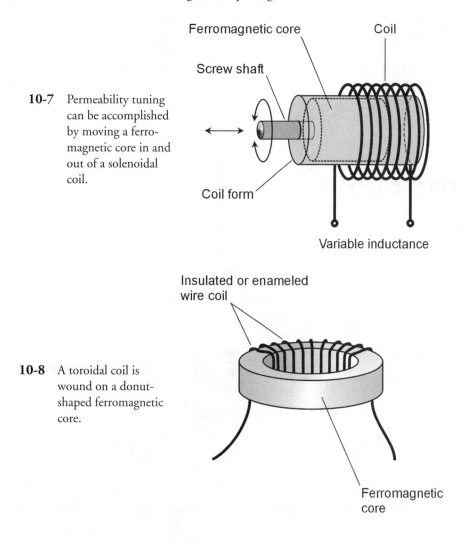

10-7 Permeability tuning can be accomplished by moving a ferromagnetic core in and out of a solenoidal coil.

10-8 A toroidal coil is wound on a donut-shaped ferromagnetic core.

There are several advantages to toroidal coils over solenoidal, or cylindrical, ones. First, fewer turns of wire are needed to get a certain inductance with a toroid compared to a solenoid. Second, a toroid can be physically smaller for a given inductance and current-carrying capacity. Third, practically all the flux is contained within the core material. This reduces unwanted mutual inductances with components near the toroid.

Toroidal coils have limitations, too. It is more difficult to permeability-tune a toroidal coil than it is to tune a solenoidal one. Toroidal coils are harder to wind than solenoidal ones. Sometimes, mutual inductance between or among physically separate coils is actually desired; with a toroid, the coils have to be wound on the same form for this to be possible.

Pot Cores

There is another way to confine the magnetic flux in a coil so that unwanted mutual inductance does not occur: wrap ferromagnetic core material around a coil (Fig. 10-9). A wraparound core of this sort is known as a *pot core.*

A typical pot core comes in two halves, inside one of which the coil is wound. Then the parts are assembled and held together by a bolt and nut. The entire assembly looks like a miniature oil tank. The wires come out of the core through small holes or slots.

Pot cores have the same advantages as toroids. The core tends to prevent the magnetic flux from extending outside the physical assembly. Inductance is greatly increased compared to solenoidal windings having a comparable number of turns. In fact, pot cores are even better than toroids if the main objective is to get a large inductance in a small space. The main disadvantage of a pot core is that tuning, or adjustment of the inductance, is all but impossible. The only way to do it is by switching in different numbers of turns, using taps at various points on the coil.

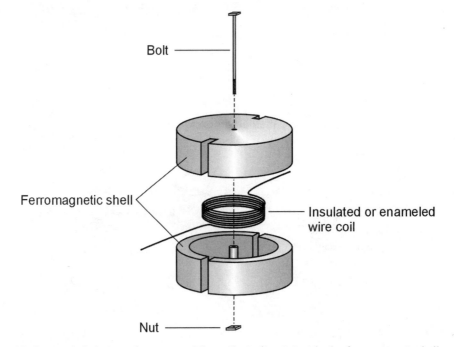

10-9 Exploded view of a pot core. The coil winding is inside the ferromagnetic shell.

Filter Chokes

The largest values of inductance that can be obtained in practice are on the order of several henrys. The primary use of a coil this large is to smooth out the pulsations in direct current that result when ac is *rectified* in a power supply. This type of coil is known as a *filter choke.* You'll learn more about power supplies later in this book.

Inductors at AF

Inductors for audio frequency (AF) applications range in value from a few millihenrys up to about 1 H. They are almost always toroidally wound, or are wound in a pot core, or comprise part of an audio transformer. Ferromagnetic cores are the rule.

Inductors can be used in conjunction with moderately large values of capacitance in order to obtain *AF-tuned circuits.* However, in recent years, audio tuning has been largely taken over by active components, particularly *integrated circuits.*

Inductors at RF

The radio frequency (RF) spectrum ranges from a few kilohertz to well above 100 GHz. At the low end of this range, inductors are similar to those at AF. As the frequency increases, cores having lower permeability are used. Toroids are common up through about 30 MHz. Above that frequency, air-core coils are more often used.

In RF applications, coils are routinely connected in series or in parallel with capacitors to obtain tuned circuits. Other arrangements yield various characteristics of *attenuation versus frequency,* serving to let signals at some frequencies pass through, while rejecting signals at other frequencies. You'll learn more about this in the discussion about *resonance* in Chap. 17.

Transmission-Line Inductors

At frequencies about 100 MHz, another type of inductor becomes practical. This is the type formed by a length of *transmission line.* A transmission line is generally used to get energy from one place to another. In radio communications, transmission lines get energy from a transmitter to an antenna, and from an antenna to a receiver.

Most transmission lines are found in either of two geometries, the *parallel-wire* type or the *coaxial* type. A parallel-wire transmission line consists of two wires running alongside each other with constant spacing (Fig. 10-10). The spacing is maintained by polyethylene rods molded at regular in-

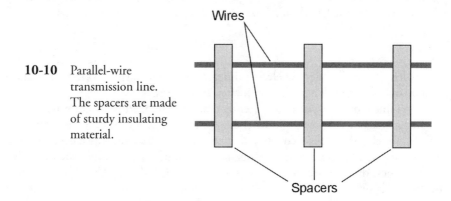

10-10   Parallel-wire transmission line. The spacers are made of sturdy insulating material.

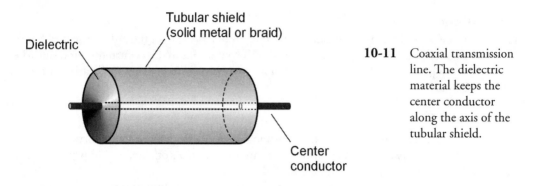

Dielectric

Tubular shield
(solid metal or braid)

Center
conductor

10-11 Coaxial transmission line. The dielectric material keeps the center conductor along the axis of the tubular shield.

tervals to the wires, or by a solid web of polyethylene. The substance separating the wires is called the *dielectric* of the transmission line. A coaxial transmission line has a wire conductor surrounded by a tubular braid or pipe (Fig. 10-11). The wire is kept at the center of this tubular *shield* by means of polyethylene beads, or more often, by solid or foamed polyethylene, all along the length of the line.

Line Inductance

Short lengths of any type of transmission line behave as inductors, as long as the line length is less than 90° (¼ of a wavelength). At 100 MHz, 90° in free space is 75 cm, or a little more than 2 ft. In general, if f is the frequency in megahertz, then ¼ wavelength in free space, expressed in centimeters (s_{cm}), is given by this formula:

$$s_{cm} = 7500/f$$

The length of a quarter-wavelength section of transmission line is shortened from the free-space quarter wavelength by the effects of the dielectric. In practice, ¼ wavelength along the line can be anywhere from about 0.66 (or 66 percent) of the free-space length for coaxial lines with solid polyethylene dielectric to about 0.95 (or 95 percent) of the free-space length for parallel-wire line with spacers molded at intervals of several centimeters. The factor by which the wavelength is shortened is called the *velocity factor* of the line.

The shortening of the wavelength in a transmission line, compared with the wavelength in free space, is a result of a slowing down of the speed with which the radio signals move in the line compared with their speed in space (the speed of light). If the velocity factor of a line is given by v, then the preceding formula for the length of a quarter-wave line, in centimeters, becomes:

$$s_{cm} = 7500v/f$$

Very short lengths of line—a few electrical degrees—produce small values of inductance. As the length approaches ¼ wavelength, the inductance increases.

Transmission line inductors behave differently than coils in one important way: the inductance of a coil, particularly an air-core coil, is independent of the frequency. But the inductance of a transmission-line section changes as the frequency changes. At first, the inductance becomes larger as the frequency increases. At a certain limiting frequency, the inductance becomes theoretically infinite. Above that frequency, the line becomes *capacitive* rather than inductive. You'll learn about capacitance in the next chapter.

Unwanted Inductances

Any length of wire has some inductance. As with a transmission line, the inductance of a wire increases as the frequency increases. Wire inductance is more significant at RF than at AF.

In some cases, especially in radio communications equipment, the inductance of, and among, wires can become a major problem. Circuits can oscillate when they should not. A receiver might respond to signals that it's not designed to intercept. A transmitter can send out signals on unauthorized and unintended frequencies. The frequency response of any circuit can be altered, degrading the performance of the equipment. Sometimes the effects of this *stray inductance* are so small that they are not important; this might be the case in a stereo hi-fi set located at a distance from other electronic equipment. But in some situations, stray inductance can cause serious equipment malfunctions.

A good way to minimize stray inductance is to use coaxial cables between and among sensitive circuits or components. The shield of the cable is connected to the *common ground* of the apparatus. In some cases, enclosing individual circuits in metal boxes can prevent stray inductance from causing feedback and other problems.

Quiz

Refer to the text in this chapter if necessary. A good score is 18 correct. Answers are in the back of the book.

1. An inductor works by
   (a) charging a piece of wire.
   (b) storing energy as a magnetic field.
   (c) choking off dc.
   (d) introducing resistance into a circuit.

2. Which of the following does *not* affect the inductance of an air-core coil, if all other factors are held constant?
   (a) The frequency
   (b) The number of turns
   (c) The diameter of the coil
   (d) The length of the coil

3. In a small inductance
   (a) energy is stored and released slowly.
   (b) the current flow is always large.
   (c) the current flow is always small.
   (d) energy is stored and released quickly.

4. A ferromagnetic core is placed in an inductor mainly to
   (a) increase the current carrying capacity.
   (b) increase the inductance.
   (c) limit the current.
   (d) reduce the inductance.

5. Inductors in series, assuming there is no mutual inductance, combine
   (a) like resistors in parallel.
   (b) like resistors in series.
   (c) like batteries in series with opposite polarities.
   (d) in a way unlike any other type of component.

6. Suppose two inductors are connected in series, without mutual inductance. Their values are 33 mH and 55 mH. What is the net inductance of the combination?
   (a) 1.8 H
   (b) 22 mH
   (c) 88 mH
   (d) 21 mH

7. If the same two inductors (33 mH and 55 mH) are connected in parallel without mutual inductance, the combination will have a value of
   (a) 1.8 H.
   (b) 22 mH.
   (c) 88 mH.
   (d) 21 mH.

8. Suppose three inductors are connected in series without mutual inductance. Their values are 4.00 nH, 140 μH, and 5.07 H. For practical purposes, the net inductance will be very close to
   (a) 4.00 nH.
   (b) 140 μH.
   (c) 5.07 H.
   (d) none of the above.

9. Suppose the three inductors mentioned above are connected in parallel without mutual inductance. The net inductance will be close to
   (a) 4.00 nH.
   (b) 140 μH.
   (c) 5.07 H.
   (d) none of the above.

10. Suppose two inductors, each of 100 μH, are connected in series, and the coefficient of coupling is 0.40. The net inductance, if the coil fields reinforce each other, is
   (a) 50.0 μH.
   (b) 120 μH.
   (c) 200 μH.
   (d) 280 μH.

11. If the coil fields oppose in the foregoing series-connected arrangement, assuming the coefficient of coupling does not change, the net inductance is

    (a) 50.0 μH.

    (b) 120 μH.

    (c) 200 μH.

    (d) 280 μH.

12. Suppose two inductors, having values of 44.0 mH and 88.0 mH, are connected in series with a coefficient of coupling equal to 1.0 (the maximum possible mutual inductance). If their fields reinforce, the net inductance is approximately

    (a) 7.55 mH.

    (b) 132 mH.

    (c) 194 mH.

    (d) 256 mH.

13. If the fields in the previous situation oppose, assuming the coefficient of coupling does not change, the net inductance will be approximately

    (a) 7.55 mH.

    (b) 132 mH.

    (c) 194 mH.

    (d) 256 mH.

14. With permeability tuning, moving the core further into a solenoidal coil

    (a) increases the inductance.

    (b) reduces the inductance.

    (c) has no effect on the inductance, but increases the current-carrying capacity of the coil.

    (d) raises the frequency.

15. A significant advantage, in some situations, of a toroidal coil over a solenoid is the fact that

    (a) the toroid is easier to wind.

    (b) the solenoid cannot carry as much current.

    (c) the toroid is easier to tune.

    (d) the magnetic flux in a toroid is practically all within the core.

16. A major feature of a pot core inductor is

    (a) high current capacity.

    (b) large inductance in small volume.

    (c) excellent efficiency at very high frequencies.

    (d) ease of inductance adjustment.

17. As an inductor core material, air

   (a) has excellent efficiency.

   (b) has high permeability.

   (c) allows large inductance to exist in a small volume.

   (d) has permeability that can vary over a wide range.

18. At a frequency of 400 Hz, which is in the AF range, the most likely form for an inductor would be

   (a) air-core.

   (b) solenoidal.

   (c) toroidal.

   (d) transmission-line.

19. At a frequency of 95.7 MHz, which is in the frequency-modulation (FM) broadcast band and is considered part of the very high frequency (VHF) radio spectrum, a good form for an inductor would be

   (a) air-core.

   (b) pot core.

   (c) either (a) or (b).

   (d) neither (a) nor (b).

20. A transmission-line inductor made from coaxial cable having velocity factor of 0.66 and working at 450 MHz, which is in the ultrahigh frequency (UHF) radio spectrum, should, in order to measure less than ¼ electrical wavelength, be cut shorter than

   (a) 16.7 m.

   (b) 11 m.

   (c) 16.7 cm.

   (d) 11 cm.

11
CHAPTER

Capacitance

ELECTRICAL COMPONENTS CAN OPPOSE THE FLOW OF AC IN THREE WAYS, TWO OF WHICH YOU'VE learned about. *Resistance* slows the flow of ac or dc charge carriers (usually electrons) by brute force. *Inductance* impedes the flow of ac charge carriers by temporarily storing the energy as a magnetic field. *Capacitance*, about which you'll learn in this chapter, impedes the flow of ac charge carriers by temporarily storing the energy as an *electric field*.

The Property of Capacitance

Imagine two huge, flat sheets of metal that are excellent electrical conductors. Suppose they are each the size of the state of Nebraska, and are placed one over the other, separated by only 1 foot of space. If these two sheets of metal are connected to the terminals of a battery, as shown in Fig. 11-1, they will become charged electrically, one positively and the other negatively.

If the plates were small, they would both become charged almost instantly, attaining a relative voltage equal to the voltage of the battery. But because the plates are gigantic, it will take a little time for the negative plate to reach full negative potential, and an equal time for the other plate to reach full positive potential. Eventually, the voltage between the two plates will equal the battery voltage,

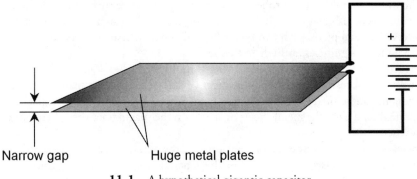

Narrow gap   Huge metal plates

11-1   A hypothetical gigantic capacitor.

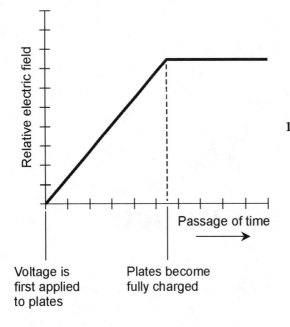

11-2 Relative electric field intensity between metal plates connected to a voltage source, as a function of time.

(graph: vertical axis "Relative electric field"; horizontal axis "Passage of time" →; lower-left note "Voltage is first applied to plates"; note at dashed line "Plates become fully charged")

and an electric field will exist in the space between the plates. This electric field will be small at first, because the plates don't charge up right away. But the charge will increase over a period of time, depending on how large the plates are, and also depending on how far apart they are. Figure 11-2 is a relative graph showing the intensity of the electric field between the plates as a function of time, elapsed from the instant the plates are connected to the battery terminals.

Energy will be stored in this electric field. The ability of the plates, and of the space between them, to store this energy is the property of *capacitance.* As a quantity or variable, capacitance is denoted by the uppercase italic letter *C.*

Practical Capacitors

It's out of the question to make a capacitor of the preceding dimensions. But two sheets, or strips, of foil can be placed one on top of the other, separated by a thin, nonconducting sheet such as paper, and then the whole assembly can be rolled up to get a large effective surface area. When this is done, the electric flux becomes great enough so that the device exhibits significant capacitance. Alternatively, two sets of several plates each can be meshed together with air in between them, and the resulting capacitance is significant at high ac frequencies.

In a capacitor, the electric flux concentration is multiplied when a *dielectric* of a certain type is placed between the plates. This increases the effective surface area of the plates, so that a physically small component can be made to have a large capacitance. The voltage that a capacitor can handle depends on the thickness of the metal sheets or strips, on the spacing between them, and on the type of dielectric used.

In general, capacitance is directly proportional to the surface area of the conducting plates or sheets. Capacitance is *inversely proportional* to the separation between conducting sheets. In other words, the closer the sheets are to each other, the greater the capacitance. The capacitance also depends on the *dielectric constant* of the material between the plates. A vacuum has a dielectric constant of 1; some substances have dielectric constants that multiply the effective capacitance many times.

The Unit of Capacitance

When a battery is connected between the plates of a capacitor, the potential difference between the plates builds up at a rate that depends on the capacitance. The greater the capacitance, the slower the rate of change of voltage in the plates. The unit of capacitance is an expression of the ratio between the current that flows and the rate of voltage change between the plates as the plates become charged. A capacitance of 1 *farad* (1 F) represents a current flow of 1 A while there is a voltage increase of 1 V/s. A capacitance of 1 F also results in 1 V of potential difference for an electric charge of 1 C.

The farad is a huge unit of capacitance. You'll almost never see a capacitor with a value of 1 F. Commonly employed units of capacitance are the *microfarad* (μF) and the *picofarad* (pF). A capacitance of 1 μF represents 0.000001 (10^{-6}) F, and 1 pF is a millionth of a microfarad, or 0.000000000001 (10^{-12}) F.

Physically small components can be made to have fairly large capacitance values. Conversely, some capacitors with small values take up large physical volumes. The physical size of a capacitor, if all other factors are held constant, is proportional to the voltage that it can handle. The higher the rated voltage, the bigger the component.

Capacitors in Series

With capacitors, there is rarely any mutual interaction. This makes capacitors easier to work with than inductors. We don't have to worry about *mutual capacitance* very often, the way we have to be concerned about mutual inductance when working with wire coils.

Capacitors in series add together like resistors or inductors in parallel. Suppose you have several capacitors with values C_1, C_2, C_3, . . . , C_n connected in series. You can find the reciprocal of the total capacitance, $1/C$, using the following formula:

$$1/C = 1/C_1 + 1/C_2 + 1/C_3 + \ldots + 1/C_n$$

The net capacitance of the series combination, C, is found by taking the reciprocal of the number you get for $1/C$.

If two or more capacitors are connected in series, and one of them has a value that is tiny compared with the values of all the others, the net capacitance is roughly equal to the smallest capacitance.

Problem 11-1

Suppose two capacitors, with values of $C_1 = 0.10$ μF and $C_2 = 0.050$ μF, are connected in series (Fig. 11-3). What is the net capacitance?

Using the preceding formula, first find the reciprocals of the values. They are $1/C_1 = 10$ and $1/C_2 = 20$. Then $1/C = 10 + 20 = 30$, and $C = 1/30 = 0.033$ μF. Note that we can work with reciprocal capacitances in this calculation only because the values of the components are specified in the same units.

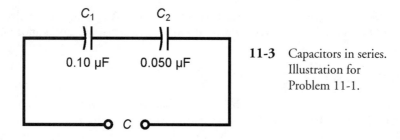

C_1 C_2

0.10 µF   0.050 µF

11-3 Capacitors in series.
Illustration for
Problem 11-1.

Problem 11-2

Suppose two capacitors with values of 0.0010 µF and 100 pF are connected in series. What is the net capacitance?

In this case, you must convert to the same size units before doing any calculations. A value of 100 pF represents 0.000100 µF. Thus, $C_1 = 0.0010$ µF and $C_2 = 0.000100$ µF. The reciprocals are $1/C_1 = 1000$ and $1/C_2 = 10,000$. Therefore, $1/C = 1000 + 10,000 = 11,000$, so $C = 1/11,000 = 0.000091$ µF. (You might rather say it's 91 pF.)

Problem 11-3

Suppose five capacitors, each of 100 pF, are in series. What is the total capacitance?

If there are n capacitors in series, all of the same value so that $C_1 = C_2 = C_3 = \ldots = C_n$, the net capacitance C is equal to $1/n$ of the capacitance of any of the components alone. Because there are five 100-pF capacitors here, the total is $C = 100/5 = 20.0$ pF.

Capacitors in Parallel

Capacitances in parallel add like resistances in series. The total capacitance is the sum of the individual component values. If two or more capacitors are connected in parallel, and one of the capacitances is far larger than any of the others, the total capacitance can be taken as approximately the value of the biggest one.

Problem 11-4

Suppose three capacitors are in parallel, having values of $C_1 = 0.100$ µF, $C_2 = 0.0100$ µF, and $C_3 = 0.001000$ µF, as shown in Fig. 11-4. What is the total capacitance?

Add them up: $C = 0.100 + 0.0100 + 0.001000 = 0.111000$. Because two of the values are given to only three significant figures, the final answer should be stated as $C = 0.111$ µF.

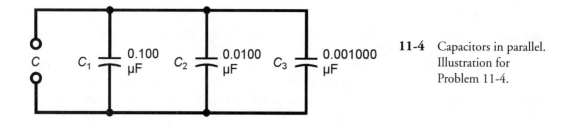

C C_1 0.100 µF C_2 0.0100 µF C_3 0.001000 µF

11-4 Capacitors in parallel.
Illustration for
Problem 11-4.

Problem 11-5

Suppose two capacitors are in parallel, one with a value of 100 μF and one with a value of 100 pF. What is the net capacitance?

In this case, you can say right away that the net capacitance is 100 μF for practical purposes. The 100-pF capacitor has a value that is only one-millionth of the capacitance of the 100-μF component. The smaller capacitance contributes essentially nothing to the net capacitance of this combination.

Fixed Capacitors

A *fixed capacitor* has a value that cannot be adjusted, and that (ideally) does not vary when environmental or circuit conditions change. Here are some of the characteristics, and common types, of fixed capacitors.

Dielectric Materials

Just as certain solids can be placed within a coil to increase the inductance, materials exist that can be sandwiched in between the plates of a capacitor to increase the capacitance. The substance between the plates is called the *dielectric* of the capacitor. Air is an efficient dielectric; it has almost no loss. But it is difficult to get very much capacitance using air as the dielectric. Some kind of solid material is usually employed as the dielectric for most fixed capacitors.

Dielectric materials accommodate electric *fields* well, but they are poor conductors of electric *currents*. In fact, dielectric materials are known as good insulators. Solid dielectrics increase the capacitance for a given surface area and spacing of the plates. Solid dielectrics also allow the plates to be rolled up, squashed, and placed very close together (Fig. 11-5). This geometry acts to maximize the capacitance per unit volume.

Paper Capacitors

In the early days of electronics, capacitors were commonly made by placing paper, soaked with mineral oil, between two strips of foil, rolling the assembly up, attaching wire leads to the two pieces of foil, and enclosing the rolled-up foil and paper in an airtight cylindrical case. *Paper capacitors* can still sometimes be found in older electronic equipment. They have values ranging from about 0.001 μF to 0.1 μF, and can handle low to moderate voltages, usually up to about 1000 V.

11-5 A cross-sectional drawing of a capacitor consisting of two foil sheets rolled up, and two sheets of dielectric material rolled up between them.

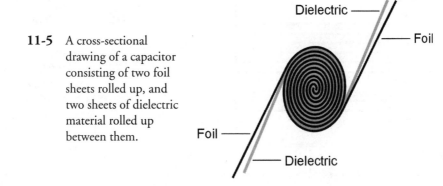

Mica Capacitors

Mica is a naturally occurring, solid, transparent mineral substance that flakes off in thin sheets. It makes an excellent dielectric for capacitors. *Mica capacitors* can be manufactured by alternately stacking metal sheets and layers of mica, or by applying silver ink to sheets of mica. The metal sheets are wired together into two meshed sets, forming the two terminals of the capacitor. This scheme is shown in Fig. 11-6.

Mica capacitors have low loss, and are therefore highly efficient, provided their voltage rating is not exceeded. Voltage ratings can be up to several thousand volts if thick sheets of mica are used. But mica capacitors are large physically in proportion to their capacitance. The main application for mica capacitors is in radio receivers and transmitters. Their capacitances are a little lower than those of paper capacitors, ranging from a few tens of picofarads up to about 0.05 μF.

Ceramic Capacitors

Ceramic materials work well as dielectrics. Sheets of metal are stacked alternately with wafers of ceramic to make these capacitors. The meshing/layering geometry of Fig. 11-6 is used. Ceramic, like mica, has low loss and allows for high efficiency.

For small values of capacitance, only one layer of ceramic is needed, and two metal plates can be glued to the disk-shaped material, one on each side. This type of component is known as a *disk-ceramic* capacitor. Alternatively, a tube or cylinder of ceramic can be employed, and metal ink applied to the inside and outside of the tube. Such units are called *tubular* capacitors. Ceramic capacitors have values ranging from a few picofarads to about 0.5 μF. Their voltage ratings are comparable to those of paper capacitors.

Plastic-Film Capacitors

Plastics make good dielectrics for the manufacture of capacitors. *Polyethylene* and *polystyrene* are commonly used. The method of manufacture is similar to that for paper capacitors. Stacking methods can be used if the plastic is rigid. The geometries can vary, and these capacitors are therefore found in various shapes.

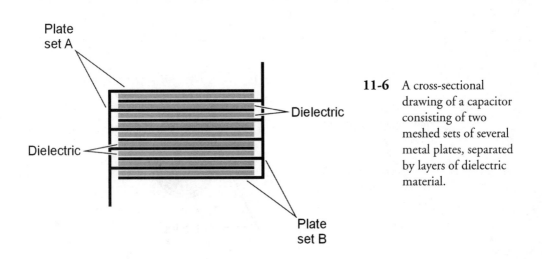

11-6 A cross-sectional drawing of a capacitor consisting of two meshed sets of several metal plates, separated by layers of dielectric material.

Capacitance values for plastic-film units range from about 50 pF to several tens of microfarads. Most often they are in the range of 0.001 μF to 10 μF. Plastic capacitors are employed at AF and RF, and at low to moderate voltages. The efficiency is good, although not as high as that for mica-dielectric or air-dielectric units.

Electrolytic Capacitors

All of the aforementioned types of capacitors provide relatively small values of capacitance. They are also *nonpolarized*, meaning that they can be hooked up in a circuit in either direction. An *electrolytic* capacitor provides greater capacitance than any of the preceding types, but it must be connected in the proper direction in a circuit to work right. An electrolytic capacitor is a *polarized* component.

Electrolytic capacitors are made by rolling up aluminum foil strips, separated by paper saturated with an *electrolyte* liquid. The electrolyte is a conducting solution. When dc flows through the component, the aluminum oxidizes because of the electrolyte. The oxide layer is nonconducting, and forms the dielectric for the capacitor. The layer is extremely thin, and this results in a high capacitance per unit volume. Electrolytic capacitors can have values up to thousands of microfarads, and some can handle thousands of volts. These capacitors are most often seen in AF circuits and in dc power supplies.

Tantalum Capacitors

Another type of electrolytic capacitor uses tantalum rather than aluminum. The tantalum can be foil, as is the aluminum in a conventional electrolytic capacitor. It can also take the form of a porous pellet, the irregular surface of which provides a large area in a small volume. An extremely thin oxide layer forms on the tantalum.

Tantalum capacitors have high reliability and excellent efficiency. They are often used in military applications because they almost never fail. They can be used in AF and digital circuits in place of aluminum electrolytics.

Semiconductor Capacitors

Later in this book, you'll learn about *semiconductors*. These materials have revolutionized electrical and electronic circuit design in the past several decades.

Semiconductor materials can be employed to make capacitors. A semiconductor *diode* conducts current in one direction, and refuses to conduct in the other direction. When a voltage source is connected across a diode so that it does not conduct, the diode acts as a capacitor. The capacitance varies depending on how much of this *reverse voltage* is applied to the diode. The greater the reverse voltage, the smaller the capacitance. This makes the diode act as a *variable capacitor*. Some diodes are especially manufactured to serve this function. Their capacitances fluctuate rapidly along with pulsating dc. They are called *varactor diodes* or simply *varactors*.

Capacitors can be formed in the semiconductor materials of an integrated circuit (also called an *IC* or *chip*) in much the same way. Sometimes, IC diodes are fabricated to serve as varactors. Another way to make a capacitor in an IC is to sandwich an oxide layer into the semiconductor material, between two layers that conduct well. Most ICs look like little boxes with protruding metal prongs (Fig. 11-7). The prongs provide the electrical connections to external circuits and systems.

Semiconductor capacitors usually have small values of capacitance. They are physically tiny, and can handle only low voltages. The advantages are miniaturization and an ability, in the case of the varactor, to change in value at a rapid rate.

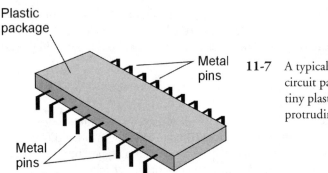

Plastic
package

Metal
pins

Metal
pins

11-7   A typical integrated-circuit package is a tiny plastic box with protruding metal pins.

Variable Capacitors

The capacitance of a component can be varied at will by adjusting the mutual surface area between the plates, or by changing the spacing between the plates. The two most common types of variable capacitors (besides varactors) are the *air variable* and the *trimmer*. You will also sometimes encounter *coaxial capacitors*.

Air Variables

By connecting two sets of metal plates so that they mesh, and by affixing one set to a rotatable shaft, a variable capacitor is made. The rotatable set of plates is called the *rotor*, and the fixed set is called the *stator*. This is the type of component you might have seen in older radio receivers, used to tune the frequency. Such capacitors are still used in transmitter output tuning networks. Figure 11-8 is a functional drawing of an air-variable capacitor.

Air variables have maximum capacitance that depends on the number of plates in each set, and also on the spacing between the plates. Common maximum values are 50 to 500 pF; minimum values are a few picofarads. The voltage-handling capability depends on the spacing between the plates. Some air variables can handle many kilovolts.

Air variables are used primarily in RF applications. They are highly efficient, and are nonpolarized, although the rotor is usually connected to common ground (the chassis or circuit board).

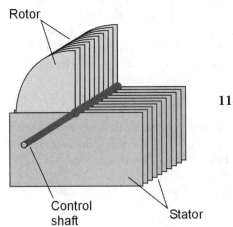

Rotor

Control
shaft

Stator

11-8   A simplified drawing of an air-variable capacitor.

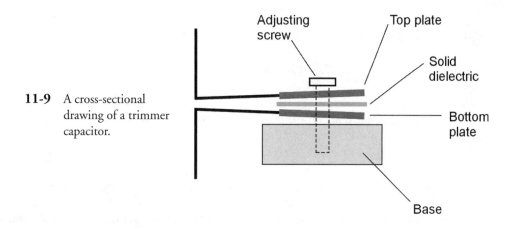

11-9 A cross-sectional drawing of a trimmer capacitor.

Trimmer Capacitors

When it is not necessary to change the value of a capacitor very often, a trimmer can be used. It consists of two plates, mounted on a ceramic base and separated by a sheet of plastic, mica, or some other solid dielectric. The plates are flexible, and can be squashed together more or less by means of a screw (Fig. 11-9). Sometimes two sets of several plates are interleaved to increase the capacitance.

Trimmers can be connected in parallel with an air variable, so that the range of the air variable can be adjusted. Some air-variable capacitors have trimmers built in. Typical maximum values for trimmers range from a few picofarads up to about 200 pF. They handle low to moderate voltages, are highly efficient, and are nonpolarized.

Coaxial Capacitors

You recall from the previous chapter that sections of transmission lines can work as inductors. They can act as capacitors, too. If a section of transmission line is less than ¼ wavelength long, and is left open at the far end (rather than shorted out), it behaves as a capacitor. The capacitance increases with length.

The most common transmission-line capacitor uses two telescoping sections of metal tubing. This is called a *coaxial capacitor*. It works because there is a certain effective surface area between the inner and the outer tubing sections. A sleeve of plastic dielectric is placed between the sections of tubing, as shown in Fig. 11-10. This allows the capacitance to be adjusted by sliding the inner section in or out of the outer section.

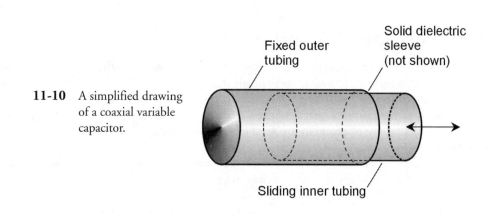

11-10 A simplified drawing of a coaxial variable capacitor.

Coaxial capacitors are used in RF applications, particularly in antenna systems. Their values are generally from a few picofarads up to about 100 pF.

Capacitor Specifications

When you are looking for a capacitor for a particular application, it's important to find a component that has the right specifications for the job. Here are two of the most important specifications to watch for.

Tolerance

Capacitors are rated according to how nearly their values can be expected to match the rated capacitance. The most common tolerance is ±10%; some capacitors are rated at ±5% or even at ±1%.

The lower (or *tighter*) the tolerance number, the more closely you can expect the actual component value to match the rated value. For example, a ±10% capacitor rated at 100 pF can range from 90 to 110 pF. But if the tolerance is ±1%, the manufacturer guarantees that the capacitance will be between 99 and 101 pF.

Problem 11-6

A capacitor is rated at 0.10 μF ±10%. What is its guaranteed range of capacitance?

First, multiply 0.10 by 10 percent to get the plus-or-minus variation. This is 0.10 × 0.10 = 0.010 μF. Then add and subtract this from the rated value to get the maximum and minimum possible capacitances. The result is a range of 0.09 to 0.11 μF.

Temperature Coefficient

Some capacitors increase in value as the temperature increases. These components have a *positive temperature coefficient.* Some capacitors decrease in value as the temperature rises; these have a *negative temperature coefficient.* Some capacitors are manufactured so that their values remain constant over a certain temperature range. Within this span of temperatures, such capacitors have *zero temperature coefficient.*

The temperature coefficient is specified in percent per degree Celsius (%/°C). Sometimes, a capacitor with a negative temperature coefficient can be connected in series or parallel with a capacitor having a positive temperature coefficient, and the two opposite effects cancel out over a range of temperatures. In other instances, a capacitor with a positive or negative temperature coefficient can be used to cancel out the effect of temperature on other components in a circuit, such as inductors and resistors.

Interelectrode Capacitance

Any two pieces of conducting material, when they are brought near each other, can act as a capacitor. Often, this *interelectrode capacitance* is so small that it can be neglected. It rarely amounts to more than a few picofarads. In utility circuits and at AF, interelectrode capacitance is not usually significant. But it can cause problems at RF. The chances for trouble increase as the frequency increases. The most common phenomena are *feedback*, and/or a change in the frequency characteristics of a circuit.

Interelectrode capacitance can be minimized by keeping wire leads as short as possible, by using shielded cables, and by enclosing sensitive circuits in metal housings.

Quiz

Refer to the text in this chapter if necessary. A good score is 18 correct. Answers are in the back of the book.

1. Capacitance acts to store electrical energy as
   (a) current.
   (b) voltage.
   (c) a magnetic field.
   (d) an electric field.

2. As capacitor plate area increases, all other things being equal,
   (a) the capacitance increases.
   (b) the capacitance decreases.
   (c) the capacitance does not change.
   (d) the current-handling ability decreases.

3. As the spacing between plates in a capacitor is made smaller, all other things being equal,
   (a) the capacitance increases.
   (b) the capacitance decreases.
   (c) the capacitance does not change.
   (d) the resistance increases.

4. A material with a high dielectric constant
   (a) acts to increase capacitance per unit volume.
   (b) acts to decrease capacitance per unit volume.
   (c) has no effect on capacitance.
   (d) causes a capacitor to become polarized.

5. A capacitance of 100 pF is the same as which of the following?
   (a) 0.01 μF
   (b) 0.001 μF
   (c) 0.0001 μF
   (d) 0.00001 μF

6. A capacitance of 0.033 μF is the same as which of the following?
   (a) 33 pF
   (b) 330 pF
   (c) 3300 pF
   (d) 33,000 pF

7. If five 0.050-μF capacitors are connected in parallel, what is the net capacitance of the combination?
   (a) 0.010 μF
   (b) 0.25 μF

(c) 0.50 μF

(d) 0.025 μF

8. If five 0.050-μF capacitors are connected in series, what is the net capacitance of the combination?

(a) 0.010 μF

(b) 0.25 μF

(c) 0.50 μF

(d) 0.025 μF

9. Suppose that two capacitors are connected in series, and their values are 47 pF and 33 pF. What is the net capacitance of this combination?

(a) 80 pF

(b) 47 pF

(c) 33 pF

(d) 19 pF

10. Suppose that two capacitors are in parallel. Their values are 47.0 pF and 470 μF. What is the net capacitance of this combination?

(a) 47.0 pF

(b) 517 pF

(c) 517 μF

(d) 470 μF

11. Suppose that three capacitors are in parallel. Their values are 0.0200 μF, 0.0500 μF, and 0.10000 μF. What is the net capacitance of this combination?

(a) 0.0125 μF

(b) 0.1700 μF

(c) 0.1000 μF

(d) 0.1250 μF

12. The main advantage of air as a dielectric material for capacitors is the fact that it

(a) has a high dielectric constant.

(b) is not physically dense.

(c) has low loss.

(d) allows for large capacitance in a small volume.

13. Which of the following is *not* a characteristic of mica capacitors?

(a) Excellent efficiency

(b) Small size, even for large values of capacitance

(c) High voltage-handling capacity

(d) Low loss

14. Which of the following capacitance values is most typical of a disk-ceramic capacitor?

(a) 100 pF

(b) 33 μF

(c) 470 μF

(d) 10,000 μF

15. Which of the following capacitance values is most typical of a paper capacitor?

(a) 0.001 pF

(b) 0.01 μF

(c) 100 μF

(d) 3300 μF

16. Which of the following capacitance ranges is most typical of an air-variable capacitor?

(a) 0.01 μF to 1 μF

(b) 1 μF to 100 μF

(c) 1 pF to 100 pF

(d) 0.001 pF to 0.1 pF

17. Which of the following types of capacitors is polarized?

(a) Paper

(b) Mica

(c) Interelectrode

(d) Electrolytic

18. If a capacitor has a negative temperature coefficient, then

(a) its capacitance decreases as the temperature rises.

(b) its capacitance increases as the temperature rises.

(c) its capacitance does not change with temperature.

(d) it will not work if the temperature is below freezing.

19. Suppose that a capacitor is rated at 33 pF ±10%. Which of the following actual capacitance values is outside the acceptable range?

(a) 30 pF

(b) 37 pF

(c) 35 pF

(d) 31 pF

20. Suppose that a capacitor, rated at 330 pF, shows an actual value of 317 pF. By how many percent does its actual capacitance differ from its rated capacitance?

(a) −0.039%

(b) −3.9%

(c) −0.041%

(d) −4.1%

12
CHAPTER

Phase

IN ALTERNATING CURRENT, EACH 360° CYCLE IS EXACTLY THE SAME AS EVERY OTHER. IN EVERY CYCLE, the waveform of the previous cycle is repeated. In this chapter, you'll learn about the most common type of ac waveform: the *sine wave*.

Instantaneous Values

An ac sine wave has a characteristic shape, as shown in Fig. 12-1. This is the way the graph of the function $y = \sin x$ looks on an (x,y) coordinate plane. (The abbreviation *sin* stands for *sine* in trigonometry.) Suppose that the peak voltage is ± 1 V, as shown. Further imagine that the period is 1 s, so the frequency is 1 Hz. Let the wave begin at time $t = 0$. Then each cycle begins every time the value of t is a whole number. At every such instant, the voltage is zero and *positive-going*.

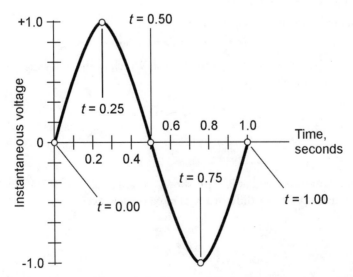

12-1 A sine wave with a period of 1 second. It thus has a frequency of 1 Hz.

If you freeze time at, say, $t = 446.00$, the voltage is zero. Looking at the diagram, you can see that the voltage will also be zero every so-many-and-a-*half* seconds, so it will be zero at $t = 446.5$. But instead of getting more positive at these instants, the voltage will be *negative-going*.

If you freeze time at so-many-and-a-*quarter* seconds, say $t = 446.25$, the voltage will be +1 V. The wave will be exactly at its positive peak. If you stop time at so-many-and-*three-quarter* seconds, say $t = 446.75$, the voltage will be exactly at its negative peak, −1 V. At intermediate times, say, so-many-and-*three-tenths* seconds, the voltage will have intermediate values.

Instantaneous Rate of Change

Figure 12-1 shows that there are times the voltage is increasing, and times it is decreasing. *Increasing*, in this context, means "getting more positive," and *decreasing* means "getting more negative." The most rapid increase in voltage occurs when $t = 0.0$ and $t = 1.0$. The most rapid decrease takes place when $t = 0.5$.

When $t = 0.25$, and also when $t = 0.75$, the instantaneous voltage neither increases nor decreases. But this condition exists only for a vanishingly small moment, a single point in time.

Suppose n is some whole number. Then the situation at $t = n.25$ is the same as it is for $t = 0.25$; also, for $t = n.75$, things are the same as they are when $t = 0.75$. The single cycle shown in Fig. 12-1 represents every possible condition of the ac sine wave having a frequency of 1 Hz and a peak value of ±1 V. The whole wave recurs, over and over, for as long as the ac continues to flow in the circuit.

Now imagine that you want to observe the *instantaneous rate of change* in the voltage of the wave in Fig. 12-1, as a function of time. A graph of this turns out to be a sine wave, too—but it is displaced to the left of the original wave by ¼ of a cycle. If you plot the instantaneous rate of change of a sine wave against time (Fig. 12-2), you get the *derivative* of the waveform. The derivative of a sine wave is a *cosine wave*. This wave has the same shape as the sine wave, but the *phase* is different by ¼ of a cycle.

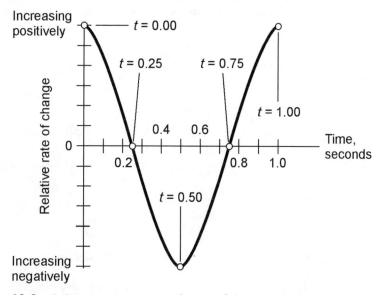

12-2 A sine wave representing the rate of change in the instantaneous voltage of the wave shown in Fig. 12-1.

Circles and Vectors

An ac sine wave represents the most efficient possible way that an electrical quantity can alternate. It has only one frequency component. All the wave energy is concentrated into this smoothly see-sawing variation. It is like a pure musical note.

Circular Motion

Suppose that you swing a ball around and around at the end of a string, at a rate of one revolution per second (1 rps). The ball describes a circle in space (Fig. 12-3A). If a friend stands some distance away, with his or her eyes in the plane of the ball's path, your friend sees the ball oscillating back and forth (Fig. 12-3B) with a frequency of 1 Hz. That is one complete cycle per second, because you swing the ball around at 1 rps.

   If you graph the position of the ball, as seen by your friend, with respect to time, the result is a sine wave (Fig. 12-4). This wave has the same fundamental shape as all sine waves. Some sine waves are taller than others, and some are stretched out horizontally more than others. But the general waveform is the same in every case. By multiplying or dividing the amplitude and the wavelength of any sine wave, it can be made to fit exactly along the curve of any other sine wave. The standard sine wave is the function $y = \sin x$ in the coordinate plane.

   You might whirl the ball around faster or slower than 1 rps. The string might be made longer or shorter. This would alter the height and/or the frequency of the sine wave graphed in Fig. 12-4. But the sine wave can always be reduced to the equivalent of constant, smooth motion in a circular orbit. This is known as the *circular motion model* of a sine wave.

Rotating Vectors

Back in Chapter 9, *degrees of phase* were discussed. If you wondered then why phase is spoken of in terms of angular measure, the reason should be clearer now. A circle has 360°. A sine wave can be represented as circular motion. Points along a sine wave thus correspond to angles, or positions, around a circle.

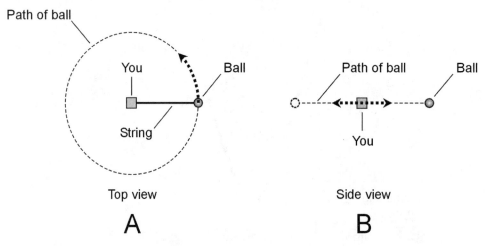

12-3   Swinging ball and string as seen from above (A) and from the side (B).

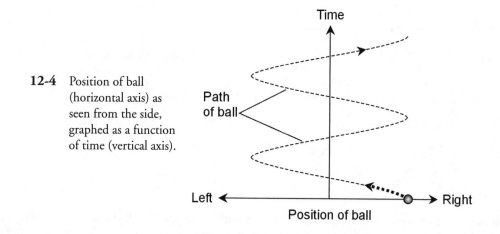

12-4 Position of ball (horizontal axis) as seen from the side, graphed as a function of time (vertical axis).

Figure 12-5 shows the way a rotating *vector* can be used to represent a sine wave. A vector is a quantity with two independent properties, called *magnitude* (or amplitude) and *direction*. At A, the vector points east, and this is assigned the value of 0°, where the wave amplitude is zero and is increasing positively. At B, the vector points north; this is the 90° instant, where the wave has attained its maximum positive amplitude. At C, the vector points west. This is 180°, the instant where the wave has gone back to zero amplitude and is getting more negative. At D, the wave points south. This is 270°, and it represents the maximum negative amplitude. When a full circle (360°) has been completed, the vector once again points east.

The four points in Fig. 12-5 are shown on a sine wave graph in Fig. 12-6. Think of the vector as revolving counterclockwise at a rate that corresponds to one revolution per cycle of the wave. If the wave has a frequency of 1 Hz, the vector goes around at a rate of 1 rps. If the wave has a frequency of

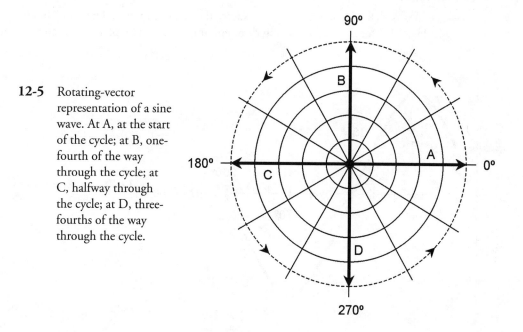

12-5 Rotating-vector representation of a sine wave. At A, at the start of the cycle; at B, one-fourth of the way through the cycle; at C, halfway through the cycle; at D, three-fourths of the way through the cycle.

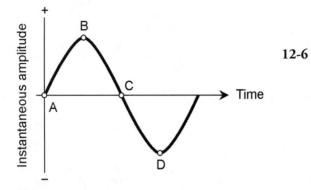

12-6 The four points for the vector model of Fig. 12-5, shown in the standard amplitude-versus-time graphical manner.

100 Hz, the speed of the vector is 100 rps, or a revolution every 0.01 s. If the wave is 1 MHz, then the speed of the vector is 1 million rps (10^6 rps), and it goes once around every 0.000001 s (10^{-6} s).

The peak amplitude of a pure ac sine wave corresponds to the length of its vector. In Fig. 12-5, time is shown by the angle counterclockwise from due east. Amplitude is independent of time. The vector length never changes, but its direction does.

Expressions of Phase Difference

The *phase difference*, also called the *phase angle*, between two waves can have meaning only when those two waves have identical frequencies. If the frequencies differ, even by just a little bit, the relative phase constantly changes, and it's impossible to specify a value for it. In the following discussions of phase angle, let's assume that the two waves always have identical frequencies.

Phase Coincidence

Phase coincidence means that two waves begin at exactly the same moment. They are "lined up." This is shown in Fig. 12-7 for two waves having different amplitudes. The phase difference in this

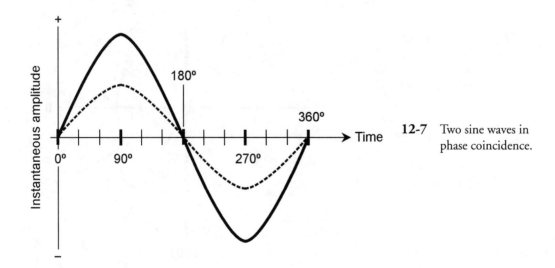

12-7 Two sine waves in phase coincidence.

case is 0°. You could say it's some whole-number multiple of 360°, too—but engineers and technicians rarely speak of any phase angle of less than 0° or more than 360°.

If two sine waves are in phase coincidence, and if neither wave has dc superimposed, then the resultant is a sine wave with positive or negative peak amplitudes equal to the sum of the positive and negative peak amplitudes of the composite waves. The phase of the resultant is the same as that of the composite waves.

Phase Opposition

When two sine waves begin exactly $\frac{1}{2}$ cycle, or 180°, apart, they are said to be in *phase opposition*. This is illustrated by the drawing of Fig. 12-8. In this situation, engineers sometimes say that the waves are *out of phase*, although this expression is a little nebulous because it could be taken to mean some phase difference other than 180°.

If two sine waves have the same amplitudes and are in phase opposition, they cancel each other out. This is because the instantaneous amplitudes of the two waves are equal and opposite at every moment in time.

If two sine waves are in phase opposition, and if neither wave has dc superimposed, then the resultant is a sine wave with positive or negative peak amplitudes equal to the difference between the positive and negative peak amplitudes of the composite waves. The phase of the resultant is the same as the phase of the stronger of the two composite waves.

Any sine wave without superimposed dc has the unique property that, if its phase is shifted by 180°, the resultant wave is the same as turning the original wave upside down. Not all waveforms have this property. Perfect square waves do, but some rectangular and sawtooth waves don't, and irregular waveforms almost never do.

Intermediate Phase Differences

Two sine waves can differ in phase by any amount from 0° (phase coincidence), through 90° (*phase quadrature*, meaning a difference a quarter of a cycle), 180° (phase opposition), 270° (phase quadrature again), to 360° (phase coincidence again).

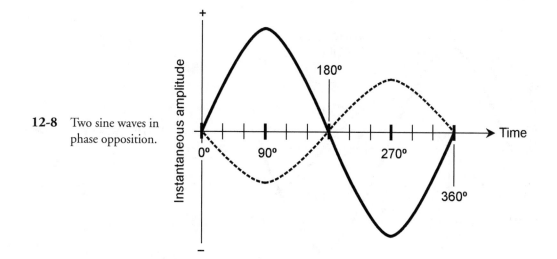

12-8   Two sine waves in phase opposition.

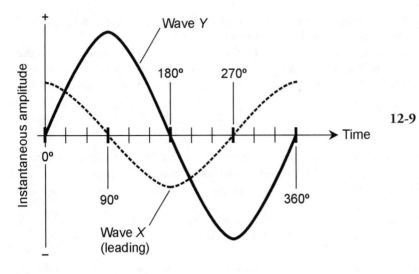

12-9   Wave *X* leads wave *Y* by 90° of phase (¼ of a cycle).

Leading Phase

Imagine two sine waves, called wave *X* and wave *Y*, with identical frequency. If wave *X* begins a fraction of a cycle *earlier* than wave *Y*, then wave *X* is said to be *leading* wave *Y* in phase. For this to be true, *X* must begin its cycle less than 180° before *Y*. Figure 12-9 shows wave *X* leading wave *Y* by 90°.

Note that if wave *X* (the dashed line in Fig. 12-9) is leading wave *Y* (the solid line), then wave *X* is displaced to the *left* of wave *Y*. In a time-domain graph or display, displacement to the left represents earlier moments in time, and displacement to the right represents later moments in time.

Lagging Phase

Suppose that some sine wave *X* begins its cycle more than 180°, but less than 360°, ahead of wave *Y*. In this situation, it is easier to imagine that wave *X* starts its cycle *later* than wave *Y*, by some value between 0° and 180°. Then wave *X* is not leading, but *lagging*, wave *Y*. Figure 12-10 shows wave *X* lagging wave *Y* by 90°.

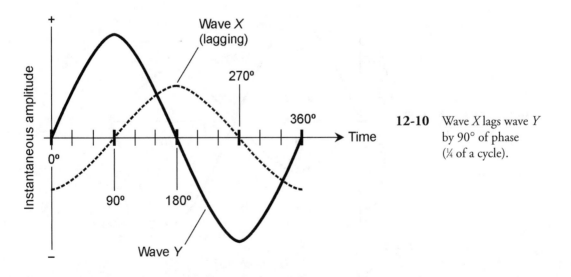

12-10   Wave *X* lags wave *Y* by 90° of phase (¼ of a cycle).

Vector Diagrams of Phase Difference

The vector renditions of sine waves, such as are shown in Fig. 12-5, are well suited to showing phase relationships.

If a sine wave X leads a sine wave Y by some number of degrees, then the two waves can be drawn as vectors, with vector **X** being that number of angular degrees *counterclockwise* from vector **Y**. If a sine wave X lags a sine wave Y by some number of degrees, then **X** appears to point in a direction that is *clockwise* from **Y** by that number of angular degrees. If two waves are in phase coincidence, then their vectors point in exactly the same direction. If two waves are in phase opposition, then their vectors point in exactly opposite directions.

The drawings of Fig. 12-11 show four phase relationships between two sine waves X and Y. At A, X is in phase with Y. At B, X leads Y by 90°. At C, X and Y are 180° apart in phase. At D, X lags Y by 90°. In all of these examples, think of the vectors rotating *counterclockwise* as time passes, but always maintaining the same angle with respect to each other, and always staying at the same lengths. If the frequency in hertz is f, then the pair of vectors rotates together, counterclockwise, at an angular speed of f, expressed in complete 360° revolutions per second.

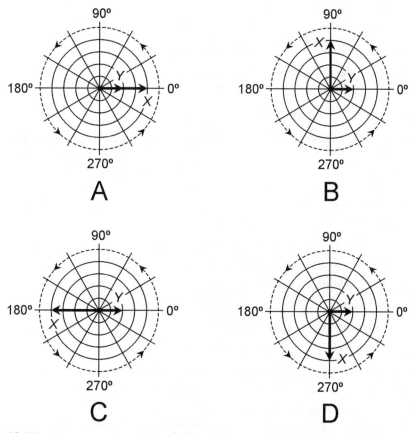

12-11   Vector representations of phase difference. At A, waves X and Y are in phase. At B, X leads Y by 90°. At C, X and Y are 180° out of phase. At D, X lags Y by 90°. Time is represented by counterclockwise motion of both vectors at a constant angular speed.

Quiz

Refer to the text in this chapter if necessary. A good score is 18 correct. Answers are in the back of the book.

1. Which of the following is *not* a general characteristic of an ac wave?
   (a) The wave shape is identical for each cycle.
   (b) The polarity reverses periodically.
   (c) The electrons always flow in the same direction.
   (d) There is a definite frequency.

2. All sine waves
   (a) have similar general appearance.
   (b) have instantaneous rise and fall times.
   (c) are in the same phase as cosine waves.
   (d) rise instantly, but decay slowly.

3. The derivative of a sine wave
   (a) is shifted in phase by $\frac{1}{2}$ cycle from the sine wave.
   (b) is the rate of change in the instantaneous value.
   (c) has instantaneous rise and decay times.
   (d) rises instantly, but decays slowly.

4. A phase difference of 180° in the circular motion model of a sine wave represents
   (a) $\frac{1}{4}$ revolution.
   (b) $\frac{1}{2}$ revolution.
   (c) a full revolution.
   (d) two full revolutions.

5. You can add or subtract a certain number of degrees of phase to or from a wave, and end up with exactly the same wave again. This number is
   (a) 90, or any whole-number multiple of it.
   (b) 180, or any whole-number multiple of it.
   (c) 270, or any whole-number multiple of it.
   (d) 360, or any whole-number multiple of it.

6. You can add or subtract a certain number of degrees of phase to or from a sine wave, and end up with an inverted (upside-down) representation of the original. This number is
   (a) 90, or any odd whole-number multiple of it.
   (b) 180, or any odd whole-number multiple of it.
   (c) 270, or any odd whole-number multiple of it.
   (d) 360, or any odd whole-number multiple of it.

7. Suppose a wave has a frequency of 300 kHz. How long does one complete cycle take?
   (a) 1.300 s
   (b) 0.00333 s
   (c) 1/3000 s
   (d) 3.33×10^{-6} s

8. If a wave has a frequency of 440 Hz, how long does it take for 10° of a cycle to occur?
   (a) 0.00273 s
   (b) 0.000273 s
   (c) 0.0000631 s
   (d) 0.00000631 s

9. Suppose two waves are in phase coincidence. One has peak values of ± 3 V and the other has peak values of ± 5 V. The resultant has voltages of
   (a) ± 8 V pk, in phase with the composites.
   (b) ± 2 V pk, in phase with the composites.
   (c) ± 8 V pk, in phase opposition with respect to the composites.
   (d) ± 2 V pk, in phase opposition with respect to the composites.

10. As shown on a graph, shifting the phase of an ac sine wave by 90° is the same thing as
    (a) moving it to the right or left by a full cycle.
    (b) moving it to the right or left by $\frac{1}{4}$ cycle.
    (c) turning it upside down.
    (d) leaving it alone.

11. Two pure sine waves that differ in phase by 180° can be considered to
    (a) be offset by two full cycles.
    (b) be in phase opposition.
    (c) be separated by less than $\frac{1}{4}$ cycle.
    (d) have a frequency of $\frac{1}{2}$ cycle.

12. Suppose two sine waves are in phase opposition. Wave X has a peak amplitude of ± 4 V and wave Y has a peak amplitude of ± 8 V. The resultant has voltages of
    (a) ± 4 V pk, in phase with the composites.
    (b) ± 4 V pk, out of phase with the composites.
    (c) ± 4 V pk, in phase with wave X.
    (d) ± 4 V pk, in phase with wave Y.

13. If wave X leads wave Y by 45°, then
    (a) wave Y is $\frac{1}{4}$ cycle ahead of wave X.
    (b) wave Y is $\frac{1}{4}$ cycle behind wave X.
    (c) wave Y is $\frac{1}{8}$ cycle behind wave X.
    (d) wave Y is 1.16 cycle ahead of wave X.

14. If wave *X* lags wave *Y* by ⅓ cycle, then
    (a) wave *Y* is 120° ahead of wave *X*.
    (b) wave *Y* is 90° ahead of wave *X*.
    (c) wave *Y* is 60° ahead of wave *X*.
    (d) wave *Y* is 30° ahead of wave *X*.

15. Refer to Fig. 12-12. In this example,
    (a) *X* lags *Y* by 45°.
    (b) *X* leads *Y* by 45°.
    (c) *X* lags *Y* by 135°.
    (d) *X* leads *Y* by 135°.

16. Which of the drawings in Fig. 12-13 represents the situation of Fig. 12-12?
    (a) Drawing A
    (b) Drawing B
    (c) Drawing C
    (d) Drawing D

17. In vector diagrams such as those of Fig. 12-13, the length of the vector represents
    (a) the average amplitude of a sine wave.
    (b) the frequency of a sine wave.
    (c) the phase of a sine wave.
    (d) the peak amplitude of a sine wave.

18. In vector diagrams such as those of Fig. 12-13, the angle between two vectors represents
    (a) the average of the peak amplitudes of two sine waves.
    (b) the frequency difference between two sine waves.
    (c) the phase difference between two sine waves.
    (d) the difference between the peak amplitudes of two sine waves.

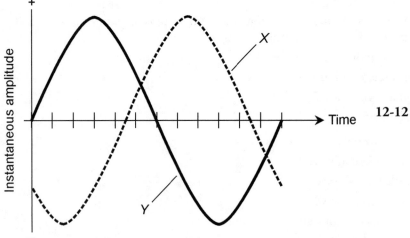

12-12   Illustration for Quiz Question 15.

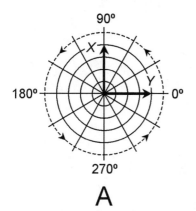

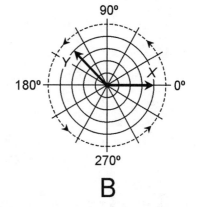

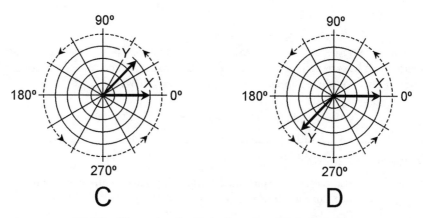

12-13 Illustration for Quiz Questions 16 through 20.

19. In vector diagrams such as those of Fig. 12-13, the distance from the center of the graph represents
    (a) average amplitude.
    (b) frequency.
    (c) phase.
    (d) peak amplitude.

20. In diagrams like those of Fig. 12-13, the progression of time is sometimes depicted as
    (a) movement of a vector to the right.
    (b) movement of a vector to the left.
    (c) counterclockwise rotation of a vector.
    (d) clockwise rotation of a vector.

<h1 style="text-align:center">13
CHAPTER</h1>

<h1 style="text-align:center">Inductive Reactance</h1>

IN DC CIRCUITS, RESISTANCE CAN BE EXPRESSED AS A NUMBER RANGING FROM ZERO (REPRESENTING a perfect conductor) to extremely large values. Physicists call resistance a *scalar* quantity, because it can be expressed on a one-dimensional *scale*, as shown in Fig. 13-1.

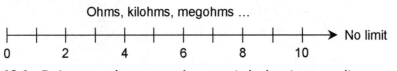

Ohms, kilohms, megohms …

13-1 Resistance can be represented as numerical values (corresponding to ohms) along a half line or ray.

Coils and Direct Current

Suppose you have some wire that conducts electricity very well. If you wind a length of the wire into a coil and connect it to a source of dc (Fig. 13-2), the wire draws a large current. It doesn't matter

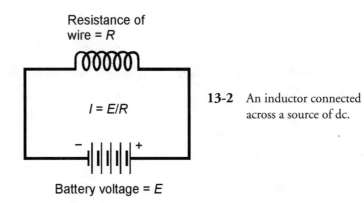

Resistance of wire = R

$I = E/R$

Battery voltage = E

13-2 An inductor connected across a source of dc.

whether the wire is a single-turn loop, or whether it's lying haphazardly on the floor, or whether it's wrapped around a stick. The current amperes is equal to the applied voltage in volts divided by the wire resistance in ohms. It's that simple.

You can make an electromagnet, as you've already seen, by passing dc through a coil wound around an iron rod. Electromagnets are known for the high current they draw from batteries or power supplies. The coil of an electromagnet heats up as energy is dissipated in the resistance of the wire. If the voltage of the battery or power supply increases, the wire in the coil gets hotter. Ultimately, if the supply can deliver enough current, the wire will melt.

Coils and Alternating Current

Suppose you change the voltage source, connected across the coil, from dc to ac (Fig. 13-3). Imagine that you can vary the frequency of the ac, from a few hertz to hundreds of hertz, then kilohertz, then megahertz.

At first, the current will be high, just as it is with dc. But the coil has a certain amount of inductance, and it takes some time for current to establish itself in the coil. Depending on how many turns there are and on whether the core is air or a ferromagnetic material, you'll reach a point, as the ac frequency increases, when the coil starts to get sluggish. That is, the current won't have time to get established in the coil before the polarity of the ac voltage reverses. At high ac frequencies, the current through the coil will have difficulty following the voltage placed across the coil. This sluggishness in a coil for ac is, in effect, similar to dc resistance. As the frequency is raised, the effect gets more pronounced. Eventually, if you keep increasing the frequency of the ac source, the coil will not even come near establishing a current with each cycle. Then the coil will act like a high resistance.

The opposition that the coil offers to ac is called *inductive reactance*. It, like resistance, is measured in ohms. It can vary, just as resistance does, from near zero (a short piece of wire) to a few ohms (a small coil) to kilohms or megohms (bigger and bigger coils). Like resistance, inductive reactance affects the current in an ac circuit. But, unlike simple resistance, reactance changes with frequency. This effect is not merely a decrease in the current, although in practice this does happen. Inductive reactance produces a change in the way the current flows with respect to the voltage.

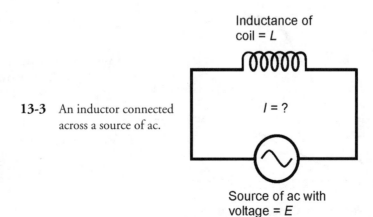

Inductance of
coil = L

$I = ?$

13-3   An inductor connected across a source of ac.

Source of ac with
voltage = E

Reactance and Frequency

Inductive reactance is one of two form of reactance. (The other form, called *capacitive reactance*, will be discussed in the next chapter.) Reactance in general is symbolized by the italic uppercase letter X. Inductive reactance is symbolized X_L.

If the frequency of an ac source is given, in hertz, as f, and the inductance of a coil in henrys is given as L, then the inductive reactance in ohms, X_L, is calculated as follows:

$$X_L = 2\pi f L$$

In this formula, the symbol π stands for the mathematical constant *pi*, which is the number of diameters around the circumference of a circle. It is equal to approximately 3.14. We can consider the value of 2π to be equal to 6.28 in most practical situations. Therefore, the preceding formula can be written a little more simply as:

$$X_L = 6.28 f L$$

This same formula applies if the frequency, f, is in kilohertz and the inductance, L, is in millihenrys. And it also applies if f is in megahertz and L is in microhenrys. Just remember that if frequency is in thousands, inductance must be in thousandths, and if frequency is in millions, inductance must be in millionths.

Inductive reactance increases *linearly* with increasing ac frequency. This means that the function of X_L versus f is a straight line when graphed. Inductive reactance also increases linearly with inductance. Therefore, the function of X_L versus L also appears as a straight line on a graph. The value of X_L is *directly proportional* to f, and is also directly proportional to L. These relationships are graphed, in relative form, in Fig. 13-4.

Problem 13-1

Suppose a coil has an inductance of 0.500 H, and the frequency of the ac passing through it is 60.0 Hz. What is the inductive reactance?

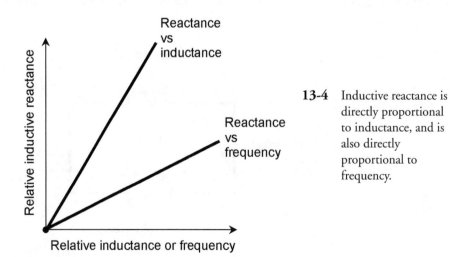

13-4   Inductive reactance is directly proportional to inductance, and is also directly proportional to frequency.

Using the preceding formula, calculate $X_L = 6.28 \times 60.0 \times 0.500 = 188 \ \Omega$. This is rounded to three significant figures.

Problem 13-2

What will be the inductive reactance of the preceding coil if the supply is a battery that supplies pure dc?

Because dc has a frequency of zero, $X_L = 6.28 \times 0 \times 0.500 = 0 \ \Omega$. That is, there will be no inductive reactance. Inductance doesn't have any practical effect with pure dc.

Problem 13-3

If a coil has an inductive reactance of 100 Ω at a frequency of 5.00 MHz, what is its inductance?

In this case, you need to plug numbers into the formula and solve for the unknown L. Start out with the equation $100 = 6.28 \times 5.00 \times L = 31.4 \times L$. Because the frequency is given in megahertz, the inductance will come out in microhenrys. You can divide both sides of the equation by 31.4, getting $L = 100/31.4 = 3.18 \ \mu H$.

Points in the *RL* Plane

Inductive reactance can be plotted along a half line, just as can resistance. In a circuit containing both resistance and inductance, the characteristics become two-dimensional. You can orient the resistance and reactance half lines perpendicular to each other to make a quarter-plane coordinate system, as shown in Fig. 13-5. Resistance is plotted horizontally, and inductive reactance is plotted vertically upward.

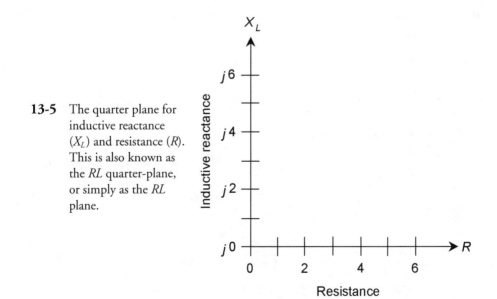

13-5 The quarter plane for inductive reactance (X_L) and resistance (R). This is also known as the *RL* quarter-plane, or simply as the *RL* plane.

In this scheme, resistance-inductance (*RL*) combinations form *complex impedances.* (The term *impedance* comes from the root *impede*, and fully describes how electrical components impede, or inhibit, the flow of ac. You'll learn all about this in Chap. 15.) Each point on the *RL plane* corresponds to one unique complex impedance value. Conversely, each complex impedance value corresponds to one unique point on the *RL* plane.

You might ask, "What's the little *j* doing in Fig. 13-5?" For reasons that will be made clear in Chap. 15, impedances on the *RL* plane are written in the form $R + jX_L$, where *R* is the resistance in ohms, and X_L is the inductive reactance in ohms. The little *j* is called a *j operator* and is a mathematical way of expressing the fact that reactance is denoted at right angles to resistance in complex-impedance graphs.

If you have a pure resistance, say $R = 5 \, \Omega$, then the *complex* impedance is $5 + j0$, and is at the point (5,0) on the *RL* plane. If you have a pure inductive reactance, such as $X_L = 3 \, \Omega$, then the complex impedance is $0 + j3$, and is at the point (0,j3) on the *RL* plane. These points, and a couple of others, are shown in Fig. 13-6.

In real life, all coils have some resistance, because no wire is a perfect conductor. All resistors have at least a tiny bit of inductive reactance, because they take up some physical space and they have wire leads. So there is really no such thing as a mathematically perfect pure resistance such as $5 + j0$, or a mathematically perfect pure reactance like $0 + j3$. But sometimes you can get extremely close to theoretical ideals in real life.

Often, resistance and inductive reactance are both deliberately placed in a circuit. Then you get impedances values such as $2 + j3$ or $4 + j1.5$. These are shown in Fig. 13-6 as points on the *RL* plane.

Remember that values for X_L are *reactances*, not actual inductances. Because of this, they vary with the frequency in an *RL* circuit. Changing the frequency has the effect of making complex impedance points move around in the *RL* plane. They move vertically, going upward as the ac frequency increases, and downward as the ac frequency decreases. If the ac frequency goes down to zero, the inductive reactance vanishes. Then $X_L = 0$, we have pure dc, and the point is right on the resistance axis.

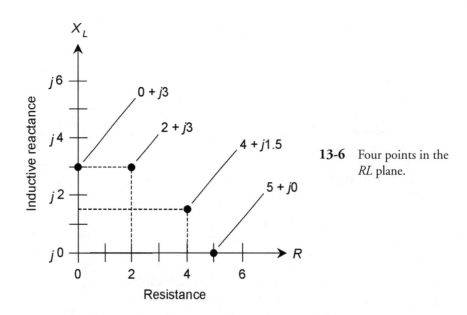

13-6 Four points in the *RL* plane.

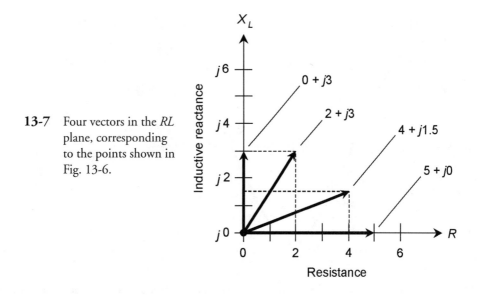

13-7 Four vectors in the *RL* plane, corresponding to the points shown in Fig. 13-6.

Vectors in the *RL* Plane

Engineers sometimes represent points in the *RL* plane as vectors. Recall that a vector is a mathematical quantity that has a defined magnitude (length) and defined direction (orientation). Expressing a point in the *RL* plane as a vector thus gives that point a unique magnitude and a unique direction.

In Fig. 13-6, four different points are shown. Each point is represented by a certain distance to the right of the *origin* $(0,j0)$, and a certain distance upward from the origin. The first of these is the resistance, R, and the second is the inductive reactance, X_L. Thus, the *RL* combination is a two-dimensional quantity. There is no way to uniquely define *RL* combinations as single numbers, or *scalars*, because there are two different quantities that can vary independently.

Another way to depict these points is to draw lines from the origin out to them. Then you can think of the points as *rays*, each having a certain length, or magnitude, and a certain direction, or angle counterclockwise from the resistance axis. These rays, going out to the points, are *complex impedance vectors* (Fig. 13-7).

Current Lags Voltage

When an ac voltage is placed across an inductor and starts to increase (either positively or negatively) from zero, it takes a fraction of a cycle for the current to follow. Once the voltage starts decreasing from its maximum peak (either positive or negative) in the cycle, it again takes a fraction of a cycle for the current to follow. The instantaneous current can't quite keep up with the instantaneous voltage, as it does in a pure resistance. Thus, in a circuit containing inductive reactance, the current is said to *lag* the voltage in phase.

Pure Inductance

Suppose that you place an ac voltage across a coil, with a frequency high enough so that the inductive reactance, X_L, is much larger than the resistance, R. In this situation, the current is $\frac{1}{4}$ of a cycle behind the voltage. That is, the current lags the voltage by 90°, as shown in Fig. 13-8.

At very low frequencies, large inductances are normally needed in order for the current lag to be

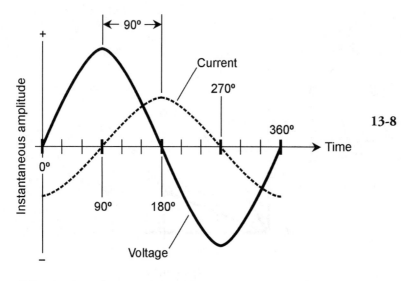

13-8   In a pure inductive reactance, the current lags the voltage by 90°.

a full 90°. This is because any coil has some resistance; no wire is a perfect conductor. If some wire were found that had a mathematically zero resistance, and if a coil of any size were wound from this wire, then the current would lag the voltage by 90° in this inductor, no matter what the ac frequency.

When the value of X_L is very large compared with the value of R in a circuit—that is, when there is an essentially pure inductive reactance—the vector in the *RL* plane points straight up along the X_L axis. Its angle is 90° from the R axis, which is considered the *zero line* in the *RL* plane.

Inductance with Resistance

When the resistance in a resistance-inductance (*RL*) circuit is significant compared with the inductive reactance, the current lags the voltage by something less than 90° (Fig. 13-9). If R is small compared with X_L, the current lag is almost 90°, but as R gets larger relative to X_L, the lag decreases.

The value of R in an *RL* circuit can increase relative to X_L because resistance is deliberately placed in series with the inductance. It can also happen because the ac frequency gets so low that X_L

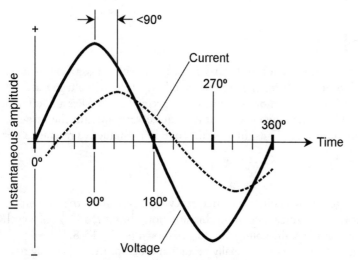

13-9   In a circuit with inductive reactance and resistance, the current lags the voltage by less than 90°.

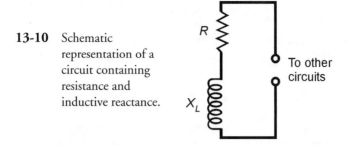

13-10 Schematic representation of a circuit containing resistance and inductive reactance.

decreases until it is comparable to the loss resistance *R* in the coil winding. In either case, the situation can be schematically represented by an inductance in series with a resistance (Fig. 13-10).

If you know the values of X_L and *R*, you can find the *angle of lag*, also called the *RL phase angle*, by plotting the point $R + jX_L$ on the *RL* plane, drawing the vector from the origin out to that point, and then measuring the angle of the vector, counterclockwise from the resistance axis. You can use a protractor to measure this angle, or you can compute its value using trigonometry.

Actually, you don't have to know the actual values of X_L and *R* in order to find the angle of lag. All you need to know is their ratio. For example, if $X_L = 5 \ \Omega$ and $R = 3 \ \Omega$, you get the same *RL* phase angle that you get if $X_L = 50 \ \Omega$ and $R = 30 \ \Omega$, or if $X_L = 20 \ \Omega$ and $R = 12 \ \Omega$. The angle of lag is the same for any values of X_L and *R* in the ratio 5:3.

Pure Resistance

As the resistance in an *RL* circuit becomes large with respect to the inductive reactance, the angle of lag gets small. The same thing happens if the inductive reactance gets small compared with the resistance. When *R* is many times greater than X_L, the vector in the *RL* plane lies almost on the *R* axis, going east (to the right). The *RL* phase angle in this case is close to 0°. The current is nearly in phase with the voltage.

In a pure resistance, with no inductance at all, the current is precisely in phase with the voltage (Fig. 13-11). A pure resistance doesn't store and release energy as an inductive circuit does, so there is no sluggishness in it.

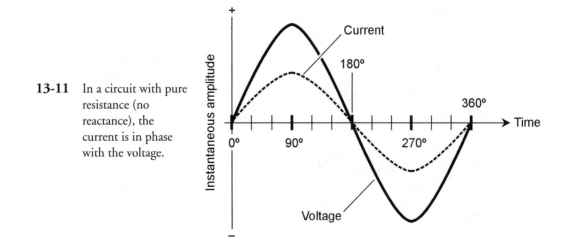

13-11 In a circuit with pure resistance (no reactance), the current is in phase with the voltage.

How Much Lag?

If you know the ratio of the inductive reactance to the resistance (X_L/R) in an *RL* circuit, then you can find the phase angle. Of course, you can also find the phase angle if you know the actual values of X_L and R.

Pictorial Method

It isn't necessary to construct an entire *RL* plane to find phase angles. You can use a ruler that has centimeter (cm) and millimeter (mm) markings, and a protractor. First, draw a line a little more than 10 cm long, going from left to right on a sheet of paper. Use the ruler and a sharp pencil. Then, with the protractor, construct a line off the left end of this first line, going vertically upward. Make this line at least 10 cm long. The horizontal line, or the one going to the right, is the *R* axis of a coordinate system. The vertical line, or the one going upward, is the X_L axis.

If you know the values of X_L and *R*, divide them down or multiply them up so they're both between 0 and 100. For example, if $X_L = 680$ Ω and $R = 840$ Ω, you can divide them both by 10 to get $X_L = 68$ and $R = 84$. Plot these points lightly by making hash marks on the vertical and horizontal lines you've drawn. The *R* mark in this example will be 84 mm to the right of the origin, and the X_L mark will be 68 mm up from the origin.

Next, draw a line connecting the two hash marks, as shown in Fig. 13-12. This line will run at a slant, and will form a triangle along with the two axes. Your hash marks, and the origin of the coordinate system, form the three *vertices* of a *right triangle*. The triangle is called *right* because one of its angles is a right angle (90°). Measure the angle between the slanted line and the *R* axis. Extend one or both of the lines if necessary in order to get a good reading on the protractor. This angle will be between 0 and 90°, and represents the phase angle in the *RL* circuit.

The complex impedance vector, $R + jX_L$, is found by constructing a rectangle using the origin and your two hash marks as three of the four vertices, and drawing new horizontal and vertical lines to complete the figure. The vector is the diagonal of this rectangle, as shown in Fig. 13-13. The

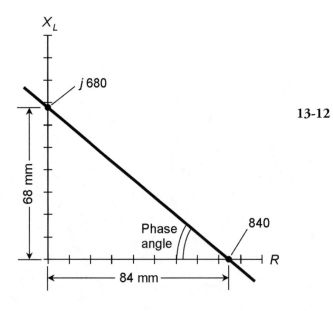

13-12 Pictorial method of finding phase angle in a circuit containing resistance and inductive reactance.

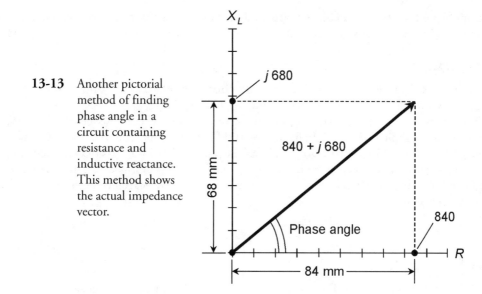

13-13 Another pictorial method of finding phase angle in a circuit containing resistance and inductive reactance. This method shows the actual impedance vector.

phase angle is the angle between this vector and the *R* axis. It will be the same as the angle of the slanted line in Fig. 13-12.

Trigonometric Method

If you have a good scientific calculator that can find the *arctangent* of a number (also called the *inverse tangent* and symbolized either as arctan or $\tan^{-1}$), you can determine the *RL* phase angle more precisely than the pictorial method allows. Given the values of X_L and *R*, the *RL* phase angle is the arctangent of their ratio. Phase angle is symbolized by the lowercase Greek letter phi (pronounced "fie" or "fee" and written ϕ). Therefore:

$$\phi = \tan^{-1} (X_L/R) \qquad \text{or} \qquad \phi = \arctan (X_L/R)$$

Problem 13-4

Suppose the inductive reactance in an *RL* circuit is 680 Ω and the resistance is 840 Ω. What is the phase angle?

The ratio X_L/R is 680/840. A calculator will display this quotient as something like 0.8095 and some more digits. Find the arctangent of this number. You should get 38.99 and some more digits. This can be rounded off to 39.0°.

Problem 13-5

Suppose an *RL* circuit operates at a frequency of 1.0 MHz with a resistance of 10 Ω and an inductance of 90 μH. What is the phase angle? What does this tell us about the nature of this *RL* circuit at this frequency?

Find the inductive reactance using the formula $X_L = 6.28fL = 6.28 \times 1.0 \times 90 = 565$ Ω. Then find the ratio $X_L/R = 565/10 = 56.5$. The phase angle is equal to arctan 56.5, which, rounded to two significant figures, is 89°. The circuit contains an almost pure inductive reactance, because the phase angle is close to 90°. The resistance contributes little to the behavior of this *RL* circuit at 1.0 MHz.

Problem 13-6

What is the phase angle for the preceding circuit at a frequency of 10 kHz? With that information, what can we say about the behavior of the circuit at 10 kHz?

This requires that X_L be calculated again, for the new frequency. Let's use megahertz, so it goes in the formula with microhenrys. A frequency of 10 kHz is the same as 0.010 MHz. Calculating, we get $X_L = 6.28fL = 6.28 \times 0.010 \times 90 = 5.65\ \Omega$. The ratio X_L/R is $5.65/10 = 0.565$. Therefore, the phase angle is arctan 0.565, which, rounded to two significant figures, is 29°. This is not close to either 0° or 90°. Thus, at 10 kHz, the resistance and the inductive reactance both play significant roles in the behavior of the circuit.

Quiz

Refer to the text in this chapter if necessary. A good score is 18 correct. Answers are in the back of the book.

1. As the number of turns in a coil that carries ac increases without limit, the current in the coil will
   - (a) eventually become very large.
   - (b) stay the same.
   - (c) decrease, approaching zero.
   - (d) be stored in the core material.

2. As the number of turns in a coil increases, the reactance at a constant frequency
   - (a) increases.
   - (b) decreases.
   - (c) stays the same.
   - (d) is stored in the core material.

3. As the frequency of an ac wave gets lower, the value of X_L for a particular coil of wire
   - (a) increases.
   - (b) decreases.
   - (c) stays the same.
   - (d) depends on the voltage.

4. Suppose a coil has an inductance of 100 mH. What is the reactance at a frequency of 1000 Hz?
   - (a) 0.628 Ω
   - (b) 6.28 Ω
   - (c) 62.8 Ω
   - (d) 628 Ω

5. Suppose a coil shows an inductive reactance of 200 Ω at 500 Hz. What is its inductance?
   (a) 0.637 H
   (b) 628 H
   (c) 63.7 mH
   (d) 628 mH

6. Imagine a 400-μH inductor with a reactance of 33 Ω. What is the frequency?
   (a) 13 kHz
   (b) 0.013 kHz
   (c) 83 kHz
   (d) 83 MHz

7. Suppose an inductor has $X_L = 555$ Ω at $f = 132$ kHz. What is L?
   (a) 670 mH
   (b) 670 μH
   (c) 460 mH
   (d) 460 μH

8. Suppose a coil has $L = 689$ μH at $f = 990$ kHz. What is X_L?
   (a) 682 Ω
   (b) 4.28 Ω
   (c) 4.28 kΩ
   (d) 4.28 MΩ

9. Suppose an inductor has $L = 88$ mH with $X_L = 100$ Ω. What is f?
   (a) 55.3 kHz
   (b) 55.3 Hz
   (c) 181 kHz
   (d) 181 Hz

10. Each point in the *RL* plane
    (a) corresponds to a unique resistance.
    (b) corresponds to a unique inductance.
    (c) corresponds to a unique combination of resistance and inductive reactance.
    (d) corresponds to a unique combination of resistance and inductance.

11. If the resistance R and the inductive reactance X_L both are allowed to vary from zero to unlimited values, but are always in the ratio 3:1, the points in the *RL* plane for all the resulting impedances will lie along
    (a) a vector pointing straight up.
    (b) a vector pointing east.
    (c) a circle.
    (d) a ray of indefinite length, pointing outward from the origin.

12. Each specific complex impedance value defined in the form $R + jX_L$

    (a) corresponds to a specific point in the RL plane.

    (b) corresponds to a specific inductive reactance.

    (c) corresponds to a specific resistance.

    (d) All of the above are true.

13. A vector is defined as a mathematical quantity that has

    (a) magnitude and direction.

    (b) resistance and inductance.

    (c) resistance and reactance.

    (d) inductance and reactance.

14. In an RL circuit, as the ratio of inductive reactance to resistance (X_L/R) decreases, the phase angle

    (a) increases.

    (b) decreases.

    (c) stays the same.

    (d) becomes alternately positive and negative.

15. In a circuit containing inductive reactance but no resistance, the phase angle is

    (a) constantly increasing.

    (b) constantly decreasing.

    (c) equal to $0°$.

    (d) equal to $90°$.

16. If the inductive reactance and the resistance in an RL circuit are equal (as expressed in ohms), then what is the phase angle?

    (a) $0°$

    (b) $45°$

    (c) $90°$

    (d) It depends on the actual values of the resistance and the inductive reactance.

17. In Fig. 13-14, the impedance shown is which of the following?

    (a) $8.0 \ \Omega$

    (b) $90 \ \Omega$

    (c) $90 + j8.0$

    (d) $8.0 + j90$

18. Note that in the diagram of Fig. 13-14, the R and X_L scale divisions are of different sizes. The phase angle can nevertheless be determined. It is

    (a) about $50°$, from the looks of it.

    (b) $48°$, as measured with a protractor.

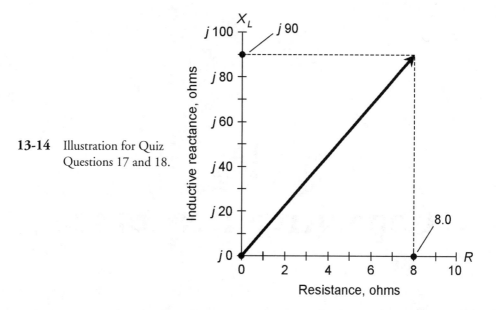

13-14 Illustration for Quiz Questions 17 and 18.

(c) 85°, as calculated using trigonometry.

(d) 6.5°, as calculated using trigonometry.

19. Consider an *RL* circuit that consists of a 100-μH inductor and a 100-Ω resistor. What is the phase angle at a frequency of 200 kHz?

(a) 45.0°

(b) 51.5°

(c) 38.5°

(d) There isn't enough data given to calculate it.

20. Suppose an *RL* circuit has an inductance of 88 mH, and the resistance is 95 Ω. At 800 Hz, what is the phase angle?

(a) 78°

(b) 12°

(c) 43°

(d) 47°

14
CHAPTER

Capacitive Reactance

CAPACITIVE REACTANCE IS THE NATURAL COUNTERPART OF INDUCTIVE REACTANCE. IT, LIKE INDUCTIVE reactance, can be represented as a ray. The capacitive-reactance ray goes in a negative direction and is assigned negative ohmic values. When the capacitive-reactance and inductive-reactance rays are joined at their endpoints (both of which correspond to a reactance of zero), a complete number line is the result, as shown in Fig. 14-1. This line depicts all possible values of reactance.

Capacitors and Direct Current

Suppose you have two big, flat metal plates, both of which are excellent electrical conductors. Imagine that you stack them one on top of the other, with only air in between. If you connect a source of dc across the plates (Fig. 14-2), the plates will become electrically charged, and will reach a potential difference equal to the dc source voltage. It won't matter how big or small the plates are; their mutual voltage will always be the same as that of the source, although, if the plates are huge, it will take awhile for them to become fully charged. Once the plates are fully charged, the current will drop to zero.

If you put some insulating material, such as glass, between the plates, their mutual voltage will not change, although the charging time will increase. If you increase the source voltage, the poten-

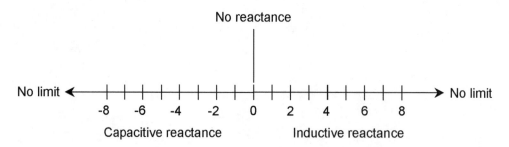

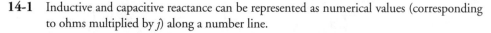

14-1 Inductive and capacitive reactance can be represented as numerical values (corresponding to ohms multiplied by *j*) along a number line.

214

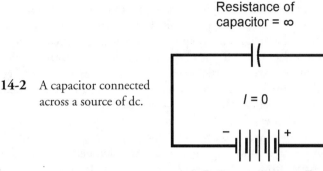

14-2 A capacitor connected across a source of dc.

Resistance of capacitor = ∞

$I = 0$

Battery voltage = E

tial difference between the plates will follow along, more or less rapidly, depending on how large the plates are and on what is between them. If the voltage is increased without limit, arcing will eventually take place. That is, sparks will begin to jump between the plates.

Capacitors and Alternating Current

Now, imagine that the voltage source connected across the plates is changed from dc to ac (Fig. 14-3). Imagine that you can adjust the frequency of this ac from a low value of a few hertz, to hundreds of hertz, to many kilohertz, megahertz, and gigahertz.

At first, the voltage between the plates will follow just about exactly along as the ac source polarity reverses. But the set of plates has a certain amount of capacitance. Perhaps they can charge up fast, if they are small and if the space between them is large, but they can't charge instantaneously. As you increase the frequency of the ac voltage source, there will come a point at which the plates do not get charged up very much before the source polarity reverses. The charge won't have time to get established with each ac cycle. At high ac frequencies, the voltage between the plates will have trouble following the current that is charging and discharging them. Just as the plates begin to get a good charge, the ac current will pass its peak and start to discharge them, pulling electrons out of the negative plate and pumping electrons into the positive plate.

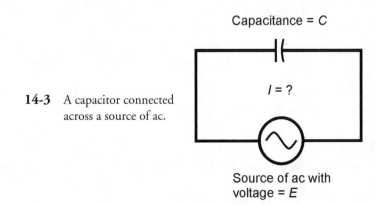

Capacitance = C

$I = ?$

14-3 A capacitor connected across a source of ac.

Source of ac with voltage = E

As the frequency is raised without limit, the set of plates starts to act more and more like a short circuit. When the frequency is low, there is a small charging current, but this quickly drops to zero as the plates become fully charged. As the frequency becomes high, the current flows for more and more of every cycle before dropping off; the charging time remains constant while the *period* of the charging/discharging wave is getting shorter. Eventually, if you keep on increasing the frequency, the period of the wave will be much shorter than the charging/discharging time, and current will flow in and out of the plates in just about the same way as it would flow if the plates were shorted out.

The opposition that the set of plates offers to ac is the *capacitive reactance*. It is measured in ohms, just like inductive reactance, and just like resistance. But it is, by convention, assigned negative values rather than positive ones. Capacitive reactance, denoted X_C, can vary, just as resistance and inductive reactance do, from near zero (when the plates are huge and close together, and/or the frequency is very high) to a few negative ohms, to many negative kilohms or megohms.

Capacitive reactance, like inductive reactance, varies with frequency. But X_C gets larger (negatively) as the frequency goes down. This is the opposite of what happens with inductive reactance, which gets larger (positively) as the frequency goes up.

Often, capacitive reactance is talked about in terms of its *absolute value*, with the minus sign removed. Then we say that the absolute value of X_C increases as the frequency goes down, or that the absolute value of X_C is decreases as the frequency goes up.

Capacitive Reactance and Frequency

In one sense, capacitive reactance behaves like a reflection of inductive reactance. But looked at another way, X_C is an extension of X_L into negative values.

If the frequency of an ac source (in hertz) is given as f, and the capacitance (in farads) is given as C, then the capacitive reactance in ohms, X_C, is calculated as follows:

$$X_C = -1/(2\pi f C)$$

Again, we meet our friend π! And again, for most practical purposes, we can take 2π to be equal to 6.28. Thus, the preceding formula can be expressed like this:

$$X_C = -1/(6.28 f C)$$

This same formula applies if the frequency, f, is in megahertz and the capacitance, C, is in microfarads.

Capacitive reactance varies *inversely* with the frequency. This means that the function X_C versus f appears as a curve when graphed, and this curve "blows up" as the frequency gets close to zero. Capacitive reactance also varies inversely with the actual value of capacitance, given a fixed frequency. Therefore, the function of X_C versus C also appears as a curve that blows up as the capacitance approaches zero.

The negative of X_C is *inversely proportional* to frequency, and also to capacitance. Relative graphs of these functions are shown in Fig. 14-4.

Problem 14-1

Suppose a capacitor has a value of 0.00100 μF at a frequency of 1.00 MHz. What is the capacitive reactance?

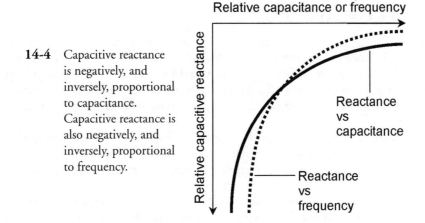

Relative capacitance or frequency

Relative capacitive reactance

Reactance
vs
capacitance

Reactance
vs
frequency

14-4 Capacitive reactance is negatively, and inversely, proportional to capacitance. Capacitive reactance is also negatively, and inversely, proportional to frequency.

Use the formula and plug in the numbers. You can do this directly, because the data is specified in microfarads (millionths) and in megahertz (millions):

$$X_C = -1/(6.28 \times 1.0 \times 0.00100) = -1/(0.00628) = -159 \ \Omega$$

This is rounded to three significant figures, because all the data is given to that many digits.

Problem 14-2

What is the capacitive reactance of the preceding capacitor if the frequency decreases to zero (that is, if the voltage source is pure dc)?

In this case, if you plug the numbers into the formula, you get a zero denominator. Mathematicians will tell you that such a quantity is undefined. But we can say that the reactance is negative infinity for all practical purposes.

Problem 14-3

Suppose a capacitor has a reactance of $-100 \ \Omega$ at a frequency of 10.0 MHz. What is its capacitance?

In this problem, you need to put the numbers in the formula and solve for the unknown C. Begin with this equation:

$$-100 = -1/(6.28 \times 10.0 \times C)$$

Dividing through by -100, you get:

$$1 = 1/(628 \times 10.0 \times C)$$

Multiply each side of this by C, and you obtain $C = 1/(628 \times 10.0)$. This can be worked out with a calculator. You should find that $C = 0.000159$ to three significant figures. Because the frequency is given in megahertz, the capacitance comes out in microfarads. That means $C = 0.000159 \ \mu F$. You can also say it is 159 pF. (Remember that 1 pF = 0.000001 μF.)

Points in the *RC* Plane

In a circuit containing resistance and capacitive reactance, the characteristics are two-dimensional in a way that is analogous to the situation with the *RL* plane from the previous chapter. The resistance ray and the capacitive-reactance ray can be placed end to end at right angles to make a quarter plane called the *RC plane* (Fig. 14-5). Resistance is plotted horizontally, with increasing values toward the right. Capacitive reactance is plotted downward, with increasingly negative values as you go down.

The combinations of R and X_C in this *RC* plane form impedances. You'll learn about impedance in greater detail in the next chapter. Each point on the *RC* plane corresponds to one and only one impedance. Conversely, each specific impedance coincides with one and only one point on the plane.

Any impedance that consists of a resistance R and a capacitive reactance X_C can be written in the form $R + jX_C$. Remember that X_C is always negative or zero. Because of this, engineers will often write $R - jX_C$ instead.

If an impedance is a pure resistance R with no reactance, then the complex impedance is $R - j0$ (or $R + j0$; it doesn't matter if j is multiplied by 0!). If $R = 3\ \Omega$ with no reactance, you get an impedance of $3 - j0$, which corresponds to the point $(3,j0)$ on the *RC* plane. If you have a pure capacitive reactance, say $X_C = -4\ \Omega$, then the complex impedance is $0 - j4$, and this is at the point $(0,-j4)$ on the *RC* plane. Again, it's important, for completeness, to write the "0" and not just the "$-j4$." The points for $3 - j0$ and $0 - j4$, and two others, are plotted on the *RC* plane in Fig. 14-6.

In practical circuits, all capacitors have some *leakage resistance*. If the frequency goes to zero (pure dc), a tiny current always flows, because no capacitor has a perfect insulator between its plates. In addition to this, all resistors have a little capacitive reactance because they occupy a finite physical space. So there is no such thing as a mathematically perfect resistor, either. The points $3 - j0$ and

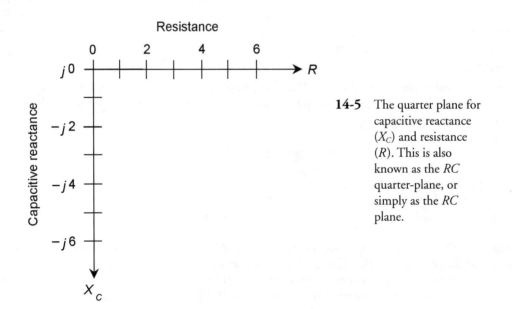

14-5 The quarter plane for capacitive reactance (X_C) and resistance (R). This is also known as the *RC* quarter-plane, or simply as the *RC* plane.

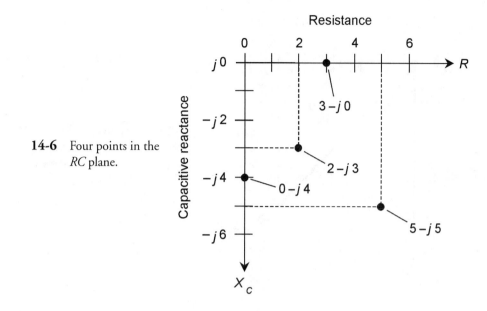

14-6 Four points in the *RC* plane.

$0 - j4$ represent an *ideal resistor* and an *ideal capacitor,* respectively—components that can be worked with in theory, but that you will never see in the real world.

Sometimes, resistance and capacitive reactance are both placed in a circuit deliberately. Then you get impedances such as $2 - j3$ and $5 - j5$, both shown in Fig. 14-6.

Remember that the values for X_C are reactances, not the actual capacitances. If you raise or lower the frequency, the value of X_C will change. A higher frequency causes X_C to get smaller negatively (closer to zero). A lower frequency causes X_C to get larger negatively (farther from zero, or lower down on the *RC* plane). If the frequency goes to zero, then the capacitive reactance drops off the bottom of the *RC* plane to negative infinity!

Vectors in the *RC* Plane

Recall from the last chapter that *RL* impedances can be represented as vectors. The same is true for *RC* impedances.

In Fig. 14-6, four different complex impedance points are shown. Each point is represented by a certain distance to the right of the origin $(0, j0)$, and a certain displacement downward. The first of these is the resistance, R, and the second is the capacitive reactance, X_C. The complex *RC* impedance is a two-dimensional quantity.

Impedance points in the *RC* plane can be rendered as vectors, just as they can in the *RL* plane. Then the points become rays, each with a certain length and direction. The magnitude and direction for a vector, and the coordinates for the point, both uniquely define the same complex impedance. The length of the vector is the distance of the point from the origin, and the direction is the angle measured *clockwise* from the resistance (R) line, and specified in *negative degrees.* The equivalent vectors, for the points in Fig. 14-6, are shown in Fig. 14-7.

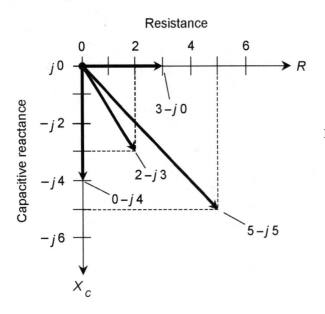

14-7   Four vectors in the *RC* plane, corresponding to the points shown in Fig. 14-6.

Current Leads Voltage

When ac is driven through a capacitor and starts to increase (in either direction), it takes a fraction of a cycle for the voltage between the plates to follow. Once the current starts decreasing from its maximum peak (in either direction) in the cycle, it again takes a fraction of a cycle for the voltage to follow. The instantaneous voltage can't quite keep up with the instantaneous current, as it does in a pure resistance. Thus, in a circuit containing capacitive reactance, the voltage lags the current in phase. Another, and more often used, way of saying this is that the current *leads* the voltage.

Pure Capacitance

Suppose an ac voltage source is connected across a capacitor. Imagine that the frequency is low enough, and/or the capacitance is small enough, so the absolute value of the capacitive reactance, X_C, is extremely large compared with the resistance, *R*. Then the current leads the voltage by just about 90° (Fig. 14-8).

The situation depicted in Fig. 14-8 represents a pure capacitive reactance. The vector in the *RC* plane in this situation points straight down. Its angle is −90° from the *R* axis.

Capacitance and Resistance

When the resistance in a resistance-capacitance circuit is significant compared with the absolute value of the capacitive reactance, the current leads the voltage by something less than 90° (Fig. 14-9). If *R* is small compared with the absolute value of X_C, the difference is almost a quarter of a cycle. As *R* gets larger, or as the absolute value of X_C becomes smaller, the phase difference decreases. A circuit containing resistance and capacitance is called an *RC circuit.*

The value of *R* in an *RC* circuit might increase relative to the absolute value of X_C because resistance is deliberately put into a circuit. It can also happen if the frequency becomes so high that the absolute value of the capacitive reactance drops to a value comparable with the loss resistance in

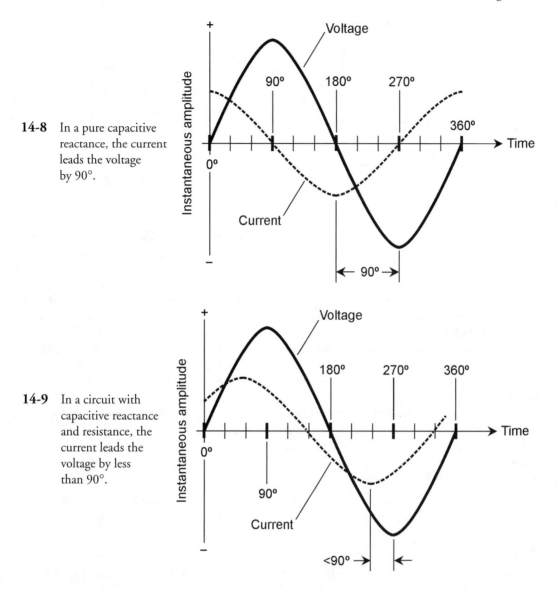

14-8 In a pure capacitive reactance, the current leads the voltage by 90°.

14-9 In a circuit with capacitive reactance and resistance, the current leads the voltage by less than 90°.

the circuit conductors. In either case, the situation can be represented by a resistance, R, in series with a capacitive reactance, X_C (Fig. 14-10).

If you know the values of X_C and R, you can find the *angle of lead*, also called the *RC phase angle*, by plotting the point $R - jX_C$ on the RC plane, drawing the vector from the origin $0 - j0$ out to that point, and then measuring the angle of the vector clockwise from the R axis. You can use a protractor to measure this angle, as you did in the previous chapter for RL phase angles. Or you can use trigonometry to calculate the angle.

As with RL circuits, you need only know the ratio of X_C to R to determine the phase angle. For example, if $X_C = -4\ \Omega$ and $R = 7\ \Omega$, you'll get the same angle as with $X_C = -400\ \Omega$ and $R = 700\ \Omega$, or with $X_C = -16\ \Omega$ and $R = 28\ \Omega$. The phase angle will be the same whenever the ratio of X_C to R is equal to $-4:7$.

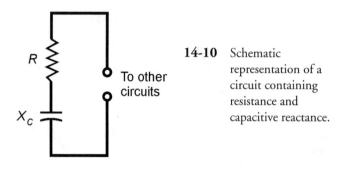

14-10 Schematic representation of a circuit containing resistance and capacitive reactance.

Pure Resistance

As the resistance in an *RC* circuit gets large compared with the absolute value of the capacitive reactance, the angle of lead becomes smaller. The same thing happens if the absolute value of X_C gets small compared with the value of *R*.

When *R* is many times larger than the absolute value of X_C, whatever their actual values, the vector in the *RC* plane points almost along the *R* axis. Then the *RC* phase angle is close to 0°. The voltage comes nearly into phase with the current. The plates of the capacitor do not come anywhere near getting fully charged with each cycle. The capacitor is said to "pass the ac" with very little loss, as if it were shorted out. But it will still have an extremely high X_C for any ac signals at much lower frequencies that might exist across it at the same time. (This property of capacitors can be put to use in electronic circuits. An example is when an engineer wants to let radio-frequency signals get through while blocking signals at audio frequencies.)

Ultimately, if the absolute value of the capacitive reactance gets small enough, the circuit acts as a pure resistance, and the current is in phase with the voltage.

How Much Lead?

If you know the ratio of capacitive reactance to resistance, or X_C/R, in an *RC* circuit, then you can find the phase angle. Of course, you can find this angle if you know the precise values, too.

Pictorial Method

You can use a protractor and a ruler to find phase angles for *RC* circuits, just as you did with *RL* circuits in the previous chapter, as long as the angles aren't too close to 0° or 90°. First, draw a line somewhat longer than 10 cm, going from left to right on the paper. Then, use the protractor to construct a line going somewhat more than 10 cm vertically downward, starting at the left end of the horizontal line. The horizontal line is the *R* axis of an *RC* plane. The line going down is the X_C axis.

If you know the actual values of X_C and *R*, divide or multiply them by a constant, chosen to make both values fall between −100 and 100. For example, if $X_C = -3800\ \Omega$ and $R = 7400\ \Omega$, divide them both by 100, getting −38 and 74. Plot these points on the lines. The X_C point goes 38 mm down from the intersection point between your two axes. The *R* point goes 74 mm to the right of the intersection point. Next, draw a line connecting the two points, as shown in Fig. 14-11. This line will be at a slant and will form a triangle along with the two axes. This is a right triangle, with the right angle at the origin of the *RC* plane. Measure the angle between the slanted line and the *R* axis. Use the protractor for this. Extend the lines, if necessary, using the ruler, to get a good reading

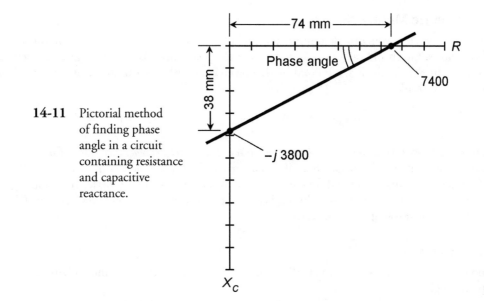

14-11 Pictorial method of finding phase angle in a circuit containing resistance and capacitive reactance.

on the protractor. This angle will be between 0 and 90°. Multiply this reading by −1 to get the *RC* phase angle. That is, if the protractor shows 27°, the *RC* phase angle is −27°.

The actual vector is found by constructing a rectangle using the origin and your two points, making new perpendicular lines to complete the figure. The vector is the diagonal of this rectangle, running out from the origin (Fig. 14-12). The phase angle is the angle between the *R* axis and this vector, multiplied by −1. It will have the same measure as the angle of the slanted line you constructed in Fig. 14-11.

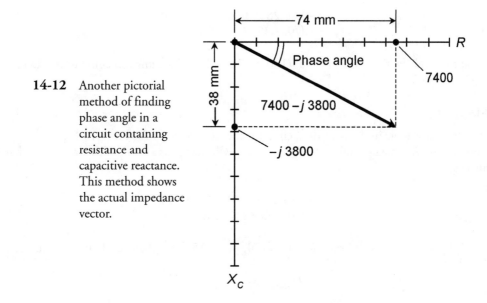

14-12 Another pictorial method of finding phase angle in a circuit containing resistance and capacitive reactance. This method shows the actual impedance vector.

Trigonometric Method

Using trigonometry, you can determine the *RC* phase angle more precisely than the pictorial method allows. Given the values of X_C and *R*, the *RC* phase angle is the arctangent of their ratio. Phase angle in *RC* circuits is symbolized by the lowercase Greek letter ϕ, just as it is in *RL* circuits. Here are the formulas:

$$\phi = \tan^{-1} (X_C/R) \qquad \text{or} \qquad \phi = \arctan (X_C/R)$$

When doing problems of this kind, remember to use the *capacitive reactance* values for X_C, and not the capacitance values. This means that, if you are given the capacitance, you must use the formula for X_C in terms of capacitance and frequency and then calculate the phase angle. You should get angles that come out negative or zero. This indicates that they're *RC* phase angles rather than *RL* phase angles (which are always positive or zero).

Problem 14-4

Suppose the capacitive reactance in an *RC* circuit is −3800 Ω and the resistance is 7400 Ω. What is the phase angle?

Find the ratio X_C/R = −3800/7400. The calculator display should show you something like −0.513513513. Find the arctangent, or $\tan^{-1}$, getting a phase angle of −27.18111109° on the calculator display. Round this off to −27.18°.

Problem 14-5

Suppose an *RC* circuit works at a frequency of 3.50 MHz. It has a resistance of 130 Ω and a capacitance of 150 pF. What is the phase angle?

First, find the capacitive reactance for a capacitor of 150 pF at 3.50 MHz. Convert the capacitance to microfarads, getting *C* = 0.000150 μF. Remember that *micro*farads go with *mega*hertz (millionths go with millions to cancel each other out). Then:

$$X_C = -1/(6.28 \times 3.50 \times 0.000150)$$
$$= -1/0.003297 = -303 \text{ Ω}$$

Now you can find the ratio X_C/R = −303/130 = −2.33. The phase angle is equal to the arctangent of −2.33, or −66.8°.

Problem 14-6

What is the phase angle in the preceding circuit if the frequency is raised to 7.10 MHz?

You need to find the new value for X_C, because it will change as a result of the frequency change. Calculating:

$$X_C = -1/(6.28 \times 7.10 \times 0.000150)$$
$$= -1/0.006688 = -150 \text{ Ω}$$

The ratio X_C/R in this case is equal to −150/130, or −1.15. The phase angle is the arctangent of −1.15, which turns out to be −49.0°.

Quiz

Refer to the text in this chapter if necessary. A good score is at least 18 correct. Answers are in the back of the book.

1. As the size of the plates in a capacitor increases, all other things being equal,
   - (a) the value of X_C increases negatively.
   - (b) the value of X_C decreases negatively.
   - (c) the value of X_C does not change.
   - (d) we cannot say what happens to X_C without more data.

2. If the dielectric material between the plates of a capacitor is changed, all other things being equal,
   - (a) the value of X_C increases negatively.
   - (b) the value of X_C decreases negatively.
   - (c) the value of X_C does not change.
   - (d) we cannot say what happens to X_C without more data.

3. As the frequency of a wave gets lower, all other things being equal, the value of X_C for a capacitor
   - (a) increases negatively.
   - (b) decreases negatively.
   - (c) does not change.
   - (d) depends on the current.

4. What is the reactance of a 330-pF capacitor at 800 kHz?
   - (a) $-1.66\ \Omega$
   - (b) $-0.00166\ \Omega$
   - (c) $-603\ \Omega$
   - (d) $-603\ \text{k}\Omega$

5. Suppose a capacitor has a reactance of $-4.50\ \Omega$ at 377 Hz. What is its capacitance?
   - (a) $9.39\ \mu\text{F}$
   - (b) $93.9\ \mu\text{F}$
   - (c) $7.42\ \mu\text{F}$
   - (d) $74.2\ \mu\text{F}$

6. Suppose a 47-μF capacitor has a reactance of $-47\ \Omega$. What is the frequency?
   - (a) $72\ \text{Hz}$
   - (b) $7.2\ \text{MHz}$
   - (c) $0.000072\ \text{Hz}$
   - (d) $7.2\ \text{Hz}$

7. Suppose a capacitor has $X_C = -8800 \; \Omega$ at $f = 830$ kHz. What is C?
   (a) 2.18 μF
   (b) 21.8 pF
   (c) 0.00218 μF
   (d) 2.18 pF

8. Suppose a capacitor has $C = 166$ pF at $f = 400$ kHz. What is X_C?
   (a) -2.4 kΩ
   (b) $-2.4 \; \Omega$
   (c) $-2.4 \times 10^{-6} \; \Omega$
   (d) -2.4 MΩ

9. Suppose a capacitor has $C = 4700$ μF and $X_C = -33 \; \Omega$. What is f?
   (a) 1.0 Hz
   (b) 10 Hz
   (c) 1.0 kHz
   (d) 10 kHz

10. Each point in the RC plane
   (a) corresponds to a unique inductance.
   (b) corresponds to a unique capacitance.
   (c) corresponds to a unique combination of resistance and capacitance.
   (d) corresponds to a unique combination of resistance and reactance.

11. If R increases in an RC circuit, but X_C is always zero, the vector in the RC plane will
   (a) rotate clockwise.
   (b) rotate counterclockwise.
   (c) always point straight toward the right.
   (d) always point straight down.

12. If the resistance R increases in an RC circuit, but the capacitance and the frequency are nonzero and constant, then the vector in the RC plane will
   (a) get longer and rotate clockwise.
   (b) get longer and rotate counterclockwise.
   (c) get shorter and rotate clockwise.
   (d) get shorter and rotate counterclockwise.

13. Each complex impedance value $R - jX_C$
   (a) represents a unique combination of resistance and capacitance.
   (b) represents a unique combination of resistance and reactance.
   (c) represents a unique combination of resistance and frequency.
   (d) All of the above are true.

14. In an *RC* circuit, as the ratio X_C/R approaches zero, the phase angle
    (a) approaches −90°.
    (b) approaches 0°.
    (c) stays the same.
    (d) cannot be found.

15. In a purely resistive circuit, the phase angle is
    (a) increasing.
    (b) decreasing.
    (c) 0°.
    (d) −90°.

16. If $X_C/R = -1$, then what is the phase angle?
    (a) 0°
    (b) −45°
    (c) −90°
    (d) Impossible to find because there's not enough data given

17. In Fig. 14-13, the impedance shown is
    (a) $8.02 + j323$.
    (b) $323 + j8.02$.
    (c) $8.02 - j323$.
    (d) $323 - j8.02$.

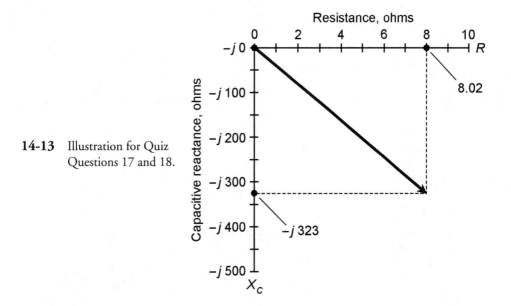

14-13 Illustration for Quiz Questions 17 and 18.

18. In Fig. 14-13, note that the R and X_C scale divisions are not the same size. What is the actual phase angle?

    (a) −1.42°

    (b) About −60°, from the looks of it

    (c) −58.9°

    (d) −88.6°

19. Suppose an *RC* circuit consists of a 150-pF capacitor and a 330-Ω resistor in series. What is the phase angle at a frequency of 1.34 MHz?

    (a) −67.4°

    (b) −22.6°

    (c) −24.4°

    (d) −65.6°

20. Suppose an *RC* circuit has a capacitance of 0.015 μF. The resistance is 52 Ω. What is the phase angle at 90 kHz?

    (a) −24°

    (b) −0.017°

    (c) −66°

    (d) None of the above

15
CHAPTER

Impedance and Admittance

IN THIS CHAPTER, A COMPLETE, WORKING DEFINITION OF COMPLEX IMPEDANCE IS DEVELOPED. YOU'LL also get acquainted with *admittance*, the extent to which an ac circuit allows (or admits) current flow, rather than impeding it. As we develop these concepts, let's review, and then expand on, some of the material presented in the previous couple of chapters.

Imaginary Numbers

Have you been wondering what *j* actually means in expressions of impedance? Well, *j* is nothing but a number: the positive square root of −1. There's a negative square root of −1, too, and it is equal to −*j*. When either *j* or −*j* is multiplied by itself, the result is −1. (Pure mathematicians often denote these same numbers as *i* or −*i*.)

The positive square root of −1 is known as the *unit imaginary number*. The *set of imaginary numbers* is composed of real-number multiples of *j* or −*j*. Some examples are *j*4, *j*35.79, −*j*25.76, and −*j*25,000.

The square of an imaginary number is always negative. Some people have trouble grasping this, but when you think long and hard about it, all numbers are abstractions. Imaginary numbers are no more imaginary (and no less real) than so-called real numbers such as 4, 35.79, −25.76, or −25,000.

The unit imaginary number *j* can be multiplied by any real number on a conventional *real number line*. If you do this for all the real numbers on the real number line, you get an *imaginary number line* (Fig. 15-1). The imaginary number line should be oriented at a right angle to the real number line when you want to graphically portray real and imaginary numbers at the same time.

In electronics, real numbers represent resistances. Imaginary numbers represent reactances.

Complex Numbers

When you add a real number and an imaginary number, you get a *complex number*. In this context, the term *complex* does not mean "complicated." A better word would be *composite*. Examples are

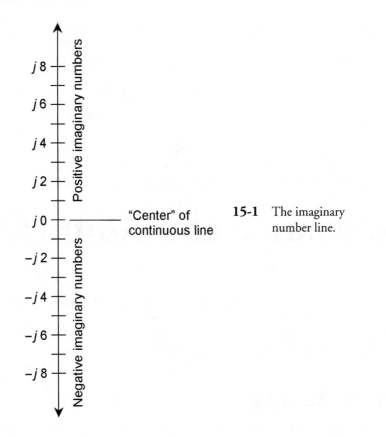

"Center" of
continuous line

15-1   The imaginary
number line.

$4 + j5$, $8 - j7$, $-7 + j13$, and $-6 - j87$. The set of complex numbers needs two dimensions—a *plane*—to be graphically defined.

Adding and Subtracting Complex Numbers

Adding complex numbers is just a matter of adding the real parts and the complex parts separately. For example, the sum of $4 + j7$ and $45 - j83$ works out like this:

$$(4 + 45) + j(7 - 83)$$
$$= 49 + j(-76)$$
$$= 49 - j76$$

Subtracting complex numbers is a little more involved; it's best to convert a difference to a sum. For example, the difference $(4 + j7) - (45 - j83)$ can be found by multiplying the second complex number by -1 and then adding the result:

$$(4 + j7) - (45 - j83)$$
$$= (4 + j7) + [-1(45 - j83)]$$
$$= (4 + j7) + (-45 + j83)$$
$$= -41 + j90$$

Multiplying Complex Numbers

When you multiply these numbers, you should treat them as sums of number pairs, that is, as *binomials*. It's easier to give the general formula than to work with specifics here. If *a*, *b*, *c*, and *d* are real numbers (positive, negative, or zero), then:

$$(a + jb)\,(c + jd)$$
$$= ac + jad + jbc + j^2bd$$
$$= (ac - bd) + j(ad + bc)$$

Fortunately, you won't encounter complex number multiplication problems very often in electronics. Nevertheless, a working knowledge of how complex numbers multiply can help you get a solid grasp of them.

The Complex Number Plane

A complete *complex number plane* is made by taking the real and imaginary number lines and placing them together, at right angles, so that they intersect at the zero points, 0 and *j*0. This is shown in Fig. 15-2. The result is a *Cartesian coordinate plane*, just like the ones people use to make graphs of everyday things such as stock price versus time.

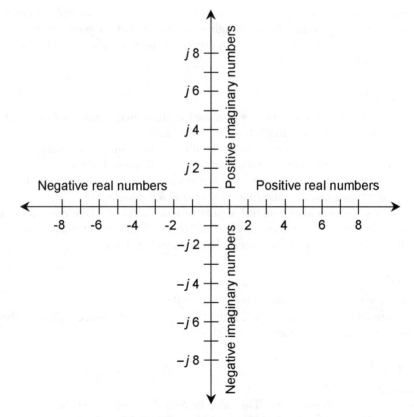

15-2 The complex number plane.

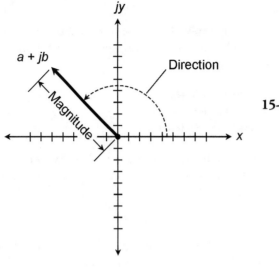

15-3   Magnitude and direction of a vector in the complex number plane.

Complex Number Vectors

Complex numbers can also be represented as vectors. This gives each complex number a unique *magnitude* and a unique *direction*. The magnitude is the distance of the point $a + jb$ from the origin $0 + j0$. The direction is the angle of the vector, expressed counterclockwise from the positive real-number axis. This is shown in Fig. 15-3.

Absolute Value

The *absolute value* of a complex number $a + jb$ is the length, or magnitude, of its vector in the complex plane, measured from the origin $(0,0)$ to the point (a,b).

In the case of a *pure real number* $a + j0$, the absolute value is simply the real number itself, a, if a is positive. If a is negative, then the absolute value of $a + j0$ is equal to $-a$.

In the case of a *pure imaginary number* $0 + jb$, the absolute value is equal to b, if b (a real number) is positive. If b is negative, the absolute value of $0 + jb$ is equal to $-b$.

If the number $a + jb$ is neither pure real or pure imaginary, the absolute value must be found by using a formula. First, square both a and b. Then add them. Finally, take the square root. This is the length, c, of the vector $a + jb$. The situation is illustrated in Fig. 15-4.

Problem 15-1

Find the absolute value of the complex number $-22 - j0$.

This is a pure real number. Actually, it is the same as $-22 + j0$, because $j0 = 0$. Therefore, the absolute value of this complex number is $-(-22) = 22$.

Problem 15-2

Find the absolute value of $0 - j34$.

This is a pure imaginary number. The value of b in this case is -34, because $0 - j34 = 0 + j(-34)$. Therefore, the absolute value is $-(-34) = 34$.

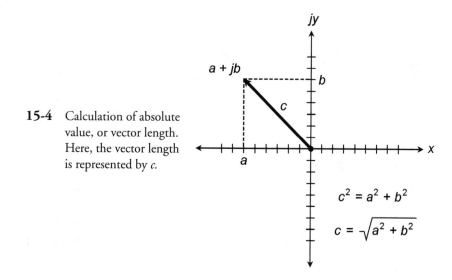

15-4 Calculation of absolute value, or vector length. Here, the vector length is represented by *c*.

$$c^2 = a^2 + b^2$$

$$c = \sqrt{a^2 + b^2}$$

Problem 15-3

Find the absolute value of $3 - j4$.

In this number, $a = 3$ and $b = -4$. Squaring both of these, and adding the results, gives us $3^2 + (-4)^2 = 9 + 16 = 25$. The square root of 25 is 5. Therefore, the absolute value of this complex number is 5.

The *RX* Plane

Recall the planes for resistance (R) and inductive reactance (X_L) from Chap. 13. This is the same as the upper-right quadrant of the complex number plane shown in Fig. 15-2. Similarly, the plane for resistance and capacitive reactance (X_C) is the same as the lower-right quadrant of the complex number plane. Resistances are represented by nonnegative real numbers. Reactances, whether they are inductive (positive) or capacitive (negative), correspond to imaginary numbers.

No Negative Resistance

There is no such thing, strictly speaking, as negative resistance. You cannot have anything better than a perfect conductor. In some cases, a supply of direct current, such as a battery, can be treated as a negative resistance; in other cases, you can have a device that acts as if its resistance were negative under certain changing (or dynamic) conditions. But for most practical applications in the *RX plane*, the resistance value is always positive. You can remove the negative axis, along with the upper-left and lower-left quadrants, of the complex number plane, obtaining a half plane, as shown in Fig. 15-5, and still get a complete set of coordinates for depicting complex impedances.

"Negative Inductors" and "Negative Capacitors"

Capacitive reactance, X_C, is effectively an extension of inductive reactance, X_L, into the realm of negatives. Capacitors act like "negative inductors." It's equally true to say that inductors act like "negative capacitors," because the negative of a negative number is a positive number. Reactance can vary from extremely large negative values, through zero, to extremely large positive values.

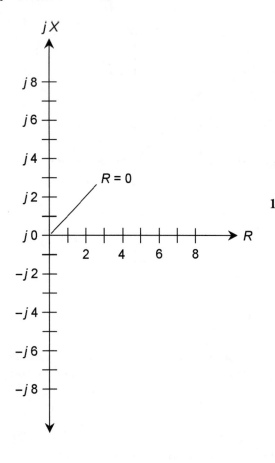

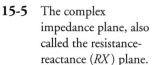

15-5 The complex impedance plane, also called the resistance-reactance (*RX*) plane.

Vector Representation of Impedance

Any impedance $R + jX$ can be represented by a complex number of the form $a + jb$. Just let $R = a$ and $X = b$. Now try to envision how the impedance vector changes as either R or X, or both, are varied. If X remains constant, an increase in R causes the vector to get longer. If R remains constant and X_L gets larger, the vector grows longer. If R stays the same but X_C gets larger negatively, the vector grows longer.

Think of the point $R + jX$ moving around in the *RX* plane, and imagine where the corresponding points on the axes lie. These points can be found by drawing dashed lines from the point $R + jX$ to the R and X axes, so that the dashed lines intersect the axes at right angles. Some examples are shown in Fig. 15-6.

Now think of the points for R and X moving toward the right and left, or up and down, on their axes. Imagine what happens to the point $R + jX$ in various scenarios. This is how impedance changes as the resistance and reactance in a circuit are varied.

Resistance is one-dimensional. Reactance is also one-dimensional. But impedance is two-dimensional. To fully define impedance, you must render it on a two-dimensional coordinate system such as the *RX* plane. The resistance and the reactance can change independently of one another.

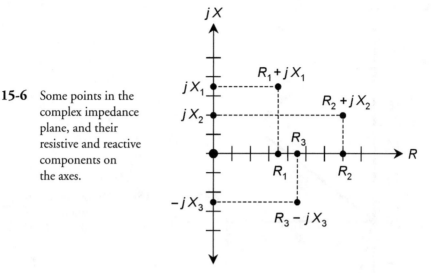

15-6 Some points in the complex impedance plane, and their resistive and reactive components on the axes.

Absolute-Value Impedance

You'll occasionally read or hear that the "impedance" of some device or component is a certain number of ohms. For example, in audio electronics, there are "8-Ω" speakers and "600-Ω" amplifier inputs. How, you ask, can manufacturers quote a single number for a quantity that is two-dimensional and needs two numbers to be completely expressed?

That's a good question, and there are two answers. First, figures like this refer to devices that have *purely resistive impedances*, also known as *nonreactive impedances*. Thus, the 8-Ω speaker really has a complex impedance of $8 + j0$, and the 600-Ω input circuit is designed to operate with a complex impedance at, or near, $600 + j0$. Second, you can talk about the length of the impedance vector (that is, the absolute value of the complex impedance), calling this a certain number of ohms. If you talk about impedance this way, however, you are being ambiguous. There can exist an infinite number of different vectors of any given length in the *RX* plane.

Sometimes, the uppercase italic letter Z is used in place of the word *impedance* in general discussions. This is what engineers mean when they say things like "$Z = 50\ \Omega$" or "$Z = 300\ \Omega$ nonreactive." In this context, if no specific impedance is given, "$Z = 8\ \Omega$" can theoretically refer to $8 + j0$, $0 + j8$, $0 - j8$, or any other complex impedance point on a half circle consisting of all points 8 units from $0 + j0$. This is shown in Fig. 15-7.

Problem 15-4

Name seven different complex impedances that can theoretically be meant by the expression "$Z = 10\ \Omega$."

It's easy name three: $0 + j10$, $10 + j0$, and $0 - j10$. These represent pure inductance, pure resistance, and pure capacitance, respectively.

A right triangle can exist having sides in a ratio of 6:8:10 units. This is true because $6^2 + 8^2 = 10^2$. (Check it and see!) Therefore, you can have $6 + j8$, $6 - j8$, $8 + j6$, and $8 - j6$, all complex impedances whose absolute value is 10.

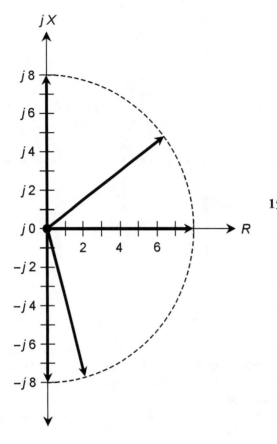

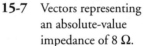

15-7 Vectors representing an absolute-value impedance of 8 Ω.

Characteristic Impedance

There is a rather exotic property of certain electronic components that you'll sometimes hear or read about. It is called *characteristic impedance* or *surge impedance*, and is symbolized Z_o. It is a specification of an important property of *transmission lines*. It can always be expressed as a positive real number, in ohms.

Transmission Lines

When it is necessary to get energy or signals from one place to another, a transmission line is required. These almost always take either of two forms, *coaxial* or *two-wire* (also called *parallel-wire*). Cross-sectional renditions of both types are shown in Fig. 15-8. Examples of transmission lines include the "ribbon" that goes from a television antenna to the receiver, the cable running from a hi-fi amplifier to the speakers, and the set of wires that carries electricity over the countryside.

Factors Affecting Z_o

The Z_o of a parallel-wire transmission line depends on the diameter of the wires, on the spacing between the wires, and on the nature of the insulating material separating the wires. In general, the Z_o

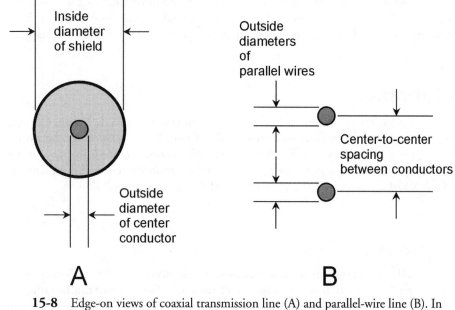

15-8 Edge-on views of coaxial transmission line (A) and parallel-wire line (B). In either type of line, Z_o depends on the conductor diameters and spacing, and on the nature of the dielectric material between the conductors. See text for discussion.

increases as the wire diameter gets smaller, and decreases as the wire diameter gets larger, all other things being equal.

In a coaxial line, as the center conductor gets thicker, the Z_o decreases if the shield stays the same size. If the center conductor stays the same size and the *shield* increases in diameter, the Z_o increases.

For either type of line, the Z_o increases as the spacing between wires, or between the center conductor and the shield, gets larger. The Z_o decreases as the spacing is reduced. Solid dielectric materials such as polyethylene reduce the Z_o of a transmission line, compared with air or a vacuum, when placed between the conductors.

An Example of Z_o in Practice

In rigorous terms, the ideal characteristic impedance for a transmission line is determined according to the nature of the *load* with which the line works.

For a system having a purely resistive impedance of a certain number of ohms, the best line Z_o value is that same number of ohms. If the load impedance is much different from the characteristic impedance of the transmission line, excessive power is wasted in heating up the transmission line.

Imagine that you have a so-called 300-Ω frequency-modulation (FM) receiving antenna, such as the folded-dipole type that you can mount indoors. Suppose that you want the best possible reception. Of course, you should choose a good location for the antenna. You should make sure that the transmission line between your radio and the antenna is as short as possible. But you should also be sure that you purchase 300-Ω TV ribbon. It has a value of Z_o that has been optimized for use with antennas whose impedances are close to $300 + j0$.

   Impedance matching is the process of making sure that the impedance of a load (such as an antenna) is purely resistive, with an ohmic value equal to the characteristic impedance of the transmission line connected to it. This concept will be discussed in more detail in the next chapter.

Conductance

In an ac circuit, electrical *conductance* works the same way as it does in a dc circuit. Conductance is symbolized by the capital letter G. It was introduced in Chap. 2. The relationship between conductance and resistance is simple: $G = 1/R$. The standard unit of conductance is the *siemens*. The larger the value of conductance, the smaller the resistance, and the more current will flow. Conversely, the smaller the value of G, the greater the value of R, and the less current will flow.

Susceptance

Sometimes, you'll come across the term *susceptance* in reference to ac circuits. Susceptance is symbolized by the capital letter B. It is the reciprocal of reactance. Susceptance can be either capacitive or inductive. These quantities are symbolized as B_C and B_L, respectively. Therefore we have these two relations:

$$B_C = 1/X_C$$
$$B_L = 1/X_L$$

   All values of B theoretically contain the j operator, just as do all values of X. But when it comes to finding reciprocals of quantities containing j, things get tricky. The reciprocal of j is equal to its negative! Expressed mathematically, we have these two facts:

$$1/j = -j$$
$$1/(-j) = j$$

As a result of these properties of j, the sign reverses whenever you find a susceptance value in terms of a reactance value. When expressed in terms of j, inductive susceptance is negative imaginary, and capacitive susceptance is positive imaginary—just the opposite situation from inductive reactance and capacitive reactance.

   Suppose you have an inductive reactance of 2 Ω. This is expressed in imaginary terms as $j2$. To find the inductive susceptance, you must find $1/(j2)$. Mathematically, this expression can be converted to a real-number multiple of j in the following manner:

$$1/(j2) = (1/j)(\tfrac{1}{2})$$
$$= (1/j)0.5$$
$$= -j0.5$$

   Now suppose you have a capacitive reactance of 10 Ω. This is expressed in imaginary terms as $-j10$. To find the capacitive susceptance, you must find $1/(-j10)$. Here's how this can be converted to the straightforward product of j and a real number:

$$1/(-j10) = (1/-j)(^{1}/_{10})$$
$$= (1/-j)0.1$$
$$= j0.1$$

When you want to find an imaginary value of susceptance in terms of an imaginary value of reactance, first take the reciprocal of the real-number part of the expression, and then multiply the result by −1.

Problem 15-5

Suppose you have a capacitor of 100 pF at a frequency of 3.00 MHz. What is B_C?

First, find X_C by the formula for capacitive reactance:

$$X_C = -1/(6.28fC)$$

Note that 100 pF = 0.000100 μF. Therefore:

$$X_C = -1/(6.28 \times 3.00 \times 0.000100)$$
$$= -1/0.001884 = -531 \ \Omega$$

The imaginary value of X_C is equal to $-j531$. The susceptance, B_C, is equal to $1/X_C$. Thus, $B_C = 1/(-j531) = j0.00188$, rounded to three significant figures.

The general formula for capacitive susceptance in siemens, in terms of frequency in hertz and capacitance in farads, is:

$$B_C = 6.28fC$$

This formula also works for frequencies in megahertz and capacitances in microfarads.

Problem 15-6

Suppose an inductor has $L = 163$ μH at a frequency of 887 kHz. What is B_L?

Note that 887 kHz = 0.887 MHz. You can calculate X_L from the formula for inductive reactance:

$$X_L = 6.28fL$$
$$= 6.28 \times 0.887 \times 163$$
$$= 908 \ \Omega$$

The imaginary value of X_L is equal to $j908$. The susceptance, $B_L =$ is equal to $1/X_L$. It follows that $B_L = -1/j908 = -j0.00110$.

The general formula for inductive susceptance in siemens, in terms of frequency in hertz and inductance in henrys, is:

$$B_L = -1/(6.28fL)$$

This formula also works for frequencies in kilohertz and inductances in millihenrys, and for frequencies in megahertz and inductances in microhenrys.

Admittance

Real-number conductance and imaginary-number susceptance combine to form *complex admittance*, symbolized by the capital letter *Y*. This is a complete expression of the extent to which a circuit allows ac to flow.

As the absolute value of complex impedance gets larger, the absolute value of complex admittance becomes smaller, in general. Huge impedances correspond to tiny admittances, and vice versa.

Admittances are written in complex form just like impedances. But you need to keep track of which quantity you're talking about! This will be obvious if you use the symbol, such as $Y = 3 - j0.5$ or $Y = 7 + j3$. When you see *Y* instead of *Z*, you know that negative *j* factors (such as in the quantity $3 - j0.5$) mean there is a net inductance in the circuit, and positive *j* factors (such as in the quantity $7 + j3$) mean there is net capacitance.

Admittance is the complex composite of conductance and susceptance. Thus, complex admittance values always take the form $Y = G + jB$. When the *j* factor is negative, a complex admittance may appear in the form $Y = G - jB$.

Do you remember how resistances combine with reactances in series to form complex impedances? In Chaps. 13 and 14, you saw series *RL* and *RC* circuits. Did you wonder why parallel circuits were ignored in those discussions? The reason was the fact that admittance, not impedance, is best for working with parallel ac circuits. Resistance and reactance combine in a messy fashion in parallel circuits. But conductance (*G*) and susceptance (*B*) merely add together in parallel circuits, yielding admittance (*Y*). Parallel circuit analysis is covered in detail in the next chapter.

The *GB* Plane

Admittance can be depicted on a plane similar to the complex impedance (*RX*) plane. Actually, it's a half plane, because there is ordinarily no such thing as negative conductance. (You can't have a component that conducts worse than not at all.) Conductance is plotted along the horizontal, or *G*, axis on this coordinate half plane, and susceptance is plotted along the *B* axis. The *GB plane* is shown in Fig. 15-9, with several points plotted.

It's Inside Out

The *GB* plane looks superficially identical to the *RX* plane. But mathematically, the two could not be more different! The *GB* plane is mathematically inside out with respect to the *RX* plane. The center, or origin, of the *GB* plane represents the point at which there is no conduction for dc or for ac. It is the zero-admittance point, rather than the zero-impedance point. In the *RX* plane, the origin represents a perfect short circuit, but in the *GB* plane, the origin corresponds to a perfect open circuit.

As you move out toward the right (east) along the *G*, or conductance, axis of the *GB* plane, the conductance improves, and the current gets greater. When you move upward (north) along the *jB* axis from the origin, you have ever-increasing positive (capacitive) susceptance. When you go down (south) along the *jB* axis from the origin, you encounter increasingly negative (inductive) susceptance.

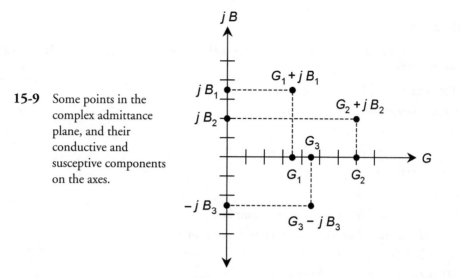

15-9 Some points in the complex admittance plane, and their conductive and susceptive components on the axes.

Vector Representation of Admittance

Complex admittances can be shown as vectors, just as can complex impedances. In Fig. 15-10, the points from Fig. 15-9 are rendered as vectors.

Generally, long vectors in the *GB* plane indicate large currents, and short vectors indicate small currents. Imagine a point moving around on the *GB* plane, and think of the vector getting longer and shorter and changing direction. Vectors pointing generally northeast, or upward and to the right, correspond to conductances and capacitances in parallel. Vectors pointing in a more or less southeasterly direction, or downward and to the right, are conductances and inductances in parallel.

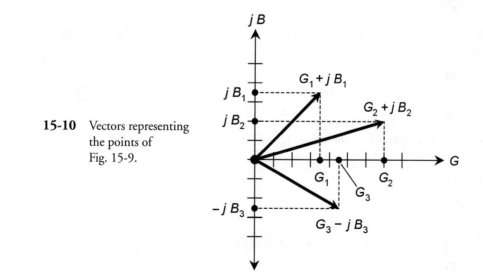

15-10 Vectors representing the points of Fig. 15-9.

Quiz

Refer to the text in this chapter if necessary. A good score is 18 or more correct. Answers are in the back of the book.

1. The square of an imaginary number
   (a) can never be negative.
   (b) can never be positive.
   (c) can be either positive or negative.
   (d) is equal to j.

2. A complex number
   (a) is the same thing as an imaginary number.
   (b) has a real-number part and an imaginary-number part.
   (c) is one-dimensional.
   (d) is a concept reserved for elite mathematicians.

3. What is the sum of $3 + j7$ and $-3 - j7$?
   (a) $0 + j0$
   (b) $6 + j14$
   (c) $-6 - j14$
   (d) $0 - j14$

4. What is $(-5 + j7) - (4 - j5)$?
   (a) $-1 + j2$
   (b) $-9 - j2$
   (c) $-1 - j2$
   (d) $-9 + j12$

5. What is the product $(-4 - j7)(6 - j2)$?
   (a) $24 - j14$
   (b) $-38 - j34$
   (c) $-24 - j14$
   (d) $-24 + j14$

6. What is the magnitude of the vector $18 - j24$?
   (a) 6
   (b) 21
   (c) 30
   (d) 52

7. The complex impedance value $5 + j0$ represents
   (a) a pure resistance.
   (b) a pure inductance.

(c) a pure capacitance.

(d) an inductance combined with a capacitance.

8. The complex impedance value $0 - j22$ represents

    (a) a pure resistance.

    (b) a pure inductance.

    (c) a pure capacitance.

    (d) an inductance combined with a resistance.

9. What is the absolute-value impedance of $3.0 - j6.0$?

    (a) $Z = 9.0 \ \Omega$

    (b) $Z = 3.0 \ \Omega$

    (c) $Z = 45 \ \Omega$

    (d) $Z = 6.7 \ \Omega$

10. What is the absolute-value impedance of $50 - j235$?

    (a) $Z = 240 \ \Omega$

    (b) $Z = 58{,}000 \ \Omega$

    (c) $Z = 285 \ \Omega$

    (d) $Z = -185 \ \Omega$

11. If the center conductor of a coaxial cable is made to have a smaller diameter, all other things being equal, what will happen to the Z_o of the transmission line?

    (a) It will increase.

    (b) It will decrease.

    (c) It will not change.

    (d) There is no way to determine this without knowing the actual dimensions.

12. If a device is said to have an impedance of $Z = 100 \ \Omega$, you can reasonably expect that this indicates

    (a) $R + jX = 100 + j0$.

    (b) $R + jX = 0 + j100$.

    (c) $R + jX = 100 + j100$.

    (d) the reactance and the resistance add up to $100 \ \Omega$.

13. Suppose a capacitor has a value of $0.050 \ \mu F$ at $665 \ kHz$. What is the capacitive susceptance, stated as an imaginary number?

    (a) $B_C = j4.79$

    (b) $B_C = -j4.79$

    (c) $B_C = j0.209$

    (d) $B_C = -j0.209$

14. An inductor has a value of 44 mH at 60 Hz. What is the inductive susceptance, stated as an imaginary number?

   (a) $B_L = -j0.060$

   (b) $B_L = j0.060$

   (c) $B_L = -j17$

   (d) $B_L = j17$

15. Susceptance and conductance add to form

   (a) complex impedance.

   (b) complex inductance.

   (c) complex reactance.

   (d) complex admittance.

16. Absolute-value impedance is equal to the square root of which of the following?

   (a) $G^2 + B^2$

   (b) $R^2 + X^2$

   (c) Z_o

   (d) $Y^2 + R^2$

17. Inductive susceptance is defined in

   (a) imaginary ohms.

   (b) imaginary henrys.

   (c) imaginary farads.

   (d) imaginary siemens.

18. Capacitive susceptance values can be defined by

   (a) positive real numbers.

   (b) negative real numbers.

   (c) positive imaginary numbers.

   (d) negative imaginary numbers.

19. Which of the following is false?

   (a) $B_C = 1/X_C$.

   (b) Complex impedance can be depicted as a vector.

   (c) Characteristic impedance is complex.

   (d) $G = 1/R$.

20. In general, as the absolute value of the impedance in a circuit increases,

   (a) the flow of ac increases.

   (b) the flow of ac decreases.

   (c) the reactance decreases.

   (d) the resistance decreases.

16
CHAPTER

RLC and *GLC*
Circuit Analysis

WHEN YOU SEE AN AC CIRCUIT THAT CONTAINS COILS AND/OR CAPACITORS, YOU SHOULD ENVISION a complex-number plane, either *RX* (resistance-reactance) or *GB* (conductance-admittance). The *RX* plane applies to series circuit analysis. The *GB* plane applies to parallel circuit analysis.

Complex Impedances in Series

When you see resistors, coils, and capacitors in series, each component has an impedance that can be represented as a vector in the *RX* plane. The vectors for resistors are constant, regardless of the frequency. But the vectors for coils and capacitors vary with frequency.

Pure Reactances

Pure inductive reactances (X_L) and capacitive reactances (X_C) simply add together when coils and capacitors are in series. Thus, $X = X_L + X_C$. In the *RX* plane, their vectors add, but because these vectors point in exactly opposite directions—inductive reactance upward and capacitive reactance downward (Fig. 16-1)—the resultant sum vector inevitably points either straight up or straight down, unless the reactances are equal and opposite, in which case they cancel and the result is the zero vector.

Problem 16-1

Suppose a coil and capacitor are connected in series, with $jX_L = j200$ and $jX_C = -j150$. What is the net reactance?

Just add the values: $jX = jX_L + jX_C = j200 + (-j150) = j(200 - 150) = j50$. This is a pure inductive reactance, because it is positive imaginary.

Problem 16-2

Suppose a coil and capacitor are connected in series, with $jX_L = j30$ and $jX_C = -j110$. What is the net reactance?

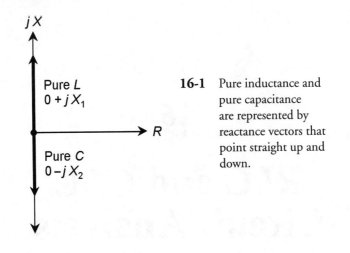

16-1 Pure inductance and pure capacitance are represented by reactance vectors that point straight up and down.

Again, add the values: $jX = j30 + (-j110) = j(30 - 110) = -j80$. This is a pure capacitive reactance, because it is negative imaginary.

Problem 16-3

Suppose a coil of inductance $L = 5.00\ \mu H$ and a capacitor of capacitance $C = 200$ pF are connected in series. Suppose the frequency is $f = 4.00$ MHz. What is the net reactance?

First, calculate the reactance of the inductor at 4.00 MHz. Proceed as follows:

$$
\begin{aligned}
jX_L &= j6.28fL \\
&= j(6.28 \times 4.00 \times 5.00) \\
&= j126
\end{aligned}
$$

Next, calculate the reactance of the capacitor at 4.00 MHz. Proceed as follows:

$$
\begin{aligned}
jX_C &= -j[1/(6.28fC)] \\
&= -j[1/(6.28 \times 4.00 \times 0.000200)] \\
&= -j199
\end{aligned}
$$

Finally, add the inductive and capacitive reactances to obtain the net reactance:

$$
\begin{aligned}
jX &= jX_L + jX_C \\
&= j126 + (-j199) \\
&= -j73
\end{aligned}
$$

This is a pure capacitive reactance.

Problem 16-4

What is the net reactance of the aforementioned inductor and capacitor combination at the frequency $f = 10.0$ MHz?

First, calculate the reactance of the inductor at 10.0 MHz. Proceed as follows:

$$jX_L = j6.28fL$$
$$= j(6.28 \times 10.0 \times 5.00)$$
$$= j314$$

Next, calculate the reactance of the capacitor at 10.00 MHz. Proceed as follows:

$$jX_C = -j[1/(6.28fC)]$$
$$= -j[1/(6.28 \times 10.0 \times 0.000200)]$$
$$= -j79.6$$

Finally, add the inductive and capacitive reactances to obtain the net reactance:

$$jX = jX_L + jX_C$$
$$= j314 + (-j79.6)$$
$$= j234$$

This is a pure inductive reactance. For series-connected components, the condition in which the capacitive and inductive reactances cancel is known as *series resonance*. We'll deal with this in more detail in the next chapter.

Adding Impedance Vectors

In the real world, there is resistance, as well as reactance, in an ac series circuit containing a coil and capacitor. This occurs because the coil wire has some resistance (it's never a perfect conductor). It can also be the case because a resistor is deliberately connected into the circuit.

Whenever the resistance in a series circuit is significant, the impedance vectors no longer point straight up and straight down. Instead, they run off toward the northeast (for the inductive part of the circuit) and southeast (for the capacitive part). This is illustrated in Fig. 16-2.

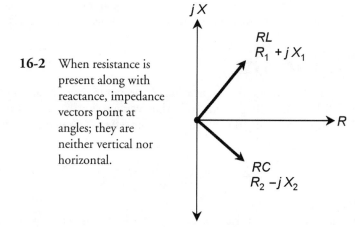

16-2 When resistance is present along with reactance, impedance vectors point at angles; they are neither vertical nor horizontal.

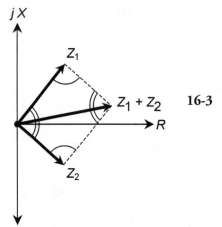

16-3 Parallelogram method of complex-impedance vector addition.

When two impedance vectors don't lie along a single line, you must use *vector addition* to be sure that you get the correct net impedance. In Fig. 16-3, the geometry of vector addition is shown. Construct a *parallelogram*, using the two vectors $Z_1 = R_1 + jX_1$ and $Z_2 = R_2 + jX_2$ as two adjacent sides of the figure. The diagonal of the parallelogram is the vector representing the net complex impedance. (Note that in a parallelogram, pairs of opposite angles have equal measures. These equalities are indicated by single and double arcs in Fig. 16-3.)

Formula for Complex Impedances in Series

Suppose you are given two complex impedances, $Z_1 = R_1 + jX_1$ and $Z_2 = R_2 + jX_2$. The net impedance, Z, of these in series is their vector sum, given by the following formula:

$$Z = (R_1 + jX_1) + (R_2 + jX_2)$$
$$= (R_1 + R_2) + j(X_1 + X_2)$$

Calculating a vector sum using the formula is easier than doing it geometrically with a parallelogram. The arithmetic method is also more exact. The resistance and reactance components add separately. Just remember that if a reactance is capacitive, then it is negative imaginary in this formula.

Series *RLC* Circuits

When an inductance, capacitance, and resistance are connected in series (Fig. 16-4), the resistance R can be imagined as belonging entirely to the coil, when you use the preceding formulas. Then you have two vectors to add, when finding the impedance of the series *RLC* circuit containing three such components:

$$Z = (R + jX_L) + (0 + jX_C)$$
$$= R + j(X_L + X_C)$$

Again, remember that X_C is never positive! So, although the formulas here have addition symbols in them, you're adding a negative number when you add in a capacitive reactance.

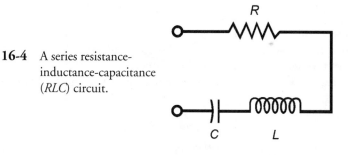

16-4 A series resistance-inductance-capacitance (*RLC*) circuit.

Problem 16-5

Suppose a resistor, a coil, and a capacitor are connected in series with $R = 50\ \Omega$, $X_L = 22\ \Omega$, and $X_C = -33\ \Omega$. What is the net impedance, Z?

Consider the resistor to be part of the coil, obtaining two complex vectors, $50 + j22$ and $0 - j33$. Adding these gives the resistance component of $50 + 0 = 50$, and the reactive component of $j22 - j33 = -j11$. Therefore, $Z = 50 - j11$.

Problem 16-6

Consider a resistor, a coil, and a capacitor that are connected in series with $R = 600\ \Omega$, $X_L = 444\ \Omega$, and $X_C = -444\ \Omega$. What is the net impedance, Z?

Again, imagine the resistor to be part of the inductor. Then the complex impedance vectors are $600 + j444$ and $0 - j444$. Adding these, the resistance component is $600 + 0 = 600$, and the reactive component is $j444 - j444 = j0$. Thus, $Z = 600 + j0$. This is a purely resistive impedance, and you can rightly call it $600\ \Omega$.

Problem 16-7

Suppose a resistor, a coil, and a capacitor are connected in series. The resistor has a value of $330\ \Omega$, the capacitance is 220 pF, and the inductance is 100 µH. The frequency is 7.15 MHz. What is the complex impedance of this series *RLC* circuit at this frequency?

First, calculate the inductive reactance. Remember that $X_L = 6.28fL$ and that megahertz and microhenrys go together in the formula. Multiply to obtain the following:

$$jX_L = j(6.28 \times 7.15 \times 100)$$
$$= j4490$$

Next, calculate the capacitive reactance using the formula $X_C = -1/(6.28fC)$. Convert 220 pF to microfarads to obtain $C = 0.000220$ µF. Then calculate:

$$jX_C = -j[1/(6.28 \times 7.15 \times 0.000220)]$$
$$= -j101$$

Now, lump the resistance and the inductive reactance together, so one of the impedance vectors is $330 + j4490$. The other is $0 - j101$. Adding these gives $Z = 330 + j4389$; this rounds off to $Z = 330 + j4390$.

Problem 16-8

Suppose a resistor, a coil, and a capacitor are connected in series. The resistance is 50.0 Ω, the inductance is 10.0 μH, and the capacitance is 1000 pF. The frequency is 1592 kHz. What is the complex impedance of this series *RLC* circuit at this frequency?

First, calculate $X_L = 6.28fL$. Convert the frequency to megahertz; 1592 kHz = 1.592 MHz. Then:

$$jX_L = j(6.28 \times 1.592 \times 10.0)$$
$$= j100$$

Then calculate $X_C = -1/(6.28fC)$. Let's convert picofarads to microfarads, and use megahertz for the frequency. Therefore:

$$jX_C = -j[1/(6.28 \times 1.592 \times 0.001000)]$$
$$= -j100$$

Let the resistance and inductive reactance go together as one vector, $50.0 + j100$. Let the capacitive reactance be represented as $0 - j100$. The sum is $Z = 50.0 + j100 - j100 = 50.0 + j0$. This is a pure resistance of 50.0 Ω. You can correctly say that the impedance is 50.0 Ω in this case.

Complex Admittances in Parallel

When you see resistors, coils, and capacitors in parallel, remember that each component, whether it is a resistor, an inductor, or a capacitor, has an admittance that can be represented as a vector in the *GB* plane. The vectors for pure conductances are constant, even as the frequency changes. But the vectors for the coils and capacitors vary with frequency.

Pure Susceptances

Pure inductive susceptances (B_L) and capacitive susceptances (B_C) add together when coils and capacitors are in parallel. Thus, $B = B_L + B_C$. Remember that B_L is never positive, and B_C is never negative. This is just the opposite situation from reactances.

In the *GB* plane, pure jB_L and jB_C vectors add. Because such vectors always point in exactly opposite directions—inductive susceptance down and capacitive susceptance up—the sum, jB, inevitably points either straight down or straight up (Fig. 16-5), unless the susceptances are equal and opposite, in which case they cancel and the result is the zero vector.

Problem 16-9

Suppose a coil and capacitor are connected in parallel, with $jB_L = -j0.05$ and $jB_C = j0.08$. What is the net susceptance?

Just add the values as follows: $jB = jB_L + jB_C = -j0.05 + j0.08 = j0.03$. This is a capacitive susceptance, because it is positive imaginary.

Problem 16-10

Suppose a coil and capacitor are connected in parallel, with $jB_L = -j0.60$ and $jB_C = j0.25$. What is the net susceptance?

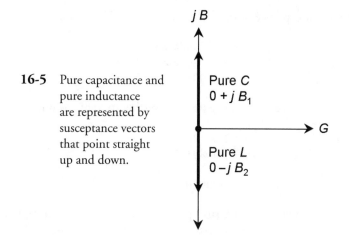

16-5 Pure capacitance and pure inductance are represented by susceptance vectors that point straight up and down.

Again, add the values: $jB = -j0.60 + j0.25 = -j0.35$. This is an inductive susceptance, because it is negative imaginary.

Problem 16-11

Suppose a coil of $L = 6.00$ µH and a capacitor of $C = 150$ pF are connected in parallel. The frequency is $f = 4.00$ MHz. What is the net susceptance?

First calculate the susceptance of the inductor at 4.00 MHz, as follows:

$$jB_L = -j[1/(6.28fL)]$$
$$= -j[1/(6.28 \times 4.00 \times 6.00)]$$
$$= -j0.00663$$

Next, calculate the susceptance of the capacitor (converting its value to microfarads) at 4.00 MHz, as follows:

$$jB_C = j(6.28fC)$$
$$= j(6.28 \times 4.00 \times 0.000150)$$
$$= j0.00377$$

Finally, add the inductive and capacitive susceptances to obtain the net susceptance:

$$jB = jB_L + jB_C$$
$$= -j0.00663 + j0.00377$$
$$= -j0.00286$$

This is a pure inductive susceptance.

Problem 16-12

What is the net susceptance of the above parallel-connected inductor and capacitor at a frequency of $f = 5.31$ MHz?

First calculate the susceptance of the inductor at 5.31 MHz, as follows:

$$jB_L = -j[1/(6.28fL)]$$
$$= -j[1/(6.28 \times 5.31 \times 6.00)]$$
$$= -j0.00500$$

Next calculate the susceptance of the capacitor (converting its value to microfarads) at 5.31 MHz, as follows:

$$jB_C = j(6.28fC)$$
$$= j(6.28 \times 5.31 \times 0.000150)$$
$$= j0.00500$$

Finally, add the inductive and capacitive susceptances to obtain the net susceptance:

$$jB = jB_L + jB_C$$
$$= -j0.00500 + j0.00500$$
$$= j0$$

This means that the circuit has no susceptance at 5.31 MHz. The situation in which there is no susceptance in an *LC* circuit is known as *parallel resonance*. It is discussed in the next chapter.

Adding Admittance Vectors

In real life, there is a small amount of conductance, as well as susceptance, in an ac parallel circuit containing a coil and capacitor. This occurs when the capacitor lets a little bit of current leak through. More often, though, it is the case because a *load* is connected in parallel with the coil and capacitor. This load can be an antenna, the input to an amplifier circuit, a test instrument, a transducer, or some other device.

When the conductance in a parallel circuit containing inductance and capacitance is significant, the admittance vectors do not point straight up and down. Instead, they run off toward the northeast (for the capacitive part of the circuit) and southeast (for the inductive part). This is illustrated in Fig. 16-6.

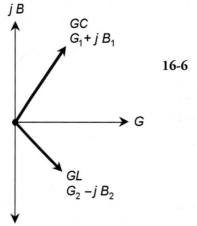

16-6   When conductance is present along with susceptance, admittance vectors point at angles; they are neither vertical nor horizontal.

You've seen how vectors add in the *RX* plane. In the *GB* plane, the principle is the same. The net admittance vector is the sum of the component admittance vectors.

Formula for Complex Admittances in Parallel

Given two admittances, $Y_1 = G_1 + jB_1$ and $Y_2 = G_2 + jB_2$, the net admittance Y of these in parallel is their vector sum, as follows:

$$Y = (G_1 + jB_1) + (G_2 + jB_2)$$
$$= (G_1 + G_2) + j(B_1 + B_2)$$

The conductance and susceptance components add separately. Just remember that if a susceptance is inductive, then it is negative imaginary in this formula.

Parallel *GLC* Circuits

When a coil, capacitor, and resistor are connected in parallel (Fig. 16-7), the resistance should be thought of as a *conductance*, whose value in siemens (symbolized S) is equal to the reciprocal of the value in ohms. Think of the conductance as all belonging to the inductor. Then you have two vectors to add, when finding the admittance of a parallel *GLC* (conductance-inductance-capacitance) circuit:

$$Y = (G + jB_L) + (0 + jB_C)$$
$$= G + j(B_L + B_C)$$

Again, remember that B_L is never positive! So, although the formulas here have addition symbols in them, you're adding a negative number when you add in an inductive susceptance.

Problem 16-13

Suppose a resistor, a coil, and a capacitor are connected in parallel. Suppose the resistor has a conductance $G = 0.10$ S, and the susceptances are $jB_L = -j0.010$ and $jB_C = j0.020$. What is the complex admittance of this combination?

Consider the resistor to be part of the coil. Then there are two complex admittances in parallel: $0.10 - j0.010$ and $0.00 + j0.020$. Adding these gives a conductance component of $0.10 + 0.00 = 0.10$ and a susceptance component of $-j0.010 + j0.020 = j0.010$. Therefore, the complex admittance is $0.10 + j0.010$.

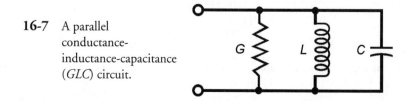

16-7   A parallel conductance-inductance-capacitance (*GLC*) circuit.

Problem 16-14

Suppose a resistor, a coil, and a capacitor are connected in parallel. Suppose the resistor has a con-ductance $G = 0.0010$ S, and the susceptances are $jB_L = -j0.0022$ and $jB_C = j0.0022$. What is the complex admittance of this combination?

Again, consider the resistor to be part of the coil. Then the complex admittances are $0.0010 - j0.0022$ and $0.0000 + j0.0022$. Adding these, the conductance component is $0.0010 + 0.0000 = 0.0010$, and the susceptance component is $-j0.0022 + j0.0022 = j0$. Thus, the admittance is $0.0010 + j0$. This is a purely conductive admittance.

Problem 16-15

Suppose a resistor, a coil, and a capacitor are connected in parallel. The resistor has a value of 100 Ω, the capacitance is 200 pF, and the inductance is 100 μH. The frequency is 1.00 MHz. What is the net complex admittance?

First, you need to calculate the inductive susceptance. Recall the formula, and plug in the num-bers as follows:

$$jB_L = -j[1/(6.28fL)]$$
$$= -j[1/(6.28 \times 1.00 \times 100)]$$
$$= -j0.00159$$

Megahertz and microhenrys go together in the formula. Next, you must calculate the capacitive susceptance. Convert 200 pF to microfarads to go with megahertz in the formula; thus $C = 0.000200$ μF. Then:

$$jB_C = j(6.28fC)$$
$$= j(6.28 \times 1.00 \times 0.000200)$$
$$= j0.00126$$

Finally, consider the conductance, which is $1/100 = 0.0100$ S, and the inductive susceptance as exist-ing together in a single component. That means that one of the parallel-connected admittances is $0.0100 - j0.00159$. The other is $0.0000 + j0.00126$. Adding these gives $0.0100 - j0.00033$.

Problem 16-16

Suppose a resistor, a coil, and a capacitor are in parallel. The resistance is 10.0 Ω, the inductance is 10.0 μH, and the capacitance is 1000 pF. The frequency is 1592 kHz. What is the complex admit-tance of this circuit at this frequency?

First, calculate the inductive susceptance. Convert the frequency to megahertz; 1592 kHz = 1.592 MHz. Plug in the numbers as follows:

$$jB_L = -j[1/(6.28fL)]$$
$$= -j[1/(6.28 \times 1.592 \times 10.0)]$$
$$= -j0.0100$$

Next, calculate the capacitive susceptance. Convert 1000 pF to microfarads to go with megahertz in the formula; thus $C = 0.001000$ μF. Then:

$$jB_C = j(6.28fC)$$
$$= j(6.28 \times 1.592 \times 0.001000)$$
$$= j0.0100$$

Finally, consider the conductance, which is $1/10.0 = 0.100$ S, and the inductive susceptance as existing together in a single component. That means that one of the parallel-connected admittances is $0.100 - j0.0100$. The other is $0.0000 + j0.0100$. Adding these gives $0.100 + j0$.

Converting Complex Admittance to Complex Impedance

The *GB* plane is, as you have seen, similar in appearance to the *RX* plane, although mathematically they are different. Once you've found a complex admittance for a parallel *RLC* circuit, you will usually want to transform this back to a complex impedance.

The transformation from a complex admittance $G + jB$ to a complex impedance $R + jX$ can be carried out using the following two formulas, one for R and the other for X:

$$R = G/(G^2 + B^2)$$
$$X = -B/(G^2 + B^2)$$

If you know the complex admittance, first find the resistance and reactance components individually using the preceding formulas. Then assemble the two components into the complex impedance, $R + jX$.

Problem 16-17

Suppose the complex admittance of a certain parallel circuit is $0.010 - j0.0050$. What is the complex impedance of this same circuit, assuming the frequency does not change?

In this case, $G = 0.010$ S and $B = -0.0050$ S. First find $G^2 + B^2$, as follows:

$$G^2 + B^2 = 0.010^2 + (-0.0050)^2$$
$$= 0.000100 + 0.000025$$
$$= 0.000125$$

Now it is easy to calculate R and X, like this:

$$R = G/0.000125$$
$$= 0.010/0.000125$$
$$= 80 \ \Omega$$

$$X = -B/0.000125$$
$$= 0.0050/0.000125$$
$$= 40 \ \Omega$$

The complex impedance is therefore $80 + j40$.

Putting It All Together

When you're confronted with a parallel circuit containing resistance, inductance, and capacitance, and you want to determine the complex impedance of the combination, do these things:

1. Find the conductance $G = 1/R$ for the resistor. (It will be positive or zero.)
2. Find the susceptance B_L of the inductor using the appropriate formula. (It will be negative or zero.)
3. Find the susceptance B_C of the capacitor using the appropriate formula. (It will be positive or zero.)
4. Find the net susceptance $B = B_L + B_C$. (It might be positive, negative, or zero.)
5. Compute R and X in terms of G and B using the appropriate formulas.
6. Assemble the complex impedance $R + jX$.

Problem 16-18

Suppose a resistor of 10.0 Ω, a capacitor of 820 pF, and a coil of 10.0 μH are in parallel. The frequency is 1.00 MHz. What is the complex impedance?
    Proceed according to the above steps, as follows:

1. Calculate $G = 1/R = 1/10.0 = 0.100$.
2. Calculate $B_L = -1/(6.28fL) = -1/(6.28 \times 1.00 \times 10.0) = -0.0159$.
3. Calculate $B_C = 6.28fC = 6.28 \times 1.00 \times 0.000820 = 0.00515$. (Remember to first convert the capacitance to microfarads, to go with megahertz.)
4. Calculate $B = B_L + B_C = -0.0159 + 0.00515 = -0.0108$.
5. Define $G^2 + B^2 = 0.100^2 + (-0.0108)^2 = 0.010117$. Then $R = G/0.010117 = 0.100/0.010117 = 9.88\ \Omega$, and $X = -B/0.010117 = 0.0108/0.010117 = 1.07\ \Omega$.
6. The complex impedance is $R + jX = 9.88 + j1.07$.

Problem 16-19

Suppose a resistor of 47.0 Ω, a capacitor of 500 pF, and a coil of 10.0 μH are in parallel. What is their complex impedance at a frequency of 2.252 MHz?
    Proceed as before:

1. Calculate $G = 1/R = 1/47.0 = 0.021277$.
2. Calculate $B_L = -1/(6.28fL) = -1/(6.28 \times 2.252 \times 10.0) = -0.00707$.
3. Calculate $B_C = 6.28fC = 6.28 \times 2.252 \times 0.000500 = 0.00707$. (Remember to first convert the capacitance to microfarads, to go with megahertz.)
4. Calculate $B = B_L + B_C = -0.00707 + 0.00707 = 0.00000$.
5. Define $G^2 + B^2 = 0.021277^2 + 0.00000^2 = 0.00045271$. Then $R = G/0.00045271 = 0.021277/0.00045271 = 46.999\ \Omega$, and $X = -B/0.00045271 = 0.00000/0.00045271 = 0.00000$.
6. The complex impedance is $R + jX = 46.9999 + j0.00000$. When we round it off to three significant figures, we get $47.0 + j0.00$. This a pure resistance equal to the value of the resistor in the circuit.

Reducing Complicated *RLC* Circuits

Sometimes you'll see circuits in which there are several resistors, capacitors, and/or coils in series and parallel combinations. Such a circuit can be reduced to an equivalent series or parallel *RLC* circuit that contains one resistance, one capacitance, and one inductance.

Series Combinations

Resistances in series simply add. Inductances in series also add. Capacitances in series combine in a somewhat more complicated way. If you don't remember the formula, here it is:

$$1/C = 1/C_1 + 1/C_2 + \cdots + 1/C_n$$

where $C_1, C_2, \ldots,$ and C_n are the individual capacitances, and C is the total capacitance. Once you've found $1/C$, take its reciprocal to obtain C. Figure 16-8A shows an example of a complicated series *RLC* circuit. The equivalent circuit, with one resistance, one capacitance, and one inductance, is shown in Fig. 16-8B.

Parallel Combinations

In parallel, resistances and inductances combine the way capacitances do in series. Capacitances simply add up. An example of a complicated parallel *RLC* circuit is shown in Fig. 16-9A. The equivalent circuit, with one resistance, one capacitance, and one inductance, is shown in Fig. 16-9B.

16-8 At A, a complicated series circuit containing multiple resistances and reactances. At B, the same circuit simplified. Resistances are in ohms; inductances are in microhenrys (μH); capacitances are in picofarads (pF).

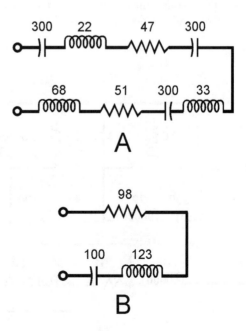

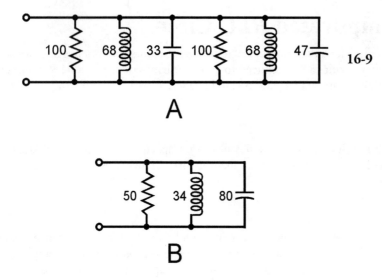

A

B

16-9 At A, a complicated parallel circuit containing multiple resistances and reactances. At B, the same circuit simplified. Resistances are in ohms; inductances are in microhenrys (μH); capacitances are in picofarads (pF).

Nightmare Scenarios

Imagine an *RLC* circuit like the one shown in Fig. 16-10. How would you find the complex impedance of this circuit at some particular frequency, such as 8.54 MHz? Don't waste much time worrying about circuits like this. You'll rarely encounter them. But rest assured that, given a frequency, a complex impedance does exist, no matter how complicated an *RLC* circuit happens to be.

An engineer could use a computer to find the theoretical complex impedance of a circuit such as the one in Fig. 16-10 at a specific frequency, or as a function of the frequency. The experimental approach would be to build the circuit, connect a signal generator to it, and then measure R and X at various frequencies with a device called an *impedance bridge*.

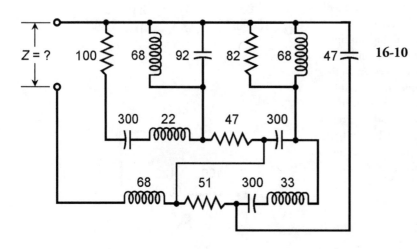

16-10 A series-parallel nightmare circuit containing multiple resistances and reactances. Resistances are in ohms; inductances are in microhenrys (μH); capacitances are in picofarads (pF).

Ohm's Law for AC Circuits

Ohm's Law for a dc circuit is a simple relationship among three variables: the current I (in amperes), the voltage E (in volts), and the resistance R (in ohms). Here are the formulas, in case you don't recall them:

$$E = IR$$
$$I = E/R$$
$$R = E/I$$

In ac circuits containing no reactance, these same formulas apply, as long as you work with root-mean-square (rms) voltages and currents. If you need a refresher concerning the meaning of rms, refer to Chapter 9.

Purely Resistive Impedances

When the impedance Z in an ac circuit contains no reactance, so that all of the current and voltage exist through and across a pure resistance R, Ohm's Law for an ac circuit is expressed as follows:

$$E = IZ$$
$$I = E/Z$$
$$Z = E/I$$

where $Z = R$, and the values I and E are rms current and voltage.

Complex Impedances

When you want to determine the relationship among current, voltage, and resistance in an ac circuit that contains resistance and reactance, things get interesting. Recall the formula for the square of the absolute-value impedance in a series RLC circuit:

$$Z^2 = R^2 + X^2$$

This means that Z is equal to the square root of the quantity $R^2 + X^2$, as follows:

$$Z = (R^2 + X^2)^{1/2}$$

This is the length of the vector $R + jX$ in the complex impedance plane. You learned this in Chap. 15. This formula applies only for series RLC circuits.

The square of the absolute-value impedance for a parallel RLC circuit, in which the resistance is R and the reactance is X, is defined this way:

$$Z^2 = R^2X^2/(R^2 + X^2)$$

This means that the absolute-value impedance, Z, must be calculated using the rather arcane formula:

$$Z = [R^2X^2/(R^2 + X^2)]^{1/2}$$

The $1/2$ power of a quantity represents the positive square root of that quantity.

Problem 16-20

Suppose a series RX circuit (shown by the generic block diagram of Fig. 16-11) has a resistance of $R = 50.0\ \Omega$ and a capacitive reactance of $X = -50.0\ \Omega$. Suppose 100-V rms ac is applied to this circuit. What is the current?

First, calculate $Z^2 = R^2 + X^2 = 50.0^2 + (-50.0)^2 = 2500 + 2500 = 5000$. Then Z is the square root of 5000, or 70.7. Therefore, $I = E/Z = 100/70.7 = 1.41$ A rms.

Problem 16-21

What are the rms ac voltages across the resistance and the reactance, respectively, in the circuit described in Problem 16-20?

The Ohm's Law formulas for dc will work here. Because the current is $I = 1.41$ A rms, the voltage drop across the resistance is equal to $E_R = IR = 1.41 \times 50.0 = 70.5$ V rms. The voltage drop across the reactance is the product of the current and the reactance: $E_X = IX = 1.41 \times (-50.0) = -70.5$ V rms. This is an rms ac voltage of equal magnitude to that across the resistance. But the phase is different.

Note that voltages across the resistance and the reactance—a capacitive reactance in this case, because it's negative—don't add up to 100 V rms, which is placed across the whole circuit. This is because, in an RX ac circuit, there is always a difference in phase between the voltage across the resistance and the voltage across the reactance. The voltages across the components always add up to the applied voltage *vectorially*, but not always *arithmetically*.

Problem 16-22

Suppose a series RX circuit (Fig. 16-11) has $R = 10.0\ \Omega$. and $X = 40.0\ \Omega$. The applied voltage is 100-V rms ac. What is the current?

Calculate $Z^2 = R^2 + X^2 = 100 + 1600 = 1700$. This means that Z is the square root of 1700, or 41.2. Therefore, $I = E/Z = 100/41.2 = 2.43$ A rms.

Problem 16-23

What are the rms ac voltages across the resistance and the reactance, respectively, in the circuit described in Problem 16-22?

Knowing the current, calculate $E_R = IR = 2.43 \times 10.0 = 24.3$ V rms. Also, $E_X = IX = 2.43 \times 40.0 = 97.2$ V rms. If you add $E_R + E_X$ arithmetically, you get $24.3 + 97.2 = 121.5$ V as the total

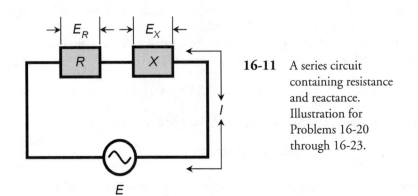

16-11 A series circuit containing resistance and reactance. Illustration for Problems 16-20 through 16-23.

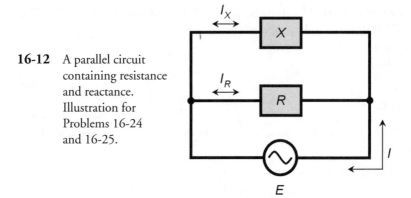

16-12 A parallel circuit containing resistance and reactance. Illustration for Problems 16-24 and 16-25.

across *R* and *X*. Again, this differs from the applied voltage! The simple dc rule does not work here, for the same reason it didn't work in the scenario of Problem 16-21.

Problem 16-24

Suppose a parallel *RX* circuit (shown by the generic block diagram of Fig. 16-12) has $R = 30.0$ Ω and $X = -20.0$ Ω. The ac supply voltage is 50.0 V rms. What is the total current drawn from the ac supply?

First, find the square of the absolute-value impedance, remembering the formula for parallel circuits: $Z^2 = R^2X^2/(R^2 + X^2) = 360{,}000/1300 = 277$. The absolute-value impedance *Z* is the square root of 277, or 16.6. The total current is therefore $I = E/Z = 50/16.6 = 3.01$ A rms.

Problem 16-25

What are the rms currents through the resistance and the reactance, respectively, in the circuit described in Problem 16-24?

The Ohm's Law formulas for dc will work here. For the resistance, $I_R = E/R = 50.0/30.0 = 1.67$ A rms. For the reactance, $I_X = E/X = 50.0/(-20.0) = -2.5$ A rms. Note that these currents don't add up to 3.01 A, the total current. The reason for this is the same as the reason ac voltages don't add arithmetically in ac circuits that contain reactance. The constituent currents, I_R and I_X, differ in phase. Vectorially, they add up to 3.01 A rms, but arithmetically, they don't.

Quiz

Refer to the text in this chapter if necessary. A good score is 18 correct. Answers are in the back of the book.

1. Suppose a coil and capacitor are connected in series. The inductive reactance is 250 Ω, and the capacitive reactance is −300 Ω. What is the complex impedance?

    (a) $0 + j550$

    (b) $0 - j50$

    (c) $250 - j300$

    (d) $-300 + j250$

2. Suppose a coil of 25.0 μH and capacitor of 100 pF are connected in series. The frequency is 5.00 MHz. What is the complex impedance?

(a) $0 + j467$

(b) $25 + j100$

(c) $0 - j467$

(d) $25 - j100$

3. When $R = 0$ in a series *RLC* circuit, but the net reactance is not zero, the impedance vector

(a) always points straight up.

(b) always points straight down.

(c) always points straight toward the right.

(d) None of the above is correct.

4. Suppose a resistor of 150 Ω, a coil with a reactance of 100 Ω, and a capacitor with a reactance of −200 Ω are connected in series. What is the complex impedance?

(a) $150 + j100$

(b) $150 - j200$

(c) $100 - j200$

(d) $150 - j100$

5. Suppose a resistor of 330 Ω, a coil of 1.00 μH, and a capacitor of 200 pF are in series. What is the complex impedance at 10.0 MHz?

(a) $330 - j199$

(b) $300 + j201$

(c) $300 + j142$

(d) $330 - j16.8$

6. Suppose a coil has an inductance of 3.00 μH and a resistance of 10.0 Ω in its winding. A capacitor of 100 pF is in series with this coil. What is the complex impedance at 10.0 MHz?

(a) $10 + j3.00$

(b) $10 + j29.2$

(c) $10 - j97$

(d) $10 + j348$

7. Suppose a coil has a reactance of 4.00 Ω. What is the complex admittance, assuming there is nothing else is in the circuit?

(a) $0 + j0.25$

(b) $0 + j4.00$

(c) $0 - j0.25$

(d) $0 - j4.00$

8. What will happen to the susceptance of a capacitor if the frequency is doubled and all other factors remain constant?

(a) It will decrease to half its former value.

(b) It will not change.

(c) It will double.

(d) It will quadruple.

9. Suppose a coil and capacitor are in parallel, with $jB_L = -j0.05$ and $jB_C = j0.03$. What is the complex admittance, assuming that nothing is in series or parallel with these components?

(a) $0 - j0.02$

(b) $0 - j0.07$

(c) $0 + j0.02$

(d) $-0.05 + j0.03$

10. Imagine a coil, a resistor, and a capacitor connected in parallel. The resistance is 1.0 Ω, the capacitive susceptance is 1.0 S, and the inductive susceptance is −1.0 S. Then, suddenly, the frequency is cut to half its former value. What is the complex admittance at the new frequency?

(a) $1.0 + j0.0$

(b) $1.0 + j1.5$

(c) $1.0 - j1.5$

(d) $1.0 - j2.0$

11. Suppose a coil of 3.50 µH and a capacitor of 47.0 pF are in parallel. The frequency is 9.55 MHz. There is nothing else in series or parallel with these components. What is the complex admittance?

(a) $0 + j0.00282$

(b) $0 - j0.00194$

(c) $0 + j0.00194$

(d) $0 - j0.00758$

12. A vector pointing southeast in the *GB* plane would indicate

(a) pure conductance with zero susceptance.

(b) conductance and inductive susceptance.

(c) conductance and capacitive susceptance.

(d) pure susceptance with zero conductance.

13. Suppose a resistor with conductance 0.0044 S, a capacitor with susceptance 0.035 S, and a coil with susceptance −0.011 S are all connected in parallel. What is the complex admittance?

(a) $0.0044 + j0.024$

(b) $0.035 - j0.011$

(c) $-0.011 + j0.035$

(d) $0.0044 + j0.046$

14. Suppose a resistor of 100 Ω, a coil of 4.50 µH, and a capacitor of 220 pF are in parallel. What is the complex admittance at a frequency of 6.50 MHz?

(a) $100 + j0.00354$

(b) $0.010 + j0.00354$

(c) $100 - j0.0144$

(d) $0.010 + j0.0144$

15. Suppose the complex admittance of a circuit is $0.02 + j0.20$. What is the complex impedance, assuming the frequency does not change?

   (a) $50 + j5.0$

   (b) $0.495 - j4.95$

   (c) $50 - j5.0$

   (d) $0.495 + j4.95$

16. Suppose a resistor of $51.0\ \Omega$, an inductor of $22.0\ \mu H$, and a capacitor of $150\ pF$ are in parallel. The frequency is 1.00 MHz. What is the complex impedance?

   (a) $51.0 - j14.9$

   (b) $51.0 + j14.9$

   (c) $46.2 - j14.9$

   (d) $46.2 + j14.9$

17. Suppose a series circuit has $99.0\ \Omega$ of resistance and $88.0\ \Omega$ of inductive reactance. An ac rms voltage of 117 V is applied to this series network. What is the current?

   (a) 1.18 A

   (b) 1.13 A

   (c) 0.886 A

   (d) 0.846 A

18. What is the voltage across the reactance in the preceding example?

   (a) 78.0 V

   (b) 55.1 V

   (c) 99.4 V

   (d) 74.4 V

19. Suppose a parallel circuit has $10\ \Omega$ of resistance and $15\ \Omega$ of reactance. An ac rms voltage of 20 V is applied across it. What is the total current?

   (a) 2.00 A

   (b) 2.40 A

   (c) 1.33 A

   (d) 0.800 A

20. What is the current through the resistance in the preceding example?

   (a) 2.00 A

   (b) 2.40 A

   (c) 1.33 A

   (d) 0.800 A

Power and Resonance in Alternating-Current Circuits

ONE OF THE BIGGEST CHALLENGES IN ELECTRICITY AND ELECTRONICS IS OPTIMIZING THE EFFICIENCY with which power is transferred from one place to another, or converted from one form to another. Also important, especially for the radio-frequency (RF) engineer, is the phenomenon of *resonance*. Power and resonance are closely related.

Forms of Power

What is power, exactly? Here is an all-encompassing definition: *Power is the rate at which energy is expended, radiated, or dissipated.* This definition can be applied to mechanical motion, chemical effects, dc and ac electricity, sound waves, radio waves, sound, heat, infrared (IR), visible light, ultraviolet (UV), X rays, gamma rays, and high-speed subatomic particles. In all cases, the energy is converted from one form into another form at a certain rate.

Units of Power

The standard unit of power is the *watt*, abbreviated W. A watt is equivalent to a *joule per second* (J/s). Sometimes power is given as *kilowatts* (kW or thousands of watts), *megawatts* (MW or millions of watts), or *gigawatts* (GW or billions of watts). It is also sometimes expressed as *milliwatts* (mW or thousandths of watts), *microwatts* (μW or millionths of watts), or *nanowatts* (nW or billionths of watts).

Volt-Amperes

In dc circuits, and also in ac circuits having no reactance, power can be defined this way: *Power is the product of the voltage across a circuit or component and the current through that same circuit or component.* Mathematically this is written $P = EI$. If E is in volts and I is in amperes, then P is in *volt-amperes* (VA). This translates into watts when there is no reactance in the circuit (Fig. 17-1). The root-mean-square (rms) values for voltage and current are always used to derive the effective, or average, power.

    Volt-amperes, also called *VA power* or *apparent power*, can take various forms. A resistor converts electrical energy into heat energy, at a rate that depends on the value of the resistance and the current through it. A light bulb converts electricity into light and heat. A radio antenna converts high-

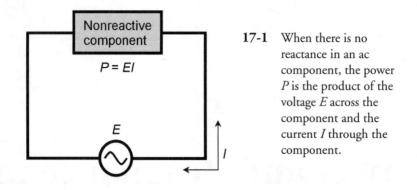

17-1 When there is no reactance in an ac component, the power *P* is the product of the voltage *E* across the component and the current *I* through the component.

frequency ac into radio waves. A speaker converts low-frequency ac into sound waves. The power in these forms is a measure of the intensity of the heat, light, radio waves, or sound waves.

Instantaneous Power

Usually, but not always, engineers think of power based on the rms, or effective, ac value. But for VA power, peak values are sometimes used instead. If the ac is a sine wave, the peak current is 1.414 times the rms current, and the peak voltage is 1.414 times the rms voltage. If the current and the voltage are exactly in phase, the product of their peak values is twice the product of their rms values.

There are instants in time when the VA power in a reactance-free, sine-wave ac circuit is twice the effective power. There are other instants in time when the VA power is zero; at still other moments, the VA power is somewhere between zero and twice the effective power level (Fig. 17-2). This constantly changing power is called *instantaneous power.*

In some situations, such as with a voice-modulated radio signal or a fast-scan television signal, the instantaneous power varies in an extremely complicated fashion. Have you ever seen the *modulation envelope* of such a signal displayed on an oscilloscope?

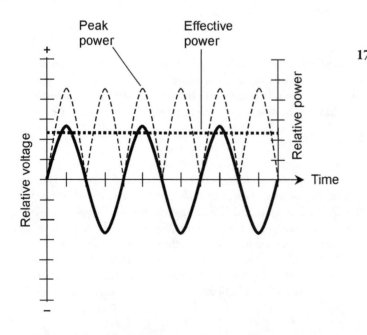

17-2 Peak versus effective power for a sine wave. The left-hand vertical scale shows relative voltage. The right-hand vertical scale shows relative power. The solid curve represents the voltage as a function of time. The light and heavy dashed waves show peak and effective power, respectively, as functions of time.

Imaginary Power

If an ac circuit contains reactance, things get interesting. In a pure resistance, the rate of energy expenditure per unit time (or *true power*) is the same as the VA power (also known as *apparent power*). But when inductance and/or capacitance exists in an ac circuit, the VA power is greater than the power actually manifested as heat, light, radio waves, or whatever. The apparent power is then greater than the true power! The extra power is called *imaginary power*, because it exists in the reactance, and reactance can be, as you have learned, rendered in mathematically imaginary numerical form. Imaginary power is also known as *reactive power*.

Inductors and capacitors store energy and then release it a fraction of a cycle later. This phenomenon, like true power, is expressible as the rate at which energy is changed from one form to another. But rather than existing as a usable form of power, such as heat, light, radio waves, sound waves, or mechanical motion, imaginary power is stored up as a magnetic or electric field, and then released back into the circuit or system. This storage and release of power takes place over and over with each repeating ac cycle.

True Power Does Not Travel

A common and usually harmless misconception about true power is the notion that it can travel. For example, if you connect a radio transmitter to a cable that runs outdoors to an antenna, you might say you're "feeding power" through the cable to the antenna. Everybody says this, even engineers and technicians. But true power always involves a change in form, such as from electrical current and voltage into radio waves. It doesn't go from place to place. It simply happens in a specific place. It's the imaginary power that moves in situations like this, especially in transmission lines between power stations and power users, or between radio transmitters and radio antennas.

In a real-life radio antenna system, some true power is dissipated as heat in the transmitter amplifiers and in the feed line (Fig. 17-3). The useful dissipation of true power occurs when the imaginary power, in the form of electric and magnetic fields, gets to the antenna, where it is changed into *electromagnetic waves*.

You will often hear expressions such as "forward power" and "reflected power," or "power is fed from this amplifier to these speakers." It is all right to talk like this, but it can sometimes lead to

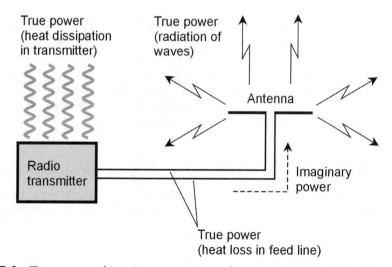

17-3   True power and imaginary power in a radio transmitter and antenna system.

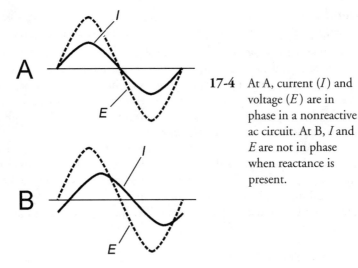

A

B

17-4   At A, current (I) and voltage (E) are in phase in a nonreactive ac circuit. At B, I and E are not in phase when reactance is present.

wrong conclusions, especially concerning impedance and standing waves. Then, you need to be keenly aware of the distinction among true, imaginary, and apparent power.

Reactance Does Not Consume Power

A pure inductance or a pure capacitance cannot dissipate any power. The only thing that such a component can do is store energy and then give it back to the circuit a fraction of a cycle later. In real life, the dielectrics or wires in coils and capacitors dissipate some power as heat, but ideal components would not do this.

A capacitor, as you have learned, stores energy as an electric field. An inductor stores energy as a magnetic field.

A component that contains reactance causes ac to shift in phase, so that the current is no longer exactly in step with the voltage. In a circuit with inductive reactance, the current lags the voltage by up to 90°, or one-quarter cycle. In a circuit with capacitive reactance, the current leads the voltage by up to 90°.

In a resistance-reactance circuit, true power is dissipated only in the resistive components. The reactive components exaggerate the VA power compared with the true power. Why, you ask, does reactance cause this discrepancy? In a circuit that is purely resistive, the voltage and current march right along in step with each other, and therefore, they combine in the most efficient possible way (Fig. 17-4A). But in a circuit containing reactance, the voltage and current are out of step with each other (Fig. 17-4B) because of their phase difference. Therefore, the actual energy expenditure, or true power, is not as great as the product of the voltage and the current.

True Power, VA Power, and Reactive Power

In an ac circuit or system containing nonzero resistance and nonzero reactance, the relationships among true power P_T, apparent (VA) power P_{VA}, and imaginary (reactive) power P_X are as follows:

$$P_{VA}^2 = P_T^2 + P_X^2$$
$$P_T < P_{VA}$$
$$P_X < P_{VA}$$

If there is no reactance in the circuit or system, then $P_{VA} = P_T$, and $P_X = 0$. Engineers strive to minimize, and if possible eliminate, the reactance in power-transmission systems.

Power Factor

In an ac circuit, the ratio of the true power to the VA power, P_T/P_{VA}, is called the *power factor*. If there is no reactance, the ideal case, then $P_T = P_{VA}$, and the power factor (PF) is equal to 1. If the circuit contains all reactance and no resistance of any significance (that is, zero or infinite resistance), then $P_T = 0$, and therefore $PF = 0$.

When a *load*, or a circuit in which you want power to be dissipated, contains resistance and reactance, then PF is between 0 and 1. That is, $0 < PF < 1$. The power factor can also be expressed as a percentage between 0 and 100, written $PF_\%$. Mathematically, we have these formulas for the power factor:

$$PF = P_T/P_{VA}$$
$$PF_\% = 100 P_T/P_{VA}$$

When a load has some resistance and some reactance, then some of the power is dissipated as true power, and some is rejected by the load as imaginary power. In a sense, this imaginary power is sent back to the power source.

There are two ways to determine the power factor in an ac circuit that contains reactance and resistance. One method is to find the cosine of the phase angle. The other method involves the ratio of the resistance to the absolute-value impedance.

Cosine of Phase Angle

Recall that in a circuit having reactance and resistance, the current and the voltage are not in phase. The phase angle (ϕ) is the extent, expressed in degrees, to which the current and the voltage differ in phase. If there is no reactance, then $\phi = 0°$. If there is a pure reactance, then either $\phi = +90°$ (if the reactance is inductive) or else $\phi = -90°$ (if the reactance is capacitive). The power factor is equal to the cosine of the phase angle:

$$PF = \cos \phi$$

Problem 17-1

Suppose a circuit contains no reactance, but a pure resistance of 600 Ω. What is the power factor?

Without doing any calculations, it is evident that $PF = 1$, because $P_{VA} = P_T$ in a pure resistance. That means $P_T/P_{VA} = 1$. But you can also look at this by noting that the phase angle is 0°, because the current is in phase with the voltage. Using your calculator, you can see that $\cos 0° = 1$. Therefore, $PF = 1 = 100\%$. The vector for this case is shown in Fig. 17-5.

Problem 17-2

Suppose a circuit contains a pure capacitive reactance of −40 Ω, but no resistance. What is the power factor?

Here, the phase angle is −90° (Fig. 17-6). A calculator will tell you that $\cos -90° = 0$. Therefore, $PF = 0$, and $P_T/P_{VA} = 0 = 0\%$. None of the power is true; all of it is reactive.

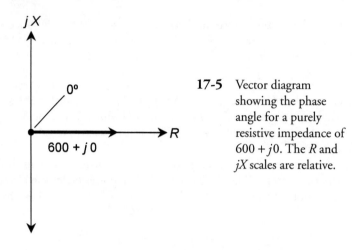

17-5 Vector diagram showing the phase angle for a purely resistive impedance of $600 + j0$. The R and jX scales are relative.

Problem 17-3

Suppose a circuit contains a resistance of 50 Ω and an inductive reactance of 50 Ω in series. What is the power factor?

The phase angle in this case is 45° (Fig. 17-7). The resistance and reactance vectors have equal lengths and form two sides of a right triangle, with the complex impedance vector forming the hypotenuse. To determine the power factor, you can use a calculator to find cos 45° = 0.707. This means that $P_T/P_{VA} = 0.707 = 70.7\%$.

The Ratio R/Z

The second way to calculate the power factor is to find the ratio of the resistance R to the absolute-value impedance Z. In Fig. 17-7, this is visually apparent. A right triangle is formed by the resistance vector R (the base), the reactance vector jX (the height), and the absolute-value impedance Z (the hypotenuse). The cosine of the phase angle is equal to the ratio of the base length to the hypotenuse length; this represents R/Z.

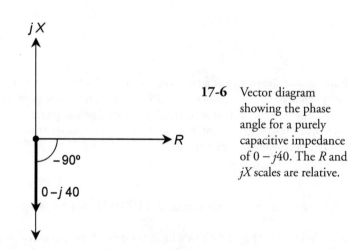

17-6 Vector diagram showing the phase angle for a purely capacitive impedance of $0 - j40$. The R and jX scales are relative.

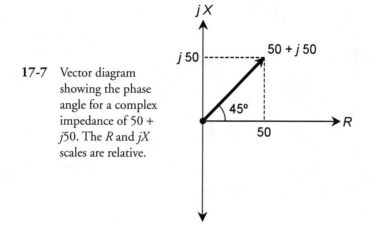

17-7 Vector diagram showing the phase angle for a complex impedance of 50 + *j*50. The *R* and *jX* scales are relative.

Problem 17-4

Suppose a circuit has an absolute-value impedance Z of 100 Ω, with a resistance $R = 80 \Omega$. What is the power factor?

Simply find the ratio $PF = R/Z = 80/100 = 0.8 = 80\%$. Note that it doesn't matter whether the reactance in this circuit is capacitive or inductive.

Problem 17-5

Suppose a circuit has an absolute-value impedance of 50 Ω, purely resistive. What is the power factor?

Here, $R = Z = 50 \Omega$. Therefore, $PF = R/Z = 50/50 = 1 = 100\%$.

Problem 17-6

Suppose a circuit has a resistance of 50 Ω and a capacitive reactance of −30 Ω in series. What is the power factor? Use the cosine method.

First, find the phase angle. Remember the formula: $\phi = \arctan(X/R)$, where X is the reactance and R is the resistance. Therefore, $\phi = \arctan(-30/50) = \arctan(-0.60) = -31°$. The power factor is the cosine of this angle; $PF = \cos(-31°) = 0.86 = 86\%$.

Problem 17-7

Suppose a circuit has a resistance of 30 Ω and an inductive reactance of 40 Ω. What is the power factor? Use the R/Z method.

Find the absolute-value impedance: $Z^2 = R^2 + X^2 = 30^2 + 40^2 = 900 + 1600 = 2500$. Therefore, $Z = 2500^{1/2} = 50 \Omega$, so $PF = R/Z = 30/50 = 0.60 = 60\%$. This problem can be represented vectorially by a 30:40:50 right triangle, as shown in Fig. 17-8.

How Much of the Power Is True?

The preceding formulas allow you to figure out, given the resistance, reactance, and VA power, how many watts are true or real power, and how many watts are imaginary or reactive power. This is important in RF equipment, because some RF wattmeters display VA power rather than true power. When there is reactance in a circuit or system, the wattage reading is therefore exaggerated.

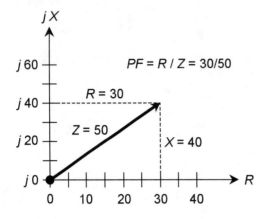

17-8   Illustration for Problem 17-7. (The vertical and horizontal scale increments differ; this is a common practice in graphs, often done for illustration convenience.)

Problem 17-8

Suppose a circuit has 50 Ω of resistance and 30 Ω of inductive reactance in series. A wattmeter shows 100 W, representing the VA power. What is the true power?

First, calculate the power factor. Suppose you use the phase-angle method. Then:

$$\phi = \arctan (X/R)$$
$$= \arctan (30/50) = 31°$$

The power factor is the cosine of the phase angle. Thus:

$$PF = \cos 31° = 0.86 = 86\%$$

Remember that $PF = P_T/P_{VA}$. This formula can be rearranged to solve for true power:

$$P_T = PF \times P_{VA}$$
$$= 0.86 \times 100$$
$$= 86 \text{ W}$$

Problem 17-9

Suppose a circuit has a resistance of 1000 Ω in parallel with a capacitance of 1000 pF. The frequency is 100 kHz. If a wattmeter designed to read VA power shows a reading of 88.0 W, what is the true power?

This problem is rather complicated because the components are in parallel. To begin, be sure the units are all in agreement so the formulas will work right. Convert the frequency to megahertz: $f = 100 \text{ kHz} = 0.100 \text{ MHz}$. Convert capacitance to microfarads: $C = 1000 \text{ pF} = 0.001000 \text{ }\mu\text{F}$. From the previous chapter, recall the formula for capacitive susceptance, and calculate it for this situation:

$$B_C = 6.28fC$$
$$= 6.28 \times 0.100 \times 0.001000$$
$$= 0.000628 \text{ S}$$

The conductance of the resistor, G, is the reciprocal of the resistance, R, as follows:

$$G = 1/R$$
$$= 1/1000$$
$$= 0.001000 \ \text{S}$$

Now, use the formulas for calculating resistance and reactance in terms of conductance and susceptance in parallel circuits. First, find the resistance:

$$R = G/(G^2 + B^2)$$
$$= 0.001000/(0.001000^2 + 0.000628^2)$$
$$= 0.001000/0.000001394$$
$$= 717 \ \Omega$$

Then, find the reactance:

$$X = -B/(G^2 + B^2)$$
$$= -0.000628/0.000001394$$
$$= -451 \ \Omega$$

Next, calculate the phase angle:

$$\phi = \arctan (X/R)$$
$$= \arctan (-451/717)$$
$$= \arctan (-0.629)$$
$$= -32.2°$$

The power factor is found from the phase angle as follows:

$$PF = \cos \phi$$
$$= \cos (-32.2°)$$
$$= 0.846 = 84.6\%$$

The VA power, P_{VA}, is given as 88.0 W. Therefore:

$$P_T = PF \times P_{VA}$$
$$= 0.846 \times 88.0$$
$$= 74.4 \ \text{W}$$

Power Transmission

Consider how electricity gets to your home. Generators produce large voltages and currents at a power plant. The problem: getting the electricity from the plant to the homes, businesses, and other facilities that need it. This process involves the use of long wire transmission lines. Transformers are also required to step the voltages up or down. As another example, consider a radio broadcast or communications station. The transmitter produces high-frequency ac. The problem is getting the power to be radiated by the antenna, located some distance from the transmitter. This involves the use of an RF transmission line. The most common type is coaxial cable. Two-wire line is also sometimes used. At ultrahigh and microwave frequencies, another kind of transmission line, known as a *waveguide*, is often employed.

Loss: The Less, The Better!

The overriding concern in any *power transmission* system is minimizing the loss. Power wastage occurs almost entirely as heat in the transmission line conductors and dielectric, and in objects near the line. Some loss can take the form of unwanted electromagnetic radiation from the line. Loss also occurs in transformers. Power loss in an electrical system is analogous to the loss of usable work produced by friction in a mechanical system. The less of it, the better!

In an ideal power transmission system, all of the power is VA power; that is, it is in the form of ac in the conductors and an alternating voltage between them. It is undesirable to have power in a transmission line or transformer exist in the form of true power, because that translates into either heat loss, or radiation loss, or both. The place for true power dissipation or radiation is in the load, such as electrical appliances or radio antennas.

Power Measurement in a Transmission Line

In an ac transmission line, power is measured by placing an ac voltmeter between the conductors, and an ac ammeter in series with one of the conductors (Fig. 17-9). Then the power P (in watts) is equal to the product of the rms voltage E (in volts) and the rms current I (in amperes). This technique can be used in any transmission line. But this is not necessarily an indication of the true power dissipated by the load at the end of the line.

Recall that any transmission line has a *characteristic impedance*. This value, Z_o, depends on the diameters of the line conductors, the spacing between the conductors, and the type of dielectric material that separates the conductors. If the load is a pure resistance R containing no reactance, and if $R = Z_o$, then the power indicated by the voltmeter/ammeter scheme will be the same as the true power dissipated by the load—provided that the voltmeter and ammeter are placed at the load end of the transmission line.

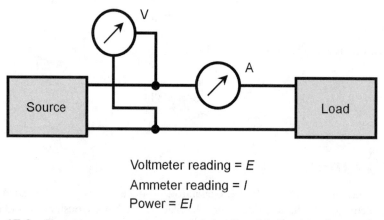

Voltmeter reading = E
Ammeter reading = I
Power = EI

17-9   Power measurement in a transmission line. Ideally, the voltage and the current should be measured at the same physical point on the line.

If the load is a pure resistance but it differs from the characteristic impedance of the line, then the voltmeter and ammeter will not give an indication of the true power. Also, if there is any reactance in the load, the voltmeter/ammeter method will not be accurate, even if the resistive component happens to be the same as the characteristic impedance of the line. The physics of this is rather complicated, and we won't get into the details here. But you should remember that it is optimum for the impedance of a load to be a pure resistance R, such that $R = Z_o$. When this is not the case, an *impedance mismatch* is said to exist.

Small impedance mismatches can often be tolerated in power transmission systems. But this is not always the case. In very high frequency (VHF), ultrahigh frequency (UHF), and microwave radio transmitting systems, even a small impedance mismatch between the load and the line can cause excessive power losses in the line. An impedance mismatch can usually be corrected by means of a *matching transformer* between a transmission line and the load, and/or the deliberate addition of reactance at the load end of the line to cancel out any existing load reactance.

Loss in a Mismatched Line

When a transmission line is terminated in a resistance $R = Z_o$, then the current and the voltage are constant all along the line, provided the line has no loss. The ratio of the voltage to the current, E/I, is equal to R and also equal to Z_o. But this is an idealized case. No line is completely *lossless*.

In a real-world transmission line, the current and voltage gradually decrease as a signal makes its way from the source to the load. But if the load is a pure resistance equal to the characteristic impedance of the line, the current and voltage remain in the same ratio at all points along the line (Fig. 17-10).

Standing Waves

If the load is not perfectly matched to the line, the current and voltage vary in a complicated way along the length of the line. In some places, the current is high; in other places it is low. The max-

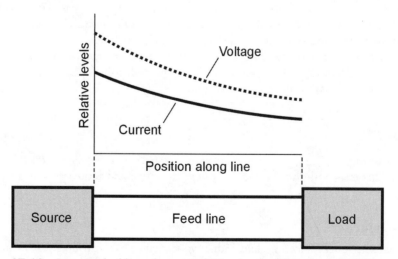

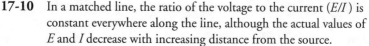

17-10   In a matched line, the ratio of the voltage to the current (E/I) is constant everywhere along the line, although the actual values of E and I decrease with increasing distance from the source.

ima and minima are called *loops* and *nodes*, respectively. At a *current loop*, the voltage is minimum (a *voltage node*), and at a *current node*, the voltage is maximum (a *voltage loop*). The current and voltage loops and nodes along a mismatched transmission line, if graphed as functions of the position on the line, form wavelike patterns that remain fixed over time. They just stand there. For this reason, they are called *standing waves*.

Standing-Wave Loss

At current loops, the loss in line conductors reaches a maximum. At voltage loops, the loss in the dielectric reaches a maximum. At current nodes, the loss in the conductors reaches a minimum. At voltage nodes, the loss in the dielectric reaches a minimum. It is tempting to suppose that everything would average out here, but it doesn't work that way! Overall, in a mismatched line, the line losses are greater than they are in a perfectly matched line. This extra line loss increases as the mismatch gets worse.

Transmission-line mismatch loss, also called *standing-wave loss*, occurs in the form of heat dissipation. It is true power. Any true power that goes into heating up a transmission line is wasted, because it cannot be dissipated in the load.

The greater the mismatch, the more severe the standing-wave loss becomes. The more loss a line has to begin with (that is, when it is perfectly matched), the more loss is caused by a given amount of mismatch. Standing-wave loss also increases as the frequency increases, if all other factors are held constant. This loss is the most significant, and the most harmful, in long lengths of transmission line, especially in RF practice at VHF, UHF, and microwave frequencies.

Line Overheating

A severe mismatch between the load and the transmission line can cause another problem: physical damage to, or destruction of, the line!

A feed line might be able to handle a kilowatt (1 kW) of power when it is perfectly matched. But if a severe mismatch exists and you try to feed 1 kW into the line, the extra current at the current loops can heat the conductors to the point where the dielectric material melts and the line shorts out. It is also possible for the voltage at the voltage loops to cause arcing between the line conductors. This perforates and/or burns the dielectric, ruining the line.

When an RF transmission line must be used with a mismatch, *derating functions* are required to determine how much power the line can safely handle. Manufacturers of prefabricated lines such as coaxial cable can supply you with this information.

Resonance

One of the most important phenomena in ac circuits, especially in RF engineering, is the property of *resonance*. This is a condition that occurs when capacitive and inductive reactance cancel each other out.

Series Resonance

Recall that capacitive reactance, X_C, and inductive reactance, X_L, can be equal in magnitude, although they are always opposite in effect. In any circuit containing an inductance and capacitance, there exists a frequency at which $X_L = -X_C$. This condition constitutes resonance. In a simple *LC* circuit, there is only one such frequency. But in some circuits involving transmission lines or antennas,

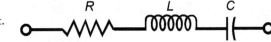

17-11 A series *RLC* circuit.

there can be many such frequencies. The lowest frequency at which resonance occurs is called the *resonant frequency*, symbolized f_o.

Refer to the schematic diagram of Fig. 17-11. You should recognize this as a series *RLC* circuit. At some particular frequency, $X_L = -X_C$. This is inevitable if L and C are finite and nonzero. This frequency is f_o for the circuit. At f_o, the effects of capacitive reactance and inductive reactance cancel out. The result is that the circuit appears as a pure resistance, with a value that is theoretically equal to R.

If $R = 0$, that is, if the resistor is a short circuit, then the circuit is called a *series LC circuit*, and the impedance at resonance will be theoretically $0 + j0$. The circuit will offer no opposition to the flow of alternating current at the frequency f_o. This condition is *series resonance*. In a practical series *LC* circuit, there is always a little bit of loss in the coil and capacitor, so the real part of the complex impedance is not exactly equal to 0 (although it can be extremely small).

Parallel Resonance

Refer to the circuit diagram of Fig. 17-12. This is a *parallel RLC circuit*. Remember that, in this sort of situation, the resistance R should be thought of as a conductance G, with $G = 1/R$. Then the circuit can be called a *parallel GLC circuit*.

At some particular frequency f_o, the inductive susceptance B_L will exactly cancel the capacitive susceptance B_C; that is, $B_L = -B_C$. This is inevitable for some frequency f_o, as long as the circuit contains finite, nonzero inductance and finite, nonzero capacitance. At the frequency f_o, the susceptances cancel each other out, leaving theoretically zero susceptance. The admittance through the circuit is then very nearly equal to the conductance, G, of the resistor.

If the circuit contains no resistor, but only a coil and capacitor, it is called a *parallel LC circuit*, and the admittance at resonance will be theoretically $0 + j0$. That means the circuit will offer great opposition to alternating current at f_o, and the complex impedance will theoretically be infinite! This condition is *parallel resonance*. In a practical parallel *LC* circuit, there is always a little bit of loss in the coil and capacitor, so the real part of the complex impedance is not infinite (although it can be extremely large).

Calculating Resonant Frequency

The formula for calculating resonant frequency f_o, in terms of the inductance L in henrys and the capacitance C in farads, is as follows:

$$f_o = 1/[2\pi (LC)^{1/2}]$$

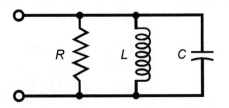

17-12 A parallel *RLC* circuit.

Considering $\pi = 3.14$ to three significant figures, this formula can be simplified to:

$$f_\circ = 0.159/(LC)^{1/2}$$

The $\frac{1}{2}$ power of a quantity represents the positive square root of that quantity. The preceding formulas are valid for series-resonant and parallel-resonant *RLC* circuits.

The formula will also work if you want to find $f_\circ$ in megahertz (MHz) when L is given in microhenrys (μH) and C is in microfarads (μF). These values are far more common than hertz, henrys, and farads in electronic circuits. Just remember that millions of hertz go with million*ths* of henrys, and with million*ths* of farads.

The Effects of *R* and *G*

Interestingly, the value of *R* or *G* does not affect the resonant frequency in either type of circuit. But these quantities are significant, nevertheless! The presence of nonzero resistance in a *series-resonant circuit*, or nonzero conductance in a *parallel-resonant circuit*, makes the resonant frequency less well-defined. Engineers say that the *resonant frequency response* becomes "more broad" or "less sharp."

In a series circuit, the resonant frequency response becomes more broad as the *resistance* increases. In a parallel circuit, the resonant frequency response becomes more broad as the *conductance* increases. The sharpest possible responses occur when $R = 0$ in a series circuit, and when $G = 0$ (that is, $R = \infty$) in a parallel circuit.

Problem 17-10

Find the resonant frequency of a series circuit with an inductance of 100 μH and a capacitance of 100 pF.

First, convert the capacitance to microfarads: 100 pF = 0.000100 μF. Then find the product $LC = 100 \times 0.000100 = 0.0100$. Take the square root of this, getting 0.100. Finally, divide 0.159 by 0.100, getting $f_\circ = 1.59$ MHz.

Problem 17-11

Find the resonant frequency of a parallel circuit consisting of a 33-μH coil and a 47-pF capacitor.

Again, convert the capacitance to microfarads: 47 pF = 0.000047 μF. Then find the product $LC = 33 \times 0.000047 = 0.00155$. Take the square root of this, getting 0.0394. Finally, divide 0.159 by 0.0394, getting $f_\circ = 4.04$ MHz.

Problem 17-12

Suppose you want to design a circuit so that it has $f_\circ = 9.00$ MHz. You have a 33-pF fixed capacitor available. What size coil will be needed to get the desired resonant frequency?

Use the formula for the resonant frequency, and plug in the values. This will allow you to use simple arithmetic to solve for L. Convert the capacitance to microfarads: 33 pF = 0.000033 μF. Then calculate as follows:

$$f_\circ = 0.159/(LC)^{1/2}$$
$$9.00 = 0.159/(L \times 0.000033)^{1/2}$$
$$9.00^2 = 0.159^2/(0.000033 \times L)$$
$$81.0 = 0.0253/(0.000033 \times L)$$

$$81.0 \times 0.000033 \times L = 0.0253$$
$$0.00267 \times L = 0.0253$$
$$L = 0.0253/0.00267$$
$$= 9.48 \ \mu H$$

Problem 17-13

Suppose a circuit must be designed to have $f_o = 455$ kHz. A coil of 100 μH is available. What size capacitor is needed?

Convert the frequency to megahertz: 455 kHz = 0.455 MHz. Then the calculation proceeds in the same way as with the preceding problem:

$$f_o = 0.159/(LC)^{1/2}$$
$$0.455 = 0.159/(100 \times C)^{1/2}$$
$$0.455^2 = 0.159^2/(100 \times C)$$
$$0.207 = 0.0253/(100 \times C)$$
$$0.207 \times 100 \times C = 0.0253$$
$$20.7 \times C = 0.0253$$
$$C = 0.0253/20.7$$
$$= 0.00122 \ \mu F$$
$$= 1220 \ pF$$

In practical circuits, variable inductors and/or variable capacitors are often placed in tuned circuits, so that small errors in the frequency can be compensated for. The most common approach is to design the circuit for a frequency slightly higher than f_o, and to use a *padder capacitor* in parallel with the main capacitor (Fig. 17-13).

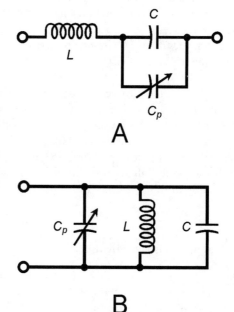

17-13 Padding capacitors (C_p) allow limited adjustment of the resonant frequency in a series *LC* circuit (as shown at A), or in a parallel *LC* circuit (as shown at B).

Resonant Devices

Resonant circuits often consist of coils and capacitors in series or parallel, but there are other kinds of hardware that exhibit resonance. Some of these are as follows.

Piezoelectric Crystals

Pieces of quartz, when cut into thin wafers and subjected to voltages, will vibrate at high frequencies. Because of the physical dimensions of such a *piezoelectric crystal*, these vibrations occur at a precise frequency f_o, and also at whole-number multiples of f_o. These multiples, $2f_o$, $3f_o$, $4f_o$, and so on, are called *harmonic frequencies* or simply *harmonics*. The frequency f_o is called the *fundamental frequency* or simply the *fundamental*. The fundamental, f_o, is defined as the lowest frequency at which resonance occurs. Quartz crystals can be made to act like LC circuits in electronic devices. A crystal exhibits an impedance that varies with frequency. The reactance is zero at f_o and the harmonic frequencies.

Cavities

Lengths of metal tubing, cut to specific dimensions, exhibit resonance at very high, ultrahigh, and microwave radio frequencies. They work in much the same way as musical instruments resonate with sound waves. But the waves are electromagnetic, rather than acoustic. Such *cavities*, also called cavity *resonators*, have reasonable physical dimensions at frequencies above about 150 MHz. Below this frequency, a cavity can be made to work, but it is long and unwieldy. Like crystals, cavities resonate at a fundamental frequency f_o, and also at harmonic frequencies.

Sections of Transmission Line

When a transmission line is cut to $\frac{1}{4}$ wavelength, or to any whole-number multiple of this, it behaves as a resonant circuit. The most common length for a transmission-line resonator is a $\frac{1}{4}$ wavelength. Such a piece of transmission line is called a *quarter-wave section*.

When a quarter-wave section is short-circuited at the far end, it acts like a parallel-resonant LC circuit, and has a high resistive impedance at the resonant frequency f_o. When it is open at the far end, it acts as a series-resonant LC circuit, and has a low resistive impedance at f_o. In effect, a quarter-wave section converts an ac short circuit into an ac open circuit and vice versa, at a specific frequency f_o.

The length of a quarter-wave section depends on the desired f_o. It also depends on how fast the electromagnetic energy travels along the line. This speed is specified in terms of a *velocity factor*, abbreviated v. The value of v is given as a fraction of the speed of light. Typical transmission lines have velocity factors ranging from about 0.66 to 0.95 (or 66 percent to 95 percent). This factor is provided by the manufacturers of prefabricated lines such as coaxial cable.

If the frequency in megahertz is f_o and the velocity factor of a line is v, then the length L_{ft} of a quarter-wave section of transmission line, in feet, is given by this formula:

$$L_{ft} = 246v/f_o$$

The length L_m in meters is given by this:

$$L_m = 75.0v/f_o$$

We use L here to stand for "length," not "inductance"!

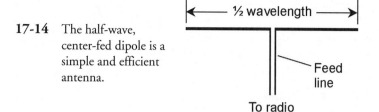

17-14 The half-wave, center-fed dipole is a simple and efficient antenna.

Antennas

Many types of antennas exhibit resonant properties. The simplest type of resonant antenna, and the only kind that will be mentioned here, is the center-fed, half-wavelength *dipole antenna* (Fig. 17-14).

The length L_{ft}, in feet, for a dipole antenna at a frequency of f_o, in megahertz, is given by the following formula:

$$L_{ft} = 468/f_o$$

This takes into account the fact that electromagnetic fields travel along a wire at about 95 percent of the speed of light. A straight, thin wire in free space has a velocity factor of approximately 0.95.

If the length of the half-wave dipole is specified in meters as L_m, then:

$$L_m = 143/f_o$$

A half-wave dipole has a purely resistive impedance of about 73 Ω at its fundamental frequency f_o. But this type of antenna is also resonant at all harmonics of f_o. The dipole is a full wavelength long at $2f_o$; it is $^3/_2$ wavelength long at $3f_o$; it is two full wavelengths long at $4f_o$, and so on.

Radiation Resistance

At f_o and all of the odd harmonics, the antenna behaves like a series-resonant *RLC* circuit with a fairly low resistance. At all even harmonics, the antenna acts like a parallel-resonant *RLC* circuit with a high resistance. Does this confuse you? There's no resistor in Fig. 17-14! Where, you ask, does the resistance come from in the half-wave dipole? The answer to this is rather esoteric, and it brings to light an interesting property that all antennas have. It is called *radiation resistance*, and is a crucial factor in the design and construction of all RF antenna systems.

When electromagnetic energy is fed into an antenna, power is radiated into space in the form of radio waves. This is a manifestation of true power, just as the dissipation of power in a pure resistance is a manifestation of true power. Although there is no physical resistor in Fig. 17-14, the radiation of radio waves is like power dissipation in a pure resistance. In fact, if a half-wave dipole antenna were replaced with a 73-Ω nonreactive resistor that could dissipate enough power without burning out, a radio transmitter connected to the opposite end of the line wouldn't know the difference. (But a receiver would!)

Problem 17-14

How many feet long is a quarter-wave section of transmission line at 7.05 MHz, if the velocity factor is 0.800?

Just use the formula:

$$L_{ft} = 246v/f_o$$
$$= (246 \times 0.800)/7.05$$
$$= 197/7.05$$
$$= 27.9 \text{ ft}$$

Quiz

Refer to the text in this chapter if necessary. A good source is 18 or more correct. Answers are in the back of the book.

1. The power in a pure reactance is
   (a) radiated.
   (b) true.
   (c) imaginary.
   (d) apparent.

2. Which of the following is *not* an example of true power?
   (a) Power in the form of heat, produced by dc flowing through a resistor
   (b) Power in the form of electromagnetic fields, radiated from a radio antenna
   (c) The product of the rms ac through a capacitor and the rms voltage across it
   (d) Power in the form of heat, produced by losses in an RF transmission line

3. Suppose the apparent power in a circuit is 100 W, and the imaginary power is 40 W. What is the true power?
   (a) 92 W
   (b) 100 W
   (c) 140 W
   (d) It is impossible to determine from this information.

4. Power factor is equal to
   (a) apparent power divided by true power.
   (b) imaginary power divided by apparent power.
   (c) imaginary power divided by true power.
   (d) true power divided by apparent power.

5. Suppose a circuit has a resistance of 300 Ω and an inductance of 13.5 μH in series, and is operated at 10.0 MHz. What is the power factor?
   (a) 0.334
   (b) 0.999
   (c) 0.595
   (d) It cannot be determined from the information given.

6. Suppose a series circuit has $Z = 88.4\ \Omega$, with $R = 50.0\ \Omega$. What is the power factor, expressed as a percentage?

(a) 99.9 percent

(b) 56.6 percent

(c) 60.5 percent

(d) 29.5 percent

7. Suppose a series circuit has $R = 53.5\ \Omega$, with $X = 75.5\ \Omega$. What is the power factor, expressed as a percentage?

(a) 70.9 percent

(b) 81.6 percent

(c) 57.8 percent

(d) 63.2 percent

8. The phase angle in an ac circuit is equal to

(a) arctan (Z/R).

(b) arctan (R/Z).

(c) arctan (R/X).

(d) arctan (X/R).

9. Suppose an ac ammeter and an ac voltmeter indicate that there are 220 W of VA power in a circuit that consists of a resistance of 50 Ω in series with a capacitive reactance of $-20\ \Omega$. What is the true power?

(a) 237 W

(b) 204 W

(c) 88.0 W

(d) 81.6 W

10. Suppose an ac ammeter and an ac voltmeter indicate that there are 57 W of VA power in a circuit. The resistance is known to be 50 Ω, and the true power is known to be 40 W. What is the absolute-value impedance?

(a) 50 Ω

(b) 57 Ω

(c) 71 Ω

(d) It is impossible to determine on the basis of this data.

11. Which of the following should be minimized in an RF transmission line?

(a) The load impedance

(b) The load resistance

(c) The line loss

(d) The transmitter power

12. Which of the following does *not* increase the loss in a transmission line?
  (a) Reducing the power output of the source
  (b) Increasing the degree of mismatch between the line and the load
  (c) Reducing the diameter of the line conductors
  (d) Raising the frequency

13. Which of the following is a significant problem that standing waves can cause in an RF transmission line?
  (a) Line overheating
  (b) Excessive power loss
  (c) Inaccuracy in power measurement
  (d) All of the above

14. Suppose a coil and capacitor are in series. The inductance is 88 mH and the capacitance is 1000 pF. What is the resonant frequency?
  (a) 17 kHz
  (b) 540 Hz
  (c) 17 MHz
  (d) 540 kHz

15. Suppose a coil and capacitor are in parallel, with $L = 10.0\ \mu H$ and $C = 10$ pF. What is f_o?
  (a) 15.9 kHz
  (b) 5.04 MHz
  (c) 15.9 MHz
  (d) 50.4 MHz

16. Suppose you want to build a series-resonant circuit with $f_o = 14.1$ MHz. A coil of 13.5 μH is available. How much capacitance is needed?
  (a) 0.945 μF
  (b) 9.45 pF
  (c) 94.5 pF
  (d) 945 pF

17. Suppose you want to build a parallel-resonant circuit with $f_o = 21.3$ MHz. A capacitor of 22.0 pF is available. How much inductance is needed?
  (a) 2.54 mH
  (b) 254 μH
  (c) 25.4 μH
  (d) 2.54 μH

18. A ¼-wave section of transmission line is cut for use at 21.1 MHz. The line has a velocity factor of 0.800. What is its physical length in meters?
  (a) 11.1 m
  (b) 3.55 m

(c) 8.87 m

(d) 2.84 m

19. What is the fourth harmonic of 800 kHz?

(a) 200 kHz

(b) 400 kHz

(c) 3.20 MHz

(d) 4.00 MHz

20. Suppose you want to build a $\frac{1}{2}$-wave dipole antenna designed to have a fundamental resonant frequency of 3.60 MHz. How long should you make it, as measured from end to end in feet?

(a) 130 ft

(b) 1680 ft

(c) 39.7 ft

(d) 515 ft

18
CHAPTER

Transformers and Impedance Matching

TRANSFORMERS ARE USED TO OBTAIN THE OPTIMUM VOLTAGE FOR THE OPERATION OF A CIRCUIT OR system. Transformers can also match impedances between a circuit and a load, or between two different circuits. Transformers can be used to provide dc isolation between electronic circuits while letting ac pass. Another application is to mate balanced and unbalanced circuits, feed systems, and loads.

Principle of the Transformer

When two wires are near each other and one of them carries a fluctuating current, a fluctuating current is induced in the other wire. This effect is known as *electromagnetic induction*. All ac transformers work according to the principle of electromagnetic induction. If the first wire carries sine-wave ac of a certain frequency, then the *induced current* is sine-wave ac of the same frequency in the second wire.

The closer the two wires are to each other, the greater is the induced current, for a given current in the first wire. If the wires are wound into coils and placed along a common axis (Fig. 18-1), the induced current will be greater than if the wires are straight and parallel. Even more *coupling*, or efficiency of induced-current transfer, is obtained if the two coils are wound one atop the other.

Primary and Secondary

The two windings, along with the core on which they are wound, constitute a transformer. The first coil is called the *primary winding*, and the second coil is known as the *secondary winding*. These are often spoken of simply as the *primary* and the *secondary*. The induced current in the secondary creates a voltage between its end terminals. In a *step-down* transformer, the secondary voltage is less than the primary voltage. In a *step-up* transformer, the secondary voltage is greater than the primary voltage. The primary voltage is abbreviated E_{pri}, and the secondary voltage is abbreviated E_{sec}. Unless otherwise stated, effective (rms) voltages are always specified.

The windings of a transformer have inductance, because they are coils. The required inductances of the primary and secondary depend on the frequency of operation, and also on the resistive part of the impedance in the circuit. As the frequency increases, the needed inductance decreases. At high resistive impedances, more inductance is generally needed than at low resistive impedances.

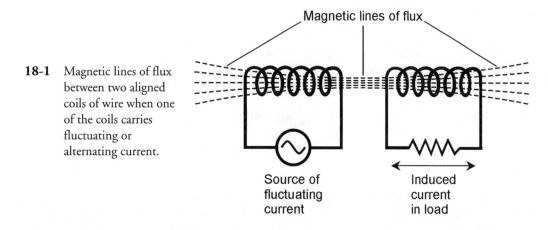

18-1 Magnetic lines of flux between two aligned coils of wire when one of the coils carries fluctuating or alternating current.

Turns Ratio

The *primary-to-secondary turns ratio* in a transformer is the ratio of the number of turns in the primary, T_{pri}, to the number of turns in the secondary, T_{sec}. This ratio is written $T_{pri}:T_{sec}$ or T_{pri}/T_{sec}. In a transformer with excellent primary-to-secondary coupling, the following relationship always holds:

$$E_{pri}/E_{sec} = T_{pri}/T_{sec}$$

That is, the primary-to-secondary voltage ratio is always equal to the primary-to-secondary turns ratio (Fig. 18-2).

Problem 18-1

Suppose a transformer has a primary-to-secondary turns ratio of exactly 9:1. The ac voltage at the primary is 117 V rms. Is this a step-up transformer or a step-down transformer? What is the voltage across the secondary?

This is a step-down transformer. Simply plug in the numbers in the preceding equation and solve for E_{sec}, as follows:

$$E_{pri}/E_{sec} = T_{pri}/T_{sec}$$
$$117/E_{sec} = 9.00$$
$$1/E_{sec} = 9.00/117$$
$$E_{sec} = 117/9.00$$
$$= 13.0 \text{ V rms}$$

18-2 The primary voltage (E_{pri}) and secondary voltage (E_{sec}) in a transformer depend on the number of turns in the primary winding (T_{pri}) versus the number of turns in the secondary winding (T_{sec}).

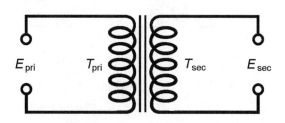

Problem 18-2

Consider a transformer with a primary-to-secondary turns ratio of exactly 1:9. The voltage at the primary is 121.4 V rms. Is this a step-up transformer or a step-down transformer? What is the voltage at the secondary?

This is a step-up transformer. Plug in numbers and solve for E_{sec}, as follows:

$$121.4/E_{sec} = 1/9.000$$
$$E_{sec}/121.4 = 9.000$$
$$E_{sec} = 9.000 \times 121.4$$
$$= 1093 \text{ V rms}$$

Sometimes the *secondary-to-primary turns ratio* is given, rather than the primary-to-secondary turns ratio. This is written T_{sec}/T_{pri}. In a step-down unit, T_{sec}/T_{pri} is less than 1. In a step-up unit, T_{sec}/T_{pri} is greater than 1. When you hear someone say that such-and-such a transformer has a certain "turns ratio," say 10:1, be sure of which ratio is meant, T_{pri}/T_{sec} or T_{sec}/T_{pri}! If you get it wrong, you'll have the secondary voltage wrong by a factor of the *square* of the turns ratio.

Ferromagnetic Cores

If a ferromagnetic substance such as laminated iron or powdered iron is placed within the pair of coils, the extent of coupling is increased far above that possible with an air core. But this improvement in coupling is obtained at a price. Some energy is invariably lost as heat in the core. Also, ferromagnetic cores limit the maximum frequency at which a transformer will work well.

The schematic symbol for an air-core transformer consists of two inductor symbols back-to-back (Fig. 18-3A). If a laminated iron core is used, two parallel lines are added to the schematic symbol (Fig. 18-3B). If the core is made of powdered iron, the two parallel lines are broken or dashed (Fig. 18-3C).

In transformers for 60-Hz utility ac, and also for low audio-frequency (AF) use, sheets of an alloy called *silicon steel*, glued together in layers, are often employed as transformer cores. The silicon steel is sometimes called *transformer iron*. The reason layering is used, rather than making the core from a single mass of metal, is that the magnetic fields from the coils cause currents to flow in a solid core. These *eddy currents* go in circles, heating up the core and wasting energy that would otherwise be transferred from the primary to the secondary. Eddy currents are choked off by breaking up the core into layers, so that currents cannot flow very well in circles.

A rather esoteric form of loss, called *hysteresis loss*, occurs in all ferromagnetic transformer cores, but especially laminated iron. Hysteresis is the tendency for a core material to be sluggish in accept-

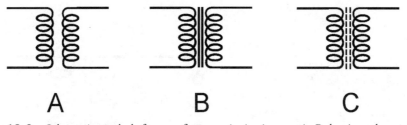

18-3   Schematic symbols for transformers. At A, air core. At B, laminated iron core. At C, ferrite or powdered iron core.

ing a fluctuating magnetic field. Laminated cores exhibit high hysteresis loss above the AF range, and are therefore not good above a few kilohertz.

At frequencies up to several tens of megahertz, *powdered iron* works well for RF transformers. This material has high magnetic permeability and concentrates the flux efficiently. High permeability cores minimize the number of turns needed in the coils, and this minimizes the loss that occurs in the wires.

Geometries

The properties of a transformer depend on the shape of its core, and on the way in which the wires are wound on it. There are several different *geometries* used with transformers.

E Core

A common core for a power transformer is the *E core*, so named because it is shaped like the capital letter E. A bar, placed at the open end of the E, completes the core assembly after the coils have been wound on the E-shaped section (Fig. 18-4A).

The primary and secondary windings can be placed on an E core in either of two ways. The simpler winding method is to put both the primary and the secondary around the middle bar of the E (Fig. 18-4B). This is called the *shell method* of transformer winding. It provides maximum coupling between the windings. However, this scheme results in considerable capacitance between the primary and the secondary. Such *interwinding capacitance* can sometimes be tolerated, but often it cannot. Another disadvantage of the shell geometry is that, when windings are placed one on top of the other, the transformer cannot handle very much voltage. High voltages cause arcing between the windings, which can destroy the insulation on the wires and lead to permanent short circuits.

Another winding method is the *core method.* In this scheme, one winding is placed at the bottom of the E section, and the other winding is placed at the top (Fig. 18-4C). The coupling occurs

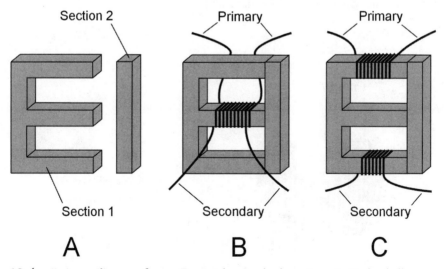

18-4   At A, a utility transformer E core, showing both sections. At B, the shell winding method. At C, the core winding method.

by means of magnetic flux in the core. The interwinding capacitance is lower than it is in a shell-wound transformer because the windings are physically farther apart. Also, a core-wound transformer can handle higher voltages than a shell-wound transformer of the same physical size. Sometimes the center part of the E is left out of the core when the core winding scheme is used.

Shell-wound and core-wound transformers are almost universally employed at 60 Hz. These configurations are also common at AF.

Solenoidal Core

A pair of cylindrical coils, wound around a rod-shaped piece of powdered iron or ferrite, was once a common configuration for RF transformers. Sometimes this type of transformer is still seen, although it is most often used as a *loopstick antenna* in portable radio receivers and in radio direction-finding equipment. The coil windings can be placed one atop the other, or they can be separated (Fig. 18-5) to reduce the capacitance between the primary and secondary.

In a loopstick antenna, the primary serves to pick up the radio signals. The secondary winding provides an optimum impedance match to the first amplifier stage, or *front end*, of the radio receiver. The use of transformers for impedance matching is discussed later in this chapter.

Toroidal Core

The *toroidal core* (or *toroid*) has become common for winding RF transformers. The core is a donut-shaped ring of powdered iron. The coils are wound around the donut. The complete assembly is called a *toroidal transformer.* The primary and secondary can be wound one over the other, or they can be wound over different parts of the core (Fig. 18-6). As with other transformers, when the windings are one on top of the other, there is more interwinding capacitance than when they are separated.

Toroids confine practically all the magnetic flux within the core material. This allows toroidal coils and transformers to be placed near other components without inductive interaction. Also, a toroidal coil or transformer can be mounted directly on a metal chassis, and the operation is not affected (assuming the wire is insulated or enameled).

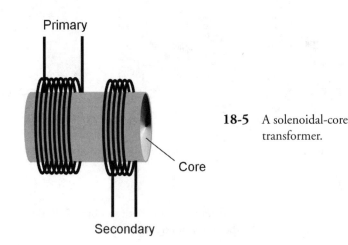

18-5   A solenoidal-core transformer.

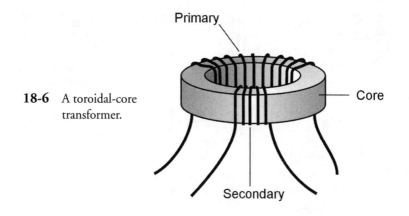

18-6 A toroidal-core transformer.

A toroidal core provides considerably more inductance per turn, for the same kind of ferromagnetic material, than a solenoidal core. It is common to see toroidal coils or transformers that have inductance values as high as 100 mH.

Pot Core

Even more inductance per turn can be obtained with a *pot core*. This is a shell of ferromagnetic material that is wrapped around a loop-shaped coil. The core is manufactured in two halves (Fig. 18-7). You wind the coil inside one of the halves, and then bolt the two together. The final core completely surrounds the loop, and the magnetic flux is confined to the core material.

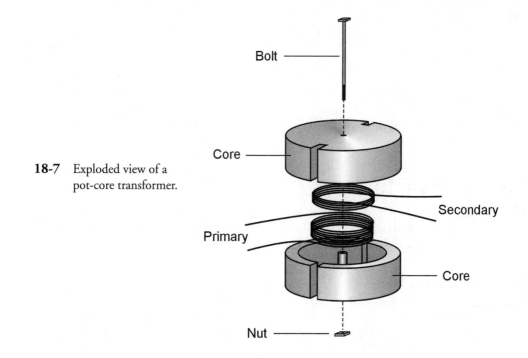

18-7 Exploded view of a pot-core transformer.

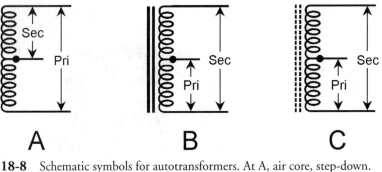

18-8   Schematic symbols for autotransformers. At A, air core, step-down.
At B, laminated iron core, step-up. At C, ferrite or powdered iron
core, step-up.

Like the toroid, the pot core is self-shielding. There is essentially no coupling to external components. A pot core can be used to wind a single, high-inductance coil. Inductance values of more than 1 H are possible with a reasonable number of wire turns.

In a pot-core transformer, the primary and secondary must be wound next to each other. This is unavoidable because of the geometry. Therefore, the interwinding capacitance of a pot-core transformer is high. Pot cores are useful at AF and the lowest-frequency parts of the RF spectrum. They are rarely employed at high radio frequencies.

Autotransformer

In some situations, there is no need to provide dc isolation between the primary and secondary windings of a transformer. In a case of this sort, an *autotransformer* can be used. It has a single, tapped winding.

Figure 18-8 shows three autotransformer configurations. The unit shown at A has an air core, and is a step-down type. The unit at B has a laminated iron core, and is a step-up type. The unit at C has a powdered iron core, and is a step-up type.

You'll sometimes see autotransformers in radio receivers or transmitters. Autotransformers work well in impedance-matching applications, and also perform well as solenoidal loopstick antennas. Autotransformers are occasionally, but not often, used in AF applications and in 60-Hz utility wiring. In utility circuits, autotransformers can step the voltage down by a large factor, but they aren't used to step voltages up by more than a few percent.

Power Transformers

Any transformer used in the 60-Hz utility line, intended to provide a certain rms ac voltage for the operation of electrical circuits, is a *power transformer*. Power transformers exist in a vast range of physical sizes, from smaller than a tennis ball to as big as a room.

At the Generating Plant

The largest transformers are employed at the places where electricity is generated. Not surprisingly, high-energy power plants have bigger transformers that develop higher voltages than low-energy, local power plants. These transformers must be able to handle high voltages and large currents simultaneously.

When electrical energy must be sent over long distances, extremely high voltages are used. This is because, for a given amount of power ultimately dissipated by the loads, the current is lower when the voltage is higher. Lower current translates into reduced loss in the transmission line.

Recall the formula $P = EI$, where P is the power (in watts), E is the voltage (in volts), and I is the current (in amperes). If you can make the voltage 10 times larger, for a given power level, then the current is reduced to $\frac{1}{10}$ as much. The *ohmic losses* in the wires are proportional to the *square* of the current. Remember that $P = I^2R$, where P is the power (in watts), I is the current (in amperes), and R is the resistance (in ohms). Engineers can't do much about the wire resistance or the power consumed by the loads, but they can adjust the voltage, and thereby the current.

Suppose the voltage in a power transmission line is increased by a factor of 10, and the load at the end of the line draws constant power. This increase in the voltage reduces the current to $\frac{1}{10}$ of its previous value. As a result, the ohmic loss is cut to $(\frac{1}{10})^2$, or $\frac{1}{100}$, of its previous amount. That's a major improvement in the efficiency of the transmission line, at least in terms of the loss caused by the resistance in the wires—and it is the reason why regional power plants have massive transformers capable of generating hundreds of thousands of volts.

Along the Line

Extreme voltage is good for *high-tension* power transmission, but it's certainly of no use to an average consumer. The wiring in a high-tension system must be done using precautions to prevent arcing (sparking) and short circuits. Personnel must be kept at least several meters away from the wires. Can you imagine trying to use an appliance, say a home computer, by plugging it into a 500,000-V rms electrical outlet?

Medium-voltage power lines branch out from the major lines, and step-down transformers are used at the branch points. These lines fan out to still lower-voltage lines, and step-down transformers are employed at these points, too. Each transformer must have windings heavy enough to withstand the product $P = EI$, the amount of VA power delivered to all the subscribers served by that transformer, at periods of peak demand.

Sometimes, such as during a heat wave, the demand for electricity rises above the normal peak level. This loads down the circuit to the point that the voltage drops several percent. This is called a *brownout*. If consumption rises further still, a dangerous current load is placed on one or more intermediate power transformers. Circuit breakers in the transformers protect them from destruction by opening the circuit. Then there is a temporary *blackout*.

At individual homes and buildings, transformers step the voltage down to either 234 V rms or 117 V rms. Usually, 234-V rms electricity is provided in the form of three sine waves, called *phases*, each separated by 120°, and each appearing at one of the three slots in the outlet (Fig. 18-9A). This voltage is commonly employed with heavy appliances, such as the kitchen oven/stove (if they are electric), heating (if it is electric), and the laundry washer and dryer. A 117-V rms outlet supplies just one phase, appearing between two of the three slots in the outlet. The third opening in the outlet leads to an earth ground (Fig. 18-9B).

In Electronic Devices

The smallest power transformers are found in electronic equipment such as television sets, ham radios, and home computers. Most solid-state devices use low voltages, ranging from about 5 V up to perhaps 50 V. This equipment needs step-down power transformers in its power supplies.

Solid-state equipment usually (but not always) consumes relatively little power, so the transformers are usually not very bulky. The exception is high-powered AF or RF amplifiers, whose tran-

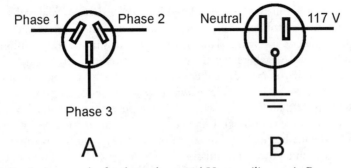

18-9   At A, an outlet for three-phase, 234-V rms utility ac. At B, a conventional single-phase utility outlet for 117-V rms utility ac.

sistors can demand more than 1000 W (1 kW) in some cases. At 12 V, this translates to a current demand of 90 A or more.

Television sets have cathode-ray tubes that need several hundred volts. This is derived by using a step-up transformer in the power supply. Such transformers don't have to supply a lot of current, though, so they are not very big or heavy. Another type of device that needs rather high voltage is a ham-radio amplifier with vacuum tubes. Such an amplifier requires from 2 kV to 5 kV.

Any voltage higher than about 12 V should be treated with respect. *Warning: The voltages in televisions and ham radios can present an electrocution hazard, even after the equipment has been switched off. Do not try to service such equipment unless you are trained to do so!*

At Audio Frequencies

Transformers for use at AF are similar to those employed for 60-Hz electricity. The differences are that the frequency is somewhat higher (up to 20 kHz), and that audio signals exist in a band of frequencies (20 Hz to 20 kHz) rather than at only one frequency.

Most AF transformers are constructed like miniature utility transformers. They have laminated E cores with primary and secondary windings wound around the crossbars, as shown in Fig. 18-4. Audio transformers can be either the step-up or the step-down type. However, rather than being made to produce a specific voltage, AF transformers are designed to match impedances.

Audio circuits, and in fact all electronic circuits that handle sine-wave or complex-wave signals, exhibit impedance at the input and output. The load has a certain impedance; a source has another impedance. Good audio design strives to minimize the reactance in the circuitry, so that the absolute-value impedance Z is close to the resistance R. This means that X must be zero or nearly zero. In the following discussion of impedance-matching transformers, for both AF and RF applications, assume that the reactance is zero, so the impedance is purely resistive with $Z = R + j0$.

Isolation and Impedance Matching

Transformers can provide *isolation* between electronic circuits. While there is *inductive coupling* in a transformer, there is comparatively little *capacitive coupling*. The amount of capacitive coupling can be reduced by using cores that minimize the number of wire turns needed in the windings, and by keeping the windings physically separated from each other (rather than overlapping).

Balanced and Unbalanced Loads and Lines

A *balanced load* is one whose terminals can be reversed without significantly affecting circuit behavior. A plain resistor is a good example. The two-wire antenna input in a television receiver is another example of a balanced load. A *balanced transmission line* is usually a two-wire line, such as old-fashioned *TV ribbon*, also called *twinlead*.

An *unbalanced load* is a load that must be connected a certain way. Switching its leads will result in improper circuit operation. In this sense, an unbalanced load is a little like a polarized component such as a battery or capacitor. Many wireless antennas are of this type. Usually, unbalanced sources and loads have one side connected to ground. The coaxial input of a television receiver is unbalanced; the shield (braid) of the cable is grounded. An *unbalanced transmission line* is usually a coaxial line, such as you find in a cable television system.

Normally, you cannot connect an unbalanced line to a balanced load, or a balanced line to an unbalanced load, and expect good performance. But a transformer can allow for mating between these two types of systems. In Fig. 18-10A, a *balanced-to-unbalanced transformer* is shown. Note that the balanced side is center-tapped, and the tap is grounded. In Fig. 18-10B, an *unbalanced-to-balanced transformer* is illustrated. Again, the balanced side has a grounded center tap.

The turns ratio of a balanced-to-unbalanced transformer (also called a *balun*) or an unbalanced-to-balanced transformer (also known as an *unbal*) can be 1:1, but this need not be the case, and often it is not. If the impedances of the balanced and unbalanced parts of the systems are the same, then a 1:1 turns ratio is ideal. But if the impedances differ, the turns ratio should be such that the impedances are matched. Shortly, we'll see how the turns ratio of a transformer can be manipulated to transform one purely resistive impedance into another.

Transformer Coupling

Transformers are sometimes used between amplifier stages in electronic equipment where a large *amplification factor* is needed. There are other methods of *coupling* from one amplifier stage to another, but transformers offer some advantages, especially in RF receivers and transmitters.

Part of the problem in getting a radio to work is that the amplifiers must operate in a stable manner. If there is too much feedback, a series of amplifiers will oscillate, and this will severely degrade the performance of the radio. Transformers that minimize the capacitance between the amplifier stages, while still transferring the desired signals, can help to prevent this oscillation.

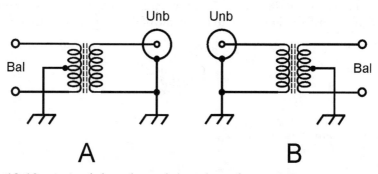

18-10 At A, a balanced-to-unbalanced transformer. At B, an unbalanced-to-balanced transformer.

Impedance Transfer Ratio

In RF and AF systems, transformers are employed to match impedances. Thus, you will sometimes hear or read about an *impedance step-up transformer* or an *impedance step-down transformer.*

The *impedance transfer ratio* of a transformer varies according to the square of the turns ratio, and also according to the square of the voltage-transfer ratio. If the primary (source) and secondary (load) impedances are purely resistive and are denoted Z_{pri} and Z_{sec}, then the following relations hold:

$$Z_{pri}/Z_{sec} = (T_{pri}/T_{sec})^2$$
$$Z_{pri}/Z_{sec} = (E_{pri}/E_{sec})^2$$

The inverses of these formulas, in which the turns ratio or voltage-transfer ratio are expressed in terms of the impedance-transfer ratio, are:

$$T_{pri}/T_{sec} = (Z_{pri}/Z_{sec})^{1/2}$$
$$E_{pri}/E_{sec} = (Z_{pri}/Z_{sec})^{1/2}$$

Problem 18-3

Consider a situation in which a transformer is needed to match an input impedance of 50.0 Ω, purely resistive, to an output impedance of 300 Ω, also purely resistive. What is the required turns ratio T_{pri}/T_{sec}?

The required transformer will have a step-up impedance ratio of $Z_{pri}/Z_{sec} = 50.0/300 = 1/6.00$. From the preceding formulas:

$$
\begin{aligned}
T_{pri}/T_{sec} &= (Z_{pri}/Z_{sec})^{1/2} \\
&= (1/6.00)^{1/2} \\
&= 0.16667^{1/2} \\
&= 0.408 \\
&= 1/2.45
\end{aligned}
$$

Problem 18-4

Suppose a transformer has a primary-to-secondary turns ratio of 4.00:1. The load, connected to the transformer output, is a pure resistance of 37.5 Ω. What is the impedance at the primary?

The impedance-transfer ratio is equal to the square of the turns ratio. Therefore:

$$
\begin{aligned}
Z_{pri}/Z_{sec} &= (T_{pri}/T_{sec})^2 \\
&= (4.00/1)^2 \\
&= 4.00^2 \\
&= 16.0
\end{aligned}
$$

We know that the secondary impedance, Z_{sec} is 37.5 Ω. Thus:

$$
\begin{aligned}
Z_{pri} &= 16.0 \times Z_{sec} \\
&= 16.0 \times 37.5 \\
&= 600 \ \Omega
\end{aligned}
$$

Radio-Frequency Transformers

In radio receivers and transmitters, transformers can be categorized generally by the method of construction used. Some have primary and secondary windings, just like utility and audio units. Others employ transmission-line sections. These are the two most common types of transformer found at radio frequencies.

Wire-Wound Types

In wire-wound RF transformers, powdered-iron cores can be used up to quite high frequencies. Toroidal cores are common, because they are self-shielding (all of the magnetic flux is confined within the core material). The number of turns depends on the frequency, and also on the permeability of the core.

In high-power applications, air-core coils are often preferred. Although air has low permeability, it has negligible hysteresis loss, and will not heat up or fracture as powdered-iron cores sometimes do. The disadvantage of air-core coils is that some of the magnetic flux extends outside of the coil. This affects the performance of the transformer when it must be placed in a cramped space, such as in a transmitter final-amplifier compartment.

A major advantage of coil-type transformers, especially when they are wound on toroidal cores, is that they can be made to work over a wide band of frequencies, such as from 3.5 MHz to 30 MHz. These are called *broadband transformers*.

Transmission-Line Types

As you recall, any transmission line has a characteristic impedance, or Z_o, that depends on the line construction. This property is sometimes used to make impedance transformers out of coaxial or parallel-wire line.

Transmission-line transformers are always made from quarter-wave sections. From the previous chapter, remember the formula for the length of a quarter-wave section:

$$L_{ft} = 246v/f_o$$

where L_{ft} is the length of the section in feet, v is the velocity factor expressed as a fraction, and f_o is the frequency of operation in megahertz. If the length L_m is specified in meters, then:

$$L_m = 75v/f_o$$

Suppose that a quarter-wave section of line, with characteristic impedance Z_o, is terminated in a purely resistive impedance R_{out}. Then the impedance that appears at the input end of the line, R_{in}, is also a pure resistance, and the following relations hold:

$$Z_o^2 = R_{in}R_{out}$$
$$Z_o = (R_{in}R_{out})^{1/2}$$

This is illustrated in Fig. 18-11. The first of the preceding formulas can be rearranged to solve for R_{in} in terms of R_{out}, or vice versa:

$$R_{in} = Z_o^2/R_{out}$$
$$R_{out} = Z_o^2/R_{in}$$

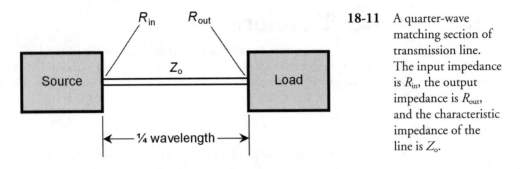

18-11   A quarter-wave matching section of transmission line. The input impedance is R_{in}, the output impedance is R_{out}, and the characteristic impedance of the line is Z_o.

These equations are valid at the frequency f_o for which the line length measures $\frac{1}{4}$ wavelength. Sometimes, the word "wavelength" is replaced by the lowercase Greek letter lambda (λ), so you will occasionally see the length of a quarter-wave section denoted as $(\frac{1}{4})\lambda$ or 0.25λ.

Neglecting line losses, the preceding relations hold at all *odd harmonics* of f_o, that is, at $3f_o$, $5f_o$, $7f_o$, and so on. At other frequencies, a quarter-wave section of line does not act as a transformer. Instead, it behaves in a complex manner that is beyond the scope of this discussion.

Quarter-wave transmission-line transformers are most often used in antenna systems, especially at the higher frequencies, where their dimensions become practical. A quarter-wave matching section should be made using unbalanced line if the load is unbalanced, and balanced line if the load is balanced.

A disadvantage of quarter-wave sections is the fact that they work only at specific frequencies. But this is often offset by the ease with which they are constructed, if radio equipment is to be used at only one frequency, or at odd-harmonic frequencies.

Problem 18-5

Suppose an antenna has a purely resistive impedance of 100 Ω. It is connected to a $\frac{1}{4}$-wave section of 75-Ω coaxial cable. What is the impedance at the input end of the section?

Use the formula from above:

$$R_{in} = Z_o^2/R_{out}$$
$$= 75^2/100$$
$$= 5625/100$$
$$= 56\ \Omega$$

Problem 18-6

Consider an antenna known to have a purely resistive impedance of 600 Ω. You want to match it to the output of a radio transmitter designed to work into a 50.0-Ω pure resistance. What is the characteristic impedance needed for a quarter-wave matching section?

Use this formula:

$$Z^2 = R_{in}R_{out}$$
$$= 600 \times 50$$
$$= 30,000$$

Therefore:

$$Z_o = (30,000)^{1/2}$$
$$= 173 \ \Omega$$

It may be difficult to find a commercially manufactured transmission line that has this particular characteristic impedance. Prefabricated lines come in standard Z_o values, and a perfect match might not be obtainable. In that case, the closest obtainable Z_o should be used. In this case, it would probably be 150 Ω. If nothing is available anywhere near the characteristic impedance needed for a quarter-wave matching section, then a coil-type transformer can be used instead.

What about Reactance?

Things are simple when there is no reactance in an ac circuit using transformers. But often, especially in RF antenna systems, pure resistance doesn't occur naturally. It has to be obtained by using inductors and/or capacitors to cancel the reactance out. The presence of reactance in a load makes a perfect match impossible with an impedance-matching transformer alone.

Recall that inductive and capacitive reactances are opposite in effect, and that their magnitudes can vary. If a load presents a complex impedance $R + jX$, it is possible to cancel the reactance X by deliberately introducing an equal and opposite reactance $-X$. This can be, and often is, done by connecting an inductor or capacitor in series with a load that contains reactance as well as resistance. The result is a pure resistance with a value equal to $(R + jX) - jX$, or simply R.

When wireless communications is contemplated over a wide band of frequencies, adjustable impedance-matching and reactance-canceling networks can be placed between the transmitter and the antenna system. Such a device is called a *transmatch* or an *antenna tuner*. These devices not only match the resistive portions of the transmitter and load impedances, but they can tune out reactances in the load. Transmatches are popular among amateur radio operators, who use equipment capable of operation from less than 2 MHz up to the highest known radio frequencies.

Quiz

Refer to the text in this chapter if necessary. A good score is 18 or more correct. Answers are in the back of the book.

1. In a step-up transformer,
   (a) the primary impedance is greater than the secondary impedance.
   (b) the secondary winding is right on top of the primary.
   (c) the primary voltage is less than the secondary voltage.
   (d) All of the above are true.

2. The capacitance between the primary and the secondary windings of a transformer can be minimized by
   (a) placing the windings on opposite sides of a toroidal core.
   (b) winding the secondary right on top of the primary.
   (c) using the highest possible frequency.
   (d) using a center tap on the balanced winding.

3. A transformer steps a voltage down from 117 V to 6.00 V. What is its primary-to-secondary turns ratio?

   (a) 1:380

   (b) 380:1

   (c) 1:19.5

   (d) 19.5:1

4. A step-up transformer has a primary-to-secondary turns ratio of 1:5.00. If 117 V rms appears at the primary, what is the ac rms voltage across the secondary?

   (a) 23.4 V rms

   (b) 585 V rms

   (c) 117 V rms

   (d) 2.93 kV rms

5. A transformer has a secondary-to-primary turns ratio of 0.167. This transformer is

   (a) a step-up unit.

   (b) a step-down unit.

   (c) neither a step-up unit nor a step-down unit.

   (d) a reversible unit.

6. Which of the following statements is false, concerning air cores compared with ferromagnetic cores?

   (a) Air concentrates the magnetic lines of flux.

   (b) Air works at higher frequencies than ferromagnetics.

   (c) Ferromagnetics are lossier than air.

   (d) A ferromagnetic-core transformer needs fewer turns of wire than an equivalent air-core transformer.

7. Eddy currents cause

   (a) an increase in efficiency.

   (b) an increase in coupling between windings.

   (c) an increase in core loss.

   (d) an increase in usable frequency range.

8. Suppose a transformer has an ac voltage of 117 V rms across its primary, and 234 V rms appears across its secondary. If this transformer is reversed (that is, connected backward), assuming that this be done without damaging the windings, what will be the voltage at the output?

   (a) 234 V rms

   (b) 468 V rms

   (c) 117 V rms

   (d) 58.5 V rms

9. The shell method of transformer winding

   (a) provides maximum coupling.

   (b) minimizes capacitance between windings.

(c) withstands more voltage than other winding methods.

(d) has windings far apart but along a common axis.

10. Which of these core types is best if you need a winding inductance of 1.5 H?

    (a) Air core

    (b) Ferromagnetic solenoid core

    (c) Ferromagnetic toroid core

    (d) Ferromagnetic pot core

11. An advantage of a toroid core over a solenoid core is the fact that

    (a) the toroid works at higher frequencies.

    (b) the toroid confines the magnetic flux.

    (c) the toroid can work for dc as well as for ac.

    (d) it is easier to wind the turns on a toroid.

12. High voltage is used in long-distance power transmission because

    (a) it is easier to regulate than low voltage.

    (b) the I^2R losses are minimized.

    (c) the electromagnetic fields are strong.

    (d) small transformers can be used.

13. In a household circuit, 234-V rms electricity usually has

    (a) one phase.

    (b) two phases.

    (c) three phases.

    (d) four phases.

14. In a transformer, a center tap often exists in

    (a) the primary winding.

    (b) the secondary winding.

    (c) an unbalanced winding.

    (d) a balanced winding.

15. An autotransformer

    (a) can be adjusted automatically.

    (b) has a center-tapped secondary.

    (c) consists of a single tapped winding.

    (d) is useful only for impedance matching.

16. Suppose a transformer has a primary-to-secondary turns ratio of 2.00:1. The input impedance is 300 Ω, purely resistive. What is the output impedance?

    (a) 75 Ω, purely resistive

    (b) 150 Ω, purely resistive

    (c) 600 Ω, purely resistive

    (d) 1200 Ω, purely resistive

17. Suppose a purely resistive input impedance of 50 Ω must be matched to a purely resistive output impedance of 450 Ω. The primary-to-secondary turns ratio of the transformer must be which of the following?

    (a) 9.00

    (b) 3.00

    (c) 1/3.00

    (d) 1/9.00

18. Suppose a quarter-wave matching section has a characteristic impedance of 75.0 Ω. The input impedance is 50.0 Ω, purely resistive. What is the output impedance?

    (a) 150 Ω, purely resistive

    (b) 125 Ω, purely resistive

    (c) 100 Ω, purely resistive

    (d) 113 Ω, purely resistive

19. Suppose a purely resistive impedance of 75 Ω must be matched to a purely resistive impedance of 300 Ω. A quarter-wave section would need to have

    (a) $Z_o = 188\ \Omega$.

    (b) $Z_o = 150\ \Omega$.

    (c) $Z_o = 225\ \Omega$.

    (d) $Z_o = 375\ \Omega$.

20. If there is reactance in the load to which a transformer is connected, then

    (a) the transformer will be destroyed.

    (b) a perfect impedance match cannot be obtained.

    (c) a center tap must be used in the secondary.

    (d) the turns ratio must be changed to obtain an impedance match.

Test: Part 2

Do not refer to the text when taking this test. A good score is at least 37 correct. Answers are in the back of the book. It's best to have a friend check your score the first time, so you won't memorize the answers if you want to take the test again.

1. Consider a series circuit that has a resistance of 100 Ω and a capacitive reactance of -200 Ω. What is the complex impedance?

    (a) $-200 + j100$

    (b) $100 + j200$

    (c) $200 - j100$

    (d) $200 + j100$

    (e) $100 - j200$

2. Mutual inductance causes the net value of a set of coils to

    (a) cancel out, resulting in zero inductance.

    (b) be greater than what it would be with no mutual coupling.

    (c) be less than what it would be with no mutual coupling.

    (d) double.

    (e) vary, depending on the extent and phase of mutual coupling.

3. Refer to Fig. Test 2-1. Wave A is

    (a) leading wave B by 90°.

    (b) lagging wave B by 90°.

    (c) leading wave B by 180°.

    (d) lagging wave B by 135°.

    (e) lagging wave B by 45°.

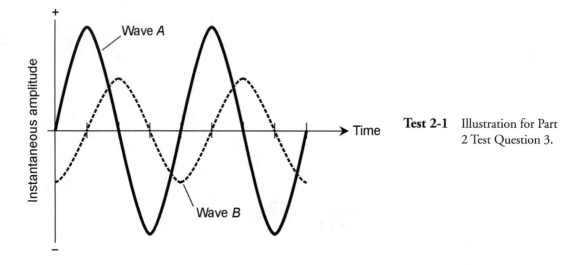

Test 2-1   Illustration for Part 2 Test Question 3.

4. If a pure sine wave with no dc component has a positive peak value of +30.0 V pk, what is its rms voltage?

   (a) 21.2 V rms

   (b) 30.0 V rms

   (c) 42.4 V rms

   (d) 60.0 V rms

   (e) 90.0 V rms

5. Suppose four capacitors are connected in parallel. Their values are 100 pF each. What is the net capacitance?

   (a) 25 pF

   (b) 50 pF

   (c) 100 pF

   (d) 200 pF

   (e) 400 pF

6. Suppose an ac transformer has a primary-to-secondary turns ratio of 8.88/1. The input voltage is 234 V rms. What is the output voltage?

   (a) 2.08 kV rms

   (b) 18.5 kV rms

   (c) 2.97 V rms

   (d) 26.4 V rms

   (e) 20.8 V rms

7. In a series *RL* circuit, as the resistance becomes small compared with the reactance, the angle of lag approaches which of the following?

   (a) 0°

   (b) 45°

(c) 90°

(d) 180°

(e) 360°

8. Suppose an ac transmission line carries 3.50 A rms and 150 V rms. Imagine that the line is perfectly lossless, and that the load impedance is a pure resistance equal to the characteristic impedance of the line. What is the *true power* in this transmission line?

(a) 525 W

(b) 42.9 W

(c) 1.84 W

(d) Nonexistent, because true power is dissipated, not transmitted

(e) Variable, depending on standing-wave effects

9. In a parallel configuration, susceptances

(a) simply add up.

(b) add like capacitances in series.

(c) add like inductances in parallel.

(d) must be changed to reactances before you can work with them.

(e) cancel out.

10. Consider a sine wave that has a frequency of 200 kHz. How many degrees of phase change occur in a microsecond (a millionth of a second)?

(a) 180°

(b) 144°

(c) 120°

(d) 90°

(e) 72°

11. At a frequency of 2.55 MHz, what is the reactance of a 330-pF capacitor?

(a) −5.28 Ω

(b) −0.00528 Ω

(c) −189 Ω

(d) −18.9 kΩ

(e) −0.000189 Ω

12. Suppose a transformer has a step-up turns ratio of 1/3.16. The impedance of the load connected to the secondary is 499 Ω, purely resistive. What is the impedance at the primary?

(a) 50.0 Ω, purely resistive

(b) 158 Ω, purely resistive

(c) 1.58 kΩ, purely resistive

(d) 4.98 kΩ, purely resistive

(e) Impossible to calculate from the data given

13. If a complex impedance is represented by $34 - j23$, what is the absolute-value impedance?

   (a) 34 Ω

   (b) 11 Ω

   (c) −23 Ω

   (d) 41 Ω

   (e) 57 Ω

14. Suppose a coil has an inductance of 750 μH. What is the inductive reactance at 100 kHz?

   (a) 75.0 Ω

   (b) 75.0 kΩ

   (c) 471 Ω

   (d) 47.1 kΩ

   (e) 212 Ω

15. If two sine waves are 180° out of phase, it represents a difference of

   (a) $\frac{1}{8}$ of a cycle.

   (b) $\frac{1}{4}$ of a cycle.

   (c) $\frac{1}{2}$ of a cycle.

   (d) 1 full cycle.

   (e) 2 full cycles.

16. If R denotes resistance and Z denotes absolute-value impedance, then R/Z represents the

   (a) true power.

   (b) imaginary power.

   (c) apparent power.

   (d) absolute-value power.

   (e) power factor.

17. Suppose two components are connected in series. One component has a complex impedance of $30 + j50$, and the other component has a complex impedance of $50 - j30$. What is the impedance of the series combination?

   (a) $80 + j80$

   (b) $20 + j20$

   (c) $20 - j20$

   (d) $-20 + j20$

   (e) $80 + j20$

18. Suppose two inductors, having values of 140 μH and 1.50 mH, are connected in series. What is the net inductance?

   (a) 141.5 μH

   (b) 1.64 μH

   (c) 0.1415 mH

(d) 1.64 mH

(e) 0.164 mH

19. Which of the following types of capacitor is polarized?

    (a) Mica

    (b) Paper

    (c) Electrolytic

    (d) Air variable

    (e) Ceramic

20. A coil with a toroidal, ferromagnetic core

    (a) has less inductance than an air-core coil with the same number of turns.

    (b) is essentially self-shielding.

    (c) works well as a loopstick antenna.

    (d) is ideal as a transmission-line transformer.

    (e) cannot be used at frequencies below 10 MHz.

21. The efficiency of an electric generator

    (a) depends on the mechanical driving power source.

    (b) is equal to the electrical output power divided by the mechanical input power.

    (c) depends on the nature of the electrical load.

    (d) is equal to driving voltage divided by output voltage.

    (e) is equal to driving current divided by output current.

22. Admittance is

    (a) the reciprocal of reactance.

    (b) the reciprocal of resistance.

    (c) a measure of the opposition a circuit offers to ac.

    (d) a measure of the ease with which a circuit passes ac.

    (e) another expression for absolute-value impedance.

23. The absolute-value impedance Z of a parallel RLC circuit, where R is the resistance and X is the net reactance, is found according to which of the following formulas?

    (a) $Z = R + X$

    (b) $Z^2 = R^2 + X^2$

    (c) $Z^2 = R^2 X^2 / (R^2 + X^2)$

    (d) $Z = 1/(R^2 + X^2)$

    (e) $Z = R^2 X^2 / (R + X)$

24. Complex numbers are used to represent impedance because

    (a) reactance cannot store power.

    (b) reactance isn't a real physical thing.

(c) they provide a way to represent what happens in resistance-reactance circuits.

(d) engineers like to work with sophisticated mathematics.

(e) Forget it! Complex numbers are never used to represent impedance.

25. Which of the following (within reason) has no effect on the value, in farads, of a capacitor?

(a) The mutual surface area of the plates

(b) The dielectric constant of the material between the plates

(c) The spacing between the plates

(d) The amount of overlap between plates

(e) The frequency

26. The $0°$ phase point in an ac sine wave is usually considered to be the point in time at which the instantaneous amplitude is

(a) zero and negative-going.

(b) at its negative peak.

(c) zero and positive-going.

(d) at its positive peak.

(e) any value; it doesn't matter.

27. The inductance of a coil can be adjusted in a practical way by

(a) varying the frequency of the signal applied to the coil.

(b) varying the number of turns using multiple taps.

(c) varying the current in the coil.

(d) varying the wavelength of the signal applied to the coil.

(e) varying the voltage across the coil.

28. Power factor is defined as the ratio of

(a) true power to VA power.

(b) true power to imaginary power.

(c) imaginary power to VA power.

(d) imaginary power to true power.

(e) VA power to true power.

29. Consider a situation in which you want to match a feed line with $Z_o = 50\ \Omega$ to an antenna with a purely resistive impedance of $200\ \Omega$. A quarter-wave matching section should have which of the following?

(a) $Z_o = 150\ \Omega$

(b) $Z_o = 250\ \Omega$

(c) $Z_o = 125\ \Omega$

(d) $Z_o = 133\ \Omega$

(e) $Z_o = 100\ \Omega$

30. The vector $40 + j30$ in the *RX* plane represents
    - (a) 40 Ω of resistance and 30 μH of inductance.
    - (b) 40 μH of inductance and 30 Ω of resistance.
    - (c) 40 Ω of resistance and 30 Ω of inductive reactance.
    - (d) 40 Ω of inductive reactance and 30 Ω of resistance.
    - (e) 40 μH of inductive reactance and 30 Ω of resistance.

31. In a series *RC* circuit where $R = 300$ Ω and $X_C = -30$ Ω,
    - (a) the current leads the voltage by a few degrees.
    - (b) the current leads the voltage by almost 90°.
    - (c) the voltage leads the current by a few degrees.
    - (d) the voltage leads the current by almost 90°.
    - (e) the voltage leads the current by 90°.

32. In a step-down transformer,
    - (a) the primary voltage is greater than the secondary voltage.
    - (b) the purely resistive impedance across the primary is less than the purely resistive impedance across the secondary.
    - (c) the secondary voltage is greater than the primary voltage.
    - (d) the output frequency is higher than the input frequency.
    - (e) the output frequency is lower than the input frequency.

33. Suppose a capacitor of 470 pF is in parallel with an inductor of 4.44 μH. What is the resonant frequency?
    - (a) 3.49 MHz
    - (b) 3.49 kHz
    - (c) 13.0 MHz
    - (d) 13.0 GHz
    - (e) It cannot be calculated from the data given.

34. A pure sine wave contains energy at
    - (a) only one specific frequency.
    - (b) a specific frequency and its even harmonics.
    - (c) a specific frequency and its odd harmonics.
    - (d) a specific frequency and all its harmonics.
    - (e) a specific frequency and its second harmonic only.

35. Inductive susceptance is
    - (a) the reciprocal of inductance.
    - (b) negative imaginary.
    - (c) equivalent to capacitive reactance.
    - (d) the reciprocal of capacitive susceptance.
    - (e) positive imaginary.

36. The rate of change (derivative) of a pure sine wave is another pure sine wave that has the same frequency as the original wave, and

    (a) is in phase with the original wave.

    (b) is 180° out of phase with the original wave.

    (c) leads the original wave by 45°.

    (d) lags the original wave by 90°.

    (e) leads the original wave by 90°.

37. True power is equal to

    (a) VA power plus imaginary power.

    (b) imaginary power minus VA power.

    (c) the vector difference between VA and reactive power.

    (d) VA power; the two are the same thing.

    (e) 0.707 times the VA power.

38. Consider a circuit in which three capacitors are connected in series. Their values are 47 μF, 68 μF, and 100 μF. The total capacitance of this combination is

    (a) 215 μF.

    (b) between 68 μF and 100 μF.

    (c) between 47 μF and 68 μF.

    (d) 22 μF.

    (e) not determinable from the data given.

39. The reactance of a section of transmission line depends on all of the following factors *except*

    (a) the velocity factor of the line.

    (b) the length of the section.

    (c) the current in the line.

    (d) the frequency of the signal in the line.

    (e) the wavelength of the signal in the line.

40. When analyzing a parallel *RLC* circuit to find the complex impedance, you should

    (a) add the resistance and reactance to get $R + jX$.

    (b) find the net conductance and susceptance, convert to resistance and reactance, and then add these to get $R + jX$.

    (c) find the net conductance and susceptance, and add these to get $R + jX$.

    (d) rearrange the components so they're connected in series, and find the complex impedance of that circuit.

    (e) subtract reactance from resistance to get $R - jX$.

41. The illustration in Fig. Test 2-2 shows a vector $R + jX$ representing

    (a) $X_C = 60\ \Omega$ and $R = 25\ \Omega$.

    (b) $X_L = 60\ \Omega$ and $R = 25\ \Omega$.

    (c) $X_L = 60\ \mu H$ and $R = 25\ \Omega$.

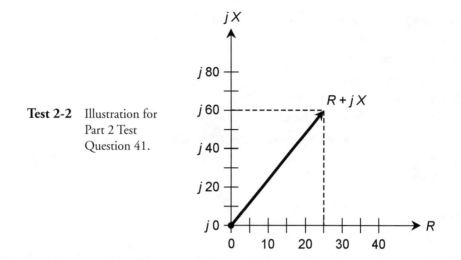

Test 2-2 Illustration for Part 2 Test Question 41.

(d) $C = 60\ \mu\text{F}$ and $R = 25\ \Omega$.

(e) $L = 60\ \mu\text{H}$ and $R = 25\ \Omega$.

42. Suppose two pure sine waves have no dc components, have the same frequency, and have the same peak-to-peak voltages, but they cancel each other out when combined. What is the phase difference between the waves?

(a) 45°

(b) 90°

(c) 180°

(d) 270°

(e) 360°

43. Suppose a series *RC* circuit has a resistance of 50 Ω and a capacitive reactance of −37 Ω. What is the phase angle?

(a) 37°

(b) 53°

(c) −37°

(d) −53°

(e) It cannot be calculated from the data given.

44. Suppose a 200-Ω resistor is in series with a coil and capacitor, such that $X_L = 200\ \Omega$ and $X_C = -100\ \Omega$. What is the complex impedance?

(a) $200 - j100$

(b) $200 - j200$

(c) $200 + j100$

(d) $200 + j200$

(e) Impossible to determine from the data given

45. The characteristic impedance of a transmission line
    (a) is negative imaginary.
    (b) is positive imaginary.
    (c) depends on the frequency.
    (d) depends on the construction of the line.
    (e) depends on the length of the line.

46. Suppose the period of a pure sine wave is 2×10^{-8}s. What is the frequency?
    (a) 2×10^8 Hz
    (b) 20 MHz
    (c) 50 kHz
    (d) 50 MHz
    (e) 500 MHz

47. Suppose a series *RC* circuit has a resistance of 600 Ω and a capacitance of 220 pF. What is the phase angle?
    (a) $-20°$
    (b) $20°$
    (c) $-70°$
    (d) $70°$
    (e) Not determinable from the data given

48. A capacitor with a negative temperature coefficient
    (a) works less well as the temperature increases.
    (b) works better as the temperature increases.
    (c) heats up as its value is made larger.
    (d) cools down as its value is made larger.
    (e) exhibits increasing capacitance as the temperature drops.

49. Suppose three coils are connected in parallel. Each has an inductance of 300 μH. There is no mutual inductance. What is the net inductance?
    (a) 100 μH
    (b) 300 μH
    (c) 900 μH
    (d) 17.3 μH
    (e) 173 μH

50. Suppose a coil has 100 Ω of inductive reactance at 30.0 MHz. What is its inductance?
    (a) 0.531 μH
    (b) 18.8 mH
    (c) 531 μH
    (d) 18.8 μH
    (e) It can't be found from the data given.

3
PART

Basic Electronics

19
CHAPTER

Introduction to Semiconductors

SINCE THE 1960S, WHEN THE TRANSISTOR BECAME COMMON IN CONSUMER DEVICES, *SEMICONDUCTORS* have acquired a dominating role in electronics. The term *semiconductor* arises from the ability of these materials to conduct some of the time, but not all the time. The conductivity can be controlled to produce effects such as amplification, rectification, oscillation, signal mixing, and switching.

The Semiconductor Revolution

Decades ago, *vacuum tubes*, also known as *electron tubes*, were the only devices available for use as amplifiers, oscillators, detectors, and other electronic circuits and systems. A typical tube (called a *valve* in England) ranged from the size of your thumb to the size of your fist. They are still used in some power amplifiers, microwave oscillators, and video display units.

Tubes generally require high voltage. Even in modest radio receivers, 100 V to 200 V dc was required when tubes were employed. This mandated bulky power supplies, and created an electrical shock hazard. Nowadays, a transistor of microscopic dimensions can perform the functions of a tube in most situations. The power supply can be a couple of AA cells or a 9-V transistor battery.

Even in high-power applications, transistors are smaller and lighter than tubes. Figure 19-1 is a size comparison drawing between a transistor and a vacuum tube for use in an AF or RF power amplifier.

Integrated circuits (ICs), hardly larger than individual transistors, can do the work of hundreds or even thousands of vacuum tubes. An excellent example of this technology is found in personal computers and the peripheral devices used with them.

19-1 A power-amplifier transistor (at left) is much smaller than a vacuum tube of comparable power-handling capacity (right).

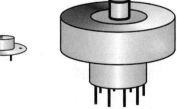

Semiconductor Materials

Various elements, compounds, and mixtures can function as semiconductors. The two most common materials are *silicon* and a compound of gallium and arsenic known as *gallium arsenide* (often abbreviated GaAs). In the early years of semiconductor technology, *germanium* formed the basis for many semiconductors; today it is seen occasionally, but not often. Other substances that work as semiconductors are *selenium*, *cadmium* compounds, *indium* compounds, and the oxides of certain metals.

Silicon

Silicon (chemical symbol Si) is widely used in diodes, transistors, and integrated circuits. Generally, other substances, or *impurities*, must be added to silicon to give it the desired properties. The best quality silicon is obtained by growing crystals in a laboratory. The silicon is then fabricated into *wafers* or *chips*.

Gallium Arsenide

Another common semiconductor is the compound gallium arsenide. Engineers and technicians call this material by its acronym-like chemical symbol, GaAs, pronounced "gas." If you hear about "gasfets" and "gas ICs," you're hearing about gallium-arsenide technology.

GaAs devices require little voltage, and will function at higher frequencies than silicon devices because the charge carriers move faster through the semiconductor material. GaAs devices are relatively immune to the effects of ionizing radiation such as X rays and gamma rays. GaAs is used in light-emitting diodes (LEDs), infrared-emitting diodes (IREDs), laser diodes, visible-light and infrared (IR) detectors, ultra-high-frequency (UHF) amplifying devices, and a variety of integrated circuits.

Selenium

Selenium exhibits conductivity that varies depending on the intensity of visible light or IR radiation that strikes it. All semiconductor materials exhibit this property, known as *photoconductivity*, to some degree; but in selenium the effect is especially pronounced. For this reason, selenium is useful for making *photocells*. Selenium is also used in certain types of *rectifiers*. A rectifier is a component or circuit that converts ac to pulsating dc.

A significant advantage of selenium is the fact that it is electrically rugged. Selenium-based components can withstand brief *transients*, or spikes, of abnormally high voltage, better than components made with most other semiconductor materials.

Germanium

Pure elemental germanium is a poor electrical conductor. It becomes a semiconductor only when *impurities* are added. Germanium was used extensively in the early years of semiconductor technology. Some diodes and transistors still use it.

A germanium diode has a low voltage drop (0.3 V, compared with 0.6 V for silicon and 1 V for selenium) when it conducts, and this makes it useful in some situations. But germanium is easily destroyed by heat. Extreme care must be used when soldering the leads of a germanium component.

Metal Oxides

Certain metal oxides have properties that make them useful in the manufacture of semiconductor devices. When you hear about MOS (pronounced "moss") or CMOS (pronounced "sea moss") technology, you are hearing about *metal-oxide semiconductor* and *complementary metal-oxide semiconductor* devices, respectively.

An advantage of MOS and CMOS devices is the fact that they need almost no power to function. They draw so little current that a battery in a MOS or CMOS device lasts just about as long as it would on the shelf. Another advantage is high speed. This allows operation at high frequencies in RF equipment, and makes it possible to perform many switching operations per second for use in computers.

Certain types of transistors, and many kinds of ICs, make use of this technology. In integrated circuits, MOS and CMOS allow for a large number of discrete diodes and transistors on a single chip. Engineers would say that MOS/CMOS has *high component density*.

The biggest problem with MOS and CMOS technology is the fact that the devices are easily damaged by static electricity. Care must be used when handling components of this type. Technicians working with MOS and CMOS components must literally ground themselves by wearing a metal wrist strap connected to a good earth ground. Otherwise, the electrostatic charges that normally build up on their bodies can destroy MOS and CMOS components when equipment is constructed or serviced.

Doping and Charge Carriers

For a semiconductor material to have the properties necessary in order to function as electronic components, impurities are usually added. The impurities cause the material to conduct currents in certain ways. The addition of an impurity to a semiconductor is called *doping*. Sometimes the impurity is called a *dopant*.

Donor Impurities

When an impurity contains an excess of electrons, the dopant is called a *donor impurity*. Adding such a substance causes conduction mainly by means of electron flow, as in an ordinary metal such as copper or aluminum. The excess electrons are passed from atom to atom when a voltage exists across the material. Elements that serve as donor impurities include antimony, arsenic, bismuth, and phosphorus. A material with a donor impurity is called an *N-type semiconductor*, because electrons have negative (N) charge.

Acceptor Impurities

If an impurity has a deficiency of electrons, the dopant is called an *acceptor impurity*. When a substance such as aluminum, boron, gallium, or indium is added to a semiconductor, the material conducts by means of *hole flow*. A *hole* is a missing electron—or more precisely, a place in an atom where an electron should be, but isn't. A semiconductor with an acceptor impurity is called a *P-type semiconductor*, because holes have, in effect, a positive (P) charge.

Majority and Minority Carriers

Charge carriers in semiconductor materials are either electrons, each of which has a unit negative charge, or holes, each of which has a unit positive charge. In any semiconductor substance, some

⟵ • = Electron

⟶ O = Hole

• • • O • • • • • • • O • • • • • O

O • • • • • • • • • O • • • • • • •

• • • • O • • • • • • • O • • • • •

• • • • • O • • • • • • • • O • •

• • O • • • • • O • • • • O • • • •

O • • • • • O • • • • O • • • • • •

• • • • O • • • • O • • • • O • • •

19-2 Pictorial representation of hole flow. Solid black dots represent electrons, moving in one direction. Open circles represent holes, moving in the opposite direction.

of the current takes the form of electrons passed from atom to atom in a negative-to-positive direction, and some of the current occurs as holes that move from atom to atom in a positive-to-negative direction.

Sometimes electrons account for most of the current in a semiconductor. This is the case if the material has donor impurities, that is, if it is of the N type. In other cases, holes account for most of the current. This happens when the material has acceptor impurities, and is thus of the P type. The dominating charge carriers (either electrons or holes) are called the *majority carriers*. The less abundant ones are called the *minority carriers*. The ratio of majority to minority carriers can vary, depending on the way in which the semiconductor material has been manufactured.

Figure 19-2 is a simplified illustration of electron flow versus hole flow in a sample of N-type semiconductor material, where the majority carriers are electrons and the minority carriers are holes. The solid black dots represent electrons. Imagine them moving from right to left in this illustration as they are passed from atom to atom. Small open circles represent holes. Imagine them moving from left to right in the illustration. In this particular example, the positive battery or power-supply terminal (or "source of holes") would be out of the picture toward the left, and the negative battery or power-supply terminal (or "source of electrons") would be out of the picture toward the right.

The P-N Junction

Merely connecting up a piece of semiconducting material, either P or N type, to a source of current can be interesting, and a good subject for science experiments. But when the two types of material are brought together, the boundary between them, called the *P-N junction*, behaves in ways that make semiconductor materials truly useful in electronic components.

The Semiconductor Diode

Figure 19-3 shows the schematic symbol for a *semiconductor diode*, formed by joining a piece of P-type material to a piece of N-type material. The N-type semiconductor is represented by the short, straight line in the symbol, and is called the *cathode*. The P-type semiconductor is represented by the arrow, and is called the *anode*.

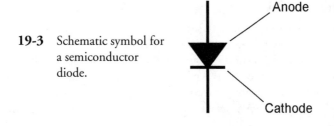

19-3 Schematic symbol for a semiconductor diode.

In the diode as shown in Figure 19-3, electrons can move easily in the direction opposite the arrow, and holes can move easily in the direction in which the arrow points. But current cannot, under most conditions, flow the other way. Electrons normally do not move with the arrow, and holes normally do not move against the arrow.

If you connect a battery and a resistor in series with the diode, you'll get a current to flow if the negative terminal of the battery is connected to the cathode and the positive terminal is connected to the anode, as shown in Fig. 19-4A. No current will flow if the battery is reversed, as shown in Fig. 19-4B. (The resistor is included in the circuit to prevent destruction of the diode by excessive current.)

It takes a specific, well-defined minimum applied voltage for conduction to occur through a semiconductor diode. This is called the *forward breakover voltage.* Depending on the type of material, the forward breakover voltage varies from about 0.3 V to 1 V. If the voltage across the junction is not at least as great as the forward breakover voltage, the diode will not conduct, even when it is connected as shown in Fig. 19-4A. This effect, known as the *forward breakover effect* or the *P-N junction threshold effect,* can be of use in circuits designed to limit the positive and/or negative peak voltages that signals can attain. The effect can also be used in a device called a *threshold detector,* in which a signal must be stronger than a certain amplitude in order to pass through.

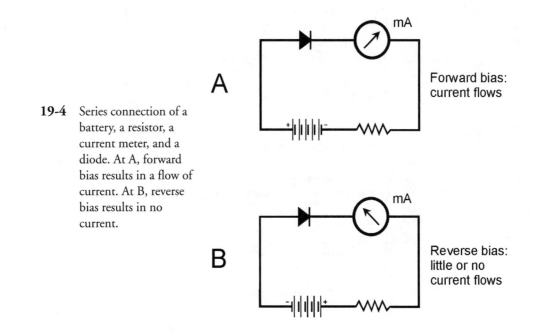

19-4 Series connection of a battery, a resistor, a current meter, and a diode. At A, forward bias results in a flow of current. At B, reverse bias results in no current.

How the Junction Works

When the N-type material is negative with respect to the P type, as in Fig. 19-4A, electrons flow easily from N to P. The N-type semiconductor, which already has an excess of electrons, receives more; the P-type semiconductor, with a shortage of electrons, has some more taken away. The N-type material constantly feeds electrons to the P type in an attempt to create an electron balance, and the battery or power supply keeps robbing electrons from the P-type material. This condition is illustrated in Fig. 19-5A, and is known as *forward bias*. Current can flow through the diode easily under these circumstances.

When the battery or dc power-supply polarity is switched so the N-type material is positive with respect to the P type, the situation is called *reverse bias*. Electrons in the N-type material are pulled toward the positive charge pole, away from the P-N junction. In the P-type material, holes are pulled toward the negative charge pole, also away from the P-N junction. The electrons are the majority carriers in the N-type material, and the holes are the majority carriers in the P-type material. The charge therefore becomes depleted in the vicinity of the P-N junction, and on both sides of it, as shown in Fig. 19-5B. This zone, where majority carriers are deficient, is called the *depletion region*. A shortage of majority carriers in any semiconductor substance means that the substance cannot conduct well. Thus, the depletion region acts like an electrical insulator. This is why a semiconductor diode will not normally conduct when it is reverse-biased. A diode is, in effect, a one-way current gate—usually!

Junction Capacitance

Some P-N junctions can alternate between conduction (in forward bias) and nonconduction (in reverse bias) millions or billions of times per second. Other junctions are slower. The main limiting

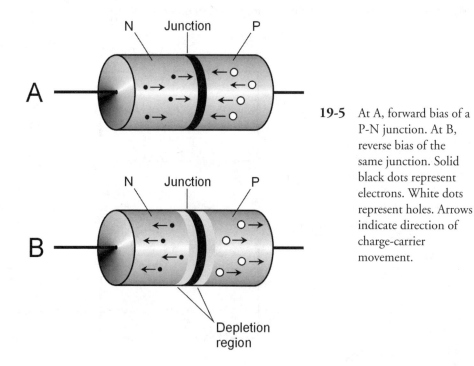

19-5   At A, forward bias of a P-N junction. At B, reverse bias of the same junction. Solid black dots represent electrons. White dots represent holes. Arrows indicate direction of charge-carrier movement.

factor is the capacitance at the P-N junction during conditions of reverse bias. As the *junction capacitance* of a diode increases, maximum frequency at which it can alternate between the conducting state and the nonconducting state decreases.

The junction capacitance of a diode depends on several factors, including the operating voltage, the type of semiconductor material, and the cross-sectional area of the P-N junction. If you examine Fig. 19-5B, you might get the idea that the depletion region, sandwiched between two semiconducting sections, can play a role similar to that of the dielectric in a capacitor. This is true! In fact, a reverse-biased P-N junction actually is a capacitor. Some semiconductor components, called *varactor diodes*, are manufactured with this property specifically in mind.

The junction capacitance of a diode can be varied by changing the reverse-bias voltage, because this voltage affects the width of the depletion region. The greater the reverse voltage, the wider the depletion region gets, and the smaller the capacitance becomes.

Avalanche Effect

Sometimes, a diode conducts when it is reverse-biased. The greater the reverse-bias voltage, the more like an electrical insulator a P-N junction gets—up to a point. But if the reverse bias rises past a specific critical value, the voltage overcomes the ability of the junction to prevent the flow of current, and the junction conducts as if it were forward-biased. This phenomenon is called the *avalanche effect* because conduction occurs in a sudden and massive way, something like a snow avalanche on a mountainside.

The avalanche effect does not damage a P-N junction (unless the voltage is extreme). It's a temporary thing. When the voltage drops back below the critical value, the junction behaves normally again.

Some components are designed to take advantage of the avalanche effect. In other cases, the avalanche effect limits the performance of a circuit. In a device designed for voltage regulation, called a *Zener diode*, you'll hear about the *avalanche voltage* or *Zener voltage* specification. This can range from a couple of volts to well over 100 V. Zener diodes are often used in voltage-regulating circuits.

For *rectifier diodes* in power supplies, you'll hear or read about the *peak inverse voltage* (PIV) or *peak reverse voltage* (PRV) specification. It's important that rectifier diodes have PIV ratings great enough so that the avalanche effect will not occur (or even come close to happening) during any part of the ac cycle.

Quiz

Refer to the text in this chapter if necessary. A good score is at least 18 correct. Answers are in the back of the book.

1. The term *semiconductor* arises from
   (a) resistor-like properties of metal oxides.
   (b) variable conductive properties of some materials.
   (c) the fact that electrons conduct better than holes.
   (d) insulating properties of silicon and GaAs.

2. Which of the following is *not* an advantage of semiconductor devices over vacuum tubes?
   (a) Smaller size
   (b) Lower working voltage

   (c) Lighter weight

   (d) Ability to withstand high voltage spikes

3. Of the following substances, which is the most commonly used semiconductor?

   (a) Germanium

   (b) Galena

   (c) Silicon

   (d) Copper

4. GaAs is

   (a) a compound.

   (b) an element.

   (c) a mixture.

   (d) a gas.

5. A disadvantage of MOS devices is the fact that

   (a) the charge carriers move fast.

   (b) the material does not react to ionizing radiation.

   (c) they can be damaged by electrostatic discharges.

   (d) they must always be used at high frequencies.

6. Selenium works especially well in

   (a) photocells.

   (b) high-frequency detectors.

   (c) RF power amplifiers.

   (d) voltage regulators.

7. Of the following, which material allows the lowest forward voltage drop in a diode?

   (a) Selenium

   (b) Silicon

   (c) Copper

   (d) Germanium

8. A CMOS integrated circuit

   (a) can only work at low frequencies.

   (b) requires very little power to function.

   (c) requires considerable power to function.

   (d) can only work at high frequencies.

9. The purpose of doping is to

   (a) make the charge carriers move faster.

   (b) cause holes to flow.

(c) give a semiconductor material specific properties.

(d) protect devices from damage in case of transients.

10. A semiconductor material is made into N type by

(a) adding an acceptor impurity.

(b) adding a donor impurity.

(c) injecting protons.

(d) taking neutrons away.

11. Which of the following does not result from adding an acceptor impurity?

(a) The material becomes P type.

(b) Current flows mainly in the form of holes.

(c) Most of the carriers have positive electric charge.

(d) The substance acquires an electron surplus.

12. In a P-type material, electrons are

(a) the majority carriers.

(b) the minority carriers.

(c) positively charged.

(d) entirely absent.

13. Holes move from

(a) minus to plus.

(b) plus to minus.

(c) P-type to N-type material.

(d) N-type to P-type material.

14. When a P-N junction does not conduct even though a voltage is applied, the junction is

(a) reverse-biased at a voltage less than the avalanche voltage.

(b) overdriven.

(c) biased past the breaker voltage.

(d) in a state of avalanche effect.

15. Holes flow the opposite way from electrons because

(a) charge carriers flow continuously.

(b) they have opposite electric charge.

(c) they have the same electric charge.

(d) Forget it! Holes flow in the same direction as electrons.

16. If an electron is considered to have a charge of −1 unit, then a hole can be considered to have

(a) a charge of −1 unit.

(b) no charge.

(c) a charge of +1 unit.

(d) a charge that depends on the semiconductor type.

17. When a P-N junction is forward-biased, conduction will not occur unless

(a) the applied voltage exceeds the forward breakover voltage.

(b) the applied voltage is less than the forward breakover voltage.

(c) the junction capacitance is high enough.

(d) the depletion region is wide enough.

18. If the reverse bias exceeds the avalanche voltage in a P-N junction,

(a) the junction will be destroyed.

(b) the junction will insulate; no current will flow.

(c) the junction will conduct current.

(d) the capacitance will become extremely low.

19. Avalanche voltage is routinely exceeded when a P-N junction acts as a

(a) current rectifier.

(b) variable resistor.

(c) variable capacitor.

(d) voltage regulator.

20. Which of the following does not affect the junction capacitance of a diode?

(a) the cross-sectional area of the P-N junction

(b) the width of the depletion region

(c) the phase of an applied ac signal

(d) the reverse-bias voltage

20

CHAPTER

How Diodes Are Used

IN THE EARLY YEARS OF ELECTRONICS, NEARLY ALL DIODES WERE VACUUM TUBES. TODAY, MOST ARE made from semiconductors. Contemporary diodes can do almost everything that the old ones could, and also some things that people in the tube era could only dream about.

Rectification

The hallmark of a *rectifier diode* is that it passes current in only one direction. This makes it useful for changing ac to dc. Generally speaking, when the cathode is negative with respect to the anode, current flows; when the cathode is positive relative to the anode, there is no current. The constraints on this behavior are the forward breakover and avalanche voltages, as you learned about in Chap. 19.

Examine the circuit shown at A in Fig. 20-1. Suppose a 60-Hz ac sine wave is applied to the input. During half the cycle, the diode conducts, and during the other half, it doesn't. This cuts off half of every cycle. Depending on which way the diode is hooked up, either the positive half or the negative half of the ac cycle will be removed. Drawing B in Fig. 20-1 shows a graph of the output

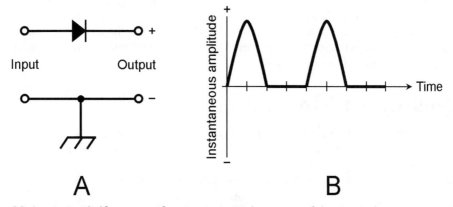

20-1 At A, a half-wave rectifier circuit. At B, the output of the circuit shown at A when an ac sine wave is applied to the input.

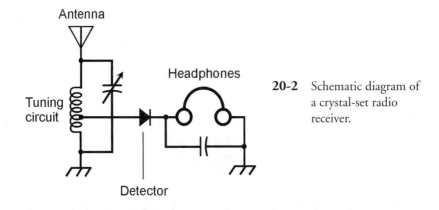

20-2 Schematic diagram of a crystal-set radio receiver.

of the circuit at A. Remember that electrons flow from negative to positive, against the arrow in the diode symbol.

The circuit and wave diagram of Fig. 20-1 show a *half-wave rectifier* circuit. This is the simplest possible rectifier. That's its chief advantage over other, more complicated rectifier circuits. You'll learn about the various types of rectifier diodes and circuits in Chap. 21.

Detection

One of the earliest diodes, existing even before vacuum tubes, was actually a primitive semiconductor device. Known as a *cat whisker*, it consisted of a fine piece of wire in contact with a small piece of the mineral *galena*. This strange-looking contraption had the ability to act as a rectifier for extremely weak RF currents. When the cat whisker was connected in a circuit such as the one shown in Fig. 20-2, the result was a device capable of picking up amplitude-modulated (AM) radio signals and producing audio output that could be heard in the headset.

The galena, sometimes called a "crystal," gave rise to the nickname *crystal set* for this primitive radio receiver. You can still build a crystal set today, using a simple RF diode, a coil, a tuning capacitor, a headset, and a long-wire antenna. Notice that there's no battery! The audio is provided by the received signal alone.

The diode in Fig. 20-2 acts to recover the audio from the radio signal. This process is called *detection*; the circuit is called a *detector* or *demodulator*. If the detector is to be effective, the diode must be of the proper type. It must have low junction capacitance, so that it can work as a rectifier (and not as a capacitor) at radio frequencies. Some modern RF diodes are microscopic versions of the old cat whisker, enclosed in a glass case with axial leads.

Frequency Multiplication

When current passes through a diode, half of the cycle is cut off, as shown in Fig. 20-1B. This occurs no matter what the frequency, from 60-Hz utility current through RF, as long as the diode capacitance is not too great.

The output wave from the diode looks much different than the input wave. This condition is known as *nonlinearity*. Whenever there is nonlinearity of any kind in a circuit—that is, whenever the output waveform is shaped differently from the input waveform—there are harmonics in the output. These are waves at integer multiples of the input frequency. (If you've forgotten what harmonics are, refer to Chap. 9.)

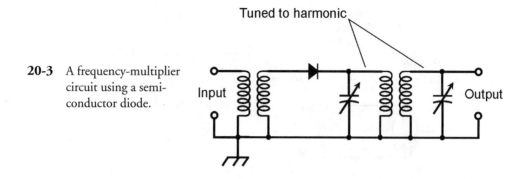

20-3 A frequency-multiplier circuit using a semiconductor diode.

Often, nonlinearity is undesirable. Then engineers strive to make the circuit *linear*, so the output waveform has exactly the same shape as the input waveform. But sometimes harmonics are desired. Then nonlinearity is introduced deliberately to produce *frequency multiplication*. Diodes are ideal for this purpose. A simple frequency-multiplier circuit is shown in Fig. 20-3. The output *LC* circuit is tuned to the desired nth harmonic frequency, nf_o, rather than to the input or fundamental frequency, f_o.

For a diode to work as a frequency multiplier, it must be of a type that would also work well as a detector at the same frequencies. This means that the component should act like a rectifier, but not like a capacitor.

Signal Mixing

When two waves having different frequencies are combined in a nonlinear circuit, new waves are produced at frequencies equal to the sum and difference of the frequencies of the input waves. Diodes can provide this nonlinearity.

Suppose there are two signals with frequencies f_1 and f_2. For mathematical convenience, let's assign f_2 to the wave with the higher frequency, and f_1 to the wave with the lower frequency. If these signals are combined in a nonlinear circuit, new waves result. One of them has a frequency of $f_2 + f_1$, and the other has a frequency of $f_2 - f_1$. These sum and difference frequencies are known as *beat frequencies*. The signals themselves are called *mixing products* or *heterodynes* (Fig. 20-4).

Figure 20-4, incidentally, is an illustration of a *frequency domain* display. The amplitude (on the vertical scale or axis) is shown as a function of the frequency (on the horizontal scale or axis). This sort of display is what engineers see when they look at the screen of a lab instrument known as a *spectrum analyzer*. In contrast, an ordinary oscilloscope displays amplitude (on the vertical scale or axis) as a function of time (on the horizontal scale or axis). The oscilloscope provides a *time domain* display.

20-4 Spectral (frequency-domain) illustration of signal mixing.

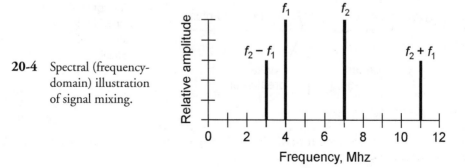

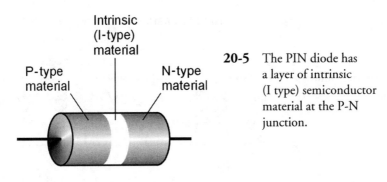

20-5   The PIN diode has a layer of intrinsic (I type) semiconductor material at the P-N junction.

Switching

The ability of diodes to conduct with forward bias, and to insulate with reverse bias, makes them useful for switching in some electronic applications. Diodes can perform switching operations much faster than any mechanical device.

One type of diode, made for use as an RF switch, has a special semiconductor layer sandwiched in between the P-type and N-type material. The material in this layer is called an *intrinsic* (or *I-type*) *semiconductor*. The *intrinsic layer* (or *I layer*) reduces the capacitance of the diode, so that it can work at higher frequencies than an ordinary diode. A diode with an I-type semiconductor layer sandwiched in between the P- and N-type layers is called a *PIN diode* (Fig. 20-5).

Direct-current bias, applied to one or more PIN diodes, allows RF currents to be effectively channeled without using relays and cables. A PIN diode also makes a good RF detector, especially at very high frequencies.

Voltage Regulation

Most diodes have an avalanche breakdown voltage that is much higher than the reverse bias ever gets. The value of the avalanche voltage depends on how a diode is manufactured. *Zener diodes* are specially made so they exhibit well-defined, constant avalanche voltages.

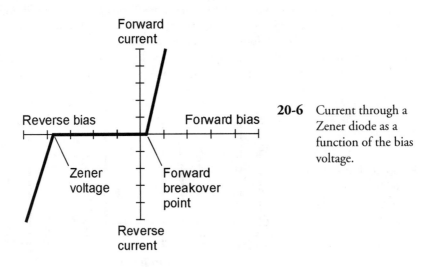

20-6   Current through a Zener diode as a function of the bias voltage.

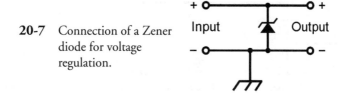

20-7 Connection of a Zener diode for voltage regulation.

Suppose a certain Zener diode has an avalanche voltage, also called the *Zener voltage*, of 50 V. If reverse bias is applied to the P-N junction, the diode acts as an open circuit as long as the bias is less than 50 V. But if the reverse-bias voltage reaches 50 V—even for a brief instant of time—the diode conducts. This effectively prevents the reverse-bias voltage from exceeding 50 V.

The current through a Zener diode, as a function of the voltage, is shown in Fig. 20-6. The Zener voltage is indicated by the abrupt rise in reverse current as the reverse-bias voltage increases. A simple Zener-diode voltage-limiting circuit is shown in Fig. 20-7. Note the polarity of the diode: the cathode is connected to the positive pole, and the anode is connected to the negative pole.

Amplitude Limiting

In Chap. 19, you learned that a diode will not conduct until the forward-bias voltage is at least as great as the forward breakover voltage. There's a corollary to this: a diode will always conduct when the forward-bias voltage reaches or exceeds the forward breakover voltage, when the device is conducting current in the forward direction. In the case of silicon diodes this is approximately 0.6 V. For germanium diodes it is about 0.3 V, and for selenium diodes it is about 1 V.

This phenomenon can be used to advantage when it is necessary to limit the amplitude of a signal, as shown in Fig. 20-8. By connecting two identical diodes back-to-back in parallel with the signal path (A), the maximum peak amplitude is limited, or *clipped*, to the forward breakover voltage of the diodes. The input and output waveforms of a clipped signal are illustrated at B. This scheme is sometimes used in radio receivers to prevent "blasting" when a strong signal comes in.

The downside of the *diode limiter* circuit, such as the one shown in Fig. 20-8, is the fact that it introduces distortion when clipping occurs. This might not be a problem for reception of digital signals, for frequency-modulated signals, or for analog signals that rarely reach the limiting voltage. But for amplitude-modulated signals with peaks that rise past the limiting voltage, it can cause trouble.

20-8 At A, connection of two diodes to act as an ac limiter. At B, illustration of sine-wave peaks cut off by the action of the diodes in an ac limiter.

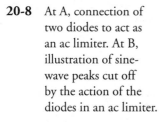

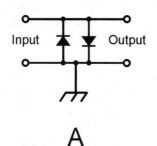

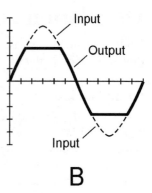

A

B

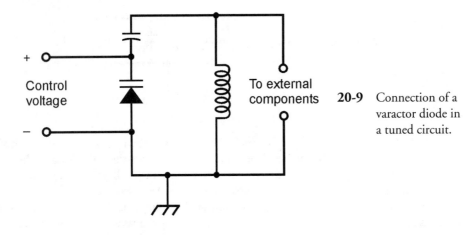

20-9 Connection of a varactor diode in a tuned circuit.

Frequency Control

When a diode is reverse-biased, there is a region at the P-N junction with dielectric (insulating) properties. As you know from Chap. 19, this is called the depletion region, because it has a shortage of majority charge carriers. The width of this zone depends on several things, including the reverse-bias voltage.

As long as the reverse bias is less than the avalanche voltage, varying the bias affects the width of the depletion region. This in turn varies the junction capacitance. This capacitance, which is always small (on the order of picofarads), varies inversely with the square root of the reverse-bias voltage, as long as the reverse bias remains less than the avalanche voltage. Thus, for example, if the reverse-bias voltage is quadrupled, the junction capacitance drops to one-half; if the reverse-bias voltage is decreased by a factor of 9, then the junction capacitance increases by a factor of 3.

Some diodes are manufactured especially for use as variable capacitors. Such a device is known as varactor diode, as you learned in Chap. 19. Varactors are used in a special type of circuit called a *voltage-controlled oscillator* (VCO). Figure 20-9 is a simple example of the *LC* circuit in a VCO, using a coil, a fixed capacitor, and a varactor. This is a parallel-tuned circuit. The fixed capacitor, whose value is large compared with that of the varactor, serves to keep the coil from short-circuiting the control voltage across the varactor. Notice that the symbol for the varactor has two lines on the cathode side.

Oscillation and Amplification

Under certain conditions, diodes can be made to produce microwave RF signals. Three types of diodes that can do this are *Gunn diodes*, *IMPATT diodes*, and *tunnel diodes*.

Gunn Diodes

A Gunn diode can produce up to 1 W of RF power output, but more commonly it works at levels of about 0.1 W. Gunn diodes are usually made from gallium arsenide. A Gunn diode oscillates because of the *Gunn effect*, named after J. Gunn of International Business Machines (IBM), who first observed it in the 1960s. A Gunn diode doesn't work like a rectifier, detector, or mixer. Instead, the oscillation takes place as a result of a quirk called *negative resistance*.

Gunn-diode oscillators are often tuned using varactor diodes. A Gunn-diode oscillator, connected directly to a microwave horn antenna, is known as a *Gunnplexer*. These devices are popular with amateur-radio experimenters at frequencies of 10 GHz and above.

IMPATT Diodes

The acronym *IMPATT* comes from the words *imp*act *a*valanche *t*ransit *t*ime. This, like negative resistance, is a rather esoteric phenomenon. An *IMPATT diode* is a microwave oscillating device like a Gunn diode, except that it uses silicon rather than gallium arsenide.

An IMPATT diode can be used as an amplifier for a microwave transmitter that employs a Gunn-diode oscillator. As an oscillator, an IMPATT diode produces about the same amount of output power, at comparable frequencies, as a Gunn diode.

Tunnel Diodes

Another type of diode that will oscillate at microwave frequencies is the *tunnel diode*, also known as the *Esaki diode*. It produces enough power so it can be used as a local oscillator in a microwave radio receiver, but not much more.

Tunnel diodes work well as amplifiers in microwave receivers, because they generate very little unwanted noise. This is especially true of gallium arsenide devices.

Energy Emission

Some semiconductor diodes emit radiant energy when a current passes through the P-N junction in a forward direction. This phenomenon occurs as electrons fall from higher to lower energy states within atoms.

LEDs and IREDs

Depending on the exact mixture of semiconductors used in manufacture, visible light of almost any color can be produced by diodes when bias is applied to them in the forward direction. Infrared-emitting devices also exist. The most common color for a *light-emitting diode* (LED) is bright red. An *infrared-emitting diode* (IRED) produces energy at wavelengths slightly longer than those of visible red light.

The intensity of the radiant energy from an LED or IRED depends to some extent on the forward current. As the current rises, the brightness increases, but only up to a certain point. If the current continues to rise, no further increase in brilliance takes place. The LED or IRED is then said to be in a state of *saturation*.

Digital Displays

Because LEDs can be made in various different shapes and sizes, they are ideal for use in digital displays. You've seen digital clock radios that use them. They are common in car radios. They make good indicators for "on/off," "a.m./p.m.," "battery low," and other conditions.

In recent years, LED displays have been largely replaced by *liquid crystal displays* (LCDs). The LCD technology has advantages over LED technology, including lower power consumption and better visibility in direct sunlight. However, LCDs require backlighting when the ambient illumination is low.

Communications

Both LEDs and IREDs are useful in communications because their intensity can be modulated to carry information. When the current through the device is sufficient to produce output, but not enough to cause saturation, the LED or IRED output follows along with rapid current changes. Analog and digital signals can be conveyed over light beams in this way. Some modern telephone systems make use of modulated light, transmitted through clear fibers. This is known as *fiber-optic* technology.

Special LEDs and IREDs produce *coherent radiation.* These are called *laser diodes.* The rays from these diodes aren't the intense, parallel beams that most people imagine when they think about lasers. A laser LED or IRED generates a cone-shaped beam of low intensity. But it can be focused into a parallel beam, and the resulting rays have some of the same advantages found in larger lasers, including the ability to travel long distances with little decrease in their intensity.

Photosensitive Diodes

Virtually all P-N junctions exhibit conductivity that varies with exposure to radiant electromagnetic energy such as IR, visible light, and UV. The reason that conventional diodes are not affected by these rays is that they are enclosed in opaque packages. Some *photosensitive diodes* have variable dc resistance that depends on the intensity of the electromagnetic rays. Other types of diodes produce their own dc in the presence of radiant energy.

Silicon Photodiodes

A silicon diode, housed in a transparent case and constructed in such a way that visible light can strike the barrier between the P-type and N-type materials, forms a *silicon photodiode.* A reverse-bias voltage is applied to the device. When radiant energy strikes the junction, current flows. The current is proportional to the intensity of the radiant energy, within certain limits.

Silicon photodiodes are more sensitive at some wavelengths than at others. The greatest sensitivity is in the *near infrared* part of the spectrum, at wavelengths just a little bit longer than the wavelength of visible red light.

When radiant energy of variable intensity strikes the P-N junction of a reverse-biased silicon photodiode, the output current follows the light-intensity variations. This makes silicon photodiodes useful for receiving modulated-light signals of the kind used in fiber-optic communications systems.

The Optoisolator

An LED or IRED and a photodiode can be combined in a single package to get a component called an *optoisolator.* This device, the schematic symbol for which is shown in Fig. 20-10, creates a modulated-light signal and sends it over a small, clear gap to a receptor. An LED or IRED converts an electrical signal to visible light or IR; a photodiode changes the visible light or infrared back into an electrical signal.

When a signal is electrically coupled from one circuit to another, the two stages interact. The input impedance of a given stage, such as an amplifier, can affect the behavior of the circuits that feed power to it. This can lead to various sorts of trouble. Optoisolators overcome this effect, because the coupling is not done electrically. If the input impedance of the second circuit changes, the impedance that the first circuit sees is not affected, because it is simply the impedance of the LED

20-10 An optoisolator has an LED or IRED at the input and a photodiode at the output.

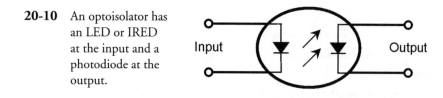

Input   Output

or IRED. That is where the "isolator" in "optoisolator" comes from. The circuits can be *electronically coupled*, and yet at the same time remain *electrically isolated*.

Photovoltaic Cells

A silicon diode, with no bias voltage applied, can generate dc all by itself if enough electromagnetic radiation hits its P-N junction. This is known as the *photovoltaic effect*. It is the principle by which solar cells work.

Photovoltaic cells are specially manufactured to have the greatest possible P-N junction surface area. This maximizes the amount of light that strikes the junction. A single silicon photovoltaic cell can produce about 0.6 V of dc electricity. The amount of current that it can deliver, and thus the amount of power it can provide, depends on the surface area of the junction.

Photovoltaic cells can be connected in series-parallel combinations to provide power for solid-state electronic devices such as portable radios. These arrays can also be used to charge batteries, allowing for use of the electronic devices when radiant energy is not available (for example, at night!). A large assembly of solar cells, connected in series-parallel, is called a *solar panel*. The power produced by a solar panel depends on the intensity of the light that strikes it, the sum total of the surface areas of all the cells, and the angle at which the light strikes the cells. Some solar panels can produce several kilowatts of electrical power in direct sunlight that shines in such a way that the sun's rays arrive perpendicular to the surfaces of all the cells.

Quiz

Refer to the text in this chapter if necessary. A good score is at least 18 correct. Answers are in the back of the book.

1. When a diode is forward-biased, the anode voltage
    (a) is negative relative to the cathode voltage.
    (b) is positive relative to the cathode voltage.
    (c) is the same as the cathode voltage.
    (d) alternates between positive and negative relative to the cathode voltage.

2. If a diode is connected in series with the secondary winding of an ac transformer, and if the peak voltage across the diode never exceeds the avalanche voltage, then the output of the complete transformer-diode circuit is
    (a) ac with half the frequency of the input.
    (b) ac with the same frequency as the input.
    (c) ac with twice the frequency of the input.
    (d) none of the above.

3. A crystal set
   (a) can be used to transmit radio signals.
   (b) requires a battery with long life.
   (c) requires no battery.
   (d) is used for rectifying 60-Hz ac.

4. A diode detector
   (a) is used in power supplies.
   (b) is employed in some radio receivers.
   (c) is used to generate microwave RF signals.
   (d) changes dc into ac.

5. If the output wave in a circuit has the same shape as the input wave, then
   (a) the circuit is operating in a linear manner.
   (b) the circuit is operating as a frequency multiplier.
   (c) the circuit is operating as a mixer.
   (d) the circuit is operating as a rectifier.

6. Suppose the two input signal frequencies to a mixer circuit are 3.522 MHz and 3.977 MHz. At which of the following frequencies can we expect a signal to exist at the output?
   (a) 455 kHz
   (b) 886 kHz
   (c) 14.00 MHz
   (d) 1.129 MHz

7. Fill in the blanks to make the following sentence correct: "A spectrum analyzer provides a display of _____ as a function of _____."
   (a) amplitude/time
   (b) time/frequency
   (c) frequency/time
   (d) amplitude/frequency

8. Zener voltage is a specialized manifestation of
   (a) forward breakover voltage.
   (b) peak forward voltage.
   (c) avalanche voltage.
   (d) reverse bias.

9. The forward breakover voltage of a silicon diode is
   (a) about 0.3 V.
   (b) about 0.6 V.
   (c) about 1.0 V.
   (d) dependent on the avalanche voltage.

10. A diode audio limiter circuit
    (a) is useful for voltage regulation.
    (b) always uses Zener diodes.
    (c) rectifies the audio to reduce distortion.
    (d) can cause distortion under some conditions.

11. The capacitance of a varactor varies with the
    (a) forward voltage.
    (b) reverse voltage.
    (c) avalanche voltage.
    (d) forward breakover voltage.

12. The purpose of the I layer in a PIN diode is to
    (a) minimize the junction capacitance.
    (b) optimize the avalanche voltage.
    (c) reduce the forward breakover voltage.
    (d) increase the current through the diode.

13. Which of these diode types can be used as the key element in the oscillator circuit of a microwave radio transmitter?
    (a) A rectifier diode
    (b) A PIN diode
    (c) An IMPATT diode
    (d) None of the above

14. A Gunnplexer is often used as a
    (a) microwave communications device.
    (b) low-frequency RF detector.
    (c) high-voltage rectifier.
    (d) signal mixer or frequency divider.

15. The most likely place you would find an LED would be in
    (a) a rectifier circuit.
    (b) a mixer circuit.
    (c) a digital frequency display.
    (d) an oscillator circuit.

16. Coherent electromagnetic radiation is produced by a
    (a) Gunn diode.
    (b) varactor diode.
    (c) rectifier diode.
    (d) laser diode.

17. Suppose you want a circuit to operate in a stable manner when the load impedance varies. You might consider a coupling method that employs

    (a) a Gunn diode.

    (b) an optoisolator.

    (c) a photovoltaic cell.

    (d) a PIN diode.

18. The electrical power that a solar panel can provide in direct sunlight depends on all of the following factors except

    (a) the ac voltage applied to the panel.

    (b) the total surface area of all the cells in the panel.

    (c) the angle at which the sunlight strikes the cells.

    (d) the intensity of the sunlight that strikes the cells.

19. Emission of energy in an IRED is caused by

    (a) high-frequency radio waves.

    (b) rectification.

    (c) changes in electron energy within atoms.

    (d) none of the above.

20. A photodiode, when not used as a photovoltaic cell, has

    (a) reverse bias.

    (b) no bias.

    (c) forward bias.

    (d) negative resistance.

21

Power Supplies

A *POWER SUPPLY* CONVERTS UTILITY AC TO DC FOR USE WITH CERTAIN ELECTRICAL AND ELECTRONIC devices. In this chapter, we'll examine the components of a typical power supply.

Power Transformers

Power transformers can be categorized as step-down or step-up. As you remember, the output, or secondary, voltage of a step-down unit is lower than the input, or primary, voltage. The reverse is true for a step-up transformer.

Step-down

Most solid-state electronic devices, such as radios, need only a few volts. The power supplies for such equipment use step-down power transformers. The physical size of the transformer depends on the current. Some devices need only a small current and a low voltage. The transformer in a radio receiver, for example, can be physically small. A ham radio transmitter or hi-fi amplifier needs more current. This means that the secondary winding of the transformer must consist of heavy-gauge wire, and the core must be bulky to contain the magnetic flux.

Step-up

Some circuits need high voltage. The cathode-ray tube (CRT) in a conventional home television set needs several hundred volts. Some ham radio power amplifiers use vacuum tubes working at more than 1 kV dc. The transformers in these appliances are step-up types. They are moderate to large in size, because of the number of turns in the secondary, and also because high voltages can spark, or *arc*, between wire turns if the windings are too tight. If a step-up transformer needs to supply only a small amount of current, it need not be big. But for ham radio transmitters and radio or television broadcast amplifiers, the transformers are large, heavy, and expensive.

Transformer Ratings

Transformers are rated according to output voltage and current. For a given unit, the *volt-ampere* (VA) capacity is often specified. This is the product of the voltage and current.

A transformer with 12-V output, capable of delivering 10 A, has 12 V × 10 A = 120 VA of capacity. The nature of power-supply filtering, to be discussed later in this chapter, makes it necessary for the power-transformer VA rating to be greater than the wattage consumed by the load.

A high-quality, rugged power transformer, capable of providing the necessary currents and/or voltages, is crucial in any power supply. The transformer is usually the most expensive component to replace.

Rectifier Diodes

Rectifier diodes are available in various sizes, intended for different purposes. Most rectifier diodes are made of silicon, and are known as *silicon rectifiers*. Some are fabricated from selenium, and are called *selenium rectifiers*. Two important features of a power-supply diode are the *average forward current* (I_o) rating and the *peak inverse voltage* (PIV) rating.

Average Forward Current

Electric current produces heat. If the current through a diode is too great, the heat will destroy the P-N junction. When designing a power supply, it is wise to use diodes with an I_o rating of at least 1.5 times the expected average dc forward current. If this current is 4.0 A, for example, the rectifier diodes should be rated at I_o = 6.0 A or more.

Note that I_o flows through the *diodes*. The current drawn by the *load* is often different from this. Also, note that I_o is an *average* figure. The *instantaneous* forward current is another thing, and can be 15 or 20 times the I_o, depending on the nature of the filtering circuit.

Some diodes have *heatsinks* to help carry heat away from the P-N junction. A selenium diode can be recognized by the appearance of its heatsink, which looks something like a baseboard radiator built around a steam pipe.

Diodes can be connected in parallel to increase the current rating over that of an individual diode. When this is done, small-value resistors should be placed in series with each diode in the set to equalize the current. Each resistor should have a value such that the voltage drop across it is about 1 V under normal operating conditions.

Peak Inverse Voltage

The PIV rating of a diode is the instantaneous reverse-bias voltage that it can withstand without the avalanche effect taking place. A good power supply has diodes whose PIV ratings are significantly greater than the peak ac input voltage. If the PIV rating is not great enough, the diode or diodes in a supply conduct for part of the reverse cycle. This degrades the efficiency of the supply because the reverse current bucks the forward current.

Diodes can be connected in series to get a higher PIV capacity than a single diode alone. This scheme is sometimes seen in high-voltage supplies, such as those needed for tube-type power amplifiers. High-value resistors, of about 500 Ω for each peak-inverse volt, are placed across each diode in the set to distribute the reverse bias equally among the diodes. In addition, each diode is shunted by (that is, connected in parallel with) a capacitor of 0.005 μF or 0.1 μF.

Half-Wave Circuit

The simplest rectifier circuit, called the *half-wave rectifier* (Fig. 21-1A), has a single diode that chops off half of the ac cycle. The effective (eff) output voltage from a power supply that uses a

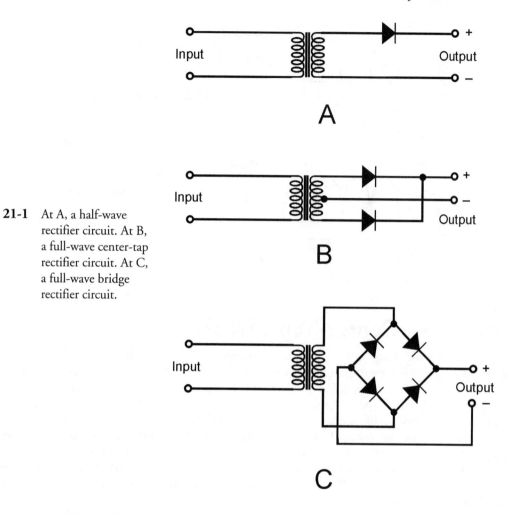

21-1 At A, a half-wave rectifier circuit. At B, a full-wave center-tap rectifier circuit. At C, a full-wave bridge rectifier circuit.

half-wave rectifier is much less than the peak transformer output voltage, as shown in Fig. 21-2A. The peak voltage across the diode in the reverse direction can be as much as 2.8 times the applied rms ac voltage.

Most engineers like to use diodes whose PIV ratings are at least 1.5 times the maximum expected peak reverse voltage. Therefore, in a half-wave rectifier circuit, the diodes should be rated for at least 2.8 × 1.5, or 4.2, times the rms ac voltage that appears across the secondary winding of the power transformer.

Half-wave rectification has shortcomings. First, the output is difficult to filter. Second, the output voltage can drop considerably when the supply is required to deliver high current. Third, half-wave rectification puts a strain on the transformer and diodes because it *pumps* them. The circuit works the diodes hard during half the ac cycle, and lets them loaf during the other half.

Half-wave rectification is usually adequate for use in a power supply that is not required to deliver much current, or when the voltage can vary without affecting the behavior of the equipment connected to it. The main advantage of a half-wave circuit is that it costs less than more sophisticated circuits.

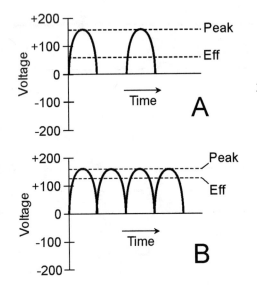

21-2 At A, the output of a half-wave rectifier. At B, the output of a full-wave rectifier. Note the difference in how the effective (eff) voltages compare with the peak voltages.

Full-Wave Center-Tap Circuit

A better scheme for changing ac to dc takes advantage of both halves of the ac cycle. A *full-wave center-tap rectifier* has a transformer with a tapped secondary (Fig. 21-1B). The center tap is connected to *electrical ground*, also called *chassis ground*. This produces voltages and currents at the ends of the winding that are in phase opposition with respect to each other. These two ac waves can be individually half-wave rectified, cutting off one half of the cycle and then the other, over and over.

The effective output voltage from a power supply that uses a full-wave center-tap rectifier is greater, relative to the peak voltage, than is the case with the half-wave rectifier (Fig. 21-2B). The PIV across the diodes can, nevertheless, be as much as 2.8 times the applied rms ac voltage. Therefore, the diodes should have a PIV rating of at least 4.2 times the applied rms ac voltage to ensure that they won't break down.

The output of a full-wave center-tap rectifier is easier to filter than that of a half-wave rectifier because the frequency of the pulsations in the dc (known as the *ripple frequency*) from a full-wave rectifier is twice the ripple frequency of the pulsating dc from a half-wave rectifier, assuming identical ac input frequency in either situation. If you compare Fig. 21-2B with Fig. 21-2A, you will see that the full-wave-rectifier output is closer to pure dc than the half-wave rectifier output. Another advantage of a full-wave center-tap rectifier is the fact that it's gentler with the transformer and diodes than a half-wave rectifier. Yet another asset: When a load is applied to the output of a power supply that uses a full-wave center-tap rectifier circuit, the voltage drops less than is the case with a half-wave supply. But because the transformer is more sophisticated, the full-wave center-tap circuit costs more than a half-wave circuit that delivers the same output voltage at the same rated maximum current.

Full-Wave Bridge Circuit

Another way to get full-wave rectification is the *full-wave bridge rectifier*, often called simply a *bridge*. It is diagrammed in Fig. 21-1C. The output waveform is similar to that of the full-wave center-tap circuit (Fig. 21-2B).

The effective output voltage from a power supply that uses a full-wave bridge rectifier is somewhat less than the peak transformer output voltage, as shown in Fig. 21-2B. The peak voltage across the diodes in the reverse direction is about 1.4 times the applied rms ac voltage. Therefore, each diode needs to have a PIV rating of at least 1.4 × 1.5, or 2.1, times the rms ac voltage that appears at the transformer secondary.

The bridge circuit does not require a center-tapped transformer secondary. It uses the entire secondary winding on both halves of the wave cycle, so it makes even more efficient use of the transformer than the full-wave center-tap circuit. The bridge is also easier on the diodes than half-wave or full-wave center-tap circuits.

Voltage-Doubler Circuit

Diodes and capacitors can be interconnected to deliver a dc output that is approximately twice the positive or negative peak ac input voltage. This is called a *voltage-doubler power supply*. This circuit works well only when the load draws low current. Otherwise, the *voltage regulation* is poor; the voltage drops a lot when the current demand is significant.

The best way to build a high-voltage power supply is to use a step-up transformer, not a voltage-doubling scheme. Nevertheless, a voltage-doubler power supply can be, and sometimes is, used when the cost of the circuit must be minimized and the demands placed on it are expected to be modest.

Figure 21-3 is a simplified diagram of a voltage-doubler power supply. It works on the entire ac cycle, so it is called a *full-wave voltage doubler*. This circuit subjects the diodes to voltage peaks in the reverse direction that are 2.8 times the applied rms ac voltage. Therefore, the diodes should be rated for PIV of at least 4.2 times the rms ac voltage that appears across the transformer secondary. When the current drawn is low, the dc output voltage of this type of power supply is approximately 2.8 times the rms ac input voltage.

Proper operation of a voltage-doubler power supply depends on the ability of the capacitors to hold a charge under maximum load. The capacitors must have large values, as well as be capable of handling high voltages. The capacitors serve two purposes: to boost the voltage and to filter the out-

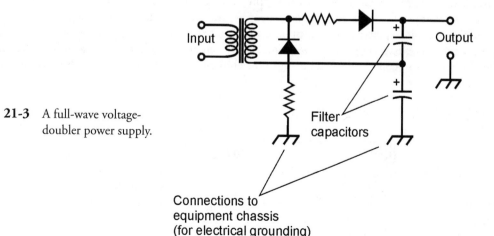

21-3 A full-wave voltage-doubler power supply.

put. The resistors, which have low ohmic values and are connected in series with the diodes, protect the diodes against *surge currents* that occur when the power supply is first switched on.

Filtering

Most dc-powered devices need something better—more pure—than the rough, pulsating dc that comes right out of a rectifier circuit. The pulsations (ripple) in the rectifier output can be eliminated by a *filter*.

Capacitors Alone

The simplest power-supply filter consists of one or more large-value capacitors, connected in parallel with the rectifier output (Fig. 21-4). A good component for this purpose is known as an *electrolytic capacitor*. This type of capacitor is *polarized*, meaning that it must be connected in the correct direction in the circuit. Each capacitor is also rated for a certain maximum voltage. Pay attention to these ratings if you ever work with electrolytic capacitors!

Filter capacitors work by trying to maintain the dc voltage at its peak level, as shown in Fig. 21-5. This is easier to do with the output of a full-wave rectifier (drawing A) than with the output of a half-wave rectifier (drawing B). With a full-wave rectifier receiving a 60-Hz ac electrical input, the ripple frequency is 120 Hz, but with a half-wave rectifier it is 60 Hz. The filter capacitors are thus recharged twice as often with a full-wave rectifier, as compared with a half-wave rectifier. This is why full-wave rectifier circuits produce more pure dc than half-wave rectifier circuits.

Capacitors and Chokes

Another way to smooth out the dc from a rectifier is to place a large-value inductor in series with the output, and a large-value capacitor in parallel. The inductor is called a *filter choke*.

In a filter that uses a capacitor and an inductor, the capacitor can be placed on the rectifier side of the choke. This is a *capacitor-input filter* (Fig. 21-6A). If the filter choke is placed on the rectifier side of the capacitor, the circuit is a *choke-input filter* (Fig. 21-6B). Capacitor-input filtering can be used when a power supply is not required to deliver much current. The output voltage, when the load is light (not much current is drawn), is higher with a capacitor-input filter than with a choke-input filter having identical input. If the supply needs to deliver large or variable amounts of current, a choke-input filter is a better choice, because the output voltage is more stable.

If the output of a power supply must have an absolute minimum of ripple, two or three capacitor/choke pairs can be connected in *cascade* (Fig. 21-7). Each pair constitutes a *section* of the filter. Multisection filters can consist of capacitor-input or choke-input sections, but the two types are never

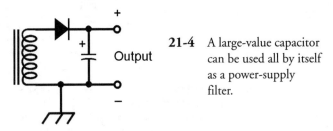

21-4 A large-value capacitor can be used all by itself as a power-supply filter.

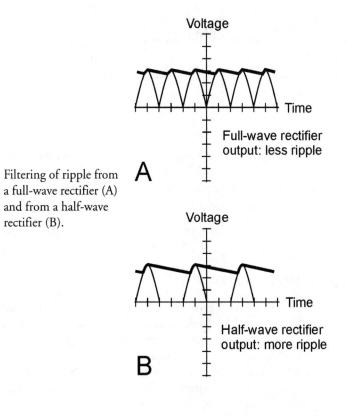

21-5 Filtering of ripple from a full-wave rectifier (A) and from a half-wave rectifier (B).

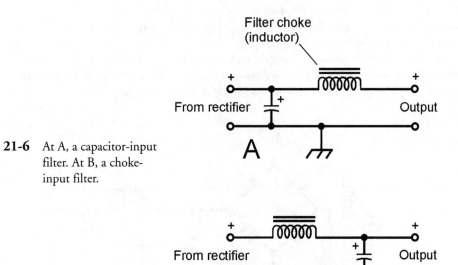

21-6 At A, a capacitor-input filter. At B, a choke-input filter.

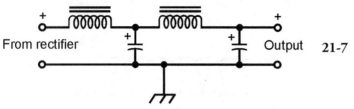

21-7 Two choke-input filter sections in cascade.

mixed. In the example of Fig. 21-7, both capacitor/choke pairs are called *L sections.* If the second capacitor is omitted, the filter becomes a *T section.* If the second capacitor is moved to the input and the second choke is omitted, the filter becomes a *pi section.* These sections are named because their schematic diagrams look something like the uppercase English L, the uppercase English T, and the uppercase Greek Π, respectively.

Voltage Regulation

If a special diode called a *Zener diode* is connected in parallel with the output of a power supply, the diode limits the output voltage. The diode must have an adequate power rating to prevent it from burning out. The limiting voltage depends on the particular Zener diode used. Zener diodes are available for any reasonable power-supply voltage.

Figure 21-8 is a diagram of a full-wave bridge dc power supply including a Zener diode for voltage regulation. Note the direction in which the Zener diode is connected in this application: with the arrow pointing from minus to plus. This is contrary to the polarity used for rectifier diodes. It's important that the polarity be correct with a Zener diode, or it will burn out.

A Zener-diode voltage regulator is inefficient when the supply is used with equipment that draws high current. When a supply must deliver a lot of current, a *power transistor* is used along with the Zener diode to obtain regulation. Figure 21-9 shows such a circuit.

Voltage regulators are available in *integrated-circuit* (IC) form. The *regulator IC*, also called a *regulator chip*, is installed in the power-supply circuit at the output of the filter. In high-voltage power supplies, *electron tubes* are sometimes used as voltage regulators. These are particularly rugged, and can withstand much higher temporary overloads than Zener diodes, transistors, or chips. However, some engineers consider such *regulator tubes* archaic.

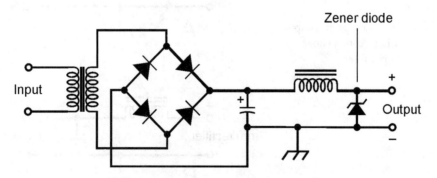

21-8 A power supply with a Zener-diode voltage regulator in the output.

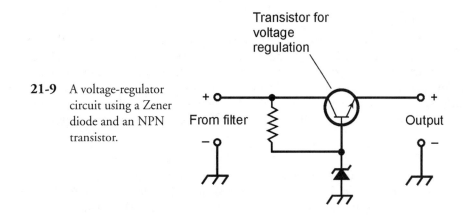

21-9 A voltage-regulator circuit using a Zener diode and an NPN transistor.

Protection of Equipment

The output of a power supply should be free of sudden changes that can damage equipment or components, or interfere with their proper performance. It is also important that voltages not appear on the external surfaces of a power supply, or on the external surfaces of any equipment connected to it.

Grounding

The best electrical ground for a power supply is the third-wire ground provided in up-to-date ac utility circuits. In an ac outlet, this connection appears as a hole shaped like an uppercase letter D turned on its side. The contacts inside this hole should be connected to a wire that ultimately terminates in a metal rod driven into the earth at the point where the electrical wiring enters the building. That constitutes an *earth ground.*

In older buildings, *two-wire ac systems* are common. These can be recognized by the presence of two slots in the utility outlets, but no ground hole. Some of these systems employ reasonable grounding by means of a scheme called *polarization*, where one slot is longer than the other, the longer slot being connected to electrical ground. But this is not as good as a *three-wire ac system*, in which the ground connection is independent of both the outlet slots.

Unfortunately, the presence of a three-wire or polarized outlet system does not always mean that an appliance connected to an outlet is well grounded. If the appliance design is faulty, or if the ground holes at the outlets were not grounded by the people who installed the electrical system, a power supply can deliver unwanted voltages to the external surfaces of appliances and electronic devices. This can present an electrocution hazard, and can also hinder the performance of equipment connected to the supply.

- **Warning: All exposed metal surfaces of power supplies should be connected to the grounded wire of a three-wire electrical cord. The third prong of the plug should never be defeated or cut off. Some means should be found to ensure that the electrical system in the building has been properly installed, so you don't work under the illusion that your system has a good ground when it actually does not. If you are in doubt about this, consult a professional electrician.**

Surge Currents

At the instant a power supply is switched on, a surge of current occurs, even with nothing connected to the supply output. This is because the filter capacitors need an initial charge, so they draw a large current for a short time. The *surge current* is far greater than the normal operating current. An extreme current surge of this sort can destroy the rectifier diodes if they are not sufficiently rated and/or protected. The phenomenon is worst in high-voltage supplies and voltage-multiplier circuits. Diode failure as a result of current surges can be prevented in at least three ways:

- Use diodes with a current rating of many times the normal operating level.
- Connect several diodes in parallel wherever a diode is called for in the circuit. *Current-equalizing resistors* are necessary (Fig. 21-10). The resistors should have small, identical ohmic values. The diodes should all be identical.
- Use an *automatic switching* circuit in the transformer primary. This type of circuit applies a reduced ac voltage to the transformer for a second or two, and then applies the full input voltage.

Transients

The ac that appears at utility outlets is a sine wave with a constant voltage near 117 V rms or 234 V rms. But there are often *voltage spikes*, known as *transients*, that can attain positive or negative peak values of several thousand volts. Transients are caused by sudden changes in the load in a utility circuit. A thundershower can produce transients throughout an entire town. Unless they are suppressed, transients can destroy the diodes in a power supply. Transients can also cause problems with sensitive electronic equipment such as computers or microcomputer-controlled appliances.

The simplest way to get rid of common transients is to place a small capacitor of about 0.01 µF, rated for 600 V or more, between each side of the transformer primary and electrical ground, as shown in Fig. 21-11. A good component for this purpose is a *disk ceramic capacitor* (not an electrolytic capacitor). Disk ceramic capacitors have no polarity issues. They can be connected in either direction to work equally well.

Commercially made *transient suppressors* are available. These devices, often mistakenly called "surge protectors," use sophisticated methods to prevent sudden voltage spikes from reaching levels where they can cause problems. It is a good idea to use transient suppressors with all sensitive electronic devices, including computers, hi-fi stereo systems, and television sets. In the event of a thundershower, the best way to protect such equipment is to physically unplug it from the wall outlets until the event has passed.

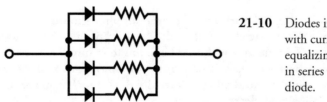

21-10 Diodes in parallel, with current-equalizing resistors in series with each diode.

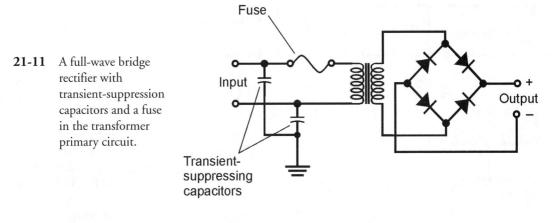

21-11 A full-wave bridge rectifier with transient-suppression capacitors and a fuse in the transformer primary circuit.

Fuses

A *fuse* is a piece of soft wire that melts, breaking a circuit if the current exceeds a certain level. A fuse is placed in series with the transformer primary, as shown in Fig. 21-11. A short circuit or overload anywhere in the power supply, or in equipment connected to it, will burn the fuse out. If a fuse blows out, it must be replaced with another of the same rating. Fuses are rated in amperes (A). Thus, a 5-A fuse will carry up to 5 A before blowing out, and a 20-A fuse will carry up to 20 A.

Fuses are available in two types: the *quick-break fuse* and the *slow-blow fuse.* A quick-break fuse is a straight length of wire or a metal strip. A slow-blow fuse usually has a spring inside along with the wire or strip. It's best to replace blown-out fuses with new ones of the same type. Quick-break fuses in slow-blow situations can burn out needlessly, causing inconvenience. Slow-blow fuses in quick-break environments might not provide adequate protection to the equipment, letting excessive current flow for too long before blowing out.

Circuit Breakers

A *circuit breaker* performs the same function as a fuse, except that a breaker can be reset by turning off the power supply, waiting a moment, and then pressing a button or flipping a switch. Some breakers reset automatically when the equipment has been shut off for a certain length of time. Circuit breakers are rated in amperes, just like fuses.

If a fuse or breaker keeps blowing out or tripping, or if it blows or trips immediately after it has been replaced or reset, then something is wrong with the power supply or with the equipment connected to it. Burned-out diodes, a bad transformer, and shorted filter capacitors in the supply can all cause this trouble. A short circuit in the equipment connected to the supply, or the connection of a device in the wrong direction (polarity), can cause repeated fuse blowing or circuit-breaker tripping.

Never replace a fuse or breaker with a larger-capacity unit to overcome the inconvenience of repeated fuse/breaker blowing/tripping. Find the cause of the trouble, and repair the equipment as needed. The "penny in the fuse box" scheme can endanger equipment and personnel, and it increases the risk of fire in the event of a short circuit.

The Complete System

Figure 21-12 is a block diagram of a complete power supply. Note the sequence in which the portions of the system, called *stages*, are connected. A final note of warning is in order here:

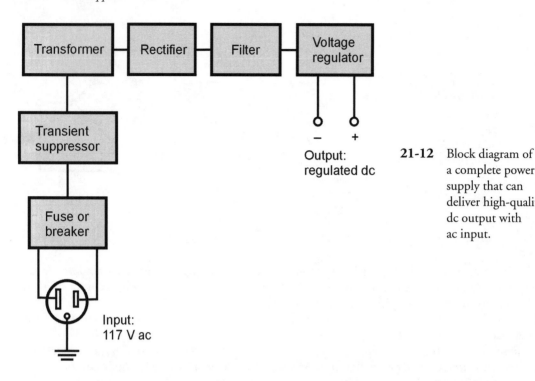

21-12 Block diagram of a complete power supply that can deliver high-quality dc output with ac input.

- **High-voltage power supplies can retain deadly voltages after they have been switched off and unplugged. This is because the filter capacitors retain their charge for some time. If you have any doubt about your ability to safely build or work with a power supply, leave it to a professional.**

Quiz

Refer to the text in this chapter if necessary. A good score is at least 18 correct. Answers are in the back of the book.

1. The output of a rectifier circuit without a filter included is

    (a) 60 Hz ac.

    (b) smooth dc.

    (c) pulsating dc.

    (d) 120 Hz ac.

2. Which of the following components is not necessarily required in a power supply designed to produce 12-V dc output with 117-V rms ac input?

    (a) The transformer

    (b) The filter

    (c) The rectifier

    (d) All of the above components are required.

3. Of the following appliances, which would need the biggest transformer?

    (a) A clock radio

    (b) A television broadcast transmitter

    (c) A shortwave radio receiver

    (d) A home television set

4. An advantage of full-wave bridge rectification is the fact that

    (a) it uses the whole transformer secondary for the entire ac input cycle.

    (b) it costs less than other rectifier types.

    (c) it cuts off half of the ac wave cycle.

    (d) it never needs a filter.

5. In a power supply designed to provide high power at low voltage, the best rectifier circuit would probably be the

    (a) half-wave arrangement.

    (b) full-wave, center-tap arrangement.

    (c) quarter-wave arrangement.

    (d) voltage doubler arrangement.

6. The part of a power supply immediately preceding the regulator is

    (a) the transformer.

    (b) the rectifier.

    (c) the filter.

    (d) the ac input.

7. If a half-wave rectifier is used with 165-V pk ac input, the effective dc output voltage is

    (a) considerably less than 165 V.

    (b) slightly less than 165 V.

    (c) exactly 165 V.

    (d) slightly more than 165 V.

8. If a full-wave bridge circuit is used with a transformer whose secondary provides 50 V rms, the peak voltage that occurs across the diodes in the reverse direction is approximately

    (a) 50 V pk.

    (b) 70 V pk.

    (c) 100 V pk.

    (d) 140 V pk.

9. What is the principal disadvantage of a voltage-doubler power supply circuit?

    (a) Excessive current

    (b) Excessive voltage

    (c) Insufficient rectification

    (d) Poor regulation under heavy loads

10. Suppose a transformer secondary provides 10-V rms ac to a voltage-doubler circuit. What is the approximate dc output voltage with no load?

  (a) 14 V

  (b) 20 V

  (c) 28 V

  (d) 36 V

11. The ripple frequency from a full-wave rectifier is

  (a) twice that from a half-wave circuit.

  (b) the same as that from a half-wave circuit.

  (c) half that from a half-wave circuit.

  (d) $\frac{1}{4}$ that from a half-wave circuit.

12. Which of the following would make the best filter for a power supply?

  (a) A capacitor in series

  (b) A choke in series

  (c) A capacitor in series and a choke in parallel

  (d) A capacitor in parallel and a choke in series

13. If you need exceptionally good ripple filtering for a power supply, which of the following alternatives will yield the best results?

  (a) Connect several capacitors in parallel.

  (b) Use a choke-input filter.

  (c) Connect several chokes in series.

  (d) Use two capacitor/choke filtering sections in cascade.

14. Voltage regulation can be accomplished by a Zener diode connected in

  (a) parallel with the filter output, forward-biased.

  (b) parallel with the filter output, reverse-biased.

  (c) series with the filter output, forward-biased.

  (d) series with the filter output, reverse-biased.

15. A current surge takes place when a power supply is first turned on because

  (a) the transformer core is suddenly magnetized.

  (b) the diodes suddenly start to conduct.

  (c) the filter capacitor(s) must be initially charged.

  (d) arcing takes place in the power switch.

16. Transient suppression is of importance mainly because it minimizes the risk of

  (a) diode failure.

  (b) transformer imbalance.

  (c) filter capacitor overcharging.

  (d) poor voltage regulation.

17. If a fuse blows, and it is replaced with one having a lower current rating, there is a good chance that

    (a) the power supply will be severely damaged.

    (b) the diodes will not rectify.

    (c) the fuse will blow out right away.

    (d) transient suppressors won't work.

18. Suppose you see a fuse with nothing but a straight wire inside. You can assume that this fuse

    (a) is a slow-blow type.

    (b) is a quick-break type.

    (c) has a low current rating.

    (d) has a high current rating.

19. In order to minimize the risk of diode destruction as a result of surge currents that can occur when a power supply is first switched on, which of the following techniques can be useful?

    (a) Connecting multiple diodes in parallel, with low-value resistors in series with each diode

    (b) Connecting multiple diodes in parallel, with low-value capacitors in series with each diode

    (c) Connecting multiple diodes in series, with low-value chokes across each diode

    (d) Connecting multiple diodes in series, with low-value resistors across each diode

20. To repair a damaged power supply with which you are not completely familiar, you should

    (a) install bleeder resistors before beginning your work.

    (b) remove the fuse before beginning your work.

    (c) leave it alone and have a professional work on it.

    (d) short out all the diodes before beginning your work.

22
CHAPTER

The Bipolar Transistor

THE WORD *TRANSISTOR* IS A CONTRACTION OF "CURRENT-*TRANS*FERRING RES*ISTOR*." A *BIPOLAR transistor* has two P-N junctions. There are two configurations: a P-type layer sandwiched between two N-type layers, or an N-type layer between two P-type layers.

NPN versus PNP

A simplified drawing of an *NPN bipolar transistor* is shown in Fig. 22-1A, and the schematic symbol is shown in Fig. 22-1B. The P-type, or center, layer is called the *base*. One of the N-type semiconductor layers is the *emitter*, and the other is the *collector*. Sometimes these are labeled *B*, *E*, and *C* in schematic diagrams. A *PNP bipolar transistor* has two P-type layers, one on either side of a thin N-type layer (Fig. 22-2A). The schematic symbol is shown in Fig. 2-22B.

It's easy to tell whether a bipolar transistor in a diagram is NPN or PNP. If the device is NPN, the arrow at the emitter points outward. If the device is PNP, the arrow at the emitter points inward.

Generally, PNP and NPN transistors can perform the same functions. The differences are the polarities of the voltages and the directions of the resulting currents. In most applications, an NPN device can be replaced with a PNP device or vice versa, the power-supply polarity can be reversed, and the circuit will work in the same way—as long as the new device has the appropriate specifications.

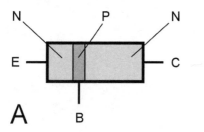

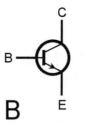

22-1 At A, pictorial diagram of an NPN transistor. At B, the schematic symbol. Electrodes are E = emitter, B = base, and C = collector.

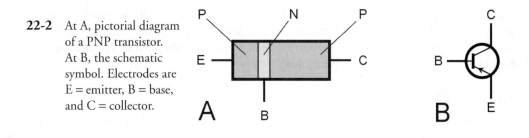

22-2 At A, pictorial diagram of a PNP transistor. At B, the schematic symbol. Electrodes are E = emitter, B = base, and C = collector.

Biasing

Imagine a bipolar transistor as consisting of two diodes in reverse series. You can't normally connect two diodes this way and get a working transistor, but the analogy is good for *modeling* the behavior of bipolar transistors. A dual-diode NPN transistor model is shown in Fig. 22-3A. The base is formed by the connection of the two anodes. The emitter is one of the cathodes, and the collector is the other cathode. Figure 22-3B shows the equivalent real-world NPN transistor circuit.

The NPN Case

The normal method of biasing an NPN transistor is to have the collector voltage positive with respect to the emitter. This is shown by the connection of the battery in Figs. 22-3A and 22-3B. Typical dc voltages for a transistor power supply range between 3 V and about 50 V. A typical voltage is 12 V.

In the model and also in the real-world transistor circuit, the base is labeled "control," because the flow of current through the transistor depends critically on what happens at this electrode.

Zero Bias for NPN

Suppose the base of a transistor is at the same voltage as the emitter. This is known as *zero bias.* When the forward bias is zero, the emitter-base current, often called simply *base current* and de-

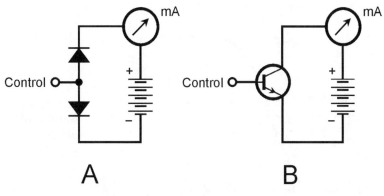

22-3 At A, the dual-diode model of a simple NPN circuit. At B, the actual transistor circuit.

noted I_B, is zero, and the emitter-base (E-B) junction does not conduct. This prevents current from flowing between the emitter and collector, unless a signal is injected at the base to change the situation. Such a signal must, at least momentarily, attain a positive voltage equal to or greater than the forward breakover voltage of the E-B junction.

Reverse Bias for NPN

Now imagine that a second battery is connected between the base and the emitter in the circuit of Fig. 22-3B, with the polarity such that E_B becomes negative with respect to the emitter. The addition of this new battery will cause the E-B junction to be *reverse-biased.* No current flows through the E-B junction in this situation (as long as the new battery voltage is not so great that avalanche breakdown occurs). A signal might be injected at the base to cause a flow of current, but such a signal must attain, at least momentarily, a positive voltage high enough to overcome both the reverse bias and the forward breakover voltage of the junction.

Forward Bias for NPN

Now suppose that E_B is made positive with respect to the emitter, starting at small voltages and gradually increasing. This is *forward bias.* If the forward bias is less than the forward breakover voltage, no current will flow. But as the base voltage E_B reaches the breakover point, the E-B junction will start to conduct.

The base-collector (B-C) junction of a bipolar transistor is normally reverse-biased. It will remain reverse-biased as long as E_B is less than the supply voltage (in this case 12 V). In practical transistor circuits, it is common for E_B to be set at a fraction of the supply voltage. Despite the reverse bias of the B-C junction, a significant emitter-collector current, called *collector current* and denoted I_C, will flow once the E-B junction conducts.

In a real transistor circuit such as the one shown in Fig. 22-3B, the meter reading will jump when the forward breakover voltage of the E-B junction is reached. Then even a small rise in E_B, attended by a rise in I_B, will cause a large increase in I_C, as shown in Fig. 22-4.

If E_B continues to rise, a point will eventually be reached where the I_C versus E_B curve levels off. The transistor is then said to be *saturated* or *in saturation.* It is wide open, conducting as much as it can.

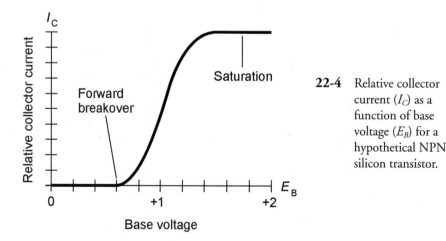

22-4   Relative collector current (I_C) as a function of base voltage (E_B) for a hypothetical NPN silicon transistor.

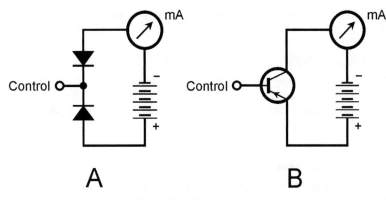

22-5   At A, the dual-diode model of a simple PNP circuit. At B, the actual
transistor circuit.

PNP Biasing

For a PNP transistor, the situation is a mirror image of the case for an NPN device. The diodes are
reversed, the arrow points inward rather than outward in the transistor symbol, and all the polarities are reversed. The dual-diode PNP model, along with the real-world transistor circuit, are shown
in Fig. 22-5. In the preceding discussion, replace every occurrence of the word "positive" with the
word "negative." Qualitatively, the same things happen: small changes in E_B cause small changes in
I_B, which in turn produce large fluctuations in I_C.

Biasing for Amplification

Because a small change in I_B causes a large variation in I_C when the bias is just right, a transistor can
operate as a *current amplifier*. If you look at Fig. 22-6, you'll see that there are some bias values at
which a transistor won't provide any current amplification. If the transistor is in saturation, the I_C

22-6   Three different
transistor bias
points. The most
amplification is
obtained when the
bias is near the middle
of the straight-line
portion of the curve.

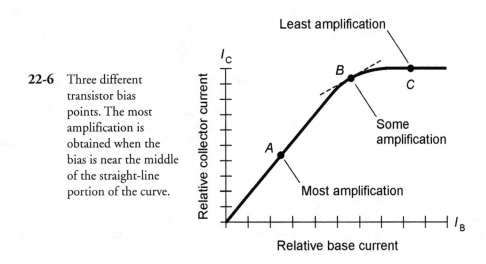

versus I_B curve is horizontal. A small change in I_B, in these portions of the curve, causes little or no change in I_C. But if the transistor is biased near the middle of the straight-line part of the curve in Fig. 22-6, the transistor will work as a current amplifier.

The same situation holds for the curve in Fig. 22-4. At some bias points, a small change in E_B does not produce much, if any, change in I_C; at other points, a small change in E_B produces a dramatic change in I_C. Whenever we want a transistor to amplify a signal, it's important that it be biased in such a way that a small change in the base current or voltage will result in a large change in the collector current.

Static Current Amplification

Current amplification is often called *beta* by engineers. It can range from a factor of just a few times up to hundreds of times. The beta of a transistor can be expressed as the *static forward current transfer ratio*, abbreviated H_{FE}. Mathematically, this is the collector current divided by the base current:

$$H_{FE} = I_C/I_B$$

For example, if a base current, I_B, of 1 mA results in a collector current, I_C, of 35 mA, then $H_{FE} = 35/1 = 35$. If $I_B = 0.5$ mA and $I_C = 35$ mA, then $H_{FE} = 35/0.5 = 70$. The H_{FE} specification for a particular transistor represents the greatest amount of current amplification that can be obtained with it.

Dynamic Current Amplification

A more practical way to define current amplification is as the ratio of the difference in I_C to the difference in I_B that occurs when a small signal is applied to the base of a transistor. Abbreviate the words "the difference in" by d. Then, according to this second definition:

$$\text{Current amplification} = dI_C/dI_B$$

Figure 22-6 is a graph of the collector current as a function of the base current (I_C versus I_B) for a hypothetical transistor. Three different points are shown, corresponding to three different bias scenarios. The ratio dI_C/dI_B is different for each of the points in this graph. Geometrically, dI_C/dI_B at a given point is the slope of a line tangent to the curve at that point. The tangent line for point B in Fig. 22-6 is a dashed straight line; the tangent lines for points A and C lie right along the curve and are therefore not shown. The steeper the slope of the line, the greater is dI_C/dI_B. Point A provides the highest value of dI_C/dI_B, provided the input signal is not too strong. This value is very close to H_{FE}.

For small-signal amplification, point A in Fig. 22-6 represents a good bias level. Engineers would say that it's a good *operating point*. At point B, dI_C/dI_B is smaller than at point A, so point B is not as good for small-signal amplification. At point C, dI_C/dI_B is practically zero. The transistor won't amplify much, if at all, if it is biased at this point.

Overdrive

Even when a transistor is biased for the greatest possible current amplification (at or near point A in Fig. 22-6), a strong ac input signal can drive it to point B or beyond during part of the signal cycle.

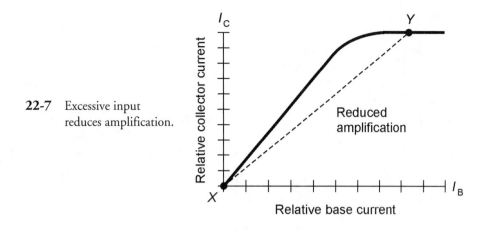

22-7   Excessive input reduces amplification.

Then, dI_C/dI_B is reduced, as shown in Fig. 22-7. Points X and Y in the graph represent the instantaneous current extremes during the signal cycle in this particular case.

When conditions are like those in Fig. 22-7, a transistor amplifier will cause *distortion* in the signal. This means that the output wave will not have the same shape as the input wave. This phenomenon is known as *nonlinearity*. It can sometimes be tolerated, but often it is undesirable. When the input signal to a transistor amplifier is too strong, the condition is called *overdrive*, and the amplifier is said to be *overdriven*.

Overdrive can cause problems other than signal distortion. An overdriven transistor is in or near saturation during part of the input signal cycle. This reduces circuit efficiency, causes excessive collector current, and can overheat the base-collector (B-C) junction. Sometimes overdrive can destroy a transistor.

Gain versus Frequency

Another important specification for a transistor is the range of frequencies over which it can be used as an amplifier. All transistors have an *amplification factor*, or *gain*, that decreases as the signal frequency increases. Some devices work well only up to a few megahertz; others can be used to several gigahertz.

Gain can be expressed in various ways. In the preceding discussion, you learned a little about *current gain*, expressed as a ratio. You will also hear about *voltage gain* or *power gain* in amplifier circuits. These, too, can be expressed as ratios. For example, if the voltage gain of a circuit is 15, then the output signal voltage (rms, peak, or peak-to-peak) is 15 times the input signal voltage. If the power gain of a circuit is 25, then the output signal power is 25 times the input signal power.

Two expressions are commonly used for the gain-versus-frequency behavior of a bipolar transistor. The *gain bandwidth product*, abbreviated f_T, is the frequency at which the gain becomes equal to 1 with the emitter connected to ground. If you try to make an amplifier using a transistor at a frequency higher than its f_T specification, you are bound to fail. The *alpha cutoff frequency* of a transistor is the frequency at which the gain becomes 0.707 times its value when the input signal frequency is 1 kHz. A transistor can have considerable gain at its alpha cutoff frequency. By looking at this

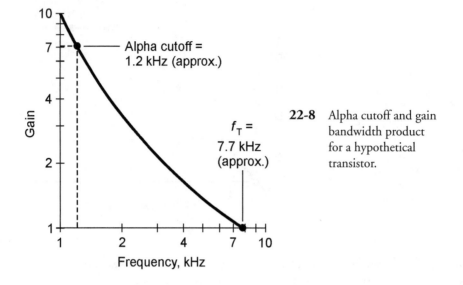

22-8 Alpha cutoff and gain bandwidth product for a hypothetical transistor.

specification for a particular transistor, you can get an idea of how rapidly it loses its ability to amplify as the frequency goes up. Some devices die off faster than others.

Figure 22-8 shows the gain bandwidth product and alpha cutoff frequency for a hypothetical transistor, on a graph of gain versus frequency. Note that the scales of this graph are not linear; that is, the divisions are not evenly spaced. This type of graph is called a *log-log* graph because both scales are *logarithmic* rather than linear.

Common Emitter Circuit

A transistor can be hooked up in three general ways. The emitter can be grounded for signal, the base can be grounded for signal, or the collector can be grounded for signal. An often-used arrangement is the *common emitter circuit.* "Common" means "grounded for the signal." The basic configuration is shown in Fig. 22-9.

A terminal can be at ground potential for a signal, and yet have a significant dc voltage. In the circuit shown, capacitor C_1 appears as a short circuit to the ac signal, so the emitter is at *signal ground.* But resistor R_1 causes the emitter to have a certain positive dc voltage with respect to ground (or a negative voltage, if a PNP transistor is used). The exact dc voltage at the emitter depends on the resistance of R_1, and on the bias. The bias is set by the ratio of the values of resistors R_2 and R_3. The bias can be anything from zero, or ground potential, to +12 V, the supply voltage. Normally it is a couple of volts.

Capacitors C_2 and C_3 block dc to or from the input and output circuitry (whatever that might be) while letting the ac signal pass. Resistor R_4 keeps the output signal from being shorted out through the power supply. A signal enters the common emitter circuit through C_2, where it causes the base current, I_B, to vary. The small fluctuations in I_B cause large changes in the collector current, I_C. This current passes through resistor R_4, causing a fluctuating dc voltage to appear across this resistor. The ac part of this passes unhindered through capacitor C_3 to the output.

The circuit of Fig. 22-9 is the basis for many amplifiers, from audio frequencies through ultrahigh radio frequencies. The common emitter configuration produces the largest gain of any arrangement. The output wave is 180° out of phase with respect to the input wave.

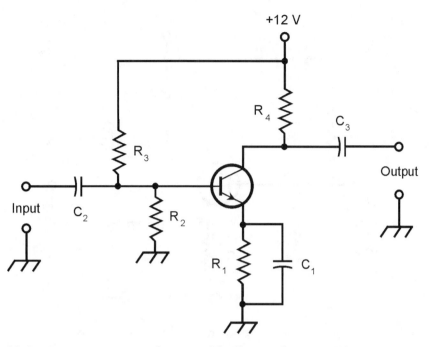

22-9   Common emitter configuration. This diagram shows an NPN transistor circuit.

Common Base Circuit

As its name implies, the *common base circuit* (Fig. 22-10) has the base at signal ground. The dc bias is the same as for the common emitter circuit, but the input signal is applied at the emitter, instead of at the base. This causes fluctuations in the voltage across R_1, causing variations in I_B. The result of these small current fluctuations is a large change in the current through R_4. Therefore, amplification occurs. The output wave is in phase with the input wave.

The signal enters through capacitor C_1. Resistor R_1 keeps the input signal from being shorted to ground. Bias is provided by R_2 and R_3. Capacitor C_2 keeps the base at signal ground. Resistor R_4 keeps the signal from being shorted out through the power supply. The output is taken through C_3.

The common base circuit provides somewhat less gain than a common emitter circuit. However, it is more stable than the common emitter configuration in some applications, especially RF power amplifiers.

Common Collector Circuit

A *common collector circuit* (Fig. 22-11) operates with the collector at signal ground. The input is applied at the base, just as it is with the common emitter circuit. The signal passes through C_2 onto the base of the transistor. Resistors R_2 and R_3 provide the correct bias for the base. Resistor R_4 limits the current through the transistor. Capacitor C_3 keeps the collector at signal ground. A fluctuating direct current flows through R_1, and a fluctuating dc voltage therefore appears across it. The ac

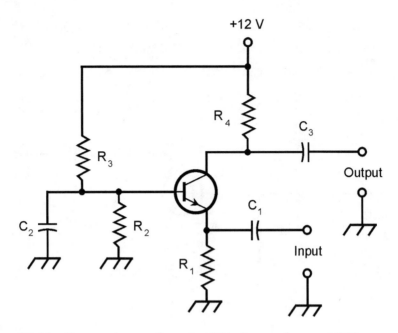

22-10   Common base configuration. This diagram shows an NPN transistor circuit.

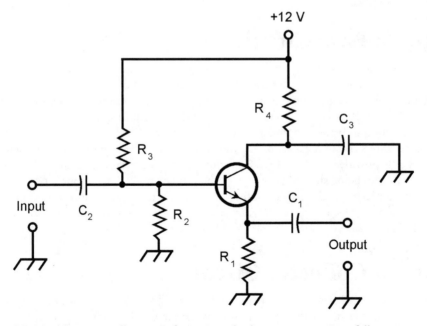

22-11   Common collector configuration, also known as an emitter follower. This diagram shows an NPN transistor circuit.

part of this voltage passes through C_1 to the output. Because the output follows the emitter current, this circuit is sometimes called an *emitter follower circuit*.

The output wave of a common collector circuit is in phase with the input wave. This circuit is unique because its input impedance is high, while its output impedance is low. For this reason, the common collector circuit can be used to match high impedances to low impedances. When well designed, an emitter follower works over a wide range of frequencies, and is a low-cost alternative to a broadband impedance-matching transformer.

Quiz

Refer to the text in this chapter if necessary. A good score is at least 18 correct. Answers are in the back of the book.

1. In a PNP circuit, the collector
    (a) has an arrow pointing inward.
    (b) is positive with respect to the emitter.
    (c) is biased at a small fraction of the base bias.
    (d) is negative with respect to the emitter.

2. In many cases, a PNP transistor can be replaced with an NPN device and the circuit will do the same thing, provided that
    (a) the power supply or battery polarity is reversed.
    (b) the collector and emitter leads are interchanged.
    (c) the arrow is pointing inward.
    (d) Forget it! A PNP transistor can never be replaced with an NPN transistor.

3. A bipolar transistor has
    (a) three P-N junctions.
    (b) three semiconductor layers.
    (c) two N-type layers around a P-type layer.
    (d) a low avalanche voltage.

4. In the dual-diode model of an NPN transistor, the emitter corresponds to
    (a) the point where the cathodes are connected together.
    (b) the point where the cathode of one diode is connected to the anode of the other.
    (c) the point where the anodes are connected together.
    (d) either of the diode cathodes.

5. The current through a transistor depends on
    (a) E_C.
    (b) E_B relative to E_C.
    (c) I_B.
    (d) more than one of the above.

6. With no signal input, a bipolar transistor would have the least I_C when
   (a) the emitter is grounded.
   (b) the E-B junction is forward-biased.
   (c) the E-B junction is reverse-biased.
   (d) the E-B current is high.

7. When a transistor is conducting as much as it can, it is said to be
   (a) in a state of cutoff.
   (b) in a state of saturation.
   (c) in a state of reverse bias.
   (d) in a state of avalanche breakdown.

8. Refer to the curve shown in Fig. 22-12. Which operating point is best if a large amplification factor is desired with a weak signal input?
   (a) Point *A*
   (b) Point *B*
   (c) Point *C*
   (d) Point *D*

9. In Fig. 22-12, the forward breakover point for the E-B junction is nearest to
   (a) no point on this graph.
   (b) point *B*.
   (c) point *C*.
   (d) point *D*.

10. In Fig. 22-12, saturation is nearest to
    (a) point *A*.
    (b) point *B*.
    (c) point *C*.
    (d) point *D*.

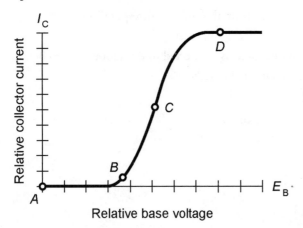

22-12 Illustration for Quiz Questions 8 through 11.

11. In Fig. 22-12, the greatest gain occurs at
    - (a) point *A*.
    - (b) point *B*.
    - (c) point *C*.
    - (d) point *D*.

12. In a common emitter circuit, the gain bandwidth product is
    - (a) the frequency at which the gain is 1.
    - (b) the frequency at which the gain is 0.707 times its value at 1 MHz.
    - (c) the frequency at which the gain is greatest.
    - (d) the difference between the frequency at which the gain is greatest, and the frequency at which the gain is 1.

13. The bipolar-transistor configuration most often used for matching a high input impedance to a low output impedance puts signal ground at
    - (a) the emitter.
    - (b) the base.
    - (c) the collector.
    - (d) any point; it doesn't matter.

14. The output is in phase with the input in
    - (a) a common emitter circuit.
    - (b) a common base circuit.
    - (c) a common collector circuit.
    - (d) more than one of the above.

15. The greatest possible amplification is obtained in
    - (a) a common emitter circuit.
    - (b) a common base circuit.
    - (c) a common collector circuit.
    - (d) more than one of the above.

16. The input is applied to the collector in
    - (a) a common emitter circuit.
    - (b) a common base circuit.
    - (c) a common collector circuit.
    - (d) none of the above.

17. The configuration noted for its stability in RF power amplifiers is the
    - (a) common emitter circuit.
    - (b) common base circuit.
    - (c) common collector circuit.
    - (d) emitter follower circuit.

18. In a common base circuit, the output is taken from
    (a) the emitter.
    (b) the base.
    (c) the collector.
    (d) more than one of the above.

19. Suppose that the input signal to a transistor amplifier results in saturation during part of the cycle. This produces
    (a) the greatest possible amplification.
    (b) reduced efficiency.
    (c) avalanche effect.
    (d) nonlinear output impedance.

20. Suppose that the gain of a transistor in a common emitter circuit is 100 at a frequency of 1 kHz, and the gain is 70.7 at 335 kHz. The gain drops to 1 at 210 MHz. The alpha cutoff frequency is
    (a) 1 kHz.
    (b) 335 kHz.
    (c) 210 MHz.
    (d) impossible to define based on this data.

23
CHAPTER

The Field Effect Transistor

THE BIPOLAR TRANSISTOR ISN'T THE ONLY SEMICONDUCTOR TRANSISTOR THAT CAN AMPLIFY. THE other major category of transistor is the *field effect transistor* (FET). There are two main types of FET: the *junction FET* (JFET) and the *metal-oxide FET* (MOSFET).

Principle of the JFET

In a JFET, the current varies because of the effects of an electric field within the device. Charge carriers (electrons or holes) flow from the *source* (S) electrode to the *drain* (D) electrode. This results in a *drain current*, I_D, that is normally the same as the *source current*, I_S. The rate of flow of charge carriers—that is, the current—depends on the voltage at a control electrode called the *gate* (G). Fluctuations in *gate voltage*, E_G, cause changes in the current through the *channel*, which is the path between the source and the drain. The current through the channel is normally equal to I_D. Small fluctuations in E_G can cause large variations in I_D. This fluctuating drain current can produce significant fluctuations in the voltage across an output resistance.

N-Channel versus P-Channel

A simplified drawing of an *N-channel JFET*, and its schematic symbol, are shown in Fig. 23-1. The N-type material forms the channel, or the path for charge carriers. The majority carriers are electrons. The drain is placed at a positive dc voltage with respect to the source.

In an N-channel device, the gate consists of P-type material. Another section of P-type material, called the *substrate*, forms a boundary on the side of the channel opposite the gate. The voltage on the gate produces an electric field that interferes with the flow of charge carriers through the channel. The more negative E_G becomes, the more the electric field chokes off the current through the channel, and the smaller I_D becomes.

A *P-channel JFET* (Fig. 23-2) has a channel of P-type semiconductor. The majority charge carriers are holes. The drain is negative with respect to the source. The more positive E_G gets, the more the electric field chokes off the current through the channel, and the smaller I_D becomes.

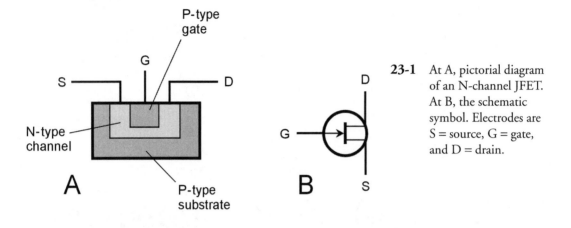

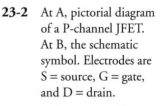

23-1 At A, pictorial diagram of an N-channel JFET. At B, the schematic symbol. Electrodes are S = source, G = gate, and D = drain.

You can recognize the N-channel JFET in schematic diagrams by the arrow pointing inward at the gate, and the P-channel JFET by the arrow pointing outward. Also, you can tell which is which (sometimes arrows are not included in schematic diagrams) by the power-supply polarity. A positive drain indicates an N-channel JFET, and a negative drain indicates a P-channel JFET.

In electronic circuits, N-channel and P-channel devices can do the same kinds of things. The main difference is the polarity. An N-channel device can almost always be replaced with a P-channel JFET, and the power-supply polarity reversed, and the circuit will still work if the new device has the right specifications. Just as there are different kinds of bipolar transistors, there are various types of JFETs, each suited to a particular application. Some JFETs work well as weak-signal amplifiers and oscillators; others are made for power amplification.

Field effect transistors have certain advantages over bipolar transistors. Perhaps the most important is that FETs are available that generate less internal noise than bipolar transistors. This makes them excellent for use as weak-signal amplifiers at very high or ultrahigh frequencies. Field effect transistors have high input impedance, which can also be an advantage in weak-signal amplifiers.

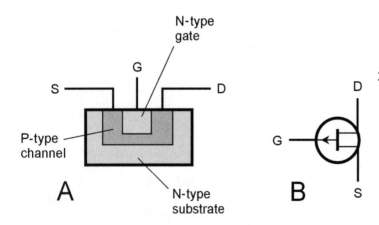

23-2 At A, pictorial diagram of a P-channel JFET. At B, the schematic symbol. Electrodes are S = source, G = gate, and D = drain.

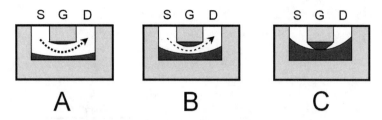

23-3 At A, the depletion region (darkest area) is narrow, the channel (white area) is wide, and many charge carriers (heavy dashed line) flow. At B, the depletion region is wider, the channel is narrower, and fewer charge carriers flow. At C, the depletion region obstructs the channel, and no charge carriers flow.

Depletion and Pinchoff

The JFET works because the voltage at the gate causes an electric field that interferes, more or less, with the flow of charge carriers along the channel. A simplified drawing of the situation for an N-channel device is shown in Fig. 23-3.

As the drain voltage E_D increases, so does the drain current I_D, up to a certain level-off value. This is true as long as the gate voltage E_G is constant, and is not too large negatively. But as E_G becomes increasingly negative (Fig. 23-3A), a *depletion region* (shown as a solid dark area) begins to form in the channel. Charge carriers cannot flow in this region; they must pass through a narrowed channel. The more negative E_G becomes, the wider the depletion region gets, as shown in drawing B. Ultimately, if the gate becomes negative enough, the depletion region completely obstructs the flow of charge carriers. This condition is called *pinchoff*, and is illustrated at C.

JFET Biasing

Two biasing methods for N-channel JFET circuits are shown in Fig. 23-4. In Fig. 23-4A, the gate is grounded through resistor R_2. The source resistor, R_1, limits the current through the JFET. The drain current, I_D, flows through R_3, producing a voltage across this resistor. The ac output signal passes through C_2. In Fig. 23-4B, the gate is connected through potentiometer R_2 to a voltage that is negative with respect to ground. Adjusting this potentiometer results in a variable negative E_G between R_2 and R_3. Resistor R_1 limits the current through the JFET. The drain current, I_D, flows through R_4, producing a voltage across it. The ac output signal passes through C_2.

In both of these circuits, the drain is positive relative to ground. For a P-channel JFET, reverse the polarities in Fig. 23-4. Typical power-supply voltages in JFET circuits are comparable to those for bipolar transistor circuits. The voltage between the source and drain, abbreviated E_D, can range from about 3 V to 150 V dc; most often it is 6 to 12 V dc. The biasing arrangement in Fig. 23-4A is preferred for weak-signal amplifiers, low-level amplifiers, and oscillators. The scheme at B is more often employed in power amplifiers having substantial input signal amplitudes.

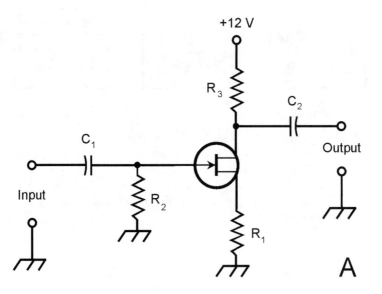

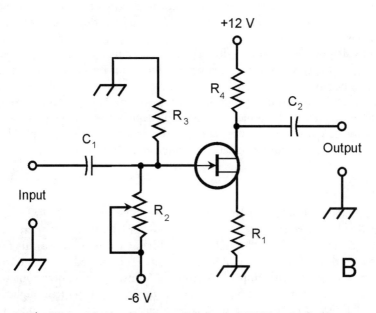

23-4   Two methods of biasing an N-channel JFET. At A, fixed gate
bias; at B, variable gate bias.

Amplification

The graph of Fig. 23-5 shows I_D as a function of E_G for a hypothetical N-channel JFET. The drain voltage, E_D, is assumed to be constant. When E_G is fairly large and negative, the JFET is pinched off, and no current flows through the channel. As E_G gets less negative, the channel opens up, and I_D begins flowing. As E_G gets still less negative, the channel gets wider and I_D increases. As E_G approaches the point where the S-G junction is at forward breakover, the channel conducts as well as it possibly can. If E_G becomes positive enough so the S-G junction conducts, some of the current in the channel leaks out through the gate. This is usually an unwanted phenomenon.

The FET Amplifies Voltage

The best amplification for weak signals is obtained when E_G is such that the slope of the curve in Fig. 23-5 is the greatest. This is shown roughly by the range marked X. For power amplification, however, results are often best when the JFET is biased at or beyond pinchoff, in the range marked Y.

In either circuit shown in Fig. 23-4, I_D passes through the drain resistor. Small fluctuations in E_G cause large changes in I_D, and these variations in turn produce wide swings in the dc voltage across R_3 (in the circuit at A) or R_4 (in the circuit at B). The ac part of this voltage goes through capacitor C_2, and appears at the output as a signal of much greater ac voltage than that of the input signal at the gate.

Drain Current versus Drain Voltage

Do you suspect that the current I_D, passing through the channel of a JFET, increases in a linear manner with increasing drain voltage E_D? This seems reasonable, but it is not what usually happens. Instead, I_D rises for awhile as E_D increases steadily, and then I_D starts to level off. The current I_D can be plotted graphically as a function of E_D for various values of E_G. When this is done, the result is a *family of characteristic curves* for the JFET. The graph of Fig. 23-6 shows a family of characteristic

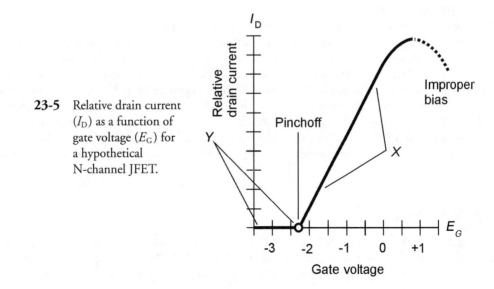

23-5 Relative drain current (I_D) as a function of gate voltage (E_G) for a hypothetical N-channel JFET.

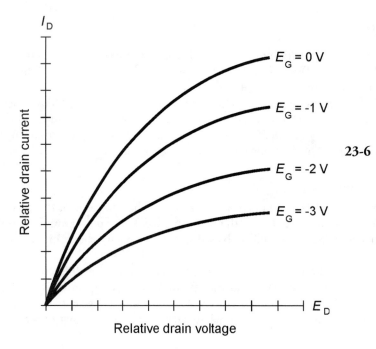

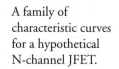

23-6 A family of characteristic curves for a hypothetical N-channel JFET.

curves for a hypothetical N-channel device. The graph of I_D versus E_G, one example of which is shown in Fig. 23-5, is also an important specification that engineers consider when choosing a JFET for a particular application.

Transconductance

Recall the discussion of *dynamic current amplification* from the last chapter. This is a measure of how well a bipolar transistor amplifies a signal. The JFET equivalent of this is called *dynamic mutual conductance* or *transconductance*.

Refer again to Fig. 23-5. Suppose that E_G is a certain value, resulting in a certain current I_D. If the gate voltage changes by a small amount dE_G, then the drain current will change by a certain increment dI_D. The transconductance is the ratio dI_D/dE_G. Geometrically, this translates to the slope of a line tangent to the curve of Fig. 23-5. The value of dI_D/dE_G is not the same at every point along the curve. When the JFET is biased beyond pinchoff, in the region marked Y, the slope of the curve is zero. Then there is no fluctuation in I_D when E_G changes by small amounts. There can be a change in I_D when there is a change in E_G only when the channel conducts current. The region where the transconductance, dI_D/dE_G, is the greatest is the region marked X, where the slope of the curve is steepest. This region of the curve represents conditions where the most gain can be obtained from the device.

The MOSFET

The acronym *MOSFET* (pronounced "*moss*-fet") stands for *metal-oxide-semiconductor field effect transistor*. A simplified cross-sectional drawing of an N-channel MOSFET, along with the schematic symbol, is shown in Fig. 23-7. The P-channel device is shown in the drawings of Fig. 23-8.

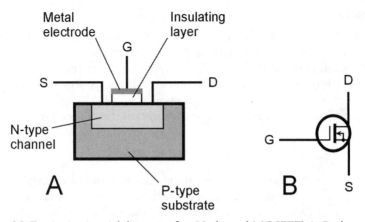

23-7 At A, pictorial diagram of an N-channel MOSFET. At B, the schematic symbol. Electrodes are S = source, G = gate, and D = drain.

Extremely High Input Impedance

When the MOSFET was first developed, it was called an *insulated-gate FET* or *IGFET.* That's still a good name for it. The gate electrode is insulated from the channel by a thin layer of dielectric material. As a result, the input impedance is even higher than that of a JFET. The gate-to-source resistance of a typical MOSFET is, as a matter of fact, comparable to that of a typical capacitor! This means that a MOSFET draws essentially no current, and therefore no power, from the signal source. This makes the device ideal for low-level and weak-signal amplifiers. But MOS devices aren't quite perfect. They have an Achilles heel. They are electrically fragile.

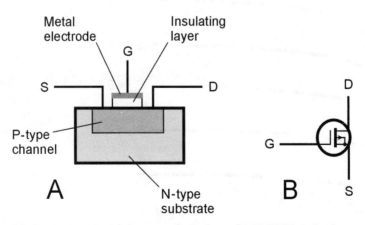

23-8 At A, pictorial diagram of a P-channel MOSFET. At B, the schematic symbol. Electrodes are S = source, G = gate, and D = drain.

Beware of Static!

The trouble with MOSFETs is that they can be easily damaged by electrostatic discharges. When building or servicing circuits containing MOS devices, technicians must use special equipment to ensure that their hands don't carry electrostatic charges that might ruin the components. If a discharge occurs through the dielectric of a MOS device, the component is permanently destroyed. Warm and humid climates do not offer total protection against the hazard. This author learned that fact by ruining several MOSFETs while designing circuits in the summertime—in Miami, Florida!

Flexibility in Biasing

In electronic circuits, an N-channel JFET can sometimes be replaced directly with an N-channel MOSFET, and P-channel devices can be similarly interchanged. But the characteristic curves for MOSFETs are not the same as those for JFETs. The main difference is that the S-G junction in a MOSFET is not a P-N junction. Therefore, forward breakover cannot occur. A gate bias voltage, E_G, more positive than +0.6 V can be applied to an N-channel MOSFET, or an E_G more negative than −0.6 V to a P-channel device, without a current leak taking place.

A family of characteristic curves for a hypothetical N-channel MOSFET is shown in the graph of Fig. 23-9.

Depletion Mode versus Enhancement Mode

Normally the channel in a JFET is wide open; as the depletion region gets wider and wider, choking off the channel, the charge carriers are forced to pass through a narrower and narrower path. This is known as the *depletion mode* of operation for a field effect transistor. A MOSFET can also

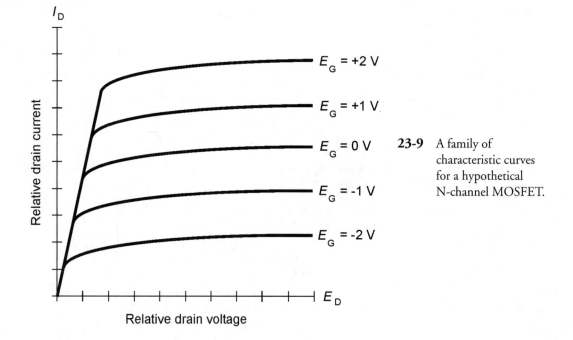

23-9 A family of characteristic curves for a hypothetical N-channel MOSFET.

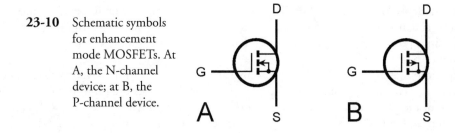

23-10 Schematic symbols for enhancement mode MOSFETs. At A, the N-channel device; at B, the P-channel device.

be made to work in the depletion mode. The drawings and schematic symbols of Figs. 23-7 and 23-8 show depletion-mode MOSFETs.

Metal-oxide semiconductor technology also allows an entirely different means of operation. An *enhancement-mode* MOSFET normally has a pinched-off channel. It is necessary to apply a bias voltage, E_G, to the gate so that a channel will form. If $E_G = 0$ in an enhancement-mode MOSFET, then $I_D = 0$ when there is no signal input. The schematic symbols for N-channel and P-channel enhancement-mode devices are shown in Fig. 23-10. Note that the vertical line is broken. This is how you can recognize an enhancement-mode device in circuit diagrams.

Common Source Circuit

There are three different circuit hookups for FETs, just as there are for bipolar transistors. These three arrangements have the source, the gate, or the drain at signal ground. In the *common source circuit*, the source is placed at signal ground. Signal input is applied to the gate. The general configuration is shown in Fig. 23-11. An N-channel JFET is used here, but the device could be an

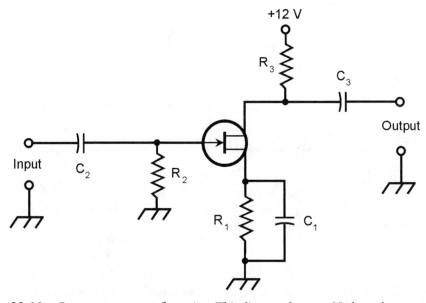

23-11 Common source configuration. This diagram shows an N-channel JFET circuit.

N-channel, depletion-mode MOSFET and the circuit diagram would be the same. For an N-channel enhancement-mode device, an extra resistor would be necessary, running from the gate to the positive power supply terminal. For P-channel devices, the supply would provide a negative, rather than a positive, voltage.

Capacitor C_1 and resistor R_1 place the source at signal ground while elevating the source above ground for dc. The ac signal enters through C_2. Resistor R_2 adjusts the input impedance and provides bias for the gate. The ac signal passes out of the circuit through C_3. Resistor R_3 keeps the output signal from being shorted out through the power supply. The circuit of Fig. 23-11 is the basis for low-level RF amplifiers and oscillators.

The common source arrangement provides the greatest gain of the three FET circuit configurations. The output is 180° out of phase with the input.

Common Gate Circuit

In the *common gate circuit* (Fig. 23-12), the gate is placed at signal ground. The input is applied to the source. The illustration shows an N-channel JFET. For other types of FETs, the same considerations apply as previously described for the common source circuit. Enhancement-mode devices would require a resistor between the gate and the positive supply terminal (or the negative terminal if the MOSFET is P-channel).

The dc bias for the common gate circuit is basically the same as that for the common source arrangement. But the signal follows a different path. The ac input signal enters through C_1. Resistor R_1 keeps the input from being shorted to ground. Gate bias is provided by R_1 and R_2. Capacitor

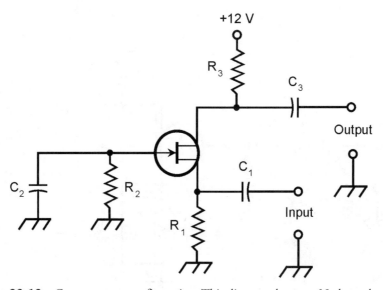

23-12 Common gate configuration. This diagram shows an N-channel JFET circuit.

C_2 places the gate at signal ground. In some common gate circuits, the gate is directly grounded, and R_2 and C_2 are not necessary. The output signal leaves the circuit through C_3. Resistor R_3 keeps the output signal from being shorted through the power supply.

The common gate arrangement produces less gain than its common source counterpart. But a common gate amplifier is not likely to break into unwanted oscillation, making it a good choice for power-amplifier circuits, especially at RF. The output is in phase with the input.

Common Drain Circuit

A *common drain circuit* is shown in Fig. 23-13. In this circuit, the collector is at signal ground. It is sometimes called a *source follower.* The FET is biased in the same way as for the common source and common gate circuits. In the illustration, an N-channel JFET is shown, but any other kind of FET could be used, reversing the polarity for P-channel devices. Enhancement-mode MOSFETs would need a resistor between the gate and the positive supply terminal (or the negative terminal if the MOSFET is P-channel).

The input signal passes through C_2 to the gate. Resistors R_1 and R_2 provide gate bias. Resistor R_3 limits the current. Capacitor C_3 keeps the drain at signal ground. Fluctuating dc (the channel current) flows through R_1 as a result of the input signal; this causes a fluctuating dc voltage to appear across R_1. The output is taken from the source, and its ac component passes through C_1.

The output of the common drain circuit is in phase with the input. This scheme is the FET analog of the bipolar common collector arrangement. The output impedance is rather low, making this circuit a good choice for broadband impedance matching.

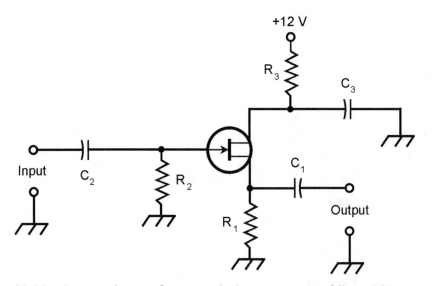

23-13 Common drain configuration, also known as a source follower. This diagram shows an N-channel JFET circuit.

Quiz

Refer to the text in this chapter if necessary. A good score is at least 18 correct. Answers are in the back of the book.

1. The current through the channel of a JFET is directly affected by all of the following, except the
   - (a) drain voltage.
   - (b) transconductance.
   - (c) gate voltage.
   - (d) gate bias.

2. In an N-channel JFET, pinchoff occurs when the gate bias is
   - (a) small and positive.
   - (b) zero.
   - (c) small and negative.
   - (d) large and negative.

3. The current consists mainly of holes when a JFET
   - (a) has a P-type channel.
   - (b) is forward-biased.
   - (c) is zero-biased.
   - (d) is reverse-biased.

4. A JFET might work better than a bipolar transistor in
   - (a) a high-voltage rectifier.
   - (b) a weak-signal RF amplifier.
   - (c) a power-supply filter.
   - (d) a power transformer.

5. In a P-channel JFET,
   - (a) the drain is forward-biased.
   - (b) the source-gate junction is forward-biased.
   - (c) the drain is negative relative to the source.
   - (d) the gate must be at dc ground.

6. A JFET is sometimes biased at or beyond pinchoff in
   - (a) a power amplifier.
   - (b) a rectifier.
   - (c) a filter.
   - (d) a weak-signal amplifier.

7. The gate of a JFET exhibits a
   - (a) forward bias.
   - (b) high impedance.

(c) low reverse resistance.

(d) low avalanche voltage.

8. Which of the following conditions is not normally desirable in a JFET?

(a) A conducting channel

(b) Holes as the majority carriers

(c) A forward-biased P-N junction

(d) A high input impedance

9. When a JFET is pinched off,

(a) the value of dI_D/dE_G is very large with no signal input.

(b) the value of dI_D/dE_G might vary considerably with no signal input.

(c) the value of dI_D/dE_G is negative with no signal input.

(d) the value of dI_D/dE_G is zero with no signal input.

10. Transconductance is the ratio of

(a) a change in drain voltage to a change in source voltage.

(b) a change in drain current to a change in gate voltage.

(c) a change in gate current to a change in source voltage.

(d) a change in drain current to a change in drain voltage.

11. Characteristic curves for JFETs generally show

(a) drain voltage as a function of source current.

(b) drain current as a function of gate current.

(c) drain current as a function of drain voltage.

(d) drain voltage as a function of gate current.

12. A disadvantage of MOS components is the fact that

(a) they can be easily damaged by static electricity.

(b) they need a high input voltage in order to amplify.

(c) they draw large amounts of current.

(d) they produce a great deal of heat.

13. The input impedance of a MOSFET is

(a) lower than that of a JFET.

(b) lower than that of a bipolar transistor.

(c) between that of a bipolar transistor and a JFET.

(d) extremely high.

14. A significant difference between MOSFETs and JFETs is the fact that

(a) MOSFETs can handle a wider range of gate bias voltages.

(b) MOSFETs can deliver greater output power.

(c) MOSFETs are more rugged.

(d) MOSFETs last longer.

15. The channel in a zero-biased JFET is normally
   (a) pinched off.
   (b) in a state of avalanche breakdown.
   (c) in a conducting state.
   (d) made of P-type semiconductor material.

16. When an enhancement-mode MOSFET is at zero bias,
   (a) the drain current is high with no signal.
   (b) the drain current fluctuates with no signal.
   (c) the drain current is low with no signal.
   (d) the drain current is zero with no signal.

17. An enhancement-mode MOSFET can be recognized in schematic diagrams by the presence of
   (a) an arrow pointing inward.
   (b) a broken vertical line inside the circle.
   (c) an arrow pointing outward.
   (d) a solid vertical line inside the circle.

18. In a source follower, which of the electrodes receives the input signal?
   (a) Any of them (doesn't matter)
   (b) The source
   (c) The gate
   (d) The drain

19. Which of the following circuits produces an output signal wave that is 180° out of phase with the input signal wave?
   (a) The common source circuit
   (b) The common gate circuit
   (c) The common drain circuit
   (d) All of the above

20. Which of the following circuits can produce the greatest signal gain (amplification factor)?
   (a) The common source circuit
   (b) The common gate circuit
   (c) The common drain circuit
   (d) All of the above circuits can amplify to the same extent.

24
CHAPTER

Amplifiers and Oscillators

NOW THAT YOU'VE LEARNED THE BASICS OF BIPOLAR AND FIELD EFFECT TRANSISTORS, YOU'RE READY to learn about amplifiers and oscillators that use these devices. As a prelude to that, let's define the most common unit that engineers use to evaluate changes in signal strength.

The Decibel

The human senses of hearing and vision don't perceive the intensity of sound and light in a linear manner. Instead, our perception of these phenomena is *logarithmic*. Electronic circuits behave in a similar way when subjected to signals that change in amplitude. Because of this, scientists and engineers invented a unit called the *decibel* (symbolized dB), in which amplitude changes are expressed according to the *base-10 logarithm* of the actual change.

More or Less?

In the decibel scheme, increases in amplitude are assigned *positive gain* values, and decreases in amplitude are assigned *negative gain* values. A negative gain figure is sometimes expressed as a *loss* by removing the minus sign.

If the output signal amplitude from a circuit is +6 dB relative to the input signal amplitude, then the output is stronger than the input. If the output is at −14 dB relative to the input, then the output is weaker than the input. In the first case, the circuit has a gain of +6 dB. In the second case, we can say that the circuit has a gain of −14 dB or a loss of 14 dB.

How large a unit is the decibel, anyway? The answer is that it has no absolute size; it is a relative unit. The decibel has meaning only when two or more signals are compared. An amplitude increase or decrease of plus or minus one decibel (±1 dB) is roughly the smallest change a listener can detect if the change is expected. If the change is not expected, the smallest difference a listener can notice is about ±3 dB.

For Voltage

Consider a circuit with an rms ac input voltage of E_{in} and an rms ac output voltage of E_{out}, specified in the same units as E_{in}. Then the *voltage gain* of the circuit, in decibels, is given by this formula:

$$\text{Gain (dB)} = 20 \log(E_{out}/E_{in})$$

The base-10 logarithm of a value x is written log x. Log functions of base 10 and base e (another type of logarithm sometimes used in physics and engineering) can be determined easily using scientific calculators. From now on, when we say "logarithm," we mean the base-10 logarithm unless otherwise specified.

Problem 24-1

Suppose a circuit has an rms ac input of 1.00 V and an rms ac output of 14.0 V. What is the gain in decibels?

First, find the ratio E_{out}/E_{in}. Because $E_{out} = 14.0$ V rms and $E_{in} = 1.00$ V rms, the ratio is 14.0/1.00, or 14.0. Next, find the logarithm of 14.0. This is about 1.146128036. Finally, multiply this number by 20, getting something like 22.92256071. Round this off to three significant figures, because that's all you're entitled to, getting 22.9 dB.

Problem 24-2

Suppose a circuit has an rms ac input voltage of 24.2 V and an rms ac output voltage of 19.9 V. What is the gain in decibels?

Find the ratio $E_{out}/E_{in} = 19.9/24.2 = 0.822\ldots$. (The three dots indicate extra digits introduced by the calculator. You can leave them in until the final roundoff.) Find the logarithm of this: log $0.822\ldots = -0.0849\ldots$. The gain is 20 times this, or $-1.699\ldots$ dB, which rounds off to -1.70 dB.

For Current

Current gain or loss is calculated in the same way as voltage gain or loss. If I_{in} is the rms ac input current and I_{out} is the rms ac output current specified in the same units as I_{in}, then:

$$\text{Gain (dB)} = 20 \log(I_{out}/I_{in})$$

For Power

The *power gain* of a circuit, in decibels, is calculated according to a slightly different formula. If P_{in} is the input signal power and P_{out} is the output signal power expressed in the same units as P_{in}, then:

$$\text{Gain (dB)} = 10 \log (P_{out}/P_{in})$$

The *coefficient* (that is, the factor by which the logarithm is to be multiplied) in the formula for power gain is 10, whereas for voltage and current gain it is 20.

Problem 24-3

Suppose a power amplifier has an input of 5.03 W and an output of 125 W. What is the gain in decibels?

Find the ratio $P_{out}/P_{in} = 125/5.03 = 24.85 \ldots$. Then find the logarithm: $\log 24.85 \ldots = 1.395 \ldots$. Finally, multiply by 10 and round off. The gain is thus $10 \times 1.395 \ldots = 14.0$ dB.

Problem 24-4

Suppose an *attenuator* (a circuit designed deliberately to produce power loss) provides 10 dB power reduction. The input power is 94 W. What is the output power?

An attenuation of 10 dB represents a gain of -10 dB. We know that $P_{in} = 94$ W, so the unknown in the power gain formula is P_{out}. We must solve for P_{out} in this formula:

$$-10 = 10 \log (P_{out}/94)$$

First, divide each side by 10, getting:

$$-1 = \log (P_{out}/94)$$

To solve this, we must take the *base-10 antilogarithm*, also known as the *antilog*, or the *inverse log*, of each side. The antilog function undoes the log function. The antilog of a value x is written antilog x. It can also be denoted as $\log^{-1}x$ or 10^x. Antilogarithms can be determined with any good scientific calculator. The solution process goes like this:

$$\text{antilog} (-1) = \text{antilog} [\log (P_{out}/94)]$$
$$0.1 = P_{out}/94$$
$$94 \times 0.1 = P_{out}$$
$$P_{out} = 9.4 \text{ W}$$

Decibels and Impedance

When determining the voltage gain (or loss) and the current gain (or loss) for a circuit in decibels, you should expect to get the same figure for both parameters only when the input impedance is identical to the output impedance. If the input and output impedances differ, the voltage gain or loss is generally not the same as the current gain or loss.

Consider how transformers work. A step-up transformer, in theory, has voltage gain, but this alone doesn't make a signal more powerful. A step-down transformer can exhibit theoretical current gain, but again, this alone does not make a signal more powerful. In order to make a signal more *power*ful, a circuit must increase the signal *power*—the *product* of the voltage and the current!

When determining power gain (or loss) for a particular circuit in decibels, the input and output impedances don't matter. In this sense, positive power gain always represents a real-world increase in signal strength. Similarly, negative power gain (or power loss) always represents a true decrease in signal strength.

Basic Bipolar Transistor Amplifier

In the previous chapters, you saw some circuits that use bipolar and field effect transistors. A signal can be applied to some control point, causing a much greater signal to appear at the output. This is the principle by which all amplifiers work.

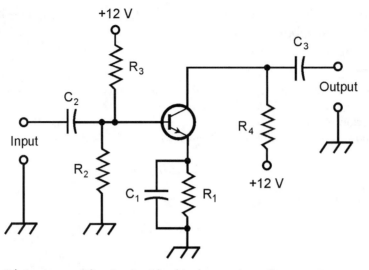

24-1 An amplifier circuit with a bipolar transistor. Component designators are discussed in the text.

In Fig. 24-1, an NPN bipolar transistor is connected as a *common emitter amplifier.* The input signal passes through C_2 to the base. Resistors R_2 and R_3 provide the base bias. Resistor R_1 and capacitor C_1 allow for the emitter to have a dc voltage relative to ground, while keeping it grounded for signals. Resistor R_1 also limits the current through the transistor. The ac output signal goes through capacitor C_3. Resistor R_4 keeps the ac output signal from being short-circuited through the power supply.

In this amplifier, the optimum capacitance values depend on the design frequency of the amplifier, and also on the impedances at the input and output. In general, as the frequency and/or circuit impedance increase, less capacitance is needed. At audio frequencies and low impedances, the capacitors might be as large as 100 μF. At radio frequencies and high impedances, values will be only a fraction of a microfarad, down to picofarads at the highest frequencies and impedances. The optimum resistor values also depend on the application. In the case of a weak-signal amplifier, typical values are 470 Ω for R_1, 4.7 kΩ for R_2, 10 kΩ for R_3, and 4.7 kΩ for R_4.

Basic JFET Amplifier

Figure 24-2 shows an N-channel JFET hooked up as a *common source amplifier.* The input signal passes through C_2 to the gate. Resistor R_2 provides the bias. Resistor R_1 and capacitor C_1 give the source a dc voltage relative to ground, while grounding it for signals. The output signal goes through C_3. Resistor R_3 keeps the output signal from being short-circuited through the power supply.

A JFET has a high input impedance, and therefore the value of C_2 should usually be small. If the device is a MOSFET, the input impedance is higher still, and C_2 will be smaller yet, sometimes 1 pF or less. The resistor values depend on the application. In some instances, R_1 and C_1 are not used, and the source is grounded directly. If R_1 is used, its optimum value will depend on the input

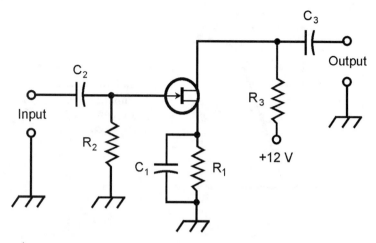

24-2   An amplifier circuit with an FET. Component designators are discussed in the text.

impedance and the bias needed for the FET. For a weak-signal amplifier, typical values are 680 Ω for R_1, 10 kΩ for R_2, and 100 Ω for R_3.

Amplifier Classes

Amplifier circuits can be categorized as *class A*, *class AB*, *class B*, and *class C*. Each class has its own special characteristics, and works best in its own unique set of applications.

The Class A Amplifier

With the previously mentioned component values, the amplifier circuits in Figs. 24-1 and 24-2 operate in class A. This type of amplifier is *linear*, meaning that the output waveform has the same shape as (although a much greater amplitude than) the input waveform.

For class A operation with a bipolar transistor, the bias must be such that, with no signal input, the device is near the middle of the straight-line portion of the I_C versus I_B (collector current versus base current) curve. This is shown for an NPN transistor in Fig. 24-3. For PNP, reverse the polarity signs. With a JFET or MOSFET, the bias must be such that, with no signal input, the device is near the middle of the straight-line part of the I_D versus E_G (drain current versus gate voltage) curve. This is shown in Fig. 24-4 for an N-channel JFET. For P channel, reverse the polarity signs.

In a class A amplifier, it is important that the input signal not be too strong. An excessively strong input signal will drive the device out of the straight-line part of the characteristic curve during part of the cycle. When this occurs, the output waveshape will not be a faithful reproduction of the input waveshape, and the amplifier will become *nonlinear*. Class A amplifiers are supposed to operate in a linear fashion at all times.

The Class AB Amplifier

Class A operation is inefficient because the transistor draws current whether there is a signal input or not. For weak-signal work, efficiency is not too critical; the things that matter are the gain and

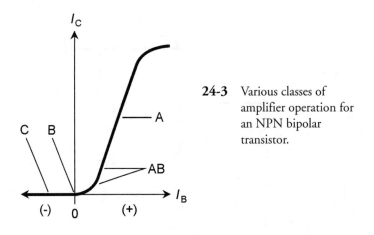

24-3 Various classes of amplifier operation for an NPN bipolar transistor.

the sensitivity. In power amplifiers, efficiency is more important, and gain and sensitivity don't matter as much.

When a bipolar transistor is biased close to cutoff under no-signal conditions, or when an FET is near pinchoff, the input signal always drives the device into the nonlinear part of the operating curve. Typical bias zones for class AB are shown in Figs. 24-3 and 24-4. A small collector or drain current flows when there is no input, but it is less than the no-signal current that flows in a class A amplifier. This is called *class AB operation*. It's more efficient than class A, but the gain and sensitivity are not as high.

There are two modes of class AB amplification. If the bipolar transistor or FET is never driven into cutoff/pinchoff during any part of the signal cycle, the amplifier is working in *class AB$_1$*. If the device goes into cutoff pinchoff for any part of the cycle (up to almost half), the amplifier is working in *class AB$_2$*.

In a class AB amplifier, the output signal waveform is not identical with the input signal waveform. But if the signal wave is *modulated,* such as in a voice radio transmitter, the waveform of the *modulating signal* comes out undistorted anyway. Thus, class AB operation is useful in RF power amplifiers.

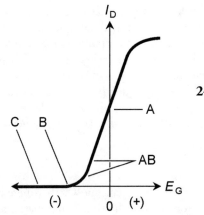
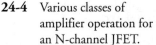

24-4 Various classes of amplifier operation for an N-channel JFET.

The Class B Amplifier

When a bipolar transistor is biased exactly at cutoff, or an FET is biased exactly at pinchoff under zero-input-signal conditions, an amplifier is working in *class B*. These operating points are labeled on the curves in Figs. 24-3 and 24-4. The class B scheme lends itself well to RF power amplification.

In class B operation, there is no collector or drain current when there is no signal. This saves energy, because the circuit does not consume power unless there is a signal going into it. (Class A and class AB amplifiers consume some power even when the input is zero.) When there is an input signal, current flows in the device during exactly half of the cycle. The output signal waveform is greatly different from the input waveshape in a class B amplifier. In fact, it is half-wave rectified.

You'll sometimes hear of class AB or class B "linear amplifiers," especially in ham radio. The term "linear" refers to the fact that the *modulation waveform* is not distorted by such an amplifier, even though the *carrier waveform* is distorted because the transistor is not biased in the straight-line part of the operating curve.

Class AB_2 and class B amplifiers draw power from the input signal source. Engineers say that such amplifiers require a certain amount of *drive* or *driving power* to function. Class A and class AB_1 amplifiers theoretically need no driving power, although there must be an input voltage.

The Class B Push-Pull Amplifier

Sometimes two bipolar transistors or FETs are used in a class B circuit, one for the positive half of the cycle and the other for the negative half. In this way, signal waveform distortion is eliminated. This is called a *class B push-pull amplifier*. This type of circuit, using two NPN bipolar transistors, is illustrated in Fig. 24-5. Resistor R_1 limits the current through the transistors. Capacitor C_1 keeps the input transformer center tap at signal ground, while allowing for some dc base bias. Resistors R_2 and R_3 bias the transistors precisely at their cutoff points. The two transistors must be identical. Not only should their part numbers be the same, but ideally they should be chosen *by experiment* to ensure that their characteristic curves are as closely matched as possible.

Class B push-pull is a popular arrangement for audio-frequency (AF) power amplification. It combines the efficiency of class B with the low distortion of class A. Its main disadvantage is that it needs two center-tapped transformers, one at the input and the other at the output. This makes push-pull amplifiers rather bulky and expensive compared to other types.

The Class C Amplifier

A bipolar transistor or FET can be biased past cutoff or pinchoff, and it will still work as a power amplifier (PA), provided that the drive is sufficient to overcome the bias during part of the cycle. This is known as *class C* operation. Bias points for class C are labeled in Figs. 24-3 and 24-4.

Class C amplifiers are nonlinear, even for amplitude modulation waveforms. Because of this, a class C circuit is useful only for signals that are either full-on or full-off. Such signals include old-fashioned Morse code, and digital schemes in which the frequency or phase (but not the amplitude) of the signal is varied.

A class C amplifier needs a lot of driving power. The gain is low. For example, it might take 300 W of signal drive to get 1 kW of signal power output. However, the efficiency is better than that of class A, AB, or B amplifiers. Let's take a closer look, now, at what *amplifier efficiency* is all about.

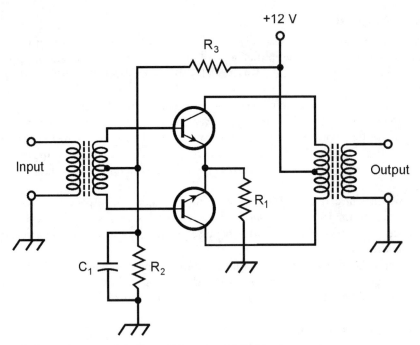

24-5 A class B push-pull amplifier using NPN bipolar transistors. Component designators are discussed in the text.

Efficiency in Power Amplifiers

In power amplification, efficiency is important. It not only provides optimum output power with minimum heat generation and minimum strain on the transistors, but it conserves energy as well. This translates into reduced cost, reduced size and weight, and longer equipment life compared with inefficient power amplifiers.

Determining dc Power Input

Suppose you connect an ammeter or milliammeter in series with the collector or drain of an amplifier and the power supply. While the amplifier is in operation, this meter will have a certain reading. The reading might appear constant, or it might fluctuate with changes in the input signal level. The dc *collector power input* to a bipolar-transistor amplifier circuit is the product of the collector current (I_C) and the collector voltage (E_C). Similarly, for an FET, the *dc drain power input* is the product of the drain current (I_D) and the drain voltage (E_D). These power figures can be further categorized as *average* or *peak* values. This discussion involves only average power.

The dc collector or drain power input can be high even when there is no input signal applied to an amplifier. A class A circuit operates this way. In fact, when a signal is applied to a class A amplifier, the meter reading, and therefore the dc collector or drain power input, does not change compared to the value under conditions of no input signal. In class AB_1 or class AB_2, there is low current (and therefore low dc collector or drain power input) with zero input signal, and a higher current (and therefore a higher dc power input) with an input signal. In class B and class C, there is no current (and therefore zero dc collector or drain power input) when there is no input signal. The current, and therefore the dc power input, increases with increasing signal input. The dc collector or drain power input is usually measured in watts, the product of amperes and volts. It can be indicated in milliwatts for low-power amplifiers, or kilowatts for high-power amplifiers.

Power Output

The *power output* of an amplifier must be measured by means of a specialized ac wattmeter. The design of AF and RF wattmeters is a sophisticated specialty in engineering.

When there is no signal input to an amplifier, there is no signal output, and therefore the power output is zero. This is true no matter what the class of amplification. The greater the signal input, in general, the greater the power output of a power amplifier, up to a certain point.

Power output, like dc input, is measured in watts. For very low power circuits, it can be in milliwatts; for high-power circuits, it is sometimes given in kilowatts.

Definition of Efficiency

The *efficiency* of a power amplifier is the ratio of the ac power output to the dc collector or drain power input.

In a bipolar-transistor amplifier, let P_C be the dc collector power input, and let P_{out} be the ac power output. For an FET amplifier, let P_D be the dc drain power input, and let P_{out} be the ac power output. Then the efficiency, eff, of the bipolar transistor amplifier is given by:

$$\text{eff} = P_{out}/P_C$$

For the FET circuit, the efficiency is:

$$\text{eff} = P_{out}/P_D$$

These are ratios, and they are always between 0 and 1. Efficiency is often expressed as a percentage instead of a ratio, so the preceding formulas are modified as follows:

$$\text{eff}_\% = 100 \; P_{out}/P_C$$
$$\text{and}$$
$$\text{eff}_\% = 100 \; P_{out}/P_D$$

Problem 24-5

Suppose a bipolar-transistor amplifier has a dc collector input of 115 W and an ac power output of 65.0 W. What is the efficiency in percent?

Use the formula for the efficiency of a bipolar transistor amplifier expressed as a percentage: $\text{eff}_\% = 100 \; P_{out}/P_C = 100 \times 65/115 = 100 \times 0.565 = 56.5\%$.

Problem 24-6

Suppose an FET amplifier is 60 percent efficient. If the power output is 3.5 W, what is the dc drain power input?

Plug in values to the formula for the efficiency of an FET amplifier expressed as a percentage. The resulting equation is solved as follows:

$$60 = 100 \times 3.5/P_D$$
$$60 = 350/P_D$$
$$60/350 = 1/P_D$$
$$P_D = 350/60$$
$$= 5.8 \text{ W}$$

Efficiency versus Class

Class A amplifiers have efficiency figures from 25 percent to 40 percent, depending on the nature of the input signal and the type of transistor used. A good class AB_1 amplifier is 35 percent to 45 percent efficient. A class AB_2 amplifier, if well designed and properly operated, can be up to about 50 percent efficient. Class B amplifiers are typically 50 percent to 65 percent efficient. Class C amplifiers can have efficiency levels as high as 75 percent.

Drive and Overdrive

Class A and AB_1 power amplifiers do not, in theory, take any power from the signal source to produce significant output power. This is one of the advantages of these classes of operation. It is only necessary that a certain voltage be present at the control electrode (the base, gate, emitter, or source) for these circuits to produce useful output signal power. Class AB_2 amplifiers need some driving power to produce ac power output. Class B amplifiers require more drive than class AB_2, and class C amplifiers need still more drive.

Whatever kind of PA is used in a given situation, it is important that the driving signal not be too strong. If *overdrive* takes place, distortion occurs in the output signal. An oscilloscope can be used to determine whether or not an amplifier is being overdriven. The scope is connected to the amplifier output terminals, and the waveform of the output signal is examined. The output waveform for a particular class of amplifier always has a characteristic shape. Overdrive is indicated by a form of distortion known as *flat topping*.

In Fig. 24-6A, the output signal waveshape for a properly operating class B amplifier is shown. In Fig. 24-6B, the output of an overdriven class B amplifier is shown. Note that the peaks are blunted or truncated. The result of this can be distortion in the modulation on a radio signal, and

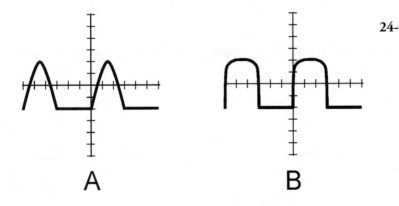

24-6 At A, an oscilloscope display of the signal output waveform from a properly operating class B power amplifier. At B, a display showing distortion in the waveform caused by overdrive.

A   B

also an excessive amount of signal output at harmonic frequencies. The efficiency of the circuit can be degraded, as well. The flat tops of the distorted waves don't contribute anything to the strength of the signal at the desired frequency, but they cause a higher-than-normal dc power input, which translates into a lower-than-normal efficiency.

Audio Amplification

The circuits you've seen so far have been general, not application-specific. With capacitors of several microfarads, and when biased for class A, these circuits are representative of audio amplifiers.

Frequency Response

High-fidelity audio amplifiers, of the kind used in music systems, must have more or less constant gain from 20 Hz to 20 kHz. This is a frequency range of 1000:1. Audio amplifiers for voice communications must work from 300 Hz to 3 kHz, a 10:1 span of frequencies. In digital communications, audio amplifiers are designed to work over a narrow range of frequencies, sometimes less than 100 Hz wide.

Hi-fi amplifiers are usually equipped with resistor-capacitor (RC) networks that tailor the frequency response. These are *tone controls*, also called *bass* and *treble* controls. The simplest hi-fi amplifiers use a single knob to control the tone. More sophisticated amplifiers have separate controls, one for bass and the other for treble. The most advanced hi-fi systems make use of *graphic equalizers*, having controls that affect the amplifier gain over several different frequency spans.

Volume Control

Audio amplifier systems usually consist of two or more *stages*. A stage is one bipolar transistor or FET (or a push-pull combination), plus peripheral resistors and capacitors. Stages can be cascaded one after the other to get high gain. In one of the stages in an audio system, a *volume control* is used. This control can be as simple as a potentiometer that allows the gain of a stage to be adjusted without affecting its linearity.

An example of a basic volume control is shown in Fig. 24-7. In this amplifier, the gain through the transistor is constant. The ac output signal passes through C_1 and appears across R_1, a potentiometer. The *wiper* (indicated by the arrow) of the potentiometer picks off more or less of the ac output signal, depending on the position of the control shaft. Capacitor C_2 isolates the potentiometer from the dc bias of the following stage.

A volume control should normally be placed in a stage where the audio power level is low. This allows the use of a low-wattage, low-cost potentiometer.

Transformer Coupling

Transformers can be used to transfer (or *couple*) signals from one stage to the next in a cascaded amplifier system (also known as an *amplifier chain*). An example of *transformer coupling* is shown in Fig. 24-8. Capacitors C_1 and C_2 keep one end of the transformer primary and secondary at signal ground. Resistor R_1 limits the current through the first transistor, Q_1. Resistors R_2 and R_3 provide the proper base bias for transistor Q_2.

The main disadvantage of this scheme is that it costs more than *capacitive coupling*. But transformer coupling can provide an optimum signal transfer between amplifier stages. By selecting a transformer with the correct turns ratio, the output impedance of Q_1 can be perfectly matched to the input impedance of Q_2.

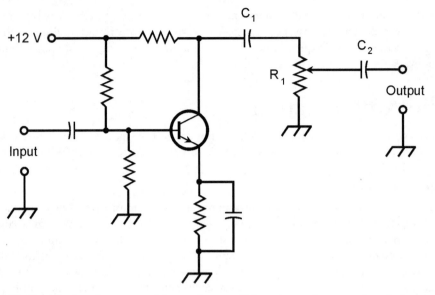

24-7   A basic volume control (potentiometer R_1) can be used to vary the gain in a low-power audio amplifier.

Tuned-Circuit Coupling

In some amplifier systems that employ transformer coupling, capacitors are added across the primary and/or secondary of the transformer. This results in resonance at a frequency determined by the capacitance and the transformer winding inductance. If the set of amplifiers is intended for use at only one frequency (and this is often the case in RF systems), this method of coupling, called

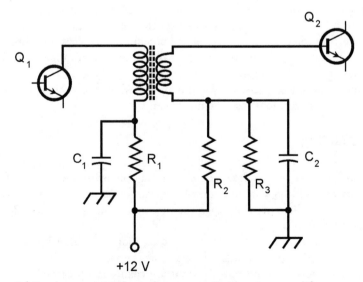

24-8   An example of transformer coupling between amplifier stages. Component designators are discussed in the text.

tuned-circuit coupling, enhances the system efficiency. But care must be taken to be sure that the amplifier chain doesn't oscillate at the resonant frequency of the tuned circuits!

Radio-Frequency Amplification

The *RF spectrum* extends upward in frequency to well over 300 GHz. The exact lower limit is a matter of disagreement in the literature. Some texts put it at 3 kHz, some at 9 kHz, some at 10 kHz, and some at the upper end of the AF range, which is normally considered to be 20 kHz.

Weak-Signal Amplifiers versus Power Amplifiers

Some RF amplifiers are designed for weak-signal work. The *front end*, or first amplifying stage, of a radio receiver requires the most sensitive possible amplifier. Sensitivity is determined by two factors: the *gain*, which has already been discussed here, and the *noise figure*, a measure of how well a circuit can amplify desired signals while generating a minimum of electronic noise.

All bipolar transistors or FETs create some *white noise* because of the movement of the charge carriers among the atoms. In general, JFETs produce less noise than bipolar transistors. Gallium arsenide FETs, also called *GaAsFETs* (pronounced "gasfets"), are the least noisy of all.

The higher the frequency at which a weak-signal amplifier is designed, the more important the noise figure gets. This is because there is less *atmospheric noise* at the higher radio frequencies, as compared with the lower frequencies. At 1.8 MHz, for example, the airwaves contain much atmospheric noise, and it doesn't make any difference if the receiver introduces a little noise itself. But at 1.8 GHz, atmospheric noise is almost nonexistent, and receiver performance depends critically on the amount of internally generated noise.

Weak-signal amplifiers almost always use resonant circuits. This optimizes the amplification at the desired frequency, while helping to cut out noise on unwanted frequencies. A typical tuned GaAsFET weak-signal RF amplifier is diagrammed in Fig. 24-9. It is designed for operation at about 10 MHz.

Broadband PAs

At RF, a PA can be either *broadband* or *tuned*. The main advantage of a broadband PA is ease of operation, because it does not need tuning. The operator need not worry about critical adjustments, nor bother to change them when changing the frequency. However, broadband PAs are slightly less efficient than tuned PAs. Another disadvantage of broadband PAs is the fact that they will amplify any signal in the design frequency range, whether or not this is desired. For example, if some earlier stage in a radio transmitter is oscillating at a frequency different from the intended signal frequency, and if this undesired energy falls within the design frequency range of the broadband PA, it will be amplified. The result will be unintended RF emission from the radio transmitter. Such unwanted signals are called *spurious emissions*.

Figure 24-10 is a schematic diagram of a typical broadband PA. The NPN bipolar transistor is a power transistor. It will reliably provide several watts of continuous RF power output over a range of frequencies from 1.5 MHz through 15 MHz. The transformers are a critical part of this circuit. They must be designed to work efficiently over a 10:1 range of frequencies.

Tuned PAs

A tuned RF power amplifier offers improved efficiency compared with broadband designs. Also, the tuning helps to reduce the chances of spurious signals being amplified and transmitted over the air. Another advantage of tuned PAs is that they can work into a wide range of load impedances. In ad-

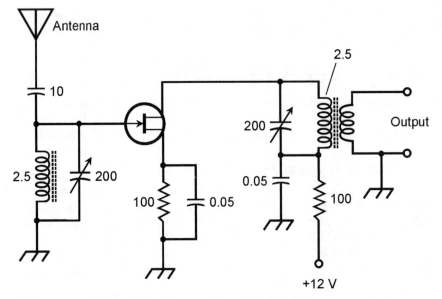

24-9 A tuned RF amplifier for use at about 10 MHz. Resistances are in ohms. Capacitances are in microfarads (μF) if less than 1, and in picofarads (pF) if more than 1. Inductances are in microhenrys (μH).

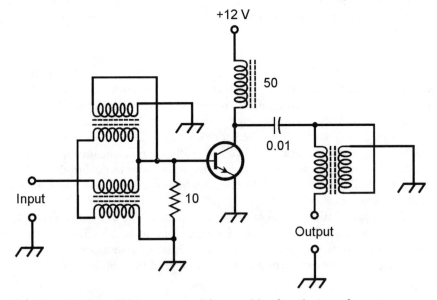

24-10 A broadband RF power amplifier, capable of producing a few watts output. Resistances are in ohms. Capacitances are in microfarads (μF). Inductances are in microhenrys (μH).

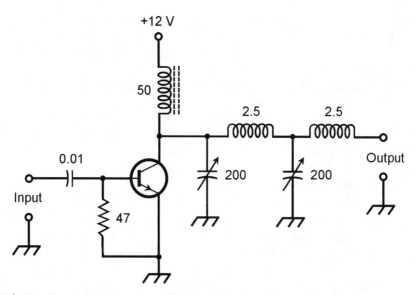

24-11 A tuned RF power amplifier, capable of producing a few watts output. Resistances are in ohms. Capacitances are in microfarads (μF) if less than 1, and in picofarads (pF) if more than 1. Inductances are in microhenrys (μH).

dition to a *tuning control*, or resonant circuit that adjusts the output of the amplifier to the operating frequency, there is a *loading control* that optimizes the signal transfer between the amplifier and the load (usually an antenna).

The main drawback of a tuned PA is that the adjustment takes time, and improper adjustment can result in damage to the transistor. If the tuning and/or loading controls are out of kilter, the efficiency of the amplifier will be extremely low (sometimes practically zero) while the dc collector or drain power input is high. Solid-state devices overheat quickly under these conditions.

A tuned RF PA, providing a few watts' output at 10 MHz or so, is shown in Fig. 24-11. The transistor is the same type as for the broadband amplifier discussed previously. The tuning and loading controls (left-hand and right-hand variable capacitors, respectively) should be adjusted for maximum RF power output as indicated by an RF wattmeter.

How Oscillators Work

Once you know how amplifiers work, it's easy to understand oscillators. All oscillators are amplifiers with positive feedback. In radio communications, oscillators generate the waves, or signals, that are ultimately sent over the air. Audio-frequency oscillators find applications in such devices as music synthesizers, modems, doorbells, sirens, alarms, and electronic toys.

Positive Feedback

Feedback can be in phase or out of phase. For a circuit to oscillate, the feedback must be in phase (positive). Out-of-phase (negative) feedback reduces the gain of an amplifier. In fact, negative feedback is used in some amplifiers to *prevent* oscillation.

The output of a common emitter or common source amplifier is out of phase from the input. If you couple the collector to the base through a capacitor, you won't get oscillation. It is necessary to reverse the phase in the feedback process in order for oscillation to occur. In addition, the amplifier gain must be high, and the coupling from the output to the input must be good. The positive feedback path must be easy for a signal to follow. Most oscillators are common emitter or common source amplifier circuits with positive feedback.

The output of a common base or common gate amplifier is in phase with the input. But these circuits have limited gain, and it's hard to make them oscillate. Common collector and common drain circuits don't have enough gain to make oscillators.

Feedback at a Single Frequency

The frequency of an oscillator is controlled by means of tuned, or resonant, circuits. These are usually inductance-capacitance (*LC*) or resistance-capacitance (*RC*) combinations. The *LC* scheme is common in radio transmitters and receivers; the *RC* method is more often used in audio work. The tuned circuit makes the feedback path easy for a signal to follow at one frequency, but hard to follow at all other frequencies. As a result, oscillation takes place at a predictable and stable frequency, determined by the inductance and capacitance, or by the resistance and capacitance.

Common Oscillator Circuits

Many circuit arrangements can be used to produce oscillation. The following several circuits are all known as *variable-frequency oscillators* (VFOs), because their frequencies are adjustable over a wide range.

The Armstrong Circuit

A common emitter or common source class A amplifier can be made to oscillate by coupling the output back to the input through a transformer that reverses the phase of the fed-back signal. The schematic diagram of Fig. 24-12 shows a common-source amplifier whose drain circuit is coupled to the gate circuit by means of a transformer. The frequency is controlled by a capacitor in series with the secondary winding. The inductance of the secondary, along with the capacitance, forms a resonant circuit that passes energy easily at one frequency while attenuating the energy at other frequencies. This circuit is known as an *Armstrong oscillator*. A bipolar transistor can be used in place of the JFET, as long as the device is biased for class A amplification.

The Hartley Circuit

Another method of obtaining controlled feedback at RF is shown in Fig. 24-13. In this circuit, a PNP bipolar transistor is used. The circuit uses a single coil with a tap on the windings to provide the feedback. A variable capacitor in parallel with the coil determines the oscillating frequency, and allows for frequency adjustment. This circuit is called a *Hartley oscillator*.

In the Hartley circuit, as well as in most other RF oscillator circuits, it is important to use only the minimum amount of feedback necessary to get oscillation. The amount of feedback is controlled by the position of the coil tap. The circuit shown in Fig. 24-13 uses about 25 percent of its amplifier power to produce feedback. The other 75 percent of the power can be used as output.

Oscillators usually produce less than 1 W of RF power output. If more power is needed, the signal can be boosted by one or more stages of amplification.

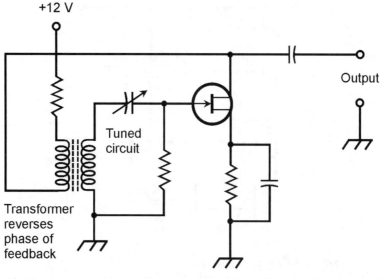

24-12 An Armstrong oscillator using an N-channel JFET. This is a common-source amplifier with positive feedback through a tuned circuit.

The Colpitts Circuit

The capacitance can be tapped, instead of the inductance, in the tuned circuit of an RF oscillator. Such a circuit is called a *Colpitts oscillator*, and a P-channel JFET version is diagrammed in Fig. 24-14. The amount of feedback is controlled by the ratio of the capacitances. A variable inductor provides for frequency adjustment. This is a matter of convenience, because it can be difficult to find a dual

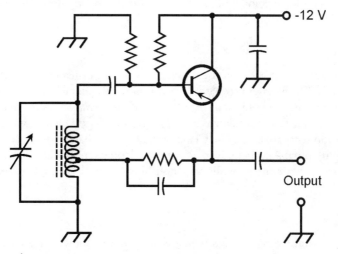

24-13 A Hartley oscillator using a PNP bipolar transistor. The Hartley circuit can be recognized by the tapped inductor in the tuned *LC* circuit.

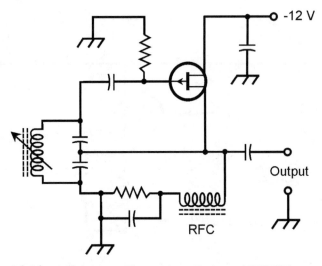

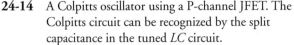

24-14 A Colpitts oscillator using a P-channel JFET. The Colpitts circuit can be recognized by the split capacitance in the tuned *LC* circuit.

variable capacitor that maintains the correct ratio of capacitances throughout its tuning range. Using fixed capacitors eliminates this problem, and it costs less, too!

Unfortunately, finding a good variable inductor for use in a Colpitts oscillator can be just about as hard as obtaining a suitable dual-gang variable capacitor. A *permeability-tuned* coil can be used, but ferromagnetic cores impair the frequency stability of an RF oscillator. A *roller inductor* can be employed, but these are bulky and expensive. An inductor with several switch-selectable taps can be used, but this does not allow for continuous frequency adjustment. Despite these shortcomings, the Colpitts circuit offers exceptional stability and reliability when properly designed, and is preferred by some engineers for this reason.

The Clapp Circuit

A variation of the Colpitts oscillator makes use of series resonance, instead of parallel resonance, in the tuned circuit. Otherwise, the circuit is basically the same as the parallel-tuned Colpitts oscillator. Figure 24-15 is a schematic diagram of a *series-tuned Colpitts oscillator* circuit that uses an NPN bipolar transistor. This circuit is also known as a *Clapp oscillator*. Its frequency won't change much when high-quality components are used. The Clapp oscillator is a reliable circuit. It isn't hard to get it to oscillate and keep it going. Another advantage of the Clapp circuit is that it allows the use of a variable capacitor for frequency control, while accomplishing feedback through a capacitive voltage divider.

Getting the Output

In the Hartley, Colpitts, and Clapp oscillators just described and shown in Figs. 24-13 through 24-15, the output is taken from the emitter or source, not from the collector or drain. There's a reason for this. The output of an oscillator can be taken from the collector or drain, just as is done in a common emitter or common-source amplifier to get maximum gain. But in an oscillator, stabil-

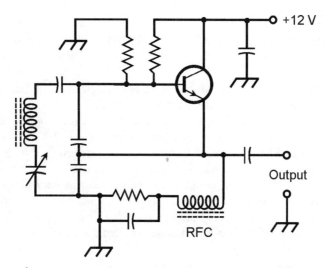

24-15 A series-tuned Colpitts oscillator, also known as a Clapp oscillator. This circuit uses an NPN bipolar transistor.

ity is more important than gain. The stability of an oscillator is better when the output is taken from the emitter or source, as compared with taking it from the collector or drain. Variations in the load impedance are less likely to affect the frequency of oscillation, and a sudden decrease in load impedance is less likely to cause the circuit to stop oscillating.

To prevent the output signal from being short-circuited to ground, an *RF choke* (RFC) is connected in series with the emitter or source in the Colpitts and Clapp oscillator circuits. The choke lets dc pass while blocking ac (just the opposite of a *blocking capacitor*). Typical values for RF chokes range from about 100 μH at high frequencies, such as 15 MHz, to 10 mH at low frequencies, such as 150 kHz.

The Voltage-Controlled Oscillator

The frequency of a VFO can be adjusted by means of a varactor diode in the tuned *LC* circuit. Recall that a varactor, also called a *varicap*, is a semiconductor diode that works as a variable capacitor when it is reverse-biased. The capacitance depends on the reverse-bias voltage. The greater this voltage, the lower the value of the capacitance.

The Hartley and Clapp oscillator circuits lend themselves well to varactor-diode frequency control. The varactor is placed in series or parallel with the tuning capacitor, and is isolated for dc by blocking capacitors. In Chap. 20, we saw an example of how a varactor can be connected in a tuned circuit (Fig. 20-9). The resulting oscillator is called a *voltage-controlled oscillator* (VCO).

Varactors are cheaper than variable capacitors or inductors. They're also less bulky. These are the chief advantages of a VCO over an old-fashioned *LC* tuned VFO.

Diode Oscillators

At ultrahigh frequencies (UHF) and microwave radio frequencies, certain types of diodes can be used as oscillators. You learned about these diodes in Chap. 20.

Crystal-Controlled Oscillators

Quartz crystals can be used in place of tuned *LC* circuits in RF oscillators, as long as it isn't necessary to change the frequency often. Crystal oscillators offer frequency stability far superior to that of *LC* tuned VFOs.

There are several ways that crystals can be connected in bipolar or FET circuits to get oscillation. One common circuit is the *Pierce oscillator*. An N-channel JFET and quartz crystal are connected in a Pierce configuration as shown in the schematic diagram of Fig. 24-16. The crystal frequency can be varied somewhat (by about 0.1 percent, or 1 part in 1000) by means of an inductor or capacitor in parallel with the crystal. But the frequency is determined mainly by the thickness of the quartz wafer, and by the angle at which it is cut from the original quartz sample.

Crystals change in frequency as the temperature changes. But they are far more stable than *LC* circuits, most of the time. Some crystal oscillators are housed in temperature-controlled chambers called *crystal ovens*. In this environment, crystals maintain their frequency so well that they are sometimes used as *frequency standards* against which other oscillators are calibrated.

The Phase-Locked Loop

One type of oscillator that combines the flexibility of a VFO with the stability of a crystal oscillator is known as a *phase-locked loop* (PLL). This makes use of a circuit called a *frequency synthesizer*.

The heart of the PLL is a VCO. The output of this oscillator passes through a *programmable multiplier/divider*, a digital circuit that divides and/or multiplies the VCO frequency by integral (whole-number) values chosen by the operator. As a result, the output frequency can be any rational-number multiple of the crystal frequency. A well-designed PLL circuit can be tuned in small digital increments over a wide range of frequencies.

The output frequency of the multiplier/divider is locked, by means of a *phase comparator*, to the signal from a crystal-controlled *reference oscillator*. As long as the output from the multiplier/divider is exactly on the reference oscillator frequency, the two signals are in phase, and the output of the phase comparator is 0 V dc. If the VCO frequency begins to drift, the output frequency of the multiplier/divider will drift, too (although at a different rate). But even a frequency change of less than 1 Hz causes the phase comparator to produce a dc *error voltage*. This error voltage is either pos-

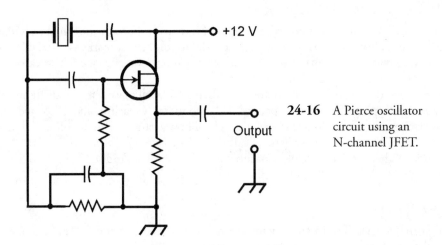

24-16 A Pierce oscillator circuit using an N-channel JFET.

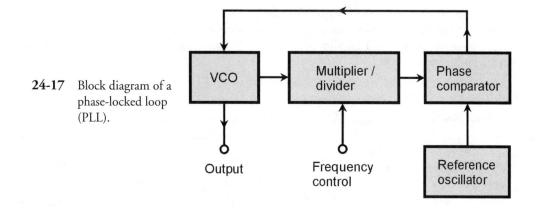

24-17   Block diagram of a phase-locked loop (PLL).

itive or negative, depending on whether the VCO has drifted higher or lower in frequency. The error voltage is applied to a varactor, causing the VCO frequency to change in a direction opposite to that of the drift. This forms a *dc feedback* circuit that maintains the VCO frequency at a precise value. It is a *loop* circuit that *locks* the VCO onto a particular frequency by means of *phase* sensing, hence the term *phase-locked loop* (PLL).

The key to the stability of the PLL lies in the fact that the reference oscillator is crystal-controlled. A block diagram of a PLL circuit is shown in Fig. 24-17. When you hear that a radio receiver, transmitter, or transceiver is synthesized, it usually means that the frequency is determined by a PLL.

The stability of a synthesizer can be enhanced by using an amplified signal from the shortwave time-and-frequency broadcast station WWV at 2.5, 5, 10, or 15 MHz, directly as the reference oscillator. These signals are frequency-exact to a minuscule fraction of 1 Hz, because they are controlled by atomic clocks. Most people don't need precision of this caliber, so you won't see consumer devices like ham radios and shortwave receivers with *primary-standard* PLL frequency synthesis. But it is employed by some corporations and government agencies.

Oscillator Stability

In an oscillator, the term *stability* has two meanings: constancy of frequency (or minimal *frequency drift*), and reliability of performance.

Constancy of Frequency

When designing a VFO of any kind, it's essential that the components maintain constant values, as much as possible, under all anticipated conditions. Some types of capacitors maintain their values better than others as the temperature rises or falls. Among the best are polystyrene capacitors. Silver-mica capacitors also work well when polystyrene units can't be found. Inductors are most temperature-stable when they have air cores. They should be wound, when possible, from stiff wire with strips of plastic to keep the windings in place. Some air-core coils are wound on hollow cylindrical cores, made of ceramic or phenolic material. Ferromagnetic solenoidal or toroidal cores aren't very good for VFO coils, because these materials change their permeability as the temperature varies. This changes the inductance, in turn affecting the oscillator frequency.

The best oscillators, in terms of frequency stability, are crystal-controlled. This includes circuits that oscillate at the fundamental frequency of the quartz crystal, circuits that oscillate at one of the crystal harmonic frequencies, or circuits that oscillate at frequencies derived from the crystal frequency by multiplier/dividers.

Reliability

An oscillator should always start working as soon as power is supplied. It should keep oscillating under all normal conditions. The failure of a single oscillator can cause an entire receiver, transmitter, or transceiver to stop working.

When an oscillator is built and put to use in a radio receiver, transmitter, or audio device, *debugging* is always necessary. This is a trial-and-error process of getting the flaws, or bugs, out of the circuit. Rarely can an engineer build something straight from the drawing board and have it work just right the first time. In fact, if two oscillators are built from the same diagram, with the same component types and values in the same geometric arrangement, one circuit might work fine, and the other might not. This usually happens because of differences in the quality of components that don't show up until the acid test.

Oscillators are designed to work into a certain range of load impedances. It's important that the load impedance not be too low. (You need never be concerned that it might be too high. In general, the higher the load impedance, the better.) If the load impedance is too low, the load will draw significant power from an oscillator. Then, even a well-designed oscillator might become unstable. Oscillators aren't meant to produce powerful signals. High power can be obtained using amplification after the oscillator.

Audio Oscillators

Audio oscillators are used in myriad electronic devices including doorbells, ambulance sirens, electronic games, telephone sets, and toys that play musical tunes. All AF oscillators are, in effect, AF amplifiers with positive feedback.

Audio Waveforms

At AF, oscillators can use *RC* or *LC* combinations to determine frequency. If *LC* circuits are used, the inductances must be large, and ferromagnetic cores are necessary.

At RF, oscillators are usually designed to produce a sine wave output. A pure sine wave represents energy at one and only one frequency. Audio oscillators, by contrast, don't always concentrate all their energy at a single frequency. (A pure AF sine wave, especially if it is continuous and frequency-constant, can be annoying.) The various musical instruments in a band or orchestra all sound different from each other, even when they play the same note (such as middle C). The reason for this is that each instrument has its own unique waveform. A clarinet sounds different than a trumpet, which in turn sounds different than a cello or piano.

Suppose you were to use an oscilloscope to look at the waveforms of musical instruments. This can be done using a high-fidelity microphone, a sensitive, low-distortion audio amplifier, and an oscilloscope. You'd see that each instrument has its own signature. Thus, each instrument's unique sound qualities can be reproduced using AF oscillators whose waveform outputs match those of the instrument. Electronic music synthesizers use audio oscillators to generate the tones you hear.

The Twin T Oscillator

An audio oscillator circuit that is popular for general-purpose use is the *twin T oscillator* (Fig. 24-18). The frequency is determined by the values of the resistors R and capacitors C. The output is a near-perfect sine wave. The small amount of distortion helps to alleviate the irritation produced by an absolutely pure sinusoid. The circuit shown in this example uses two PNP bipolar transistors. They are biased for class A amplification.

The Multivibrator

Another popular AF oscillator circuit makes use of two identical common emitter or common source amplifier circuits, hooked up so that the signal goes around and around between them. This is sometimes called a *multivibrator* circuit, although that is technically a misnomer, the term being more appropriate to various digital signal-generating circuits.

In the example of Fig. 24-19, two N-channel JFETs are connected to form a multivibrator for use at AF. Each stage amplifies the signal in class A, and reverses the phase by 180°. Therefore, the signal goes through a 360° phase shift each time it gets back to any particular point. A 360° phase shift is equivalent to no phase shift at all, so it results in positive feedback.

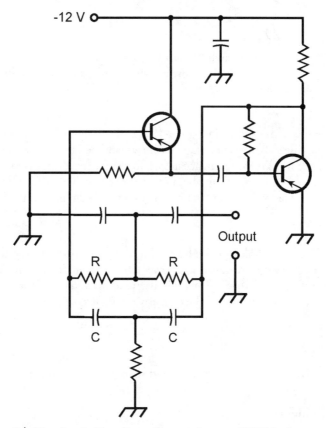

24-18 A twin T audio oscillator using two PNP bipolar transistors. The frequency is determined by the values of the resistors R and the capacitors C.

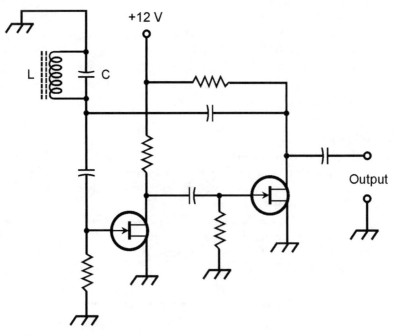

24-19   A multivibrator audio oscillator using two N-channel JFETs. The frequency is determined by the value of the inductor L and the capacitor C.

The frequency of the circuit shown in Fig. 24-19 is set by means of an *LC* circuit. The coil uses a ferromagnetic core, because stability is not of great concern and because such a core is necessary to obtain the large inductance needed for resonance at AF. Toroidal cores or pot cores are excellent in this application. The value of *L* can range from about 10 mH to as much as 1 H. The capacitance is chosen according to the formula for resonant circuits, to obtain an audio tone at the frequency desired.

IC Oscillators

In recent years, solid-state technology has advanced to the point that whole circuits can be etched onto silicon chips. Such devices are called *integrated circuits* (ICs). The *operational amplifier*, also called an *op amp*, is a type of IC that is especially useful as an audio oscillator because it has high gain, and it can easily be connected to produce positive feedback.

Quiz

Refer to the text in this chapter if necessary. A good score is at least 18 correct. Answers are in the back of the book.

1. The decibel is a unit of
    (a) relative signal strength.
    (b) voltage.

(c) power.

(d) current.

2. An oscillator at RF requires the use of

(a) a common drain or common collector circuit.

(b) a stage with gain.

(c) a tapped coil.

(d) a quartz crystal.

3. Suppose a circuit is found to have a gain figure of −15 dB. Which of the following statements is true?

(a) The output signal is stronger than the input signal.

(b) The input signal is stronger than the output signal.

(c) The input signal is 15 times as strong as the output signal.

(d) The output signal is 15 times as strong as the input signal.

4. In an oscillator circuit, the feedback should be

(a) as great as possible.

(b) kept to a minimum.

(c) just enough to sustain oscillation.

(d) done through a transformer whose wires can be switched easily.

5. A power gain of 44 dB is equivalent to which of the following output/input power ratios?

(a) 44:1

(b) 160:1

(c) 440:1

(d) 25,000:1

6. An RF choke

(a) passes RF signals but blocks dc.

(b) passes both RF signals and dc.

(c) passes dc but blocks RF signals.

(d) blocks both dc and RF signals.

7. The optimum capacitance values in an amplifier circuit depend on

(a) the power-supply voltage.

(b) the power-supply polarity.

(c) the input signal strength.

(d) the input signal frequency.

8. An oscillator might fail to start for any of the following reasons, except

(a) low battery voltage.

(b) low stage gain.

    (c) in-phase feedback.

    (d) a high-impedance load.

9. In which of the following FET amplifier types does drain current flow for exactly 50 percent of the signal cycle?

    (a) Class A

    (b) Class AB_1

    (c) Class AB_2

    (d) Class B

10. The frequency at which a quartz crystal oscillator produces energy is largely dependent on

    (a) the load impedance.

    (b) the physical thickness of the quartz wafer.

    (c) the amount of resistance through the crystal.

    (d) the power-supply voltage.

11. Which bipolar amplifier type has some distortion in the signal wave, with collector current during most, but not all, of the cycle?

    (a) Class A

    (b) Class AB_1

    (c) Class AB_2

    (d) Class B

12. An RF oscillator usually

    (a) produces an output signal with an irregular waveshape.

    (b) has most or all of its energy at a single frequency.

    (c) produces a sound that depends on its waveform.

    (d) employs an RC circuit to determine the output amplitude.

13. A class C amplifier can be made linear by

    (a) reducing the bias.

    (b) increasing the drive.

    (c) using two transistors in push-pull.

    (d) no means; a class C amplifier is always nonlinear.

14. A frequency synthesizer has

    (a) high power output.

    (b) high frequency drift rate.

    (c) exceptional stability.

    (d) an adjustable waveshape.

15. A graphic equalizer is a form of

    (a) bias control for an NPN bipolar transistor.

(b) gain control for an RF oscillator.

(c) tone control that can be used in an audio amplifier.

(d) circuit for adjusting the waveform of an RF oscillator.

16. If the impedance of the load connected to the output of an oscillator is extremely high,

(a) the frequency will drift excessively.

(b) the power output will be reduced.

(c) the oscillator might fail to start.

(d) it is no cause for concern; in fact, it is a good thing.

17. Suppose a certain bipolar-transistor PA is 66 percent efficient. The output power is 33 W. The dc collector power input is

(a) 22 W.

(b) 50 W.

(c) 2.2 W.

(d) impossible to determine without more information.

18. The arrangement in the block diagram of Fig. 24-20 represents

(a) a waveform analyzer.

(b) an audio oscillator.

(c) an RF oscillator.

(d) a sine wave generator.

19. A tuned RF PA must always be

(a) set to work over a wide range of frequencies.

(b) adjusted for maximum power output.

(c) operated at an even harmonic of the input frequency.

(d) operated in class C.

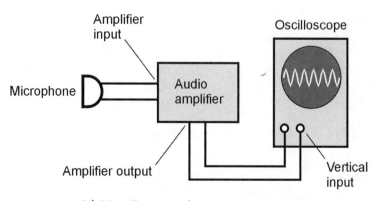

24-20 Illustration for Quiz Question 18.

20. Class B amplification can be used to obtain low distortion for audio applications
    (a) by connecting two amplifiers in cascade, thereby maximizing the gain and generating a pure sine wave output.
    (b) by biasing the bipolar transistor or FET beyond cutoff or pinchoff, thereby ensuring that the output is in phase with the input.
    (c) by connecting two identical bipolar transistors or FETs, biased exactly at cutoff or pinchoff, in a push-pull configuration.
    (d) by biasing the bipolar transistor or FET in the middle of the straight-line portion of the characteristic curve.

25
CHAPTER

Wireless Transmitters and Receivers

IN RADIO OR WIRELESS COMMUNICATIONS, A *TRANSMITTER* CONVERTS DATA INTO *ELECTROMAGNETIC* (EM) *waves* intended for recovery by one or more *receivers*. In this chapter, we'll look at how data is converted to EM waves and transmitted, and then examine how the resulting *EM fields* can be intercepted and received.

Oscillation and Amplification

A radio transmitter employs one or more oscillators to generate an RF signal, and amplifiers to generate the required power output. You just learned how these circuits work. Most transmitters have *mixers* in addition to the oscillating and amplifying stages. Signal mixing is commonly done with diodes, and was discussed in Chap. 20.

Modulation

Modulation is the process of writing data onto an electric current or EM wave. The process can be done by varying the amplitude, the frequency, or the phase of the current or wave. Another method is to transmit a series of pulses, whose duration, amplitude, or spacing is made to vary.

The Carrier

The heart of a wireless signal is a sine wave known as the *carrier*. The lowest carrier frequency used for radio communications is a few kilohertz (kHz). The highest frequency is in the hundreds of gigahertz (GHz). For efficient data transfer, the carrier frequency must be at least 10 times the highest frequency of the modulating signal.

On/Off Keying

The simplest form of modulation is *on/off keying*. This can be done in the oscillator of a radio transmitter to send Morse code, which is a binary digital mode. The duration of a Morse-code *dot* is one *bit* (binary digit). A *dash* is 3 bits long. The space between dots and dashes within a *charac-*

ter is 1 bit. The space between characters in a *word* is 3 bits. The space between words is 7 bits. The key-down (full-carrier) condition is called *mark*, and the key-up (no-signal) condition is called *space*.

Morse code is slow. Human operators use speeds ranging from about 5 words per minute (wpm) to 40 or 50 wpm. A few human operators can work at 60 to 70 wpm. These people usually copy the signals in their heads.

Frequency-Shift Keying

Digital data can be sent over wireless links by means of *frequency-shift keying* (FSK). In some FSK systems, the carrier frequency is shifted between mark and space conditions, usually by a few hundred hertz or less. In other systems, a two-tone audio-frequency (AF) sine wave modulates the carrier. This is known as *audio-frequency-shift keying* (AFSK).

The two most common codes used with FSK and AFSK are *Baudot* (pronounced "baw-DOE") and *ASCII* (pronounced "ASK-ee"). The acronym ASCII stands for *American Standard Code for Information Interchange*.

In *radioteletype* (RTTY) FSK and AFSK systems, a *terminal unit* (TU) converts the digital signals into electrical impulses that operate a teleprinter or display characters on a computer screen. The TU also generates the signals necessary to send RTTY as an operator types on a keyboard. A device that sends and receives AFSK is sometimes called a *modem*. This acronym stands for *modulator/demodulator*. A modem is basically the same as a TU. Figure 25-1 is a block diagram of an AFSK transmitter.

The main advantage of FSK or AFSK over on/off keying is the fact that there are fewer errors or misprints, because the space part of the signal is identified as such, rather than existing as a gap or pause in the data. A sudden noise burst in an on/off keyed signal can confuse a receiver into reading the space as a mark signal, but when the space is positively represented by its own signal, this is less likely to happen.

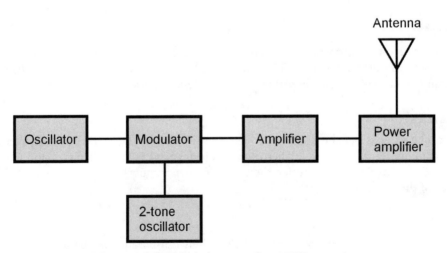

25-1   Simplified block diagram of an AFSK transmitter.

Amplitude Modulation

An AF voice signal has frequencies mostly in the range between 300 Hz and 3 kHz. Some characteristic of an RF carrier can be varied, or *modulated*, by these waveforms, thereby transmitting voice information. Figure 25-2 shows a simple circuit for obtaining *amplitude modulation* (AM). This circuit can be imagined as an RF amplifier for the carrier, with the instantaneous gain dependent on the instantaneous audio input amplitude. Another way to think of this circuit is as a mixer that combines the RF carrier and audio signals to produce sum and difference signals at frequencies just above and below that of the carrier.

The circuit shown in Fig. 25-2 works well, provided the AF input amplitude is not too great. If the AF input is excessive, then distortion occurs, intelligibility is degraded, system efficiency is reduced, and the bandwidth of the signal is increased unnecessarily.

The extent of AM is expressed as a percentage, from 0 percent (an unmodulated carrier) to 100 percent (full modulation). Increasing the modulation past 100 percent causes the same problems as excessive AF input. In an AM signal modulated 100 percent, $1/3$ of the power is used to convey the data, and the other $2/3$ is consumed by the carrier wave.

Figure 25-3 shows a *spectral display* of an AM voice radio signal. The horizontal scale is calibrated in increments of 1 kHz per division. Each vertical division represents 3 dB of change in signal strength. The maximum (reference) amplitude is 0 dB relative to 1 mW (abbreviated as 0 dBm). The data exists in *sidebands* above and below the carrier frequency. These sidebands resemble the sum and difference signals produced by a mixer. In this case the mixing occurs between the AF input signal and the RF carrier. The RF between −3 kHz and the carrier frequency constitutes the *lower*

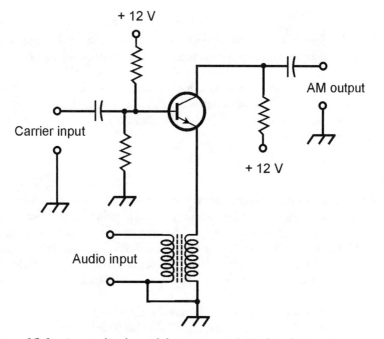

25-2   An amplitude modulator using an NPN bipolar transistor.

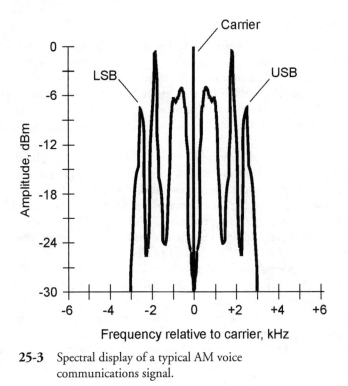

25-3 Spectral display of a typical AM voice
communications signal.

sideband (LSB); the RF from the carrier frequency to +3 kHz represents the *upper sideband* (USB). The *bandwidth* is the difference between the maximum and minimum sideband frequencies, in this case 6 kHz.

In an AM signal, the bandwidth is twice the highest audio modulating frequency. In the example of Fig. 25-3, all the voice energy is at or below 3 kHz, so the signal bandwidth is 6 kHz. This is typical of a communications signal. In standard AM broadcasting, the AF energy is spread over a wider bandwidth, nominally 10 kHz.

Single Sideband

In AM, most of the RF signal power is consumed by the carrier alone; the two sidebands are mirror-image duplicates. This is inefficient, and is also unnecessarily redundant! If the carrier and one of the sidebands is eliminated, these shortcomings can be overcome. That makes the signal stronger for a given amount of RF power, or allows the use of lower RF power in a given communications scenario. Another bonus is the fact that the bandwidth is reduced to less than half that of an AM signal modulated with the same data, so more than twice as many signals can fit into a specific range, or *band*, of frequencies.

When the carrier is removed from an AM signal along with one of the sidebands, the remaining RF energy has a spectral display resembling Fig. 25-4. This is *single sideband* (SSB) transmission. Either the LSB or the USB alone can be used, with equally good results.

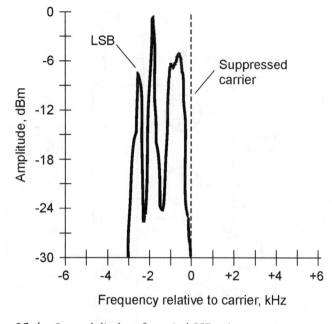

25-4 Spectral display of a typical SSB voice communications signal. In this example, the carrier and the USB energy are eliminated, leaving only the LSB energy.

An SSB signal can be obtained with a *balanced modulator*, which is an amplitude modulator/amplifier using two transistors with the inputs in push-pull and the outputs in parallel (Fig. 25-5). This cancels the carrier wave in the output, leaving only LSB and USB energy. The result is a *double sideband suppressed carrier* (DSBSC) signal, often called simply *double sideband* (DSB). At some stage following the balanced modulator, one of the sidebands is removed from the DSB signal by a *bandpass filter* to obtain an SSB signal.

Figure 25-6 is a block diagram of a simple SSB transmitter. The balanced modulator is placed in a low-power section of the transmitter. The RF amplifiers that follow any type of amplitude modulator, including a balanced modulator, must all be linear to avoid distortion and unnecessary spreading of signal bandwidth ("splatter"). They generally work in class A, except for the PA, which works in class AB or class B.

Frequency and Phase Modulation

In *frequency modulation* (FM), the instantaneous amplitude of a signal remains constant, and the instantaneous frequency is varied instead. A nonlinear PA such as a class C amplifier can be used in an FM transmitter without causing signal distortion, because the amplitude does not fluctuate.

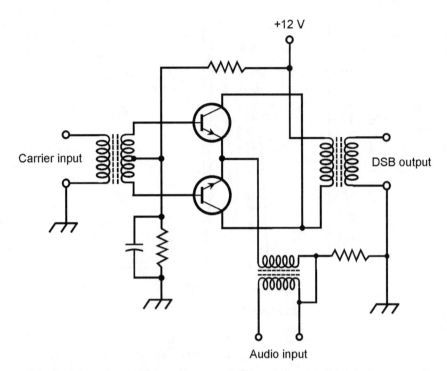

25-5   A balanced modulator using two NPN bipolar transistors. The inputs are in push-pull, but the outputs are in parallel.

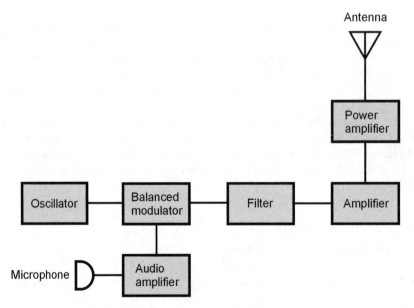

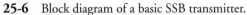

25-6   Block diagram of a basic SSB transmitter.

Frequency modulation can be obtained by applying the audio signal to a *varactor* in a tuned oscillator. An example of this scheme, known as *reactance modulation*, is shown in Fig. 25-7. The varying voltage across the varactor causes its capacitance to change in accordance with the audio waveform. The changing capacitance results in variation of the resonant frequency of the inductance-capacitance (LC) tuned circuit, causing small fluctuations in the frequency generated by the oscillator.

Another way to get FM is to modulate the phase of the oscillator signal. This causes small variations in the frequency, because any instantaneous phase change shows up as an instantaneous frequency change (and vice versa). When *phase modulation* is used, the audio signal must be processed, adjusting the frequency response of the audio amplifiers. Otherwise the signal will sound unnatural when it is received.

Deviation is the maximum extent to which the instantaneous-carrier frequency differs from the unmodulated-carrier frequency. For most FM voice communications transmitters, the deviation is standardized at ±5 kHz. This is known as *narrowband FM* (NBFM). The bandwidth of an NBFM signal is comparable to that of an AM signal containing the same modulating information. In FM hi-fi music broadcasting, and in some other applications, the deviation is much greater than ±5 kHz. This is called *wideband FM* (WBFM).

The deviation obtainable by means of direct FM is greater, for a given oscillator frequency, than the deviation that can be obtained by means of phase modulation. The deviation of a signal can be increased by a *frequency multiplier.* When an FM signal is passed through a frequency multiplier, the deviation is multiplied along with the carrier frequency.

The deviation in an FM signal should be equal to the highest modulating audio frequency if optimum fidelity is to be obtained. Thus, ±5 kHz is more than enough for voice. For music, a deviation of ±15 kHz or even ±20 kHz is required for good reproduction when the signal is received.

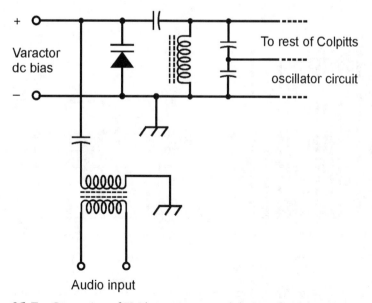

25-7   Generation of FM by reactance modulation of a Colpitts oscillator. Other oscillator types can be similarly modified.

In any FM signal, the ratio of the frequency deviation to the highest modulating audio frequency is called the *modulation index*. Ideally, this figure is between 1:1 and 2:1. If it is less than 1:1, the signal sounds muffled or distorted, and efficiency is sacrificed. Increasing it beyond 2:1 broadens the bandwidth without providing significant improvement in intelligibility or fidelity.

Pulse Modulation

Another method of modulation works by varying some aspect of a constant stream of signal pulses. Several types of *pulse modulation* (PM) are briefly described in the following sections. They are diagrammed in Fig. 25-8 as amplitude-versus-time graphs. The modulating waveform in each case is shown as a dashed curve, and the pulses are shown as vertical gray bars.

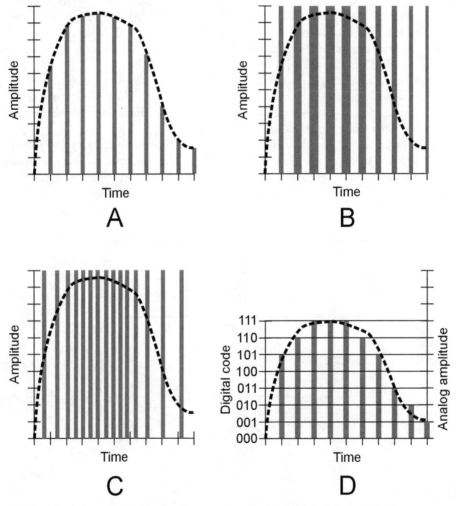

25-8 Time-domain graphs of various modes of pulse modulation. At A, pulse amplitude modulation; at B, pulse width modulation; at C, pulse interval modulation; at D, pulse code modulation.

Pulse Amplitude Modulation

In *pulse amplitude modulation* (PAM), the strength of each individual pulse varies according to the modulating waveform. In this respect, PAM resembles AM. An amplitude-versus-time graph of a hypothetical PAM signal is shown in Fig. 25-8A. Normally, the pulse amplitude increases as the instantaneous modulating-signal level increases (*positive PAM*). But this can be reversed, so higher audio levels cause the pulse amplitude to go down (*negative PAM*). Then the signal pulses are at their strongest when there is no modulation. The transmitter works a little harder if negative PAM is used.

Pulse Width Modulation

Another way to change the transmitter output is to vary the width (duration) of the pulses. This is called *pulse width modulation* (PWM) or *pulse duration modulation* (PDM), and is shown in Fig. 25-8B. Normally, the pulse width increases as the instantaneous modulating-signal level increases (*positive PWM*). But this can be reversed (*negative PWM*). The transmitter must work harder to accomplish negative PWM. Either way, the peak pulse amplitude remains constant.

Pulse Interval Modulation

Even if all the pulses have the same amplitude and the same duration, modulation can still be accomplished by varying how often they occur. In PAM and PWM, the pulses are always sent at the same time interval, known as the *sampling interval*. But in *pulse interval modulation* (PIM), also called *pulse frequency modulation* (PFM), pulses can occur more or less frequently than they do when there is no modulation. A hypothetical PIM signal is shown in Fig. 25-8C. Every pulse has the same amplitude and the same duration, but the time interval between them changes. When there is no modulation, the pulses are evenly spaced with respect to time. An increase in the instantaneous data amplitude might cause pulses to be sent more often, as is the case in Fig. 25-8C (*positive PIM*). Or, an increase in instantaneous data level might slow down the rate at which the pulses are sent (*negative PIM*).

Pulse Code Modulation

In recent years, the transmission of data has been done more and more by *digital* means. In *digital communications*, the modulating data attains only certain defined states, rather than continuously varying. Digital transmission offers better efficiency than analog transmission. With digital modes, the *signal-to-noise* (S/N) *ratio* is better, the bandwidth is narrower, and there are fewer errors. In *pulse-code modulation* (PCM), any of the above aspects—amplitude, duration, or interval—of a pulse sequence (or *pulse train*) can be varied. But rather than having infinitely many possible states, there are finitely many. The number of states is a power of 2, such as 4, 8, or 16. The greater the number of states, the better the fidelity. An example of 8-level PCM is shown in Fig. 25-8D.

Analog-to-Digital Conversion

Pulse code modulation, such as is shown at Fig. 25-8D, is one form of *analog-to-digital* (A/D) *conversion*. A voice signal, or any continuously variable signal, can be *digitized*, or converted into a train of pulses whose amplitudes can achieve only certain defined levels.

Resolution

In A/D conversion, the number of states is always a power of 2, so that it can be represented as a binary-number code. Fidelity improves as the exponent increases. The number of states is called the *sampling resolution*, or simply the *resolution*. A resolution of $2^3 = 8$ (as shown in Fig. 25-8D) is good enough for voice transmission, and is the standard for commercial digital voice circuits. A resolution of $2^4 = 16$ is adequate for high-fidelity (hi-fi) music reproduction.

Sampling Rate

The efficiency with which a signal can be digitized depends on the frequency at which sampling is done. In general, the *sampling rate* must be at least twice the highest data frequency. For an audio signal with components as high as 3 kHz, the minimum sampling rate for effective digitization is 6 kHz; the commercial voice standard is 8 kHz. For hi-fi digital transmission, the standard sampling rate is 44.1 kHz, a little more than twice the frequency of the highest audible sound (approximately 20 kHz).

Image Transmission

Nonmoving images can be sent within the same bandwidth as voice signals. For high-resolution, moving images, the necessary bandwidth is greater.

Facsimile

Nonmoving images (also called *still images*) are commonly transmitted by *facsimile*, also called *fax*. If data is sent slowly enough, any amount of detail can be transmitted within a 3-kHz-wide band, the standard for voice communications. This is why detailed fax images can be sent over a *plain old telephone service* (POTS) line.

In an electromechanical fax machine, a paper document or photo is wrapped around a *drum*. The drum is rotated at a slow, controlled rate. A spot of light scans from left to right; the drum moves the document so a single line is scanned with each pass of the light spot. This continues, line by line, until the complete *frame* (image) has been scanned. The reflected light is picked up by a *photodetector*. Dark parts of the image reflect less light than bright parts, so the current through the photodetector varies. This current modulates a carrier in one of the modes described earlier, such as AM, FM, or SSB. Typically, black is sent as a 1.5-kHz audio sine wave, and white as a 2.3-kHz audio sine wave. Gray shades produce audio sine waves having frequencies between these extremes.

At the receiver, the scanning rate and pattern can be duplicated, and a *cathode-ray tube* (CRT), *liquid crystal display* (LCD), or printer can be used to reproduce the image in *grayscale* (shades of gray ranging from black to white, without color).

Slow-Scan Television

One way to think of *slow-scan television* (SSTV) is to imagine "fast fax." An SSTV signal, like a fax signal, is sent within a band of frequencies as narrow as that of a human voice. And, like fax, SSTV transmission is of still pictures, not moving ones. The big difference between SSTV and fax is that SSTV images are sent in much less time. The time required to send a complete frame (image or scene) is 8 seconds, rather than several minutes. This speed bonus comes with a tradeoff: lower *resolution*, meaning less image detail.

Some SSTV signals are received on CRT displays. A computer can be programmed so that its monitor will act as an SSTV receiver. Converters are also available that allow SSTV signals to be viewed on a consumer-type TV set.

An SSTV frame has 120 lines. The black and white frequencies are the same as for fax transmission; the darkest parts of the picture are sent at 1.5 kHz and the brightest at 2.3 kHz. *Synchronization (sync) pulses*, that keep the receiving apparatus in step with the transmitter, are sent at 1.2 kHz. A *vertical sync pulse* tells the receiver that it's time to begin a new frame; it lasts for 30 milliseconds (ms). A *horizontal sync pulse* tells the receiver that it's time to start a new line in a frame; its duration is 5 ms. These pulses prevent *rolling* (haphazard vertical image motion) or *tearing* (lack of horizontal synchronization).

Fast-Scan Television

Conventional television is also known as *fast-scan TV* (FSTV). The frames are transmitted at the rate of 30 per second. There are 525 lines per frame. The quick frame time, and the increased resolution, of FSTV make it necessary to use a much wider frequency band than is the case with fax or SSTV. A typical video FSTV signal takes up 6 MHz of spectrum space, or 2000 times the bandwidth of a fax or SSTV signal.

Fast-scan TV is almost always sent using conventional AM. Wideband FM can also be used. With AM, one of the sidebands can be filtered out, leaving just the carrier and the other sideband. This mode is called *vestigial sideband* (VSB) transmission. It cuts the bandwidth of an FSTV signal down to about 3 MHz.

Because of the large amount of spectrum space needed to send FSTV, this mode isn't practical at frequencies below about 30 MHz. All commercial FSTV transmission is done above 50 MHz, with the great majority of channels having frequencies far higher than this. Channels 2 through 13 on your TV receiver are sometimes called the *very high frequency* (VHF) *channels;* the higher channels are called the *ultrahigh frequency* (UHF) *channels.*

Figure 25-9 is a time-domain graph of the waveform of a single line in an FSTV video signal. This represents $\frac{1}{525}$ of a complete frame. The highest instantaneous signal amplitude corresponds to the blackest shade, and the lowest amplitude to the lightest shade. Thus, the FSTV signal is sent negatively. The reason that FSTV signals are sent this way is that *retracing* (moving from the end of one line to the beginning of the next) must be synchronized between the transmitter and receiver. This is guaranteed by a defined, strong *blanking pulse.* This pulse tells the receiver when to retrace; it also shuts off the beam while the receiver display is retracing. Have you noticed that weak TV signals have poor *contrast*? (You have, if you're old enough to remember "rabbit ears"!) Weakened blanking pulses result in incomplete retrace blanking. But this is better than having the TV receiver completely lose track of when it should retrace.

Color FSTV works by sending three separate monochromatic signals, corresponding to the primary colors red, blue, and green. The signals are literally black-and-red, black-and-blue, and black-and-green. These are recombined at the receiver and displayed on the screen as a fine, interwoven matrix of red, blue, and green dots. When viewed from a distance, the dots are too small to be individually discernible. Various combinations of red, blue, and green intensities result in reproduction of all possible hues and saturations of color.

High-Definition Television

The term *high-definition television* (HDTV) refers to any of several similar methods for getting more detail into a TV picture, and for obtaining better audio quality, compared with standard FSTV.

A standard FSTV picture has 525 lines per frame, but HDTV systems have between 787 and 1125 lines per frame. The image is scanned about 60 times per second. High-definition TV is often

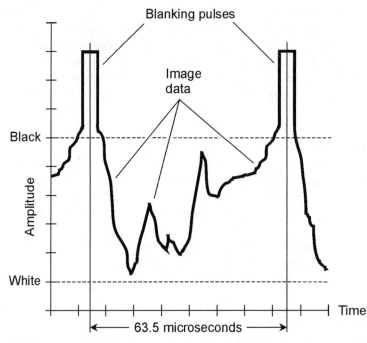

25-9 Time-domain graph of a single line in an FSTV video frame.

sent in a digital mode; this offers another advantage over conventional FSTV. Digital signals propagate better, are easier to deal with when they are weak, and can be processed in ways that analog signals cannot.

Some HDTV systems use *interlacing* in which two *rasters* are meshed together. This effectively doubles the image resolution without doubling the cost of the hardware. But it can cause annoying *jitter* in fast-moving or fast-changing images.

Digital Satellite TV

Until the early 1990s, a satellite television installation required a dish antenna several feet in diameter. A few such systems are still in use. The antennas are expensive, they attract attention (sometimes unwanted), and they are subject to damage from ice storms, heavy snows, and high winds. Digitization has changed this situation. In any communications system, digitization allows the use of smaller receiving antennas, smaller transmitting antennas, and/or lower transmitter power levels. Engineers have managed to get the diameter of the receiving dish down to about 2 ft.

A pioneer in digital TV was RCA (Radio Corporation of America), which developed the *Digital Satellite System* (DSS). The analog signal is changed into digital pulses at the transmitting station via A/D conversion. The digital signal is amplified and sent up to a geostationary satellite. The satellite has a *transponder* that receives the signal, converts it to a different frequency, and retransmits it back toward the earth. The return signal is picked up by a portable dish. A *tuner* selects the channel. *Digital signal processing* (DSP) can be used to improve the quality of reception under marginal conditions. The digital signal is changed back into analog form, suitable for viewing on a conventional FSTV set, by means of *digital-to-analog* (D/A) *conversion*.

The Electromagnetic Field

In a radio or television transmitting antenna, electrons are moving back and forth at an extreme speed. Their velocity is constantly changing as they speed up in one direction, slow down, reverse direction, speed up again, and so on. Any change of velocity (that is, of speed and/or direction) constitutes *acceleration*.

How It Happens

When electrons move, a magnetic (M) field is produced. When electrons accelerate, a changing magnetic field is produced. An alternating M field gives rise to an alternating electric (E) field, and this generates another alternating M field. This process repeats over and over, endlessly, and the effect *propagates* (travels) through space at the speed of light. The E and M fields expand alternately outward from the source in spherical wavefronts. At any given point in space, the E flux is perpendicular to the M flux. The direction of wave travel is perpendicular to both the E and M flux lines. This is an *electromagnetic* (EM) *field*.

An EM field can have any conceivable frequency, ranging from many years per cycle to quadrillions of cycles per second. The sun has a magnetic field that oscillates with a 22-year cycle. Radio waves oscillate at thousands, millions, or billions of cycles per second. Infrared (IR), visible light, ultraviolet (UV), X rays, and gamma rays are EM fields that alternate at many trillions (million millions) of cycles per second.

Frequency versus Wavelength

All EM fields have two important properties: the *frequency* and the *wavelength*. These are inversely related. You've already learned about frequency. Wavelength, for an EM field, is a rather sophisticated concept. It is measured between any two adjacent points on the wave at which the E and M fields have exactly the same amplitudes, and occur in exactly the same relative directions. The following equations relate the frequency and the wavelength of an EM field in free space (the air or a vacuum).

Let f_{MHz} be the frequency of an EM wave in megahertz, and L_{ft} be the wavelength in feet. Then the two are related as follows:

$$L_{ft} = 984 / f_{MHz}$$

If the wavelength is given as L_m in meters, then

$$L_m = 300 / f_{MHz}$$

The inverses of these formulas, for finding the frequency if the wavelength is known, are

$$f_{MHz} = 984 / L_{ft}$$
$$f_{MHz} = 300 / L_m$$

Velocity Factor

In media other than free space, the speed at which EM fields propagate is slower than the speed of light. As a result, wavelength is shortened by a factor known as the *velocity factor*, symbolized v. The value of v can be anything between 0 (representing zero speed of propagation) and 1 (representing the speed of propagation in free space, which is approximately 186,000 mi/s or 300,000 km/s). The velocity factor can also be expressed as a percentage $v_\%$. In that case, the smallest possible value is 0 percent, and the largest is 100 percent. The velocity factor in practical situations is rarely less than about 0.60, or 60 percent.

Velocity factor is important in the design of RF transmission lines and antenna systems, when sections of cable, wire, or metal tubing must be cut to specific lengths measured in wavelengths or fractions of a wavelength. Taking the velocity factor *v*, expressed as a ratio, into account, the preceding four formulas become:

$$L_{ft} = 984v\,/\,f_{MHz}$$
$$L_m = 300v\,/\,f_{MHz}$$
$$f_{MHz} = 984v\,/\,L_{ft}$$
$$f_{MHz} = 300v\,/\,L_m$$

The Electromagnetic Spectrum

The entire range of EM wavelengths is called the *electromagnetic* (EM) *spectrum.* Scientists use logarithmic scales to depict the EM spectrum, as shown in Fig. 25-10. The *RF spectrum*, which includes radio, television, and microwaves, is blown up in this illustration, and is labeled for frequency.

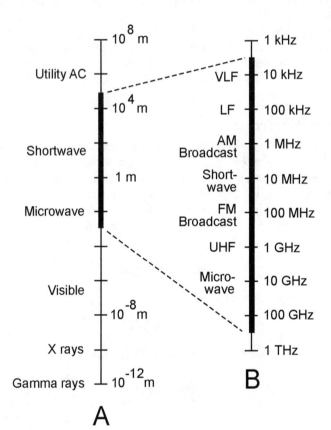

25-10   At A, the EM spectrum from 10^8 m to 10^{-12} m, with each vertical division representing two orders of magnitude (an increase or decrease of the wavelength by a factor of 100). At B, the RF spectrum, with each vertical division representing one order of magnitude (an increase or decrease of the frequency by a factor of 10).

Wave Propagation

Radio-wave propagation has been a fascinating science ever since Marconi and Tesla discovered, around the year 1900, that EM fields can travel over long distances without any supporting infrastructure whatsoever. Let's look at a few of the factors that affect wireless communications at radio frequencies.

Polarization

The orientation of the E field *lines of flux* is defined as the *polarization* of an EM wave. If the E field flux lines are parallel to the earth's surface, you have *horizontal polarization*. If the E field flux lines are perpendicular to the surface, you have *vertical polarization*. Polarization can also be slanted, of course.

In some situations, the E flux lines rotate as the wave travels through space. This is *circular polarization* if the E field intensity remains constant. If the E field intensity is more intense in some planes than in others, the polarization is *elliptical*. Rotating polarization can be *clockwise* or *counterclockwise*, viewed as the wavefronts approach. The rotational direction is called the *sense* of polarization.

The Line-of-Sight Wave

Electromagnetic waves follow straight lines unless something makes them bend. *Line-of-sight* propagation can take place even when the receiving antenna cannot be seen from the transmitting antenna. To some extent, radio waves penetrate nonconducting objects such as trees and frame houses. The line-of-sight wave consists of two components: the *direct wave* and the *reflected wave*.

The direct wave: The longest wavelengths are least affected by obstructions. At very low, low, and medium frequencies, direct waves can *diffract* around things. As the frequency rises, especially above about 3 MHz, obstructions have a greater and greater blocking effect.

The reflected wave: Electromagnetic waves reflect from the earth's surface and from conducting objects like wires and steel beams. The reflected wave always travels farther than the direct wave. The two waves are usually not in phase at the receiving antenna. If they're equally strong but 180° out of phase, a *dead spot* occurs. This phenomenon is most noticeable at the highest frequencies. At VHF and UHF, an improvement in reception can sometimes result from moving the transmitting or receiving antenna just a few inches. In mobile operation, when the transmitter and/or receiver are moving, dead spots produce rapid, repeated interruptions in the received signal. This is called *picket fencing*.

The Surface Wave

At frequencies below about 10 MHz, the earth's surface conducts ac quite well. Because of this, vertically polarized EM waves follow the surface for hundreds or even thousands of miles, with the earth actually helping to transmit the signals. The lower the frequency, the lower the *ground loss*, and the farther the waves can travel by *surface-wave propagation*. Horizontally polarized waves do not travel well in this mode, because horizontal E field flux is shorted out by the earth. Above about 10 MHz, the earth becomes lossy, and surface-wave propagation is not useful for more than a few miles.

Sky-Wave EM Propagation

Ionization in the upper atmosphere, caused by solar radiation, can return EM waves to the earth at certain frequencies. The ionization takes place at three or four distinct layers.

The lowest ionized region is called the *D layer*. It exists at an altitude of about 30 mi (50 km), and is ordinarily present only on the daylight side of the planet. This layer absorbs radio waves at some frequencies, impeding long-distance ionospheric propagation.

The *E layer*, about 50 mi (80 km) above the surface, also exists mainly during the day, although nighttime ionization is sometimes observed. The E layer can provide medium-range radio communication at certain frequencies.

The uppermost layers are called the *F_1 layer* and the *F_2 layer*. The F_1 layer, normally present only on the daylight side of the earth, forms at about 125 mi (200 km) altitude; the F_2 layer exists at about 180 mi (300 km) over most, or all, of the earth. Sometimes the distinction between the F_1 and F_2 layers is ignored, and they are spoken of together as the *F layer*. Communication by means of F-layer propagation can usually be accomplished between any two points on the earth at some frequencies between 5 MHz and 30 MHz.

Tropospheric Propagation

At frequencies above about 30 MHz, the lower atmosphere bends radio waves toward the surface. *Tropospheric banding* occurs because the *index of refraction* of air, with respect to EM waves, decreases with altitude. The effect is similar to the way sound waves sometimes travel long distances over the surface of a calm lake in the early morning or early evening, letting you hear a conversation more than a mile away. Tropospheric propagation makes it possible to communicate for hundreds of miles when the ionosphere will not return waves to the earth.

Another type of tropospheric propagation is called *ducting*. It takes place when EM waves are trapped in a layer of cool, dense air sandwiched between two layers of warmer air. Like bending, ducting occurs almost entirely at frequencies above 30 MHz.

Still another tropospheric-propagation mode is known as *troposcatter*. This takes place because air molecules, dust grains, and water droplets scatter some of the EM field. This effect is commonly seen at VHF and UHF.

Tropospheric propagation in general, without mention of the specific mode, is sometimes called *tropo*.

Auroral Propagation

In the presence of unusual solar activity, the *aurora* (northern lights or southern lights) can return radio waves to the earth. This is known as *auroral propagation*. The aurora occur at altitudes of about 40 to 250 mi (65 to 400 km). Theoretically, auroral propagation is possible, when the aurora are active, between any two points on the surface from which the same part of the aurora lie on a line of sight. Auroral propagation seldom occurs when one end of the circuit is at a latitude less than 35° north or south of the equator.

Auroral propagation is characterized by rapid and deep fading. This almost always renders analog voice and video signals unintelligible. Digital modes are most effective for communication via auroral propagation, but the carrier is often spread out over several hundred hertz as a result of phase modulation induced by auroral motion. This severely limits the maximum data transfer rate. Auroral propagation is often accompanied by deterioration in ionospheric propagation.

Meteor Scatter

Meteors produce ionized trails that persist for approximately 0.5 s up to several seconds, depending on the size of a particular meteor, its speed, and the angle at which it enters the atmosphere. This is not enough time for the transmission of much data, but during a *meteor shower*, multiple trails can result in almost continuous ionization for a period of hours. Such ionized regions reflect radio waves at certain frequencies. This is *meteor scatter propagation*. It can take place at frequencies considerably above 30 MHz, and occurs over distances ranging from just beyond the horizon up to about 1500 mi (2400 km), depending on the altitude of the ionized trail and the relative positions of the trail, the transmitting station, and the receiving station.

Moonbounce

The moon, like the earth, reflects EM fields. This makes it possible to communicate by means of *earth-moon-earth* (EME), also called *moonbounce*. High-powered transmitters, sophisticated antenna systems, and sensitive receivers are needed for EME. Some moonbounce communication is done by radio amateurs at frequencies from 50 MHz to over 2 GHz.

Transmission Media

Data can be transmitted over various *media*. The most common are *cable*, *radio* (also called *wireless*), *satellite links* (a specialized form of wireless), and *fiber optics*. Cable, radio/TV, and satellite communications use the RF spectrum. Fiber optics uses IR or visible light energy.

Cable

The earliest cables were simple wires that carried dc. Nowadays, data transmission cables more often carry ac at radio frequencies. One advantage of using RF is the fact that the signals can be amplified at intervals on a long span. This greatly increases the distances over which data can be sent by cable. Another advantage of using RF is the fact that numerous signals can be carried over a single cable, with each signal on a different frequency.

Cables can consist of pairs of wires, somewhat akin to lamp cords. But more often coaxial cable, of the type described and illustrated at the end of Chap. 10, is used. This has a center conductor surrounded by a cylindrical shield. The shield is grounded, and the center conductor carries the signals. The shield keeps signals confined to the cable, and also keeps external EM fields from interfering with the signals.

Radio

All radio and TV signals are electromagnetic waves. The radio or TV transmitter output is coupled into an *antenna system* located at some distance from the transmitter. The energy follows a *transmission line*, also called a *feed line*, from the transmitter output to the antenna itself.

Most radio antenna transmission lines are coaxial cables. There are other types, used in special applications. At microwaves, hollow tubes called *waveguides* are used to transfer the energy from a transmitter to the antenna. A waveguide is more efficient than coaxial cable at the shortest radio wavelengths. Radio amateurs sometimes use *parallel-wire* transmission lines, resembling the ribbon cable popular for use with consumer TV receiving antennas. In a parallel-wire line, the RF currents in the two conductors are always 180° out of phase, so that their EM fields cancel each other. This keeps the transmission line from radiating, guiding the EM field along toward the antenna. The energy is radiated when it reaches the antenna.

The RF bands are generally categorized from *very low frequency* (VLF) through *extremely high frequency* (EHF), according to the breakdown in Table 25-1. As noted in Chap. 24, the exact lower limit of the VLF range is a matter of disagreement in the literature. In Table 25-1, it is defined as 3 kHz, which is consistent with defining the frequency boundaries between RF bands by order of magnitude.

Table 25-1.   Bands in the RF spectrum.

Frequency designation	Frequency range	Wavelength range
Very Low (VLF)	3 kHz–30 kHz	100 km–10 km
Low (LF)	30 kHz–300 kHz	10 km–1 km
Medium (MF)	300 kHz–3 MHz	1 km–100 m
High (HF)	3 MHz–30 MHz	100 m–10 m
Very High (VHF)	30 MHz–300 MHz	10 m–1 m
Ultra High (UHF)	300 MHz–3 GHz	1 m–100 mm
Super High (SHF)	3 GHz–30 GHz	100 mm–10 mm
Extremely High (EHF)	30 GHz–300 GHz	10 mm–1 mm

Satellite Systems

At very high frequencies (VHF) and above, many communications circuits use satellites in *geostationary orbits* around the earth. If a satellite is directly over the equator at an altitude of 22,300 mi (36,000 km) and orbits from west to east, it follows the earth's rotation, thereby staying in the same spot in the sky as seen from the surface, and is thus a *geostationary satellite.*

A single geostationary satellite is on a line of sight with about 40 percent of the earth's surface. Three such satellites, placed at 120° ($\frac{1}{3}$ circle) intervals around the planet, allow coverage of all populated regions. A *dish antenna* can be aimed at a geostationary satellite, and once the antenna is in place, it need not be turned or adjusted.

Another form of satellite system uses multiple satellites in low orbits that take them over the earth's poles. These satellites are in continuous, rapid motion with respect to the surface. But if there are enough of them, they can act like repeaters in a cell phone network, and maintain reliable communications between any two points on the surface at all times. Directional antennas are not necessary in these systems, which are called *low earth orbit* (LEO) satellite networks.

Fiber Optics

Beams of IR or visible light can be modulated, just as can RF carriers. The frequency of an IR or visible light beam is higher than the frequency of any RF signal, allowing modulation by data at rates higher than anything possible with radio.

Fiber-optic technology offers several advantages over wire cables (which are sometimes called *copper* because the conductors are usually made of that metallic element). A fiber-optic cable is cheap. It is light in weight. It is immune to interference from outside EM fields. A fiber-optic cable does not corrode as metallic wires do. Fiber-optic cables are inexpensive to maintain and easy to repair. An optical fiber can carry far more signals than a cable, because the frequency bands are far wider in terms of megahertz or gigahertz.

The whole RF spectrum, from VLF through EHF, can (at least in theory) be imprinted on a single beam of light and sent through an optical fiber no thicker than a strand of hair!

Two Basic Receiver Designs

A wireless or *radio receiver* converts EM waves into the original messages sent by a distant transmitter. Let's begin our study of receivers by defining a few of the most important criteria for operation, and then we'll look at two common designs.

Specifications

The *specifications* of a receiver indicate how well it can do the functions it is designed to perform.

Sensitivity: The most common way to express receiver sensitivity is to state the number of microvolts that must exist at the antenna terminals to produce a certain *signal-to-noise* (S/N) *ratio* or *signal-plus-noise-to-noise* (S+N/N) *ratio* in decibels (dB). Sensitivity is related to the gain of the *front end* (the amplifier or amplifiers connected to the antenna), but the amount of noise this stage generates is more significant, because subsequent stages amplify the front-end noise output as well as the signal output.

Selectivity: The *passband*, or bandwidth that the receiver can hear, is established by a wideband *preselector* in the early RF amplification stages, and is honed to precision by narrowband filters in later amplifier stages. The preselector makes the receiver optimally sensitive within a range of approximately plus-or-minus 10 percent ($\pm 10\%$) of the desired signal frequency. The narrowband filter responds only to the frequency or channel of a specific signal to be received; signals in nearby channels are rejected.

Dynamic range: The signals at a receiver input vary over several orders of magnitude (multiples or powers of 10) in terms of absolute voltage. Dynamic range is the ability of a receiver to maintain a fairly constant output, and yet to maintain its rated sensitivity, in the presence of signals ranging from very weak to very strong. The dynamic range in a good receiver is in excess of 100 dB.

Noise figure: The less internal noise a receiver produces, in general, the better is the S/N ratio. Excellent S/N ratio in the presence of weak signals is only possible when the *noise figure*, a measure of internally generated receiver noise, is low. This is paramount at VHF, UHF, and microwave frequencies. Gallium-arsenide field effect transistors (GaAsFETs) are well known for the low levels of noise they generate, even at quite high frequencies. Other types of FETs can be used at lower frequencies. *Bipolar transistors* tend to be rather noisy.

Direct-Conversion Receiver

A *direct-conversion receiver* derives its output by mixing incoming signals with the output of a tunable (that is, variable frequency) *local oscillator* (LO). The received signal is fed into a mixer, along with the output of the LO. Figure 25-11 is a block diagram of a direct-conversion receiver.

For the reception of on/off keyed Morse code, also called *radiotelegraphy* or *continuous wave* (CW), the LO, also called a *beat-frequency oscillator* (BFO), is set a few hundred hertz above or below the signal frequency. This can also be done in order to receive FSK signals. The audio output has a frequency equal to the difference between the LO and incoming carrier frequencies. For reception of AM or SSB signals, the LO is set to precisely the same frequency as that of the signal carrier. This condition is known as *zero beat* because the *beat frequency*, or difference frequency, between the LO and the signal carrier is equal to zero.

A direct-conversion receiver provides rather poor *selectivity*. That means it can't separate incoming signals very well when they are close together in frequency. This is because signals on either side of the LO frequency can be heard at the same time. A *selective filter* can theoretically eliminate this. Such a filter must be designed for a fixed frequency if it is to work well. But in a direct-conversion receiver, the RF amplifier works over a wide range of frequencies.

Superheterodyne Receiver

A *superheterodyne receiver*, also called a *superhet*, uses one or more local oscillators and mixers to obtain a constant-frequency signal. A fixed-frequency signal is more easily processed than a signal that

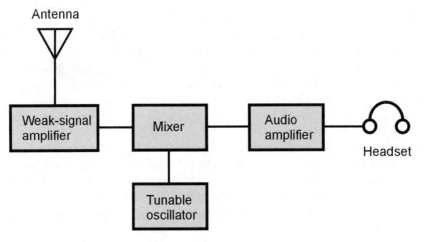

25-11 Block diagram of a direct-conversion receiver.

changes in frequency. The incoming signal is first passed through a tunable, sensitive front end. The output of the front end is mixed with the signal from a tunable, unmodulated LO. Either the sum or the difference signal is amplified. This is the *first intermediate frequency* (IF), which can be filtered to obtain a high degree of selectivity.

If the first IF signal is detected, the radio is a *single-conversion receiver.* Some receivers use a second mixer and second LO, converting the first IF to a lower-frequency *second IF.* This is a *double-conversion receiver.* The IF bandpass filter can be constructed for use on a fixed frequency, allowing superior selectivity and facilitating adjustable bandwidth. The sensitivity is enhanced because fixed IF amplifiers are easy to keep in tune.

A superheterodyne receiver can intercept or generate unwanted signals. False signals external to the receiver are called *images;* internally generated signals are called *birdies.* If the LO frequencies are carefully chosen, images and birdies do not cause problems during ordinary operation of the receiver.

Figure 25-12 is a block diagram of a generic single-conversion superheterodyne receiver. (Individual receiver designs vary somewhat.) Here's what each stage does.

Front end: The front end consists of the first RF amplifier, and often includes *LC* bandpass filters between the amplifier and the antenna. The dynamic range and sensitivity of a receiver are determined by the performance of the front end.

Mixer: A mixer stage converts the variable signal frequency to a constant IF. The output is either the sum or the difference of the signal frequency and the tunable LO frequency.

IF stages: The IF stages are where most of the gain takes place. These stages are also where optimum selectivity is obtained.

Detector: The detector extracts the information from the signal. Common circuits are the *envelope detector* for AM, the *product detector* for SSB, FSK, and CW, and the *ratio detector* for FM.

Audio amplifier: Following the detector, one or two stages of audio amplification are employed to boost the signal to a level suitable for a speaker or headset. Alternatively, the signal can be fed to a printer, facsimile machine, or computer.

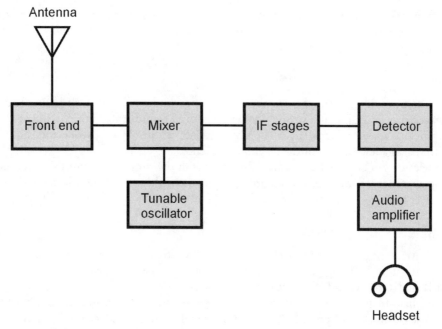

25-12 Block diagram of a single-conversion superheterodyne receiver.

Predetector Stages

In a superhet, the stages preceding the first mixer must be designed so they provide reasonable gain, but produce as little noise as possible. They must also be capable of handling strong signals without *desensitization* (losing gain), also known as *overloading*.

Preamplifier

All preamplifiers operate in class A, and most employ FETs. An FET has a high input impedance that is ideally suited to weak-signal work. Figure 25-13 shows a simple RF preamplifier circuit. Input tuning reduces noise and provides some selectivity. This circuit produces 5 dB to 10 dB gain, depending on the frequency and the choice of FET.

It is important that the preamplifier be linear, and that it remain linear in the presence of strong input signals. Nonlinearity results in unwanted mixing in RF amplifiers. The *mixing products* produce *intermodulation distortion* (IMD), or *intermod*. That can wreak havoc in a receiver, producing numerous false signals. It also degrades the S/N ratio by generating *hash*, the result of complex mixing of many false signals over a wide range of frequencies.

The Front End

At low and medium frequencies, there is considerable atmospheric noise, and the design of a front-end circuit is simple. Above 30 MHz, atmospheric noise diminishes, and the main factor that limits the sensitivity is noise generated within the receiver. For this reason, front-end design becomes increasingly critical as the frequency rises through the VHF, UHF, and microwave spectra.

The front end, like a preamplifier, must be as linear as possible; the greater the degree of non-linearity, the more susceptible the circuit is to the generation of mixing products. The front end should also have the greatest possible dynamic range.

Preselector

The preselector provides a bandpass response that improves the S/N ratio, and reduces the likelihood of receiver overloading by a strong signal far removed from the operating frequency. The preselector provides *image rejection* in a superheterodyne circuit. Most preselectors have a 3-dB bandwidth that is a few percent of the received frequency.

A preselector can be tuned by means of *tracking* with the tuning dial, but this requires careful design and alignment. Some receivers incorporate preselectors that must be adjusted independently of the receiver tuning.

IF Chain

A high IF (several megahertz) is preferable to a low IF (less than 1 MHz) for image rejection. But a low IF is better for obtaining good selectivity. Double-conversion receivers have a comparatively high first IF and a low second IF to get the best of both worlds.

Intermediate-frequency amplifiers can be cascaded with tuned-transformer coupling. The amplifiers follow the mixer and precede the detector. Double-conversion receivers have two chains of IF amplifiers. The first IF chain follows the first mixer and precedes the second mixer, and the second IF chain follows the second mixer and precedes the detector.

The selectivity of the IF chain in a superheterodyne receiver can be expressed mathematically. The bandwidths are compared for two power-attenuation values, usually 3 dB and 30 dB. This gives an indication of the shape of the bandpass response. The ratio of the 30-dB selectivity to the 3-dB selectivity is called the *shape factor*. A *rectangular response* is desirable in most applications. The smaller the shape factor, the more rectangular the response.

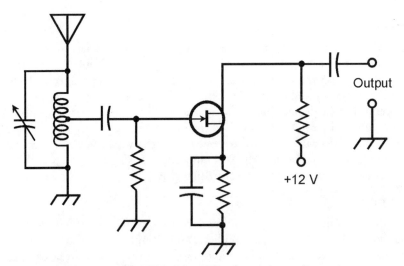

25-13   A tunable FET preamplifier for use in a radio receiver.

Detectors

Detection, also called *demodulation*, is the recovery of information such as audio, images, or printed data from a signal.

Detection of AM

The modulating waveform can be extracted from an AM signal by rectifying the carrier wave. A simplified time-domain view of this is shown in Fig. 25-14A. The rapid pulsations occur at the car-

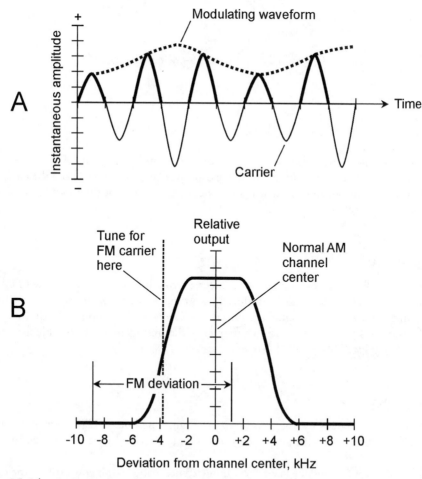

25-14 At A, envelope detection of AM, shown in the time domain. At B, slope detection of FM, shown in the frequency domain.

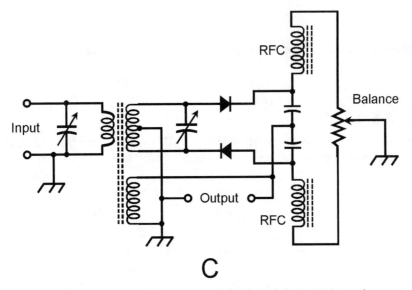

25-14   At C, a ratio detector circuit for demodulating FM signals.

rier frequency; the slower fluctuation is a duplication of the modulating data. The carrier pulsations are smoothed out by passing the output through a capacitor large enough to hold the charge for one carrier current cycle, but not so large that it smoothes out the cycles of the modulating signal. This scheme is known as *envelope detection*.

Detection of CW

For detection of CW signals, it is necessary to inject a signal into the receiver a few hundred hertz from the carrier. The injected signal is produced by a tunable *beat-frequency oscillator* (BFO). The BFO signal and the desired CW signal are mixed, or *heterodyned*, to produce audio output at the difference frequency. The BFO is tuned to a frequency that results in a comfortable listening pitch, usually 500 to 1000 Hz. This is called *heterodyne detection*.

Detection of FSK

The detection of FSK signals can be done using the same method as CW detection. The carrier beats against the BFO in the mixer, producing an audio tone that alternates between two different pitches.

With FSK, the BFO frequency is set a few hundred hertz above or below both of the carrier frequencies—that is, of both the *mark frequency* and the *space frequency*. The *frequency offset*, or difference between the BFO and signal frequencies, determines the audio output frequencies, and must

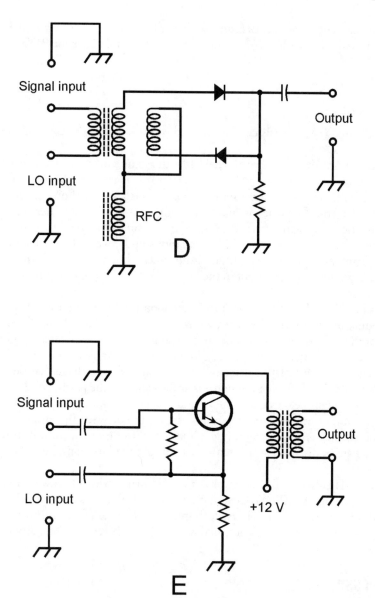

25-14 At D, a product detector using diodes. At E, a product detector using an NPN bipolar transistor biased as a class B amplifier. Both of these circuits can also be used as signal mixers.

be set so certain standard tone pitches result (such as 2125 Hz and 2295 Hz in the case of 170-Hz shift). Unlike the situation with CW reception, there is little tolerance for BFO adjustment variation or error.

Detection of FM

Frequency-modulated (FM) signals can be detected in various ways. These methods also work for phase modulation.

Slope detection: An AM receiver can detect FM in a crude manner by setting the receiver frequency near, but not on, the FM unmodulated-carrier frequency. An AM receiver has a filter with a passband of a few kilohertz, having a selectivity curve such as that shown in Fig. 25-14B. If the FM unmodulated-carrier frequency is near either edge, or *skirt*, of the filter response, frequency variations in the incoming signal cause it to swing in and out of the receiver passband. This causes the instantaneous receiver output to vary. The relationship between the instantaneous FM deviation and the instantaneous output amplitude is not linear, however, because the skirt of the passband is not a straight line, as is apparent in the figure. The result is an unnatural-sounding received signal.

PLL: If an FM signal is injected into a PLL, the loop produces an error voltage that is a duplicate of the modulating waveform. A *limiter*, which keeps the signal amplitude from varying, can be placed ahead of the PLL so the receiver does not respond to AM. Weak signals tend to abruptly appear and disappear, rather than fading, in an FM receiver that employs limiting.

Discriminator: This type of FM detector produces an output voltage that depends on the instantaneous signal frequency. When the signal is at the center of the passband, the output voltage is zero. If the frequency falls below center, the output voltage becomes positive. If the frequency rises above center, the output becomes negative. The relationship between the instantaneous FM deviation and the instantaneous output amplitude is linear, so the output is a faithful reproduction of the incoming signal data. A discriminator is sensitive to amplitude variations, but this can be overcome by a limiter.

Ratio detector: This type of FM detector is a discriminator with a built-in limiter. The original design was developed by RCA (Radio Corporation of America), and is used in high-fidelity receivers and in the audio portions of TV receivers. A simple ratio detector circuit is shown in Fig. 25-14C. The balance potentiometer should be adjusted for the best received signal quality.

Detection of SSB

For reception of SSB signals, a *product detector* is preferred, although a direct-conversion receiver can also do the job. A product detector also works well for the reception of CW and FSK. The incoming signal combines with the output of an unmodulated LO, producing audio or video. Product detection is done at a single frequency, rather than at a variable frequency as in direct-conversion reception. The single, constant frequency is obtained by mixing the incoming signal with the output of the LO.

Two product-detector circuits, which are also representative of the mixers used in superhet receivers, are shown in Fig. 25-14D and E. At D, diodes are used; there is no amplification. At E, a bipolar transistor is employed; this circuit provides some gain. The essential characteristic of either circuit is the nonlinearity of the semiconductor devices. This generates the sum and difference frequency signals that result in audio or video output.

Audio Stages

Enhanced selectivity can be obtained by tailoring the frequency response in the AF amplifier stages following the detector, in addition to the RF selectivity provided in the IF amplifier stages preceding the detector.

Filtering

A voice signal requires a band of frequencies ranging from about 300 Hz to 3000 Hz. An *audio bandpass filter*, with a passband of 300 Hz to 3000 Hz, can improve the quality of reception with some voice receivers. An ideal voice audio filter has little or no attenuation within the passband range, and high attenuation outside the range, with a near-rectangular response.

A CW or FSK signal requires only a few hundred hertz of bandwidth to be clearly read. Audio CW filters can narrow the response bandwidth to 100 Hz or less, but passbands narrower than about 100 Hz produce *ringing*, degrading the quality of reception. With FSK, the bandwidth of the filter must be at least as large as the difference (shift) between mark and space, but it need not, and should not, be much greater.

An *audio notch filter* is a *band-rejection filter* with a sharp, narrow response. An interfering carrier that produces a tone of constant frequency in the receiver output can be greatly attenuated with this type of filter. Audio notch filters are tunable from at least 300 Hz to 3000 Hz. Some sophisticated AF notch filters can tune themselves automatically. When an interfering carrier appears and remains for a few tenths of a second, the notch is activated and centers itself on the audio frequency of the detected offending signal.

Squelching

A *squelch* silences a receiver when no signal is present, allowing reception of signals when they appear. Most FM communications receivers use squelching systems. The squelch is normally closed, allowing no audio output, when no signal is present. The squelch opens, allowing everything to be heard, if the signal amplitude exceeds the *squelch threshold*, which can be adjusted by the operator.

In some systems, the squelch does not open unless the signal has certain characteristics. This is known as *selective squelching*. The most common way to achieve this is the use of subaudible (below 300 Hz) tone generators, or AF tone-burst generators. The squelch opens only when an incoming signal is modulated by a tone, or sequence of tones, having the proper characteristics. This can prevent unauthorized transmissions from accessing repeaters or being picked up by receivers.

Television Reception

A television (TV) receiver has a tunable front end, an oscillator and mixer, a set of IF amplifiers, a video demodulator, an audio demodulator and amplifier chain, a picture CRT or display with associated peripheral circuitry, and a loudspeaker.

Fast-Scan TV

Figure 25-15 is a block diagram of a receiver for conventional analog FSTV. In the United States, conventional FSTV broadcasts are made on 68 different channels numbered from 2 through 69. Each channel is 6 MHz wide, including video and audio information. Channels 2 through 13 comprise the *VHF TV broadcast channels*. Channels 14 through 69 are the *UHF TV broadcast channels*.

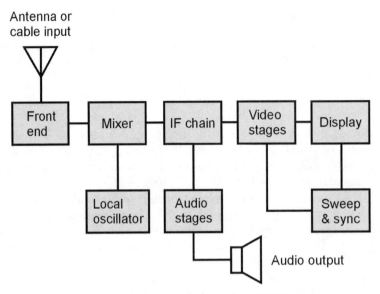

25-15 Block diagram of a conventional FSTV receiver.

In digital cable television, there are more channels. The number of possible channels is virtually unlimited, because signals are not transmitted over the air and do not consume EM spectrum space.

Slow-Scan TV

A *slow-scan television* (SSTV) communications station needs a transceiver with SSB capability, a standard TV set or personal computer, a video camera, and a *scan converter* that translates between the SSTV signal and either FSTV imagery or computer video data. The scan converter consists of two data converters (one for receiving and the other for transmitting), some digital *memory*, a *tone generator*, and a TV detector. Scan converters are commercially available. Computers can be programmed to perform this function. Some amateur radio operators build their own scan converters.

Specialized Wireless Modes

Some less common wireless communications techniques are effective under certain circumstances.

Dual-Diversity Reception

A *dual-diversity receiver* can reduce the fading that occurs in radio reception at high frequencies (approximately 3 to 30 MHz) when signals are propagated by means of the ionosphere. Two receivers are used. Both are tuned to the same signal, but they employ separate antennas, spaced several wavelengths apart. The outputs of the receiver detectors are fed into a common audio amplifier, as shown in Fig. 25-16.

Dual-diversity tuning is critical, and the equipment is expensive. In some installations, three or more antennas and receivers are employed. This provides superior immunity to fading, but it compounds the tuning difficulty and further increases the expense.

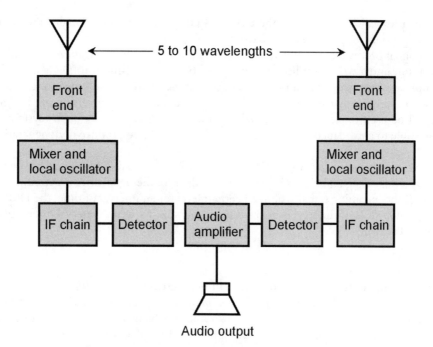

25-16   Block diagram of a dual-diversity radio receiver system.

Synchronized Communications

Digital signals require less bandwidth than analog signals to convey a given amount of information per unit time. *Synchronized communications* refers to a specialized digital mode, in which the transmitter and receiver operate from a common time standard to optimize the amount of data that can be sent in a communications channel or band.

In synchronized digital communications, also called *coherent communications*, the receiver and transmitter operate in lock-step. The receiver evaluates each transmitted binary digit, or *bit*, for a block of time lasting for the specified duration of a single bit. This makes it possible to use a receiving filter having extremely narrow bandwidth. The synchronization requires the use of an external frequency/time standard. The broadcasts of standard time-and-frequency radio stations such as WWV or WWVH can be used for this purpose. Frequency dividers are employed to obtain the necessary synchronizing frequencies. A tone or pulse is generated in the receiver output for a particular bit if, but only if, the average signal voltage exceeds a certain value over the duration of that bit. False signals, such as can be caused by filter ringing, sferics, or ignition noise, are generally ignored, because they rarely produce sufficient average bit voltage.

Experiments with synchronized communications have shown that the improvement in S/N ratio, compared with nonsynchronized systems, is several decibels at low to moderate data speeds. Further improvement can be obtained by the use of DSP.

How DSP Can Improve Reception

In DSP with analog modes such as SSB or SSTV, the signals are first changed into digital form by A/D conversion. Then the digital data is cleaned up so the pulse timing and amplitude adhere strictly to the protocol (standards) for the type of digital data being used. Finally, the digital signal is changed back to the original voice or video by D/A conversion.

Digital signal processing can extend the range of a wireless communications circuit, because it allows reception under worse conditions than would be possible without it. Digital signal processing also improves the quality of marginal signals, so that the receiving equipment or operator makes fewer errors. In circuits that use only digital modes, A/D and D/A conversion are irrelevant, but DSP can still be used to clean up the signal. This improves the accuracy of the system, and also makes it possible to copy data over and over many times (that is, to produce multigeneration duplicates).

The DSP circuit minimizes noise and interference in a received digital signal as shown in Fig. 25-17. A hypothetical signal before DSP is shown at the top; the signal after processing is shown at the bottom. If the incoming signal is above a certain level for an interval of time, the DSP output is *high* (also called *logic 1*). If the level is below the critical point for a time interval, then the output is *low* (also called *logic 0*).

Multiplexing

Signals in a communications channel or band can be intertwined, or *multiplexed*, in various ways. The most common methods are *frequency division multiplexing* (FDM) and *time division multiplexing* (TDM). In FDM, the channel is broken down into subchannels. The carrier frequencies of the signals are spaced so they do not overlap. Each signal is independent of the others. In TDM, signals are broken into segments by time, and then the segments are transferred in a rotating sequence. The receiver must be synchronized with the transmitter by means of a time standard such as WWV. Multiplexing requires an *encoder* that combines or intertwines the signals in the transmitter, and a *decoder* that separates or untangles the signals in the receiver.

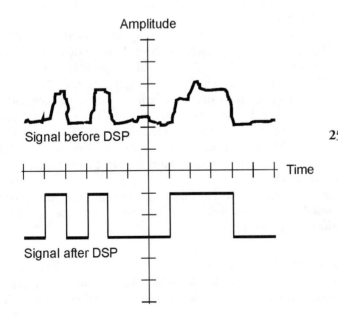

Signal before DSP

Signal after DSP

25-17   Digital signal processing can clean up a signal, improving reception.

Spread Spectrum

In *spread-spectrum communications*, the main carrier frequency is rapidly varied independently of signal modulation, and the receiver is programmed to follow. As a result, the probability of *catastrophic interference*, in which one strong interfering signal can obliterate the desired signal, is near zero. It is difficult for unauthorized people to eavesdrop on a spread-spectrum communications link unless they gain access to the *sequencing code*, also known as the *frequency spreading function.* Such a function can be complex, and can be kept secret. If the transmitting and receiving operator do not divulge the function to anyone, and if they do not tell anyone about the existence of their contact, then no one else on the band will know the contact is taking place.

During a spread-spectrum contact between a given transmitter and receiver, the operating frequency can fluctuate over a range of kilohertz, megahertz, or tens of megahertz. As a band becomes occupied with an increasing number of spread-spectrum signals, the overall noise level in the band appears to increase. Therefore, there is a practical limit to the number of spread-spectrum contacts that a band can handle. This limit is roughly the same as it would be if all the signals were constant in frequency, and had their own discrete channels.

A common method of generating spread spectrum is *frequency hopping*. The transmitter has a list of channels that it follows in a certain order. The receiver must be programmed with this same list, in the same order, and must be synchronized with the transmitter. The *dwell time* is the interval at which the frequency changes occur, which is the same as the length of time that the signal remains on any given frequency. The dwell time should be short enough so that a signal will not be noticed, and not cause interference, on any frequency. There are numerous *dwell frequencies*, so the signal energy is diluted to the extent that, if someone tunes to any particular frequency in the sequence, the signal is not noticeable.

Another way to get spread spectrum, called *frequency sweeping*, is to frequency-modulate the main transmitted carrier with a waveform that guides it up and down over the assigned band. This FM is independent of signal intelligence. A receiver can intercept the signal if, but only if, its tuning varies according to the same waveform, over the same band, at the same frequency, and in the same phase as that of the transmitter.

Quiz

Refer to the text in this chapter if necessary. A good score is at least 18 correct. Answers are in the back of the book.

1. A radio wave has a frequency of 1.55 MHz. The highest modulating frequency that can be used effectively is about
    (a) 1.55 kHz.
    (b) 15.5 kHz.
    (c) 155 kHz.
    (d) 1.55 MHz.

2. The reflected wave
    (a) arrives in phase with the direct wave.
    (b) arrives out of phase with the direct wave.
    (c) arrives in a variable phase compared with the direct wave.
    (d) is always horizontally polarized.

3. An advantage of FSK over on/off keying is the fact that FSK
   (a) offers better frequency stability.
   (b) can provide faster data rates.
   (c) reduces the number of receiving errors.
   (d) Forget it! On-off keying is just as good as FSK.

4. The highest layer of the ionosphere is
   (a) the D layer.
   (b) the E layer.
   (c) the F layer.
   (d) none of the above.

5. If an AM signal is modulated with audio having frequencies up to 5 kHz, then what is the complete signal bandwidth?
   (a) 10 kHz
   (b) 20 kHz
   (c) 50 kHz
   (d) 500 kHz

6. An LSB, suppressed-carrier signal can be demodulated by
   (a) an envelope detector.
   (b) a diode.
   (c) a ratio detector.
   (d) a product detector.

7. Which of the following modes is used to send image data over telephone lines?
   (a) On/off keying
   (b) Fax
   (c) AM
   (d) Product detection

8. The S/N ratio is a measure of
   (a) the sensitivity of a receiver.
   (b) the selectivity of a receiver.
   (c) the dynamic range of a receiver.
   (d) the efficiency of a transmitter.

9. Suppose an SSB suppressed carrier is at 14.335 MHz, and audio data is contained in a band from 14.335 to 14.338 MHz. What is this mode?
   (a) DSB
   (b) LSB
   (c) USB
   (d) AFSK

10. A receiver that responds to a desired signal, but not to another signal very close by in frequency, has good

    (a) sensitivity.

    (b) noise figure.

    (c) dynamic range.

    (d) selectivity.

11. Fill in the blank in the following sentence to make it true: "The deviation of a narrowband voice FM signal normally extends up to _____ either side of the unmodulated-carrier frequency."

    (a) 3 kHz

    (b) 5 kHz

    (c) 10 kHz

    (d) 3 MHz

12. An FM detector with built-in limiting is

    (a) a ratio detector.

    (b) a discriminator.

    (c) an envelope detector.

    (d) a product detector.

13. In which mode of PM does the peak power level of the pulses vary?

    (a) PAM

    (b) PDM

    (c) PIM

    (d) PFM

14. A continuously variable signal (such as music audio) can be recovered from a signal having only a few discrete levels or states by means of

    (a) a ratio detector.

    (b) a D/A converter.

    (c) a product detector.

    (d) an envelope detector.

15. In which mode are signals intertwined in the time domain at the transmitter, and then separated again at the receiver?

    (a) FDM

    (b) AFSK

    (c) PCM

    (d) None of the above

16. Which of the following modes can be demodulated with an envelope detector?

    (a) AM

    (b) CW

(c) FSK

(d) USB

17. The bandwidth of a fax signal is kept narrow by

(a) sending the data at a slow rate of speed.

(b) maximizing the number of digital states.

(c) optimizing the range of colors sent.

(d) using pulse modulation.

18. The dynamic range in a superhet is largely influenced by the performance of the

(a) local oscillator.

(b) product detector.

(c) front end.

(d) selectivity in the IF chain.

19. Frequency sweeping can be used to get a transmitter to produce

(a) spread-spectrum signals.

(b) time division multiplexed signals.

(c) narrowband AM signals.

(d) double sideband, suppressed carrier signals.

20. Fill in the blank to make the following sentence true: "The reception of _____ can be improved by the use of DSP."

(a) SSB signals

(b) SSTV signals

(c) synchronized communications signals

(d) any of the above

26
CHAPTER

Digital Basics

A SIGNAL IS *DIGITAL* WHEN IT CAN ATTAIN ONLY SPECIFIC LEVELS. THIS IS IN CONTRAST TO *ANALOG* signals or quantities that vary over a continuous range. A simple analog waveform is shown at Fig. 26-1A; note that the amplitude varies smoothly over time. Figure 26-1B is an example of a digital approximation of the same signal.

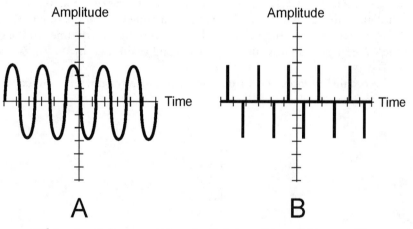

26-1   An analog wave (A), and a digital rendition of this wave (B).

Numbering Systems

People are used to dealing with the *decimal number system*, which has 10 different digits. But machines use schemes that have some power of 2 digits, such as 2 (2^1), 4 (2^2), 8 (2^3), 16 (2^4), 32 (2^5), 64 (2^6), and so on.

Decimal

The *decimal number system* is also called *base 10* or *radix 10.* The set of possible digits is {0, 1, 2, 3, 4, 5, 6, 7, 8, 9}. The first digit to the left of the radix or "decimal" point is multiplied by 10^0, or 1. The next digit to the left is multiplied by 10^1, or 10. The power of 10 increases as you move farther to the left. The first digit to the right of the decimal point is multiplied by a factor of 10^{-1}, or $\frac{1}{10}$. The next digit to the right is multiplied by 10^{-2}, or $\frac{1}{100}$. This continues as you go farther to the right. Once the process of multiplying each digit is completed, the resulting values are added up. For example:

$$
\begin{aligned}
& 2 \times 10^3 \\
+ & 7 \times 10^2 \\
+ & 0 \times 10^1 \\
+ & 4 \times 10^0 \\
+ & 5 \times 10^{-1} \\
+ & 3 \times 10^{-2} \\
+ & 8 \times 10^{-3} \\
+ & 1 \times 10^{-4} \\
+ & 6 \times 10^{-5} \\
= & 2704.53816
\end{aligned}
$$

Binary

The *binary number system* is a method of expressing numbers using only the digits 0 and 1. It is sometimes called *base 2* or *radix 2.* The digit immediately to the left of the radix point is the "ones" digit. The next digit to the left is a "twos" digit; after that comes the "fours" digit. Moving farther to the left, the digits represent 8, 16, 32, 64, and so on, doubling every time. To the right of the radix point, the value of each digit is cut in half again and again, that is, $\frac{1}{2}$, $\frac{1}{4}$, $\frac{1}{8}$, $\frac{1}{16}$, $\frac{1}{32}$, $\frac{1}{64}$, and so on.

Consider the decimal number 94. In the binary number system, this number is written as 1011110. It breaks down as follows:

$$
\begin{aligned}
& 0 \times 2^0 \\
+ & 1 \times 2^1 \\
+ & 1 \times 2^2 \\
+ & 1 \times 2^3 \\
+ & 1 \times 2^4 \\
+ & 0 \times 2^5 \\
+ & 1 \times 2^6 \\
= & 1011110
\end{aligned}
$$

When you work with a computer or calculator, you give it a decimal number that is converted into binary form. The computer or calculator does its operations entirely using the digits 0 and 1. When the process is complete, the machine converts the result back into decimal form for display.

In a communications system, binary numbers can represent alphanumeric characters, shades of color, frequencies of sound, and other variable quantities.

Octal

Another scheme, sometimes used in computer programming, is the *octal number system,* so named because it has eight symbols (according to our way of thinking), or 2^3. Every digit is an element of the set {0, 1, 2, 3, 4, 5, 6, 7}. This system is also known as *base 8* or *radix 8.*

Hexadecimal

Another system used in computer work is the *hexadecimal number system.* It has 16 (2^4) symbols. These digits are the usual 0 through 9 plus six more, represented by A through F, the first six letters of the alphabet. The digit set is {0, 1, 2, 3, 4, 5, 6, 7, 8, 9, A, B, C, D, E, F}. This system is sometimes called *base 16* or *radix 16.*

Logic

Logic refers to the reasoning used by electronic machines. The term is also used in reference to the circuits that make up digital devices and systems.

Boolean Algebra

Boolean algebra is a system of mathematical logic using the numbers 0 and 1 with the operations AND (multiplication), OR (addition), and NOT (negation). Combinations of these operations are NAND (NOT AND) and NOR (NOT OR). This peculiar form of mathematical logic, which gets its name from the nineteenth-century British mathematician George Boole, is used in the design of digital logic circuits.

Symbology

In Boolean algebra, the AND operation, also called *logical conjunction,* is written using an asterisk (*), a multiplication symbol (×), or by running two characters together, for example, X*Y.

The NOT operation, also called *logical inversion,* is denoted by placing a tilde (˜) over the quantity, as a minus sign (−) or dash (-) followed by the quantity, as a "lazy inverted L" (¬) followed by the quantity, or as the quantity followed by an accent or "prime sign" ('). An example is −X.

The OR operation, also called *logical disjunction,* is written using a plus sign (+), for example, X + Y.

The foregoing are the symbols used by engineers. Table 26-1A shows the values of these functions, where 0 indicates "falsity" and 1 indicates "truth."

In mathematics and philosophy courses involving logic, you may see other symbols used for conjunction and disjunction. The AND operation in some texts is denoted by a detached arrow-

Table 26-1A. Boolean operations.

X	Y	−X	X*Y	X+Y
0	0	1	0	0
0	1	1	0	1
1	0	0	0	1
1	1	0	1	1

**Table 26-1B. Common theorems in
Boolean algebra.**

Theorem (logic equation)	What it's called
X+0 = X	OR identity
X*1 = X	AND identity
X+1 = 1	
X*0 = 0	
X+X = X	
X*X = X	
−(−X) = X	Double negation
X+(−X) = 1	
X*(−X) = 0	Contradiction
X+Y = Y+X	Commutativity of OR
X*Y = Y*X	Commutativity of AND
X+(X*Y) = X	
X*(−Y)+Y = X+Y	
X+Y+Z = (X+Y)+Z = X+(Y+Z)	Associativity of OR
X*Y*Z = (X*Y)*Z = X*(Y*Z)	Associativity of AND
X*(Y+Z) = (X*Y)+(X*Z)	Distributivity
−(X+Y) = (−X)*(−Y)	DeMorgan's Theorem
−(X*Y) = (−X)+(−Y)	DeMorgan's Theorem

head pointing up ($\wedge$) or by an ampersand (&), and the OR operation is denoted by a detached arrowhead pointing down ($\vee$).

Theorems

Table 26-1B shows several logic equations. Such facts are called *theorems*. Statements on either side of the equals sign in each case are *logically equivalent.*

When two statements are logically equivalent, it means that one is true *if and only if* (*iff*) the other is true. For example, the statement X = Y means that X implies Y, and also that Y implies X. Logical equivalence is sometimes symbolized by a double arrow with one or two shafts ($\leftrightarrow$ or $\Leftrightarrow$).

Boolean theorems are used to analyze and simplify complicated logic functions. This makes it possible to build a circuit to perform a specific digital function, using the smallest possible number of logic switches.

Digital Circuits

All binary digital devices and systems employ high-speed electronic switches that perform Boolean operations. These switches are called *logic gates*. By combining logic gates, sophisticated digital systems can be built up. Even the most advanced computers are, at the basic level, comprised of logic gates.

Positive and Negative Logic

Usually, the binary digit 1 stands for "true" and is represented by a voltage of about +5 V. The binary digit 0 stands for "false" and is represented by about 0 V. This is *positive logic*. There are other logic forms, the most common of which is *negative logic*. In a common form of negative logic, the digit 1 (the logic high state) is represented by about 0 V, and the digit 0 (the logic low state) is represented by about +5 V. The remainder of this chapter deals with positive logic.

Basic Gates

An *inverter* or *NOT gate* has one input and one output. It reverses the state of the input. An *OR gate* can have two or more inputs. If both, or all, of the inputs are 0 (low), then the output is 0. If any of the inputs is 1 (high), then the output is 1. An *AND gate* can have two or more inputs. If both, or all, of the inputs are 1, then the output is 1. Otherwise the output is 0.

Other Gates

Sometimes an inverter and an OR gate are combined. This produces a *NOR gate*. If an inverter and an AND gate are combined, the result is a *NAND gate*.

An *exclusive OR gate*, also called an *XOR gate*, has two inputs and one output. If the two inputs are the same (either both 1 or both 0), then the output is 0. If the two inputs are different, then the output is 1.

The functions of logic gates are summarized in Table 26-2. Their schematic symbols are shown in Fig. 26-2.

Black Boxes

Logic gates can be combined to form circuits with many inputs and outputs. When two or more logic gates are combined, the outputs are always specific *logical functions* of the inputs. A complex combination of logic gates is sometimes called a *black box*.

The functions of a black box can always be determined using Boolean algebra, if the gates inside, and the way they are interconnected, is known. Conversely, if a certain complex logical function is needed for an application, a black box can be designed to perform that function by using Boolean algebra to break the function down into components of NOT, OR, AND, NOR, NAND, and XOR.

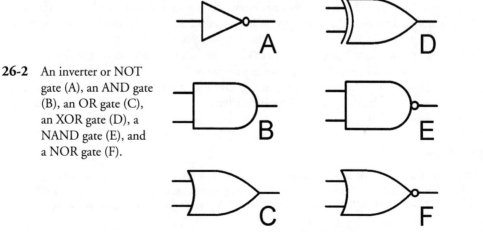

26-2 An inverter or NOT gate (A), an AND gate (B), an OR gate (C), an XOR gate (D), a NAND gate (E), and a NOR gate (F).

Table 26-2.   Logic gates and their characteristics.

Gate type	Number of inputs	Remarks
NOT	1	Changes state of input.
OR	2 or more	Output high if any inputs are high.
		Output low if all inputs are low.
AND	2 or more	Output low if any inputs are low.
		Output high if all inputs are high.
NOR	2 or more	Output low if any inputs are high.
		Output high if all inputs are low.
NAND	2 or more	Output high if any inputs are low.
		Output low if all inputs are high.
XOR	2	Output high if inputs differ.
		Output low if inputs are the same.

Forms of Binary Data

In communications, binary (two-level) data is less susceptible to noise and other interference than analog or multilevel digital data. There are several forms.

- *Morse code* is the oldest binary means of sending and receiving messages. It is a binary code because it has only two possible states: on (key-down) and off (key-up). It is used mainly by amateur radio operators in their hobby activities. A "human ear/brain machine," scrutinizing a Morse code signal, is an amazingly effective digital communications receiver.
- *Baudot*, also called the *Murray code*, is a five-unit digital code not widely used by today's digital equipment, except in some radioteletype communications.
- *ASCII* (American National Standard Code for Information Interchange) is a seven-unit code for the transmission of text and some programs. Letters, numerals, symbols, and control operations are represented. ASCII is designed for computers. There are 2^7, or 128, possible representations. Both upper- and lowercase letters can be represented, along with numerals and certain symbols.

Flip-flops

A *flip-flop* is also known as a *sequential logic gate.* In a sequential gate, the output state depends on both the inputs and the outputs. A flip-flop has two states, called *set* and *reset*. Usually, the set state is logic 1 (high), and the reset state is logic 0 (low). Here are some common types.

- *R-S flip-flop* inputs are labeled R (reset) and S (set). The outputs are Q and –Q. (Often, rather than –Q, you will see Q′, or perhaps Q with a line over it.) The outputs are always in logically opposite states. The symbol for an R-S flip-flop, also known as an *asynchronous flip-flop*, is shown in Fig. 26-3A. The *truth table* (a specialized form of table denoting logic functions) for an R-S flip-flop is at Table 26-3A.
- *Synchronous flip-flop* states change when triggered by the signal from a *clock*. In *static triggering*, the outputs change state only when the clock signal is either high or low. This type of circuit is sometimes called a *gated flip-flop*. In *positive-edge triggering*, the outputs change state at the instant the clock pulse is positive-going. The term *edge triggering* derives from the fact that the abrupt rise or fall of a pulse looks like the edge of a cliff (Fig. 26-3B). In *negative-edge triggering*, the outputs change state at the instant the clock pulse is negative-going.
- *Master/slave (M/S) flip-flop* inputs are stored before the outputs are allowed to change state. This device essentially consists of two R-S flip-flops in series. The first flip-flop is called the *master*, and the second is called the *slave*. The master flip-flop functions when the clock output is high, and the slave acts during the next low portion of the clock output. This time delay prevents confusion between the input and output.
- *J-K flip-flop* operation is similar to that of an R-S flip-flop, except that the J-K has a predictable output when the inputs are both 1. Table 26-3B shows the input and output states for this type of flip-flop. The output changes only when a triggering pulse is received. The symbol for a J-K flip-flop is shown in Fig. 26-3C.
- *R-S-T flip-flop* operation is similar to that of an R-S flip-flop, except that a high pulse at the T input causes the circuit to change state.
- The *T flip-flop* has only one input. Each time a high pulse appears at the T input, the output state is reversed.

Clocks

In electronics, the term *clock* refers to a circuit that generates pulses at high speed and at precise intervals. It sets the tempo for the operation of digital devices. In a computer, the clock acts like a metronome for the *microprocessor*. Clock speeds are measured and expressed in hertz (Hz), kilohertz (kHz), megahertz (MHz), or gigahertz (GHz).

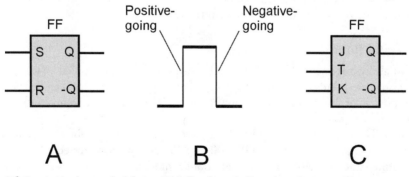

26-3 At A, the symbol for an R-S flip-flop. At B, pulse edges are either negative-going or positive-going. At C, the symbol for a J-K flip-flop.

Table 26-3. Flip-flop states.

A:
R-S Flip-flop

R	S	Q	–Q
0	0	Q	–Q
0	1	1	0
1	0	0	1
1	1	?	?

B:
J-K Flip-flop

J	K	Q	–Q
0	0	Q	–Q
0	1	1	0
1	0	0	1
1	1	–Q	Q

Counters

A *counter* consists of a set of flip-flops or equivalent circuits. Each time a pulse is received, the binary number stored by the counter increases by 1. A *frequency counter* measures the frequency of a wave by tallying the cycles in a given interval of time. The circuit consists of a *gate*, which begins and ends each counting cycle at defined intervals. The accuracy is a function of the length of the *gate time;* the longer the time base, the better the accuracy. The readout is in base-10 digital numerals.

Binary Digital Communications

The use of binary data provides excellent communications accuracy and efficiency. If multilevel signaling is required, then all the levels can be represented by groups of binary digits. A group of 3 binary digits, for example, can represent 2^3, or 8, levels. A group of 4 binary digits can represent 2^4, or 16, levels. The term *binary digit* is commonly contracted to *bit*.

Bits, Bytes, and Baud

A bit is almost always represented by either 0 or 1. A group of 8 bits is a called an *octet*, and in many systems this also corresponds to a unit called a *byte*. Large quantities of data can be expressed either according to powers of 2, or according to powers of 10. This can cause some confusion and gives rise to endless debates over semantics.

One *kilobit* (kb) is equal to 1000 bits. A *megabit* (Mb) is 1000 kilobits, or 1,000,000 bits. A *gigabit* (Gb) is 1000 megabits, or 1,000,000,000 bits. When data is expressed in bits, powers of 10 are used to define large quantities. If you hear about a modem that operates at 56 kbps, it means 56,000 *bits per second* (bps). Bits, kilobits, megabits, and gigabits per second (bps, kbps, Mbps, and Gbps) are commonly used to express data in communications.

Data quantity in storage and memory is usually specified in *kilobytes* (units of $2^{10} = 1,024$ bytes), *megabytes* (units of $2^{20} = 1,048,576$ bytes), and *gigabytes* (units of $2^{30} = 1,073,741,824$ bytes). The abbreviations for these units are KB, MB, and GB, respectively. Note that the uppercase K represents 2^{10} or 1024, while the lowercase k represents 10^3 or 1000. But M and G are always uppercase, no matter whether powers of 2 or 10 are used. (Is all of this confusing to you? Don't be discouraged. It confuses almost everybody.)

Larger data units are being used as memory and storage media continue to grow. The *terabyte* (TB) is 2^{40} bytes, or 1024 GB. The *petabyte* (PB) is 2^{50} bytes, or 1024 TB. The *exabyte* (EB) is 2^{60} bytes, or 1024 PB.

The term *baud* refers to the number of times per second that a signal changes state. The units of bps and baud are not equivalent, even though people often speak of them as if they are. These days, baud (or "baud rate") is seldom used to express data speed.

When computers are linked in a *network*, each computer has a *modem* (modulator/demodulator) connecting it to the communications medium. The slowest modem determines the speed at which the machines communicate. Table 26-4 shows common data speeds and the approximate time periods required to send 1, 10, and 100 pages of double-spaced, typewritten text at each speed.

Forms of Conversion

Any analog (continuously variable) signal can be converted into a string of pulses whose amplitudes have a finite number of states, usually some power of 2. This is analog-to-digital (A/D) conversion, and its reverse is digital-to-analog (D/A) conversion, as you've already learned. The difference between analog and digital signals can be intuitively seen by examining Fig. 26-4. This is essentially a rendition of 8-level pulse code modulation (PCM), as described in Chap. 25 and illustrated in Fig. 25-8D. Imagine the curve being sampled as a train of pulses (A/D), or the pulses being smoothed into the curve (D/A).

Table 26-4.   Time needed to send data at various speeds.
Abbreviations: s = second, ms = millisecond (0.001 s), µs = microsecond (0.000001 s).

A:

Speed, kbps	Time for one page	Time for 10 pages	Time for 100 pages
28.8	0.38 s	3.8 s	38 s
38.4	280 ms	2.8 s	28 s
57.6	190 ms	1.9 s	19 s
100	110 ms	1.1 ms	11 s
250	44 ms	440 ms	4.4 s
500	22 ms	220 ms	2.2 s

B:

Speed, Mbps	Time for one page	Time for 10 pages	Time for 100 pages
1.00	11 ms	110 ms	1.1 s
2.50	4.4 ms	44 ms	440 ms
10.0	1.1 ms	11 ms	110 ms
100	110 µs	1.1 ms	11 ms

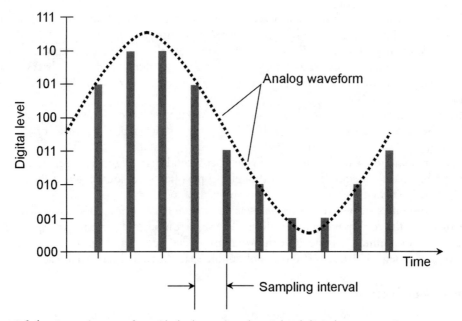

26-4   An analog waveform (dashed curve) and an 8-level digital representation (vertical bars).

Binary digital data can be sent and received one bit at a time along a single line or channel. This is *serial data transmission*. Higher data speeds can be obtained by using multiple lines or a wideband channel, sending independent sequences of bits along each line or subchannel. This is *parallel data transmission*.

Parallel-to-serial (P/S) *conversion* receives bits from multiple lines or channels, and transmits them one at a time along a single line or channel. A *buffer* stores the bits from the parallel lines or channels while they are awaiting transmission along the serial line or channel. *Serial-to-parallel* (S/P) *conversion* receives bits from a serial line or channel, and sends them in batches along several lines or channels. The output of an S/P converter cannot go any faster than the input, but the circuit is useful when it is necessary to interface between a serial-data device and a parallel-data device.

Figure 26-5 illustrates a circuit in which a P/S converter is used at the source (transmitting station), and an S/P converter is used at the destination (receiving station). In this example, the words are 8-bit bytes. However, the words could have 16, 32, 64, or even 128 bits, depending on the communications scheme.

Data Compression

Data compression is a way of maximizing the amount of digital information that can be stored in a given space, or sent in a certain period of time.

Text files can be compressed by replacing often-used words and phrases with symbols such as =, #, &, $, and @, as long as none of these symbols occurs in the uncompressed file. When the data is received, it is uncompressed by substituting the original words and phrases for the symbols.

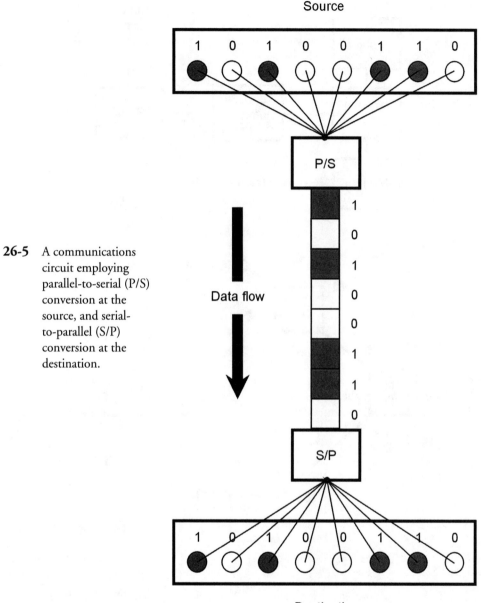

26-5 A communications circuit employing parallel-to-serial (P/S) conversion at the source, and serial-to-parallel (S/P) conversion at the destination.

Digital images can be compressed in either of two ways. In *lossless image compression*, detail is not sacrificed; only the redundant bits are eliminated. In *lossy image compression*, some detail is lost, although the loss is usually not significant.

Packet Wireless

In *packet wireless*, a computer is connected to a radio transceiver using a *terminal node controller* (TNC), which is similar to a modem. An example is shown in Fig. 26-6A. The computer has a tele-

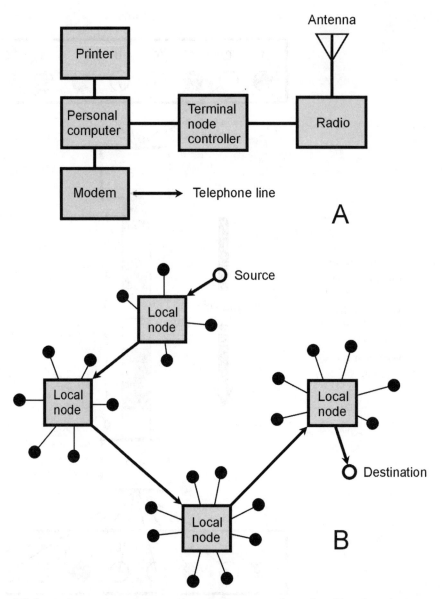

26-6   At A, a packet-wireless station. At B, passage of a packet through nodes in a
wireless communications circuit.

phone modem as well as a TNC, so messages can be sent and received using conventional online
services as well as radio.

Figure 26-6B shows how a packet wireless message is routed. Black dots represent *subscribers*. Rectangles represent local nodes, each of which serves subscribers by means of short-range links at very
high, ultra high, or microwave radio frequencies. The nodes are interconnected by terrestrial radio
links if they are relatively near each other. If the nodes are widely separated, satellite links are used.

The RGB Color Model

All visible colors can be obtained by combining red, green, and blue light. The *red/green/blue* (RGB) *color model* takes advantage of this fact.

Hue, Saturation, and Brightness

Color, as perceived by the human eye, is a function of the wavelength of optical EM energy. When most of the energy is concentrated near a single wavelength, you see an intense *hue*. The vividness of a hue is called the *saturation*. The *brightness* of a color is a function of how much total energy the light contains. In most video displays, there is a control for adjusting the brightness, also called *brilliance*.

3D Color

A color *palette* is obtained by combining pure red, green, and blue in various ratios. Assign each primary color an axis in Cartesian three-dimensional (3D) space as shown in Fig. 26-7. Call the axes R (for red), G (for green), and B (for blue). Color brightness can range from 0 to 255, or binary

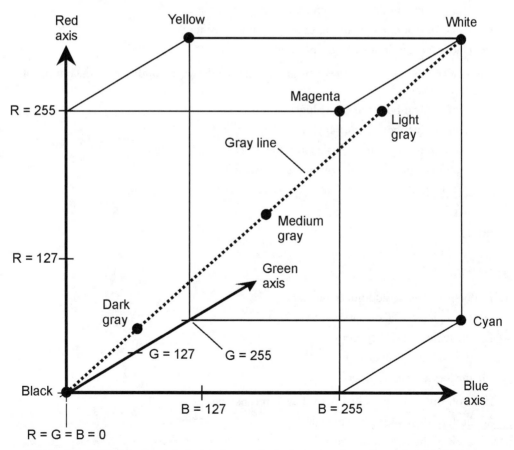

26-7   The RGB color model, depicted as a cube in Cartesian three-dimensional (3D) space.

00000000 to 11111111. The result is 16,777,216 (256^3) possible colors. Any point within the cube represents a unique color.

Colors in the RGB model can also be represented by six-digit hexadecimal numbers, such as 005CFF. In this scheme, the first two digits represent the red (R) intensity in 256 levels ranging from 00 to FF. The middle two digits represent the green (G) intensity, and the last two digits represent the blue (B) intensity.

Some RGB systems use only 16 levels for each primary color (binary 0000 through 1111). This results in 4096 possible colors.

Quiz

Refer to the text in this chapter if necessary. A good score is at least 18 correct. Answers are in the back of the book.

1. Suppose inverters are placed in series with both inputs of an AND gate. Under what conditions is the output of the resulting black box high?

    (a) If and only if both inverter inputs are high

    (b) If and only if both inverter inputs are low

    (c) If and only if one inverter input is high and the other is low

    (d) Under no conditions (the output is always low)

2. Suppose an AND gate is followed by an inverter. Under what conditions is the output of the resulting black box low?

    (a) If and only if both inputs are high

    (b) If and only if both inputs are low

    (c) If and only if one input is high and the other is low

    (d) Under no conditions (the output is always high)

3. What is the binary equivalent of decimal 29?

    (a) 10101

    (b) 11101

    (c) 10111

    (d) 11011

4. In Boolean algebra, addition represents

    (a) the logical NOT operation.

    (b) the logical AND operation.

    (c) the logical OR operation.

    (d) the logical NAND operation.

5. If the output to a logical inverter is low, it means that

    (a) both of the two inputs are high.

    (b) both of the two inputs are low.

(c) the single input is high.

(d) the single input is low.

6. What is the decimal equivalent of binary 1111?

    (a) 15

    (b) 31

    (c) 63

    (d) 127

7. The output of an AND gate is high if and only if

    (a) one input is high and the others are low.

    (b) one input is low and the others are high.

    (c) any of the inputs is low.

    (d) all inputs are high.

8. DeMorgan's Theorem states that, for all logical statements X and Y,

    (a) $-(X*Y)$ is equivalent to X+Y.

    (b) $X*Y$ is equivalent to $-(X+Y)$.

    (c) $(-X)+(-Y)$ is equivalent to $X*Y$.

    (d) $(-X)+(-Y)$ is equivalent to $-(X*Y)$.

9. If you see a number represented by FF in the documentation for an electronic circuit or system, you can be certain that the number is

    (a) radix 16.

    (b) radix 10.

    (c) radix 8.

    (d) radix 2.

10. Which of the following codes is digital?

    (a) Morse

    (b) Baudot

    (c) ASCII

    (d) All of the above

11. Which of the following voltages could normally represent a 1 in positive logic?

    (a) 0.0 V

    (b) +0.5 V

    (c) +5 V

    (d) +50 V

12. Which of the following voltages might normally represent a 1 in negative logic?

    (a) 0.0 V

    (b) +12 V

(c) +5 V

(d) +50 V

13. Fill in the blank to make the following sentence true: "_____ data transmission refers to digital bits sent along a single line, one bit after another."

(a) Serial

(b) Synchronous

(c) Parallel

(d) Analog

14. The inputs of an R-S flip-flop are known as

(a) low and high.

(b) octal and binary.

(c) synchronous and asynchronous.

(d) reset and set.

15. When both inputs of an R-S flip-flop are 0,

(a) the outputs stay as they are.

(b) $Q = 0$ and $-Q = 1$.

(c) $Q = 1$ and $-Q = 0$.

(d) the resulting outputs can be absurd.

16. The fifth digit from the right in a binary number carries which decimal value?

(a) 64

(b) 32

(c) 24

(d) 16

17. What is the octal equivalent of binary 1010?

(a) 4

(b) 10

(c) 12

(d) There is no way to tell without more information.

18. What is the largest possible radix 10 number that can be represented as a six-digit binary number?

(a) Decimal 256

(b) Decimal 128

(c) Decimal 64

(d) Decimal 63

19. If one of the inputs to a two-input NOR gate is high, what is the output state?

(a) There is not enough information to tell.

(b) Low

(c) High

(d) It depends on the state of the other input.

20. Suppose a logic circuit has four inputs W, X, Y, and Z. How many possible input combinations are there?

(a) 4

(b) 8

(c) 16

(d) 32

Test: Part 3

DO NOT REFER TO THE TEXT WHEN TAKING THIS TEST. A GOOD SCORE IS AT LEAST 37 CORRECT. Answers are in the back of the book. It's best to have a friend check your score the first time, so you won't memorize the answers if you want to take the test again.

1. In a JFET, the control electrode is usually the
   - (a) source.
   - (b) emitter.
   - (c) drain.
   - (d) base.
   - (e) gate.

2. A diode can be used as a frequency multiplier because of its
   - (a) junction capacitance.
   - (b) nonlinearity.
   - (c) avalanche voltage.
   - (d) forward breakover.
   - (e) charge carrier concentration.

3. Which of the following is not a common form of data transmission?
   - (a) Polarization modulation
   - (b) Frequency modulation
   - (c) Amplitude modulation
   - (d) Phase modulation
   - (e) Pulse modulation

4. A very brief, high-voltage spike on an ac power line is called
   - (a) a bleeder.
   - (b) an arc.

(c) a transient.

(d) an avalanche.

(e) a clipped peak.

5. Which of the following is not characteristic of an oscillator?

(a) Negative feedback

(b) Good output-to-input coupling

(c) Reasonably high transistor gain

(d) Alternating current signal output

(e) Usefulness as a signal generator

6. Which layer of the ionosphere exists at the lowest altitude?

(a) The F layer

(b) The E layer

(c) The D layer

(d) The C layer

(e) The B layer

7. The beta of a bipolar transistor is another name for its

(a) current amplification factor.

(b) voltage amplification factor.

(c) power amplification factor.

(d) maximum amplification frequency.

(e) optimum amplification frequency.

8. In a schematic diagram, the symbol for a PNP bipolar transistor can be recognized by

(a) a broken vertical line inside the symbol circle.

(b) an arrow inside the symbol circle, pointing from the base toward the emitter.

(c) an arrow inside the symbol circle, pointing outward from the collector.

(d) a collector biased at a positive voltage with respect to the emitter.

(e) an arrow inside the symbol circle, pointing from the emitter toward the base.

9. Fill in the blank in the following sentence to make it true: "In an oscillator, _____ is an expression of the extent to which the circuit maintains constant signal frequency output under variable operating conditions, and is also an expression of its overall reliability."

(a) sensitivity

(b) drift ratio

(c) gain

(d) selectivity

(e) stability

10. A Zener diode would most likely be found in

(a) the mixer in a superheterodyne receiver.

(b) the PLL in a circuit for detecting FM.

(c) the product detector in a receiver for SSB.

(d) the voltage regulator in a power supply.

(e) the AF oscillator in an AFSK transmitter.

11. When the bias in an FET stops the flow of current, the condition is called

(a) forward breakover.

(b) cutoff.

(c) reverse bias.

(d) pinchoff.

(e) avalanche.

12. The VA rating of a transformer is an expression of

(a) the maximum frequency at which it can function.

(b) the type of core material it has.

(c) the voltage step-up or step-down ratio.

(d) the impedance transfer ratio.

(e) none of the above.

13. In an N-type semiconductor, the minority carriers are

(a) electrons.

(b) protons.

(c) holes.

(d) neutrons.

(e) positrons.

14. A disadvantage of a half-wave rectifier is the fact that

(a) the output voltage is excessive compared to that of a full-wave rectifier.

(b) the output current is excessive compared to that of a full-wave rectifier.

(c) the output waveform is harder to filter than is the case with a full-wave rectifier.

(d) it requires several expensive diodes, whereas a full-wave rectifier requires only a single cheap diode.

(e) it requires an expensive center-tapped transformer, whereas a full-wave rectifier does not need a transformer at all.

15. A power gain of 30 dB is equivalent to which amplification factor?

(a) 0.001

(b) $\frac{1}{30}$

(c) 30

(d) 1000

(e) None of the above

16. Suppose an RF power amplifier has a dc collector input of 60 W, and is 75 percent efficient. What is the RF signal output power?

    (a) 80 W

    (b) 60 W

    (c) 45 W

    (d) 40 W

    (e) Impossible to determine from this data

17. If a listener is not expecting any change in the intensity of a signal or sound, the smallest change that can be noticed is approximately which of the following?

    (a) ±0.3 dB

    (b) ±1 dB

    (c) ±3 dB

    (d) ±10 dB

    (e) ±30 dB

18. A common base circuit is commonly employed as

    (a) a microwave oscillator.

    (b) a low-pass filter.

    (c) a noise generator.

    (d) a phase-locked loop.

    (e) an RF power amplifier.

19. A semiconductor material especially noted for its photoconductivity, and which is often used in making photocells, is

    (a) selenium.

    (b) lithium salt.

    (c) copper chloride.

    (d) aluminum oxide.

    (e) polyethylene.

20. Which type of PA circuit provides the highest efficiency when used with the appropriate amount of driving power?

    (a) Class A

    (b) Class AB

    (c) Class B

    (d) Class C

    (e) All of the above circuits are equally efficient.

21. Baudot is a form of

    (a) video modulation.

    (b) diode.

(c) digital code.

(d) voice modulation.

(e) AM detector.

22. The most stable type of oscillator circuit uses

(a) a tapped coil.

(b) a split capacitor.

(c) negative feedback.

(d) a common base arrangement.

(e) a quartz crystal.

23. If the source-gate junction in an FET conducts,

(a) it is a sign of improper bias.

(b) the device will work in class C.

(c) the device will oscillate.

(d) the device will work in class A.

(e) the circuit will have good stability.

24. What is the radix of the octal number system?

(a) 2

(b) 2^2

(c) 2^3

(d) 2^4

(e) 2^5

25. Signal-to-noise ratio (S/N) is often specified when stating

(a) the selectivity of a receiver.

(b) the stability of an oscillator.

(c) the modulation coefficient of a transmitter.

(d) the sensitivity of a receiver.

(e) the polarization of an EM wave.

26. In a reverse-biased semiconductor diode, the junction capacitance depends on

(a) the width of the depletion region.

(b) the reverse current.

(c) the P:N ratio.

(d) the gate bias.

(e) the avalanche voltage.

27. Suppose the bandwidth of the channel in a receiver's IF filter is 4.5 kHz at 30-dB power attenuation, and 3.0 kHz at 3-dB power attenuation. What is the shape factor?

(a) 10:1

(b) 1.5:1

(c) 1:1

(d) 1:10

(e) More information is necessary to determine this.

28. A simple power supply filter can be built with

(a) a capacitor in series with the dc output.

(b) an inductor in parallel with the dc output.

(c) a rectifier in parallel with the dc output.

(d) a resistor in series and an inductor in parallel with the dc output.

(e) a capacitor in parallel with the dc output.

29. Which of the following bipolar transistor circuits can, in theory, provide the most amplification?

(a) Common emitter

(b) Common base

(c) Common collector

(d) Common gate

(e) Common drain

30. The ratio of the difference in I_C to the difference in I_B that occurs when a small signal is applied to the base of a bipolar transistor is called the

(a) static current amplification.

(b) dynamic current amplification.

(c) power factor.

(d) efficiency.

(e) current attenuation factor.

31. An example of a device that is commonly used as an oscillator at microwave frequencies is

(a) a rectifier diode.

(b) a weak-signal diode.

(c) a Gunn diode.

(d) a Zener diode.

(e) an avalanche diode.

32. In the operation of a PNP bipolar transistor, which of the following is normal concerning the emitter and collector voltages?

(a) The collector is positive relative to the emitter.

(b) The collector is at the same voltage as the emitter.

(c) The collector is negative relative to the emitter.

(d) The collector can be either positive or negative relative to the emitter.

(e) The collector must be at ground potential.

33. What can be done to minimize the capacitance of the P-N junction in a semiconductor diode, thereby making the component effective as a high-speed RF switch?

   (a) The surface area of the P-N junction can be maximized.

   (b) A layer of intrinsic semiconductor can be placed between the P- and N-type materials.

   (c) The frequency of the applied signal can be made as high as possible.

   (d) The diode can be forward-biased with a high voltage.

   (e) A small-value resistor can be connected in series with each of several diodes, and the series diode-resistor combinations can all be connected in parallel.

34. Which type of modulation consists of one voice sideband, with a suppressed carrier?

   (a) AM

   (b) LSB

   (c) FM

   (d) RTTY

   (e) PCM

35. Suppose a series-parallel network of resistors is assembled into an *RF attenuator pad*, which is a circuit deliberately intended to produce a constant signal loss over a wide band of frequencies. If the input RF voltage to this attenuator pad is 500 μV and the output RF voltage is 100 μV, what is the loss in decibels?

   (a) 14.0 dB

   (b) 6.99 dB

   (c) 5.00 dB

   (d) 1.40 dB

   (e) 0.699 dB

36. In an AND gate, the output is high

   (a) if any input is high.

   (b) only when all inputs are low.

   (c) if any input is low.

   (d) only when all inputs are high.

   (e) only when all inputs have identical logic states.

37. A voltage-controlled oscillator makes use of

   (a) a varactor diode.

   (b) a Zener diode.

   (c) negative feedback.

   (d) a split capacitance.

   (e) adjustable gate or base bias.

38. The best electrical ground for a power supply is obtained by

   (a) connecting a small-value capacitor between the hot wire and a good earth ground, such as an 8-ft copper rod driven into the earth.

(b) ensuring that the ground wire from the power supply is connected to the third wire hole in one of the utility outlets of a properly installed three-wire electrical circuit.

(c) using a center-tapped transformer and a full-wave rectifier circuit, in conjunction with good filtering, voltage regulation, and transient suppression.

(d) connecting a large-value inductor in series with the hot wire, thereby ensuring that the voltage on that wire cannot be shunted to ground.

(e) making sure that all equipment connected to the supply is provided with fuses and/or circuit breakers of the proper ratings.

39. Suppose the output signal voltage from an amplifier is 35 times the input signal voltage. If the input and output impedances are both pure resistances of the same value, what is the gain?

(a) 15 dB

(b) 31 dB

(c) 35 dB

(d) 350 dB

(e) 700 dB

40. In an exclusive OR gate, the output is high

(a) if either input is high.

(b) only when both inputs are low.

(c) if either input is low.

(d) only when both inputs are high.

(e) only when the inputs have opposite logic states.

41. A ratio detector is a circuit for demodulating which of the following?

(a) AM

(b) PM

(c) FM

(d) SSB

(e) AFSK

42. Suppose there is a binary digital black box with two inputs, called X and Y, and an output called Z. If $X = 0$ and $Y = 1$, then $Z = 0$. In all other instances, $Z = 1$. Which of the following logical expressions represents the contents of the black box?

(a) $X + (-Y)$

(b) $X * (-Y)$

(c) $X - (*Y)$

(d) $X = (-Y)$

(e) $X - (+Y)$

43. A method of modulation in which the strength of pulses varies is called

(a) pulse ratio modulation.

(b) pulse position modulation.

(c) pulse frequency modulation.

(d) pulse amplitude modulation.

(e) pulse width modulation.

44. Boolean algebra is

(a) useful for calculating amplifier gain in decibels.

(b) a useful tool in binary digital circuit design.

(c) used to calculate the value of an unknown.

(d) used with negative pulse modulation schemes.

(e) used with positive pulse modulation schemes.

45. A voltage-doubler power supply is best for use in

(a) circuits that need high voltage but do not draw much current.

(b) low-voltage devices.

(c) high-current appliances.

(d) all kinds of electronic equipment.

(e) broadcast transmitter power amplifiers.

46. An optoisolator consists of

(a) two Zener diodes back-to-back.

(b) an LED and a photodiode.

(c) two NPN transistors in series.

(d) an NPN transistor followed by a PNP transistor.

(e) a PNP transistor followed by an NPN transistor.

47. When a semiconductor is reverse-biased with a large enough voltage, it will conduct. This phenomenon is known as

(a) bias effect.

(b) avalanche effect.

(c) forward breakover.

(d) saturation.

(e) conduction effect.

48. Synchronizing pulses in a conventional analog TV signal

(a) keep the brightness constant.

(b) keep the contrast constant.

(c) keep the image from tearing or rolling.

(d) ensure that the colors are right.

(e) keep the image in good focus.

49. In an enhancement-mode MOSFET,

(a) the channel conducts fully with zero gate bias.

(b) the channel conducts partially with zero gate bias.

(c) the channel conducts ac but not dc.

(d) the channel conducts dc but not ac.

(e) the channel does not conduct with zero gate bias.

50. In a step-up power transformer,

(a) the primary voltage is more than the secondary voltage.

(b) the secondary voltage is more than the primary voltage.

(c) the primary and secondary voltages are the same.

(d) the secondary must be center-tapped.

(e) the primary must be center-tapped.

4
PART

Specialized Devices
and Systems

27
CHAPTER

Antennas

ANTENNAS CAN BE CATEGORIZED INTO TWO MAJOR CLASSES: RECEIVING TYPES AND TRANSMITTING types. Most transmitting antennas can also function effectively for reception. Some receiving antennas can efficiently transmit EM signals; others cannot.

Radiation Resistance

When RF current flows in an electrical conductor, some EM energy is radiated into space. If a resistor is substituted for the antenna, in combination with a capacitor or inductor to mimic any inherent reactance in the antenna, the transmitter behaves in the same manner as when connected to the actual antenna. For any antenna operating at a specific frequency, there exists a specific resistance, in ohms, for which this can be done. This is known as the *radiation resistance* (R_R) of the antenna at the frequency in question. Radiation resistance is specified in ohms. This phenomenon was introduced in Chap. 17. Now you'll see why it's important!

Determining Factors

Suppose a thin, straight, lossless vertical antenna is placed over perfectly conducting ground. Then R_R is a function of the vertical-antenna conductor height in wavelengths (Fig. 27-1A). Suppose a thin, straight, lossless wire is placed in free space and fed at the center. Then R_R is a function of the overall conductor length in wavelengths (Fig. 27-1B).

Antenna Efficiency

Efficiency is rarely crucial to the performance of a receiving antenna system, but it is important in any transmitting antenna system. Radiation resistance always appears in series with *loss resistance* (R_L). The antenna efficiency, *Eff*, is given by:

$$Eff = R_R/(R_R + R_L)$$

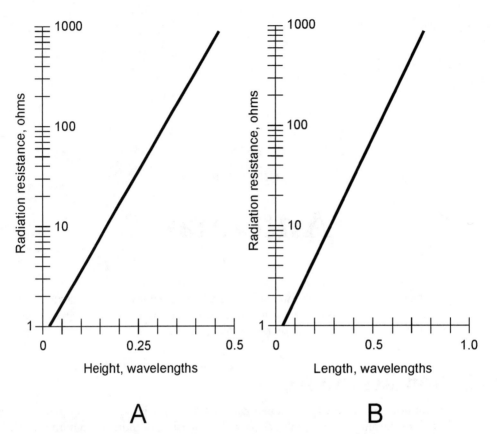

27-1   Approximate values of radiation resistance for vertical antennas over perfectly
conducting ground (A) and for center-fed antennas in free space (B).

which is the ratio of the radiation resistance to the total antenna system resistance. As a percentage,

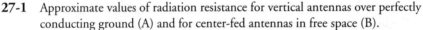

$$Eff_\% = 100 \, R_R/(R_R + R_L)$$

High efficiency in a transmitting antenna is obtained when the radiation resistance is much larger than the loss resistance. Then most of the power goes into useful EM radiation, and relatively little is dissipated as heat in the earth and in objects surrounding the antenna. When the opposite situation exists—radiation resistance comparable to, or smaller than, the loss resistance, a transmitting antenna is inefficient. This is often the case for extremely short radiators, because they always have low radiation resistance. To some extent this can be overcome by careful antenna design and location to minimize loss resistance, but even the most concerted efforts rarely reduce the loss resistance to less than a few ohms.

If an antenna system has a high loss resistance, it can nevertheless work efficiently if its radiation resistance is high enough. When an antenna radiator is just the right length at a given frequency, and if it is constructed of wire or thin tubing, its radiation resistance can be in excess of 1000 Ω. This makes it easy to construct an efficient antenna even when the loss resistance is fairly high.

Half-Wave Antennas

A half wavelength in free space is given by the equation:

$$L_{ft} = 492/f_o$$

where L_{ft} is the linear distance in feet, and f_o is the fundamental frequency, in megahertz, at which the antenna exhibits resonance. A half wavelength in meters, L_m, is given by:

$$L_m = 150/f_o$$

For ordinary wire, the results as obtained above should be multiplied by a velocity factor, v, of 0.95 (95 percent). For tubing or large-diameter wire, v can range down to about 0.90 (90 percent).

Open Dipole

An *open dipole* or *doublet* is a half-wavelength ($\lambda/2$) radiator fed at the center (Fig. 27-2A). Each leg of the antenna is a quarter wavelength ($\lambda/4$) long. For a straight wire radiator, the length L_{ft}, in feet, at a design frequency f_o, in megahertz, for a center-fed, $\lambda/2$ dipole is approximately:

$$L_{ft} = 468/f_o$$

The length in meters is approximately:

$$L_m = 143/f_o$$

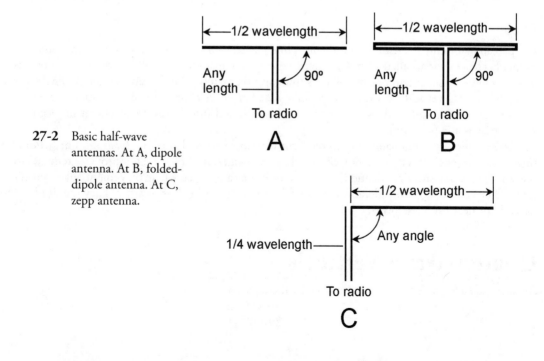

27-2 Basic half-wave antennas. At A, dipole antenna. At B, folded-dipole antenna. At C, zepp antenna.

These values assume $v = 0.95$. In free space, the impedance at the feed point is a pure resistance of approximately 73 Ω. This represents the radiation resistance alone, in the absence of reactance at the resonant frequency.

Folded Dipole

A *folded dipole antenna* is a λ/2, center-fed antenna constructed of two parallel wires with their ends connected together (Fig. 27-2B). The feed-point impedance of the folded dipole is a pure resistance of approximately 290 Ω. This makes the folded dipole ideal for use with high-impedance, parallel-wire transmission lines in applications where gain and directivity are not especially important.

Half-Wave Vertical

A radiator measuring λ/2 can be stood up, fed at the base (the bottom end) against an earth ground with a transmatch or antenna tuner designed for high impedances, and connected to a radio by a coaxial cable feed line. This type of antenna is an efficient radiator even in the presence of considerable loss resistance, because the radiation resistance is extremely high.

Zepp

A *zeppelin antenna*, also called a *zepp*, is a λ/2 radiator, fed at one end with a λ/4 section of open-wire line (Fig. 27-2C). The impedance at the feed point is an extremely high, pure resistance. At the transmitter end of the line, the impedance is a low, pure resistance. A zeppelin antenna can operate well at all harmonics of the design frequency. If an *antenna tuner*, also called a *transmatch*, is available to tune out reactance, the feed line can be of any length. Feed-line radiation can be minimized by carefully cutting the radiator to λ/2 at the fundamental frequency, and by using the antenna only at this frequency or one of its harmonics. Zepp antennas are rarely used at frequencies above 30 MHz, except when modified to form a *J pole*.

J Pole

A zepp can be oriented vertically, and the feed line placed so it lies in the same line as the radiating element. This antenna, called a J pole, radiates equally well in all horizontal directions. The J pole is used as a low-budget antenna at frequencies from approximately 10 MHz up through 300 MHz. It is, in effect, a λ/2 vertical fed with an *impedance matching section* consisting of a length of transmission line measuring λ/4. It does not require any radials, and this makes it convenient in locations where space is at a premium.

Some radio amateurs hang long J poles, cut for 3.5 MHz or 1.8 MHz, from kites or helium-filled balloons. Such antennas work well, but they are dangerous if they are not properly tethered to prevent them from breaking off and flying away, or if they are placed where they might fall on power lines. They can develop considerable electrostatic charge, even in clear weather. They are deadly if flown in or near thunderstorms.

Quarter-Wave Verticals

A *quarter wavelength* is related to frequency by the equation:

$$L_{\text{ft}} = 246v/f_{\text{o}}$$

where L_{ft} represents $\lambda/4$ in feet, f_o represents the frequency in megahertz, and v represents the velocity factor. If the length is expressed in meters as L_m, then the formula is:

$$L_m = 75v/f_o$$

For a typical wire conductor, $v = 0.95$ (95 percent); for metal tubing, v can range down to approximately 0.90 (90 percent).

A $\lambda/4$ antenna must be operated against a low-loss RF ground. The feed-point impedance over perfectly conducting ground is approximately 37 Ω. This represents radiation resistance in the absence of reactance, and provides a reasonable impedance match to most coaxial transmission lines.

Ground-Mounted Vertical

The simplest vertical antenna is a $\lambda/4$ radiator mounted at ground level. The radiator is fed with a coaxial cable. The center conductor is connected to the base of the radiator, and the shield is connected to a ground system.

Unless an extensive *ground radial* system is installed, a ground-mounted vertical is likely to be inefficient unless the surface happens to be an excellent conductor (saltwater or a salt marsh, for example). Another problem is the fact that vertically polarized antennas receive more human-made noise than horizontal antennas. In addition, the EM fields from ground-mounted transmitting antennas are more likely to cause interference to nearby electronic devices than are the EM fields from antennas installed at a height.

Ground Plane

A *ground-plane antenna* is a vertical radiator, usually measuring $\lambda/4$, operated against a system of $\lambda/4$ conductors called *radials*. The *feed point*, where the feed line joins the antenna radiator and the hub of the radial system, is elevated. When the feed point is at least $\lambda/4$ above the ground, only three or four radials are necessary to obtain a low-loss RF ground system. The radials extend outward from the base of the antenna at an angle between 0° and 45° with respect to the horizon. Figure 27-3A illustrates a typical ground-plane antenna with four horizontal radials.

A ground-plane antenna should be fed with coaxial cable. The feed-point impedance of a ground-plane antenna having a $\lambda/2$ radiator is about 37 Ω if the radials are horizontal; the impedance increases as the radials droop, reaching about 50 Ω at a *droop angle* of 45°.

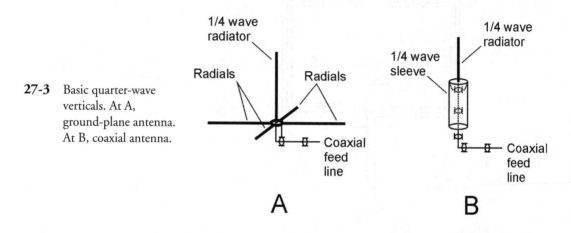

27-3 Basic quarter-wave verticals. At A, ground-plane antenna. At B, coaxial antenna.

Coaxial Antenna

The radials in a ground-plane antenna can extend straight downward, and then can be merged to form a λ/2 cylinder or sleeve that is concentric with the feed line. In this case, the feed-point impedance is approximately 73 Ω. This is known as a *coaxial antenna* (Fig. 27-3B).

Loops

Any receiving or transmitting antenna, consisting of one or more turns of wire forming a dc short circuit, comprises a *loop antenna*.

Small Loop

A *small loop antenna* has a circumference of less than 0.1λ (for each turn) and is suitable for receiving, but generally not for transmitting. It is the least responsive to signals coming from along its axis, and most responsive in the plane perpendicular to its axis. A capacitor can be connected in series or parallel with the loop to provide a resonant response. An example of such an arrangement is shown in Fig. 27-4.

Small loops are useful for *radio direction finding* (RDF), and also for reducing interference caused by human-made noise or strong local signals. The *null* along the axis is sharp and deep, and can be pointed in the direction of an offending signal or noise source.

Loopstick

For receiving applications at frequencies up to approximately 20 MHz, a *loopstick antenna* is sometimes used. This antenna, a variant of the small loop, consists of a coil wound on a solenoidal (rod-shaped), powdered-iron core. A series or parallel capacitor, in conjunction with the coil, forms a tuned circuit. A loopstick displays directional characteristics similar to those of the small loop antenna shown in Fig. 27-4. The sensitivity is maximum off the sides of the coil, and a sharp null occurs off the ends.

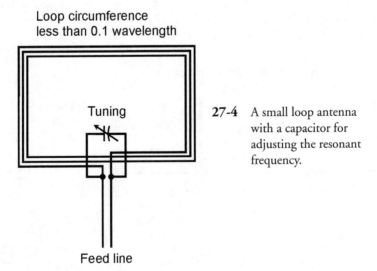

Loop circumference
less than 0.1 wavelength

Tuning

27-4   A small loop antenna with a capacitor for adjusting the resonant frequency.

Feed line

Large Loop

A *large loop antenna* usually has a circumference of either λ/2 or λ (a full wavelength), is circular or square in shape, and lies entirely in a single plane. It can work well for transmitting or receiving. The λ/2 loop presents a high impedance at the feed point, and maximum radiation/response occurs in the plane of the loop. The λ loop presents an impedance of about 100 Ω at the feed point, and the maximum radiation/response occurs along the axis (that is, perpendicular to the plane containing the loop).

The λ/2 loop exhibits a slight power loss relative to a λ/2 dipole in its favored directions. The λ loop shows a slight gain over a λ/2 dipole in its favored directions. These properties hold for loops up to several percent larger or smaller than exact λ/2 or λ circumferences (as determined for the wavelength in free space). Resonance can be obtained by means of an antenna tuner if the loop is fed with open-wire transmission line.

Sometimes, loop antennas measuring several wavelengths in circumference are strung up horizontally among multiple supports. These are technically large loops, but their gain and directional characteristics are hard to predict. If fed with open-wire line and an antenna tuner, and if placed at least λ/4 above the surface, such an antenna can be exceptionally effective.

Ground Systems

End-fed λ/4 antennas, such as the ground plane, require low-loss RF ground systems in order to perform efficiently. Center-fed λ/2 antennas, such as the dipole, do not. However, good grounding is advisable for any antenna system in order to minimize interference and electrical hazards.

Electrical versus RF Ground

Electrical grounding is important for personal safety. It can help protect equipment from damage if lightning strikes in the vicinity. It also minimizes the risk of *electromagnetic interference* (EMI) to and from radio equipment. In a three-wire electrical system, the ground prong on the plug should never be defeated, because such modification can result in dangerous voltages appearing on exposed metal surfaces.

A good RF ground system can help minimize EMI, even if it is not necessary for efficient antenna operation. Figure 27-5 shows a proper RF ground scheme (A) and an improper one (B). In a good RF ground system, each device is connected to a common *ground bus*, which in turn runs to the earth ground through a single conductor. This conductor should be as short as possible. A poor ground system contains *ground loops* that can act like loop antennas and increase the risk of EMI.

Radials and the Counterpoise

With a surface-mounted vertical antenna, there should be as many radials as possible, and they should be as long as possible. They can lie on the surface or be buried a few inches underground. The greater the number of radials of a given length, the better the antenna will work. Also, the longer the radials for a given number, the better. The radials should all converge toward, and be connected to, a ground rod at the feed point.

A *counterpoise* is a means of obtaining an RF ground or ground plane without a direct earth-ground connection. A grid of wires, a screen, or a metal sheet is placed above the surface and oriented horizontally, to provide *capacitive coupling* to the earth. This minimizes RF ground loss. Ideally, the radius of a counterpoise should be at least λ/4 at the lowest operating frequency.

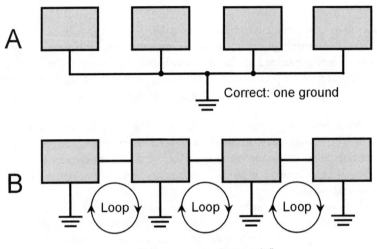

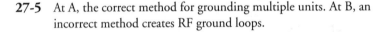

27-5 At A, the correct method for grounding multiple units. At B, an incorrect method creates RF ground loops.

Gain and Directivity

The *power gain* of a transmitting antenna is the ratio of the maximum *effective radiated power* (ERP) to the actual RF power applied at the feed point. Power gain is expressed in decibels (dB).

Suppose the ERP, in watts, for a given antenna is P_{ERP}, and the applied power, also in watts, is P. Then the following equation holds:

$$\text{Power gain (dB)} = 10 \log_{10} (P_{ERP}/P)$$

Power gain is always measured in the favored direction or directions of an antenna. These are the directions in which the antenna performs the best.

For power gain to be defined, a *reference antenna* must be chosen with a gain that is defined as 0 dB. This reference antenna is usually a $\lambda/2$ dipole in free space. Power-gain figures taken with respect to a dipole (in its favored directions) are expressed in units called dBd. The reference antenna for power-gain measurements can also be an *isotropic antenna*, which theoretically radiates and receives equally well in all directions in three dimensions. In this case, units of power gain are called dBi. For any given antenna, the power gains in dBd and dBi are different by approximately 2.15 dB:

$$\text{Power gain (dBi)} = 2.15 + \text{Power gain (dBd)}$$

Directivity Plots

Antenna radiation and response patterns are represented by plots such as those shown in Fig. 27-6. The location of the antenna is assumed to be at the center (or *origin*) of a *polar coordinate system*. The greater the radiation or reception capability of the antenna in a certain direction, the farther from the center the points on the chart are plotted.

A dipole antenna, oriented horizontally so that its conductor runs in a north-south direction, has a *horizontal plane* (or *H plane*) pattern similar to that in Fig. 27-6A. The *elevation plane* (or *E plane*) pattern depends on the height of the antenna above effective ground at the viewing angle. With the dipole oriented so that its conductor runs perpendicular to the page, and the antenna ¼ wavelength above effective ground, the E plane antenna pattern for a half-wave dipole resembles the graph shown at B.

Forward Gain

Forward gain is expressed in terms of the ERP in the *main lobe* (favored direction) of a *unidirectional* (one-directional) antenna compared with the ERP from a reference antenna, usually a half-wave dipole, in its favored directions. This gain is calculated and defined in dBd at microwave frequencies; large dish antennas can have forward gain upward of 35 dBd. In general, as the wavelength decreases (the frequency gets higher), it becomes easier to obtain high forward gain figures.

Front-to-Back Ratio

The *front-to-back* (f/b) *ratio* of a unidirectional antenna is an expression of the concentration of radiation/response in the main lobe, relative to the direction opposite the center of the main lobe. Figure 27-7 shows a hypothetical directivity plot for a unidirectional antenna pointed north. The outer circle depicts the RF *field strength* in the direction of the center of the main lobe, and represents 0 dB. The next smaller circle represents a field strength 5 dB down with respect to the main lobe. Continuing inward, circles represent 10 dB down, 15 dB down, and 20 dB down. The origin represents 25 dB down, and also shows the location of the antenna. The f/b ratio is found, in this case, by comparing the signal levels between north (azimuth 0°) and south (azimuth 180°).

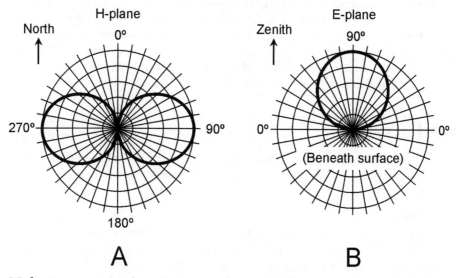

27-6   Directivity plots for a dipole antenna. At A, the H plane (horizontal plane) plot; at B, the E plane (elevation plane) plot.

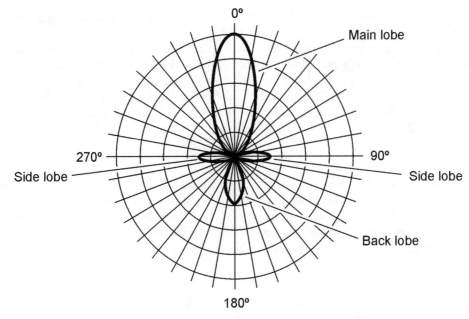

27-7   Directivity plot for a hypothetical antenna. Front-to-back and front-to-side ratios can be determined from such a graph. This plot is in the H plane.

Front-to-Side Ratio

The *front-to-side* (f/s) *ratio* is another expression of the directivity of an antenna system. The term applies to unidirectional antennas, and also to *bidirectional* antennas. The f/s ratio is expressed in decibels (dBd), just as is the f/b ratio. The EM field strength in the favored direction is compared with the field strength at right angles to the favored direction. An example is shown in Fig. 27-7. The f/s ratios are found, in this case, by comparing the signal levels between north and east (right-hand f/s), or between north and west (left-hand f/s). The right-hand and left-hand f/s ratios are usually the same in theory, although they can differ slightly in practice.

Phased Arrays

A *phased array* uses two or more *driven elements* (radiators connected directly to the feed line) to produce gain in some directions at the expense of other directions.

End-Fire Array

A typical *end-fire array* consists of two parallel λ/2 dipoles fed 90° out of phase and spaced λ/4 apart (Fig. 27-8A). This produces a unidirectional radiation pattern. Or, the two elements can be driven in phase and spaced at a separation of λ (Fig. 27-8B). This results in a bidirectional radiation pattern. In the phasing system, the branches of the transmission line must be cut to precisely the correct lengths, and the velocity factor of the line must be known and be taken into account.

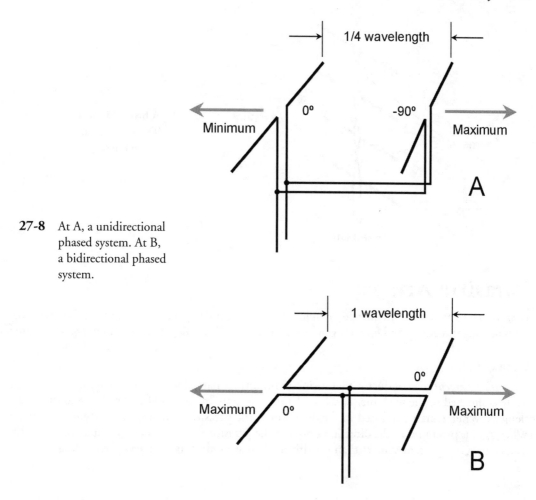

27-8 At A, a unidirectional phased system. At B, a bidirectional phased system.

Longwire

A wire antenna measuring λ or more, and fed at a high-current point or at one end, is a *longwire antenna.* A longwire antenna offers gain over a half-wave dipole. As the wire is made longer, the main lobes get more nearly in line with the antenna, and their amplitudes increase. The gain is a function of the length of the antenna; the longer the wire, the greater the gain. A longwire antenna must be as straight as possible for proper operation.

Broadside Array

Figure 27-9 shows the geometric arrangement of a *broadside array.* The driven elements can each consist of a single radiator, as shown in the figure, or they can consist of more complex antennas with directive properties. In any case, all the elements are identical. If a reflecting screen is placed behind the array of dipoles in Fig. 27-9, the system is known as a *billboard antenna.* The directional properties depend on the number of elements, whether or not the elements have gain themselves, and on the spacing among the elements. In general, the larger the number of elements, the greater the forward gain, and the greater the f/b and f/s ratios.

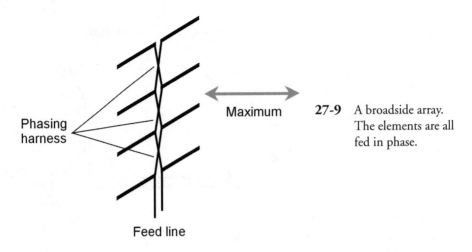

Maximum

27-9 A broadside array. The elements are all fed in phase.

Phasing harness

Feed line

Parasitic Arrays

Parasitic arrays are used at frequencies ranging from approximately 5 MHz into the microwave range for obtaining directivity and forward gain. Examples include the *Yagi antenna* and the *quad antenna*.

Concept

A *parasitic element* is a conductor that forms an important part of an antenna system, but is not directly connected to the feed line. Parasitic elements operate by means of *EM coupling* to the driven element. When gain is produced in the direction of the parasitic element, the element is a *director*. When gain is produced in the direction opposite the parasitic element, the element is a *reflector*. Directors are a few percent shorter than the driven element; reflectors are a few percent longer.

Yagi

The *Yagi antenna*, sometimes called a "beam," is an array of parallel, straight elements. A two-element Yagi is formed by placing a director or a reflector parallel to, and a specific distance away from, a single $\lambda/2$ driven element. The optimum spacing for a driven-element/director Yagi is 0.1λ to 0.2λ, with the director tuned 5 percent to 10 percent higher than the resonant frequency of the driven element. The optimum spacing for a driven-element/reflector Yagi is 0.15λ to 0.2λ, with the reflector tuned 5 percent to 10 percent lower than the resonant frequency of the driven element. The gain of a well-designed *two-element Yagi* is approximately 5 dBd.

A Yagi with one director and one reflector, along with the driven element, increases the gain and f/b ratio compared with a two-element Yagi. An optimally designed *three-element Yagi* has approximately 7 dBd gain. An example is shown in Fig. 27-10. (This drawing should not be used as an engineering blueprint.)

The gain, f/b ratio, and f/s ratio of a properly designed Yagi antenna all increase as elements are added. This is usually done by placing extra directors in front of a three-element Yagi. Each director is slightly shorter than its predecessor.

Quad

A *quad antenna* operates according to the same principles as the Yagi, except full-wavelength loops are used instead of straight half-wavelength elements.

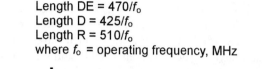

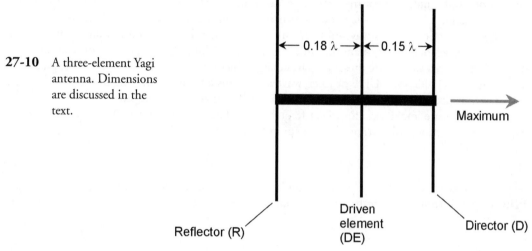

Length DE = 470/f_o
Length D = 425/f_o
Length R = 510/f_o
where f_o = operating frequency, MHz

27-10 A three-element Yagi antenna. Dimensions are discussed in the text.

← 0.18 λ → ← 0.15 λ →

Maximum

Reflector (R)

Driven element (DE)

Director (D)

A *two-element quad* can consist of a driven element and a reflector, or it can have a driven element and a director. A *three-element quad* has one driven element, one director, and one reflector. The director has a perimeter of 0.95λ to 0.97λ, the driven element has a perimeter of exactly λ, and the reflector has a perimeter of 1.03λ to 1.05λ. These dimensions, as all antenna element dimensions, are electrical dimensions (taking the velocity factor of wire or tubing into account), and not free-space dimensions.

Additional directors can be added to the basic three-element quad design to form quads having any desired numbers of elements. The gain increases as the number of elements increases. Each succeeding director is slightly shorter than its predecessor. Long quad antennas are practical at frequencies above 100 MHz. At lower frequencies, their construction tends to be unwieldy.

Antennas for Ultrahigh and Microwave Frequencies

At ultrahigh frequencies (UHF) and microwave frequencies, high-gain antennas are reasonable in size because the wavelengths are short.

Waveguides

In Chap. 25, waveguides were briefly mentioned as an option for feeding antenna systems at UHF and microwave frequencies. Here's a little more about them.

A waveguide is a hollow metal pipe having a rectangular or circular cross section. The EM field travels down the pipe, provided that the wavelength is short enough (or the cross-sectional dimensions of the pipe are large enough). In order to efficiently propagate an EM field, a *rectangular waveguide* must have height and width that both measure at least 0.5λ, and preferably more than 0.7λ. A *circular waveguide* should be at least 0.6λ in diameter, and preferably 0.7λ or more.

The characteristic impedance (Z_o) of a waveguide varies with frequency. In this sense, it differs from coaxial or parallel-wire lines, whose Z_o values are generally independent of the frequency.

A properly installed and well-maintained waveguide is an exceptional transmission line because dry air has practically no loss, even at UHF and microwave frequencies. But it is important that the interior of a waveguide be kept free from dirt, dust, insects, spiderwebs, and condensation. Even a small obstruction can seriously degrade the performance and cause the waveguide to become *lossy*.

The main limitation of a waveguide, from a practical standpoint, is inflexibility. You can't run a waveguide from one point to another in a haphazard fashion, as you can do with coaxial cable. Bends or turns in a waveguide present a particular problem, because they must be made gradually. Installing a waveguide for a UHF or microwave antenna is a little like putting in the conduit for a new electrical circuit in a home or business. It's a significant construction project!

Waveguides are impractical for use at frequencies below approximately 300 MHz, because the required cross-sectional dimensions become prohibitively large.

Dish

A *dish antenna* must be correctly shaped and precisely aligned. The most efficient shape, especially at the shortest wavelengths, is a *paraboloidal reflector*. However, a *spherical reflector*, having

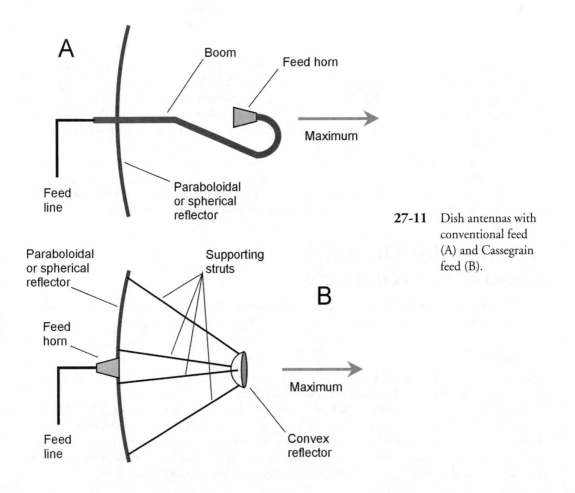

27-11 Dish antennas with conventional feed (A) and Cassegrain feed (B).

the shape of a section of a sphere, can also work well. The feed system consists of a coaxial line or waveguide from the receiver and/or transmitter, and a horn or helical driven element at the focal point of the reflector. *Conventional dish feed* is shown in Fig. 27-11A. *Cassegrain dish feed* is shown in Fig. 27-11B.

The larger the diameter of the reflector in wavelengths, the greater the gain, the f/b ratio, and the f/s ratio, and the narrower the main lobe. A dish antenna must be at least several wavelengths in diameter for proper operation. The reflecting element can be sheet metal, a screen, or a wire mesh. If a screen or mesh is used, the spacing between the wires must be a small fraction of a wavelength.

Helical

A *helical antenna* is a circularly polarized, high-gain, unidirectional antenna. A typical helical antenna is shown in Fig. 27-12. The reflector diameter should be at least 0.8λ at the lowest operating frequency. The radius of the helix should be approximately 0.17λ at the center of the intended operating frequency range. The longitudinal spacing between helix turns should be approximately 0.25λ in the center of the operating frequency range. The overall length of the helix should be at least λ at the lowest operating frequency. A helical antenna can provide about 15 dBd forward gain. Helical antennas are sometimes used in space communications systems.

Corner Reflector

A *corner reflector*, employed with a λ/2 dipole driven element, is illustrated in Fig. 27-13. This provides some gain over a λ/2 dipole by itself. The reflector is made of wire mesh, screen, or sheet metal. The *flare angle* of the reflecting element is approximately 90°. Corner reflectors are widely used in terrestrial communications at UHF and microwave frequencies. Several λ/2 dipoles can be fed in phase and placed along a common axis with a single, elongated reflector, forming a *collinear corner reflector array*.

Horn

The *horn antenna* is shaped like a squared-off trumpet or trombone horn. It provides a unidirectional radiation and response pattern, with the favored direction coincident with the opening of the horn. The feed line is a waveguide that joins the antenna at the narrowest point (throat) of the horn.

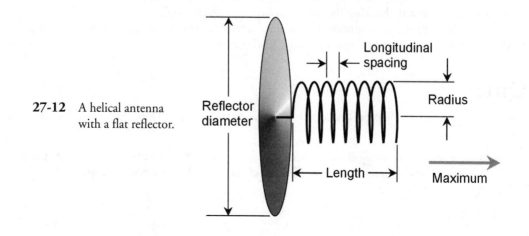

27-12 A helical antenna with a flat reflector.

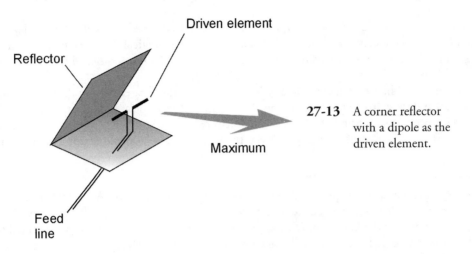

27-13 A corner reflector with a dipole as the driven element.

Horns are sometimes used all by themselves, but they are also used to feed large dish antennas. This optimizes the f/s ratio by minimizing extraneous radiation and response that occurs if a dipole is used as the driven element for the dish.

Safety

Antennas should never be placed in such a way that they can fall or blow down on power lines. Also, it should not be possible for power lines to fall or blow down on an antenna.

Wireless equipment having outdoor antennas should not be used during thundershowers, or when lightning is anywhere in the vicinity. Antenna construction and maintenance should never be undertaken when lightning is visible or thunder can be heard, even if a storm appears to be many miles away. Ideally, antennas should be disconnected from electronic equipment, and connected to a substantial earth ground, at all times when the equipment is not in use.

Tower and antenna climbing is a job for professionals. Under no circumstances should an inexperienced person attempt to climb such a structure.

Indoor transmitting antennas can expose operating personnel to EM field energy. The extent of the hazard, if any, posed by such exposure has not been firmly established. However, there is sufficient concern to warrant checking the latest publications on the topic.

For detailed information concerning antenna safety, consult a professional antenna engineer and/or a comprehensive text on antenna design and construction.

Quiz

Refer to the text in this chapter if necessary. A good score is at least 18 correct. Answers are in the back of the book.

1. Imagine a center-fed, straight wire antenna in free space, whose length can be varied at will. At which of the following lengths is the radiation resistance at the feed point the highest?

    (a) $\lambda/4$

    (b) λ

(c) λ/3

(d) λ/2

2. What is the approximate length, in meters (m), of a center-fed, λ/2 wire antenna at a frequency of 18.1 MHz? Assume the velocity factor is typical for wire.

(a) 7.90 m

(b) 8.29 m

(c) 13.6 m

(d) 25.9 m

3. What is the approximate height, in feet (ft), of a ground-mounted, λ/4 vertical antenna, made of aluminum tubing, at a frequency of 7.025 MHz? Assume the velocity factor is 92 percent.

(a) 64.43 ft

(b) 32.22 ft

(c) 19.64 ft

(d) 9.822 ft

4. In order to obtain reasonable efficiency with a λ/4 vertical antenna mounted over poorly conducting soil at a frequency of 10 MHz, it is necessary to

(a) make the antenna out of large-diameter aluminum or copper tubing.

(b) suspend the antenna from a kite or balloon.

(c) feed the antenna with a waveguide.

(d) minimize the loss resistance by installing numerous ground radials.

5. A driven element in a parasitic array is

(a) grounded for dc.

(b) a loop with a circumference of λ/2.

(c) connected to the feed line.

(d) a straight length of wire.

6. In a zepp antenna, feed-line radiation can be minimized by

(a) using the antenna only at frequencies where its length is a whole-number multiple of λ/2.

(b) using the antenna only at frequencies where its length is a whole-number multiple of λ/4.

(c) using the antenna only at frequencies where its length is an even-number multiple of λ/10.

(d) using the antenna only at frequencies where its length is an odd-number multiple of λ/4.

7. Consider a single-turn, closed loop having a circumference of 95 ft. Suppose this loop is circular, lies entirely in a single plane, is fed with open-wire transmission line and an antenna tuner, and is operated at 10 MHz. The maximum radiation and response for this loop can be expected to occur

(a) in the plane containing the loop.

(b) along a line perpendicular to the plane containing the loop.

(c) along a line coinciding with the path of the feed line.

(d) equally in all directions in three-dimensional space.

8. Suppose an antenna system has a radiation resistance of 40 Ω at 8.5 MHz. What is the efficiency of this system?

(a) 25 percent

(b) 40 percent

(c) 80 percent

(d) Impossible to determine without more information

9. Suppose an antenna system has a radiation resistance of 40 Ω and a loss resistance of 10 Ω. What is the efficiency of this system?

(a) 25 percent

(b) 40 percent

(c) 80 percent

(d) Impossible to determine without more information

10. The null in a loopstick antenna exists

(a) in the plane perpendicular to the coil axis.

(b) in the line that coincides with the coil axis.

(c) at a 45° angle with respect to the coil axis.

(d) nowhere, because a loopstick is an isotropic antenna.

11. As elements are added to a properly designed Yagi antenna,

(a) the forward gain decreases.

(b) the f/b ratio decreases.

(c) the f/s ratio increases.

(d) None of the above apply.

12. When the radiation resistance in an antenna system represents most of the total system resistance,

(a) the system can be expected to have high loss.

(b) the system can be expected to have high efficiency.

(c) the system can be expected to have a high f/s ratio.

(d) the system can be expected to have a high f/b ratio.

13. What state of affairs is optimal in a waveguide?

(a) A colony of spiders inside

(b) A cross-sectional measurement of less than λ/4

(c) Operation at frequencies below 3 MHz

(d) None of the above

14. Consider a single-turn, closed loop having a circumference of 65 ft. Suppose this loop is circular, lies entirely in a single plane, is fed with open-wire transmission line and an antenna tuner, and is operated at 7 MHz. The maximum radiation and response for this loop can be expected to occur

    (a) in the plane containing the loop.

    (b) along a line perpendicular to the plane containing the loop.

    (c) along a line coinciding with the path of the feed line.

    (d) equally in all directions in three-dimensional space.

15. Suppose the total resistance in an antenna system is 80 Ω, and the loss resistance is 20 Ω. The efficiency of the system is

    (a) 80 percent.

    (b) 75 percent.

    (c) 25 percent.

    (d) impossible to determine without more information.

16. In a parasitic array, a director is usually

    (a) tuned to a slightly higher frequency than the driven element.

    (b) connected to a system of ground radials.

    (c) a closed loop that is grounded for RF.

    (d) a straight length of wire or tubing.

17. A J pole is a modified form of

    (a) ground-plane antenna.

    (b) folded dipole antenna.

    (c) quad antenna.

    (d) zepp antenna.

18. Which of the following antenna types is well suited to use with parallel-wire transmission line?

    (a) The ground plane

    (b) The ground-mounted vertical

    (c) The folded dipole

    (d) None of the above

19. Which of the following antenna types is not designed for transmitting at 10 GHz?

    (a) A horn antenna

    (b) A dish antenna

    (c) A zepp antenna

    (d) A helical antenna

20. A vertical antenna measuring $\lambda/2$ in height does not require a ground system with low resistance because

    (a) its loss resistance is high.

    (b) its radiation resistance is low.

    (c) its main lobe is sharp.

    (d) None of the above apply.

28
CHAPTER

Integrated Circuits

INTEGRATED CIRCUITS (ICs), ALSO CALLED *CHIPS*, HAVE STIMULATED AS MUCH EVOLUTION AS ANY single development in the history of electronic technology. Most ICs look like gray or black plastic boxes with protruding pins. The schematic symbols are triangles or rectangles with lines emerging from them, labeled according to their functions.

Advantages of IC Technology

Integrated circuits have advantages over *discrete components* (individual transistors, diodes, capacitors, and resistors). Here are some of the most important considerations.

Compactness

An obvious asset of IC design is economy of space. They are far more compact than equivalent circuits made from discrete components. A corollary to this is the fact that more complex circuits can be built, and kept down to a reasonable size, using ICs as compared with discrete components.

High Speed

The interconnections among the components within an IC are physically tiny, making high switching speeds possible. Electric currents travel fast, but not instantaneously. The less time charge carriers need to get from component X to component Y, in general, the more computations are possible within a given span of time, and the less time is needed for complex operations.

Low Power Consumption

Another advantage of ICs is the fact that they use less power than equivalent discrete-component circuits. This is crucial if batteries are to be used for operation. Because ICs use so little current, they produce less heat than their discrete-component equivalents. This translates into better efficiency. It also minimizes the problems that bedevil equipment that gets hot with use, such as frequency drift and generation of internal noise.

Reliability

Integrated circuits fail less often, per component-hour of use, than systems that use discrete components. This is because all the component interconnections are sealed within the IC case, preventing corrosion or the intrusion of dust. The reduced failure rate translates into less *downtime*, or time during which the equipment is out of service for repairs.

Ease of Maintenance

Integrated-circuit technology lowers maintenance costs. Repair procedures are simplified when failures occur. Many appliances use sockets for ICs, and replacement is simply a matter of finding the faulty IC, unplugging it, and plugging in a new one. Special desoldering equipment is used with appliances having ICs soldered directly to the circuit boards.

Modular Construction

Modern IC appliances use *modular construction.* In this scheme, individual ICs perform defined functions within a circuit board; the circuit board or *card*, in turn, fits into a socket and has a specific purpose. Computers, programmed with customized software, are used by repair technicians to locate the faulty card in an appliance. The whole card can be pulled and replaced, getting the appliance back to the consumer in the shortest possible time. Then a computer can be used to troubleshoot the faulty card, getting the card ready for use in the next appliance that happens to come along with a failure in the same card.

Limitations of IC Technology

No technological advancement ever comes without a downside. Integrated circuits have some limitations that must be considered when designing an electronic system.

Inductors Impractical

While some components are easy to fabricate onto chips, other components defy the IC manufacturing process. Inductances, except for extremely low values, are an example. Devices using ICs must generally be designed to work with discrete inductors, external to the ICs themselves. However, resistance-capacitance (RC) circuits are capable of doing most things that inductance-capacitance (LC) circuits can do, and these can exist inside an IC.

High Power Impossible

The small size and low current consumption of ICs comes with an inherent limitation: high-power amplifiers cannot, in general, be fabricated onto chips. High power necessitates a certain amount of physical mass and volume. High-power devices and systems invariably generate large amounts of heat, which must be efficiently radiated or conducted away.

Linear ICs

A *linear IC* is used to process *analog signals* such as voices, music, and radio transmissions.

The term *linear* arises from the fact that, in general, the amplification factor is constant as the input amplitude varies. That is, the output signal strength is a *linear function* of the input signal strength (Fig. 28-1).

28-1 In a linear IC, the relative output is a linear (straight-line) function of the relative input. The solid lines show examples of linear IC characteristics. The dashed curves show functions not characteristic of properly operating linear ICs.

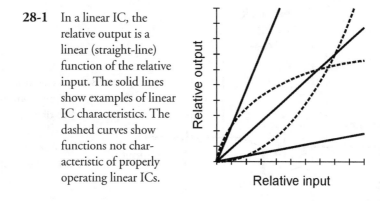

Operational Amplifier

An *operational amplifier*, or *op amp*, is a specialized linear IC that consists of several bipolar transistors, resistors, diodes, and capacitors, interconnected so that high gain is possible over a wide range of frequencies. An op amp might comprise an entire IC. Or, an IC might consist of two or more op amps. Thus, you'll sometimes hear of *dual op amps* or *quad op amps*. Some ICs have op amps in addition to other circuits.

An op amp has two inputs, one *noninverting* and one *inverting*, and one output. When a signal goes into the noninverting input, the output is in phase with the input; when a signal goes into the inverting input, the output is 180° out of phase with the input. An op amp has two power supply connections, one for the emitters of the transistors (V_{ee}) and one for the collectors (V_{cc}). The usual schematic symbol for an op amp is a triangle. The inputs, output, and power-supply connections are drawn as lines emerging from it (Fig. 28-2).

The gain of an op amp is determined by one or more external resistors. Normally, a resistor is connected between the output and the inverting input. This is called the *closed-loop configuration*. The feedback is negative (out of phase), causing the gain of the op amp to be less than it would be if there were no feedback (the *open-loop configuration*). Figure 28-3 is a schematic diagram of a noninverting closed-loop amplifier.

The reason for providing negative feedback in an op amp circuit is the fact that without it, the gain may be too great. Excessive amplifier gain can cause problems. Open-loop op amps are prone to instability, especially at low frequencies. They also generate a lot of internal noise.

When an RC combination is used in the feedback loop of an op amp, the gain depends on the input-signal frequency. It is possible to get a *lowpass response*, a *highpass response*, a *resonant peak*, or a *resonant notch* using an op amp and various RC feedback arrangements. These four types of responses are shown in Fig. 28-4.

28-2 Schematic symbol for an op amp. Connections are discussed in the text.

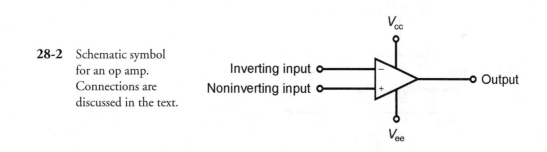

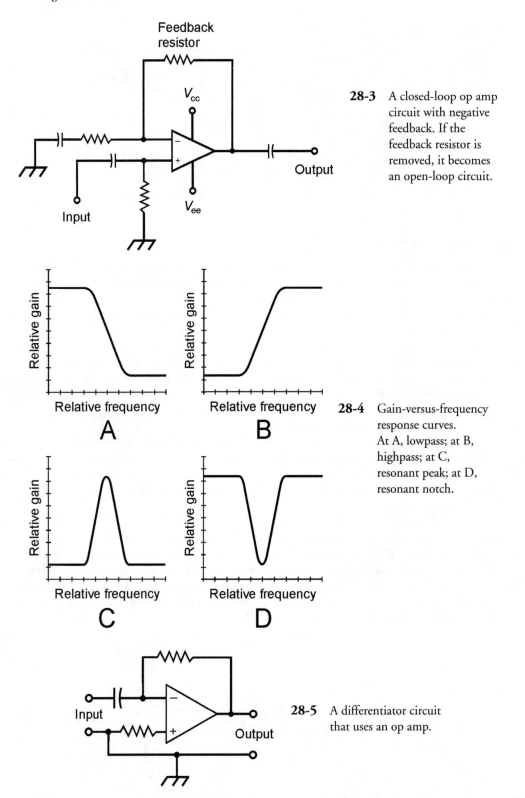

28-3   A closed-loop op amp circuit with negative feedback. If the feedback resistor is removed, it becomes an open-loop circuit.

28-4   Gain-versus-frequency response curves. At A, lowpass; at B, highpass; at C, resonant peak; at D, resonant notch.

28-5   A differentiator circuit that uses an op amp.

Op Amp Differentiator

A *differentiator* is an electronic circuit whose instantaneous output amplitude is proportional to the rate at which the input amplitude changes. The circuit mathematically *differentiates* the input signal. Op amps can be used as differentiator circuits. An example is shown in Fig. 28-5.

When the input to a differentiator is a constant dc voltage, the output is zero (no signal). When the input amplitude is increasing, the output is a positive dc voltage. When the input decreases, the output is a negative dc voltage. If the input waveform fluctuates periodically (the usual case), the output varies according to the instantaneous rate of change of the input amplitude. This results in an output signal with the same frequency as that of the input signal, although the waveform is often quite different. A pure sine wave input produces a pure sine wave at the output, but the phase is shifted 90° to the left (that is, ¼ cycle earlier in time). Complex input waveforms can produce a wide variety of outputs from a differentiator.

Op Amp Integrator

An *integrator* is an electronic circuit whose instantaneous output amplitude is proportional to the accumulated input signal amplitude as a function of time. The circuit mathematically *integrates* the input signal. The function of an integrator is basically the inverse, or opposite, of the function of a differentiator circuit. Figure 28-6 shows how an op amp can be connected to obtain an integrator.

If an integrator circuit is supplied with an input signal waveform that fluctuates periodically (the usual case), the output voltage varies according to the *integral*, or *antiderivative*, of the input voltage. This results in an output signal with the same frequency as that of the input signal, although the waveform is likely to be different. A pure sine wave input produces a pure sine wave output, but the phase is shifted 90° to the right (that is, ¼ cycle later in time). Complex input waveforms can produce many types of output waveforms.

An indefinite rise, either negatively or positively, in output voltage cannot occur in a practical integrator. If the mathematical integral (in pure theory) of an input function is an endlessly increasing output function, the actual output voltage rises to a certain maximum, either positive or negative, and stays there. This maximum is less than or equal to the power supply or battery voltage.

Voltage Regulator

A *voltage regulator IC* acts to control the output voltage of a power supply. This is important with precision electronic equipment. These ICs are available with various voltage and current ratings. Typical voltage regulator ICs have three terminals, and because of this, they are sometimes mistaken for power transistors.

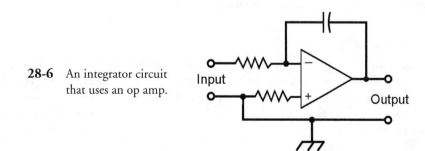

28-6 An integrator circuit that uses an op amp.

Timer

A *timer IC* is a specialized oscillator that produces a delayed output. The delay is adjustable to suit the needs of a particular device. The delay is generated by counting the number of oscillator pulses; the length of the delay can be adjusted by means of external resistors and capacitors. Timers are commonly used in circuits such as digital frequency counters, where a precise time interval or window must be provided.

Multiplexer

A *multiplexer IC* allows several different signals to be combined in a single channel by means of a process called *multiplexing*. An analog multiplexer can also be used in reverse; then it works as a *de-multiplexer*. Thus, you'll sometimes hear or read about a *multiplexer/demultiplexer IC*.

Comparator

A *comparator IC* has two inputs. It compares the voltages at the two inputs, which are called A and B. If the voltage at input A is significantly higher than the voltage at input B, the output is about +5 V. This is logic 1, or high. If the voltage at input A is lower than or equal to the voltage at input B, the output voltage is about +2 V. This is designated as logic 0, or low.

Voltage comparators are available for a variety of applications. Some can switch between low and high states at a rapid rate of speed, while others are slow. Some have low input impedance, and others have high impedance. Some are intended for AF or low-frequency RF use; others are fabricated for video or high-frequency RF applications. Voltage comparators can be used to actuate, or *trigger*, other devices such as relays and electronic switching circuits.

Digital ICs

A *digital IC*, also sometimes called a *digital-logic IC*, operates using two discrete states: high (logic 1) and low (logic 0). Digital logic is discussed in Chap. 26. Digital ICs contain massive arrays of logic gates that perform Boolean operations at high speed.

Transistor-Transistor Logic

In *transistor-transistor logic* (TTL), arrays of bipolar transistors, some with multiple emitters, operate on dc pulses. This technology has several variants, some of which date back to around 1970. A basic TTL gate is illustrated in Fig. 28-7. This gate uses two NPN bipolar transistors, one of which is a *dual-emitter* device. The transistors are always either completely cut off, or else completely saturated. Because of this, TTL is relatively immune to external noise.

Emitter-Coupled Logic

Another bipolar-transistor logic form is known as *emitter-coupled logic* (ECL). In an ECL device, the transistors are not operated at saturation, as they are with TTL. This increases the speed. But noise pulses have a greater effect in ECL, because unsaturated transistors are sensitive to, and can actually amplify, external signals and noise. The schematic of Fig. 28-8 shows a basic ECL gate using four NPN bipolar transistors.

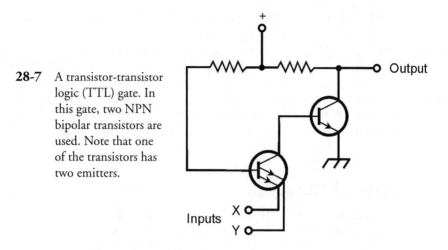

28-7 A transistor-transistor logic (TTL) gate. In this gate, two NPN bipolar transistors are used. Note that one of the transistors has two emitters.

Metal-Oxide-Semiconductor Logic

Digital ICs can also be constructed using metal-oxide-semiconductor (MOS) devices. *N-channel MOS* (NMOS) *logic*, pronounced "EN-moss logic," offers simplicity of design, along with high operating speed. *P-channel MOS logic*, pronounced "PEA-moss logic," is similar to NMOS logic, but the speed is slower. An NMOS or PMOS digital IC is like a circuit that uses only N-channel MOSFETs, or only P-channel MOSFETs, respectively.

Complementary-metal-oxide-semiconductor (CMOS) *logic*, pronounced "SEA-moss logic," employs both N-type and P-type silicon on a single chip. This is analogous to using both N-channel

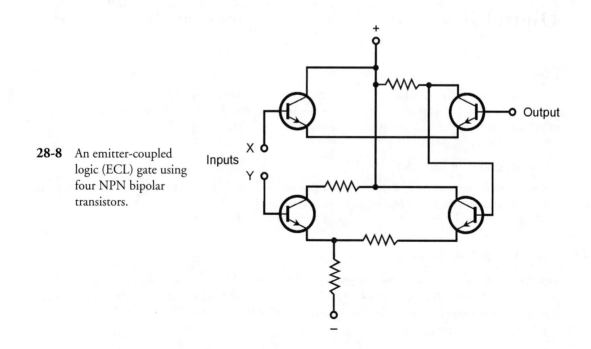

28-8 An emitter-coupled logic (ECL) gate using four NPN bipolar transistors.

and P-channel MOSFETs in a circuit. The advantages of CMOS technology include extremely low current drain, high operating speed, and immunity to noise.

All forms of MOS logic ICs require care in handling to prevent destruction by electrostatic discharges. The precautions are the same as those that are required when handling MOSFETs. All technical personnel who work with the devices should be grounded. This can be achieved by having technicians wear metal wrist straps connected to a good electrical ground, and by ensuring that the relative humidity in the lab is not allowed to get too low. When MOS ICs are stored, the pins should be pushed into special conductive foam that is manufactured for that purpose.

Component Density

The number of elements per chip in an IC is called the *component density*. There has been a steady increase in the number of components that can be fabricated on a single chip. There is an absolute limit on component density, imposed by the atomic structure of the semiconductor material— literally, the size of the atoms! Technology has begun to approach that barrier.

Small Scale

In *small scale integration* (SSI), there are fewer than 10 transistors on a chip. These types of ICs can carry the largest currents, and can be useful in voltage regulation and other moderate-power applications.

Medium Scale

In *medium scale integration* (MSI), there are 10 to 100 transistors per chip. This allows for considerable miniaturization, but it is not a high level of component density, relatively speaking. An advantage of MSI (in a few applications) is that fairly large currents can be carried by the individual gates. Both bipolar and MOS technologies can be adapted to MSI.

Large Scale

In *large scale integration* (LSI), there are 100 to 1000 transistors per semiconductor chip. This is an *order of magnitude* (a factor of 10 times) more dense than MSI. Electronic wristwatches, single-chip calculators, and simple microcomputers are examples of devices using LSI ICs.

Very-Large-Scale Devices

Very-large-scale integration (VLSI) devices have from 1000 to 1,000,000 transistors per chip. This can be up to three orders of magnitude more dense than LSI. Microcomputers and *memory* ICs are made using VLSI.

Ultra-Large-Scale Devices

You might sometimes hear of *ultra-large-scale integration* (ULSI). Devices of this kind have more than 1,000,000 transistors per chip. The principal uses for this technology are in the fields of high-level computing, supercomputing, robotics, and artificial intelligence (AI).

IC Memory

Binary digital data, in the form of high and low levels (logic ones and zeros), can be stored in *memory ICs.* These devices can take various physical forms.

Random Access

A *random access memory* (RAM) chip stores binary data in *arrays.* The data can be *addressed* (selected) from anywhere in the matrix. Data is easily changed and stored back in RAM, in whole or in part. A RAM chip is sometimes called a *read/write memory.*

An example of the use of RAM is a word-processing computer file that you are actively working on. This paragraph, this chapter, and in fact the whole text of this book was written in semiconductor RAM in small sections before being incrementally stored on the computer hard drive, and ultimately on external media.

There are two major categories of RAM: *dynamic RAM* (DRAM) and *static RAM* (SRAM). A DRAM chip contains transistors and capacitors, and data is stored as charges on the capacitors. The charge must be replenished frequently, or it will be lost through discharge. Replenishing is done automatically several hundred times per second. An SRAM chip uses a flip-flop to store the data. This gets rid of the need for constant replenishing of charge, but the tradeoff is that SRAM ICs require more elements than DRAM chips to store a given amount of data.

With any RAM chip, the data in it will vanish when power is removed, unless some provision is made for *memory backup.* The most common means of memory backup is the use of a small cell or battery with a long shelf life. Modern IC memories need so little current to store their data that a *backup battery* lasts as long in the circuit as it would on the shelf. Memory that disappears when power is removed is called *volatile memory.* If memory is retained when power is removed, it is *nonvolatile.*

Read Only

By contrast to RAM chips, the data in a *read-only memory* (ROM) chip can be easily accessed, in whole or in part, but not easily written over. A standard ROM chip is *programmed* at the factory. This permanent programming is known as *firmware.* But there are also ROM chips that you can program and reprogram yourself. The data contents of ROM chips are generally nonvolatile. That means a power failure is no cause for concern.

An *erasable programmable ROM* (EPROM) chip is an IC whose memory is of the read-only type, but that can be reprogrammed by a certain procedure. It is more difficult to rewrite data in an EPROM than in a RAM; the usual process for erasure involves exposure to ultraviolet (UV). An EPROM IC can be recognized by the presence of a transparent window with a removable cover, through which the UV is focused to erase the data. The IC must be taken from the circuit in which it is used, exposed to UV for several minutes, and then reprogrammed.

The data in some EPROM chips can be erased by electrical means. Such an IC is called an *electrically erasable programmable read-only memory* (EEPROM) chip. These do not have to be removed from the circuit for reprogramming.

Quiz

Refer to the text in this chapter if necessary. A good score is at least 18 correct. Answers are in the back of the book.

1. Because of the small size of ICs compared with equivalent circuits made from discrete components,
   - (a) more heat is generated.
   - (b) higher power output is possible.
   - (c) higher switching speeds are attainable.
   - (d) fewer calculations need to be done per unit time.

2. Which of the following is not an advantage of ICs over discrete components?
   - (a) Higher component density
   - (b) Ease of maintenance
   - (c) Lower gain
   - (d) Lower current consumption

3. In which of the following devices would you never find an IC as the main component?
   - (a) The final amplifier in a large radio broadcast transmitter
   - (b) The microprocessor in a personal computer
   - (c) A battery-powered calculator
   - (d) A low-power audio amplifier

4. Which, if any, of the following component types is not practical for direct fabrication on a chip?
   - (a) Resistors
   - (b) Capacitors
   - (c) Diodes
   - (d) All three of the above component types are practical for direct fabrication on a chip.

5. An op amp circuit usually employs negative feedback to
   - (a) maximize the gain.
   - (b) control the gain.
   - (c) allow oscillation over a wide band of frequencies.
   - (d) Forget it! Op amp circuits never have negative feedback.

6. Suppose a communications channel carries several signals at once. Which type of IC might be used to single one of the signals out for reception?
   - (a) An op amp
   - (b) A timer
   - (c) A comparator
   - (d) A multiplexer/demultiplexer

7. Which type of IC is commonly used to determine whether two dc voltages are the same or different?
   - (a) An op amp
   - (b) A timer

(c) A comparator

(d) A multiplexer/demultiplexer

8. Which type of digital IC is least susceptible to noise?

(a) Transistor-transistor logic

(b) Base-coupled logic

(c) Emitter-coupled logic

(d) N-channel-coupled logic

9. Which of the following is *not* a characteristic of CMOS chips?

(a) Sensitivity to damage by electrostatic discharge

(b) Low current demand

(c) Ability to work at high speed

(d) Ability to handle extremely high power

10. An absolute limit to the component density that can be achieved in IC technology is determined by

(a) the maximum current levels needed.

(b) the maximum attainable impedance.

(c) the size of the semiconductor atoms.

(d) Forget it! There is no limit to the component density that can be achieved in IC technology.

11. With a ROM chip,

(a) it is easier to read data from the device than to write data into it.

(b) it is more difficult to read data from the device than to write data into it.

(c) it is easy to read data from the device, and just as easy to write data into it.

(d) it is difficult to read data from the device, and equally difficult to write data into it.

12. With a RAM chip,

(a) it is easier to read data from the device than to write data into it.

(b) it is more difficult to read data from the device than to write data into it.

(c) it is easy to read data from the device, and just as easy to write data into it.

(d) it is difficult to read data from the device, and equally difficult to write data into it.

13. Which type of IC must be physically removed from the circuit to have its memory contents changed?

(a) An EEPROM chip

(b) An EPROM chip

(c) An SRAM chip

(d) A DRAM chip

14. With respect to memory, which of the following terms means that the data contents vanish if power is removed from a chip?

    (a) Volatility

    (b) Component density

    (c) Multiplexing

    (d) Corruptibility

15. In terms of the maximum number of transistors on a single chip, how many orders of magnitude larger is VLSI than MSI?

    (a) Two

    (b) Three

    (c) Four

    (d) Forget it! The maximum number of transistors on a VLSI chip is not larger, but smaller than the maximum number of transistors on an MSI chip.

16. Which of the following illustrations in Fig. 28-9 shows the input and output of a differentiator circuit?

    (a) A

    (b) B

    (c) C

    (d) None of them

17. Which of the following illustrations in Fig. 28-9 shows the input and output of an integrator circuit?

    (a) A

    (b) B

    (c) C

    (d) None of them

18. Flip-flops are used to store data in

    (a) timers.

    (b) op amps.

    (c) multiplexers.

    (d) none of the above.

19. Capacitors are used to store data in

    (a) timers.

    (b) op amps.

    (c) multiplexers.

    (d) none of the above.

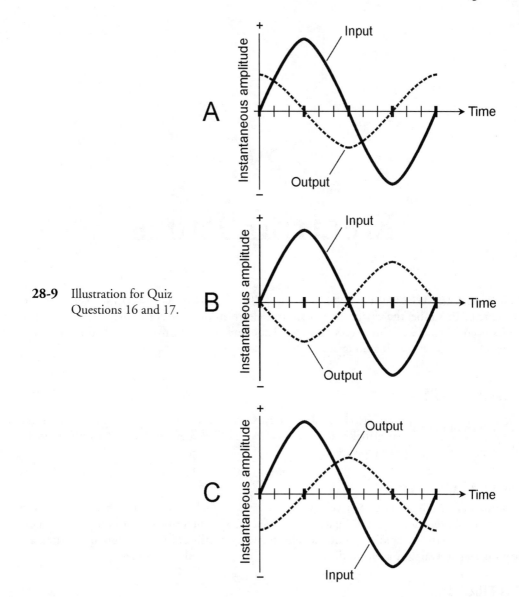

28-9 Illustration for Quiz Questions 16 and 17.

20. The phase of a sine wave signal can be shifted 180° by
    (a) applying the signal to the input of a timer.
    (b) applying the signal to the inverting input of an op amp.
    (c) applying the signal to the input of a flip-flop.
    (d) applying the signal to the input of an SRAM chip.

29

Electron Tubes

ELECTRON TUBES, ALSO CALLED *TUBES* OR *VALVES* (IN ENGLAND), ARE USED IN SOME ELECTRONIC equipment. In a tube, the *charge carriers* are free electrons that travel through space between electrodes inside the device. This makes tubes fundamentally different from semiconductor devices, in which charge carriers move among atoms in a solid medium.

Tube Forms

There are two basic types of electron tube: the *vacuum tube* and the *gas-filled tube*. As their names imply, vacuum tubes have virtually all the gases removed from their envelopes. Gas-filled tubes contain elemental vapor at low pressure.

Vacuum Tube

Vacuum tubes accelerate electrons to high speeds, resulting in large currents. This current can be focused into a beam and guided in a particular direction. The intensity and/or beam direction can be changed with extreme rapidity, producing effects such as rectification, detection, oscillation, amplification, signal mixing, waveform displays, spectral displays, and video imaging.

Gas-Filled Tube

Gas-filled tubes have a constant voltage drop, no matter what the current. This makes them useful as voltage regulators for high-voltage, high-current power supplies. Gas-filled tubes can withstand conditions that would destroy semiconductor regulating devices. Gas-filled tubes emit infrared (IR), visible light, and/or ultraviolet (UV). This property can be put to use for decorative lighting. A small *neon bulb* can be employed to construct an AF *relaxation oscillator* (Fig. 29-1).

Diode Tube

Even before the year 1900, scientists knew that electrons could carry electric current through a vacuum. They also knew that hot electrodes emit electrons more easily than cool ones. These phenom-

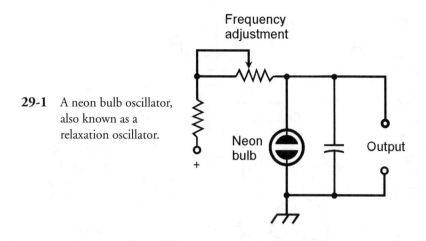

29-1   A neon bulb oscillator, also known as a relaxation oscillator.

ena were put to use in the first electron tubes, known as *diode tubes*, for the purpose of rectification. Diode tubes are rarely used nowadays, although they can still be found in some power supplies that are required to deliver several thousand volts for long periods at a 100 percent *duty cycle* (that is, continuous operation).

Electrodes in a Tube

In a tube, the electron-emitting electrode is the *cathode*. The cathode is usually heated by means of a wire *filament*, similar to the glowing element in an incandescent bulb. The heat drives electrons from the cathode. The cathode of a tube is analogous to the source of an FET, or to the emitter of a bipolar transistor. The electron-collecting electrode is the *anode*, also called the *plate*. The plate is the tube counterpart of the drain of an FET or the collector of a bipolar transistor. In most tubes, intervening *grids* control the flow of electrons from the cathode to the plate. The grids are the counterparts of the gate of an FET or the base of a bipolar transistor.

Directly Heated Cathode

In some tubes, the filament also serves as the cathode. This type of electrode is called a *directly heated cathode*. The negative supply voltage is applied directly to the filament. The filament voltage for most tubes is 6 V or 12 V dc. It is important that dc be used to heat the filament in this type of tube, because ac will tend to modulate the output. The schematic symbol for a diode tube with a directly heated cathode is shown in Fig. 29-2A.

Indirectly Heated Cathode

In many types of tubes, the filament is enclosed within a cylindrical cathode, and the cathode gets hot from the IR radiated by the filament. This is an *indirectly heated cathode*. The filament normally receives 6 V or 12 V ac or dc. In an indirectly heated cathode arrangement, ac does not cause modulation problems, as it can with a directly heated cathode tube. The schematic symbol for a diode tube with an indirectly heated cathode is shown in Fig. 29-2B.

Because the electron emission in a tube depends on the filament or *heater*, tubes need a certain amount of time to warm up before they can operate properly. This time can vary from a few seconds

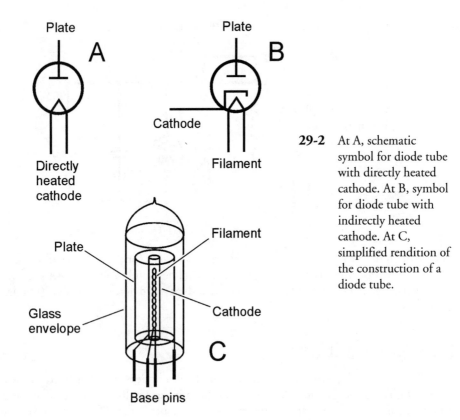

29-2 At A, schematic symbol for diode tube with directly heated cathode. At B, symbol for diode tube with indirectly heated cathode. At C, simplified rendition of the construction of a diode tube.

(for a small tube with a directly heated cathode) to a couple of minutes (for massive power-amplifier tubes with indirectly heated cathodes). The warm-up time for a large tube is about the same as the boot-up time for a personal computer.

Cold Cathode

In a gas-filled tube, the cathode does not have a filament to heat it. Such an electrode is called a *cold cathode*. Various chemical elements are used in gas-filled tubes. In fluorescent devices, neon, argon, and xenon are common. In gas-filled voltage-regulator (VR) tubes, mercury vapor is used. In a mercury-vapor VR tube, the warm-up period is the time needed for the elemental mercury, which is a liquid at room temperature, to vaporize (approximately 2 minutes).

Plate

The plate, or anode, of a tube is a cylinder concentric with the cathode and filament, as shown in Fig. 29-2C. The plate is connected to the positive dc supply voltage. Tubes operate at plate voltages ranging from 50 V to more than 3 kV. These voltages are potentially lethal. Technicians unfamiliar with vacuum tubes should not attempt to service equipment that contains them. The output of a tube-type amplifier circuit is almost always taken from the plate circuit. The plate exhibits high impedance for signal output, similar to that of an FET.

Control Grid

The flow of current can be controlled by means of an electrode between the cathode and the plate. This electrode, the *control grid* (or simply the *grid*), is a wire mesh or screen that lets electrons pass through. The grid impedes the flow of electrons if it is provided with a negative voltage relative to the cathode. The greater the negative *grid bias*, the more the grid obstructs the flow of electrons through the tube.

Triode Tube

A tube with one grid is a *triode*. The schematic symbol is shown at A in Fig. 29-3. In this case the cathode is indirectly heated; the filament is not shown. This omission is standard in schematics showing tubes with indirectly heated cathodes. When the cathode is directly heated, the filament symbol serves as the cathode symbol. The control grid is usually biased with a negative dc voltage ranging from near 0 to as much as half the positive dc plate voltage.

Tetrode Tube

A second grid can be added between the control grid and the plate. This is a spiral of wire or a coarse screen, and is called the *screen grid* or *screen*. This grid normally carries a positive dc voltage at 25 to 35 percent of the plate voltage.

The screen grid reduces the capacitance between the control grid and plate, minimizing the tendency of a tube amplifier to oscillate. The screen grid can also serve as a second control grid, allowing two signals to be injected into a tube. This tube has four elements, and is known as a *tetrode*. Its schematic symbol is shown at B in Fig. 29-3.

Pentode Tube

The electrons in a tetrode can bombard the plate with such force that some of them bounce back, or knock other electrons from the plate. This so-called secondary emission can hinder tube perfor-

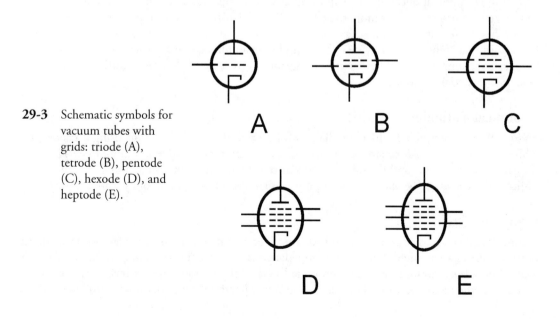

29-3 Schematic symbols for vacuum tubes with grids: triode (A), tetrode (B), pentode (C), hexode (D), and heptode (E).

mance and, at high power levels, cause screen current so high that the electrode is destroyed. This problem can be eliminated by placing another grid, called the *suppressor grid* or *suppressor*, between the screen and the plate. The suppressor repels *secondary electrons* emanating from the plate, preventing most of them from reaching the screen. The suppressor also reduces the capacitance between the control grid and the plate more than a screen grid by itself.

Greater gain and stability are possible with a *pentode*, or tube with five elements, than with a tetrode or triode. The schematic symbol for a pentode is shown at C in Fig. 29-3. The suppressor is often connected directly to the cathode.

Hexode and Heptode Tubes

In some older radio and TV receivers, tubes with four or five grids were sometimes used. These tubes had six and seven elements, respectively, and were called *hexode* and *heptode*. The usual function of such tubes was signal mixing. The schematic symbol for a hexode is shown at D in Fig. 29-3; the symbol for a heptode is at E. You will not encounter hexodes and heptodes in modern electronics, because solid-state components are used for signal mixing.

Interelectrode Capacitance

In a vacuum tube, the cathode, grid(s), and plate exhibit *interelectrode capacitance* that is the primary limiting factor on the frequency range in which the device can produce gain. The interelectrode capacitance in a typical tube is a few picofarads. This is negligible at low frequencies, but at frequencies above approximately 30 MHz, it becomes a significant consideration. Vacuum tubes intended for use as RF amplifiers are designed to minimize this capacitance.

Circuit Configurations

The most common application of vacuum tubes is in amplifiers, especially in radio and television transmitters at power levels of more than 1 kW. Some high-fidelity audio systems also employ vacuum tubes. In recent years, tubes have gained favor with some popular music bands. Some musicians insist that "tube amps" provide richer sound than amplifiers using power transistors. There are two basic vacuum-tube amplifier circuit arrangements: the *grounded-cathode* configuration and the *grounded-grid* configuration.

Grounded Cathode

Figure 29-4 is a simplified schematic diagram of a grounded-cathode circuit using a triode tube. This circuit is the basis for many tube-type RF power amplifiers and audio amplifiers. The input impedance is moderate, and the output impedance is high. Impedance matching between the amplifier and the load can be obtained by tapping a coil in the output circuit, or by using a transformer.

Grounded Grid

Figure 29-5 shows a basic grounded-grid RF amplifier circuit. The input impedance is low, and the output impedance is high. The output impedance is matched by the same means as with the grounded-cathode arrangement. The grounded-grid configuration requires more driving (input) power than the grounded-cathode scheme. A grounded-cathode amplifier might produce 1 kW of

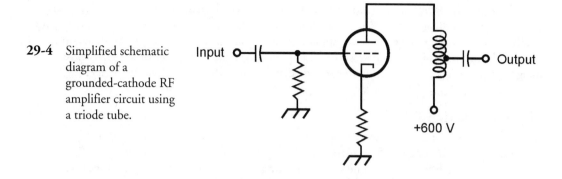

29-4 Simplified schematic diagram of a grounded-cathode RF amplifier circuit using a triode tube.

RF output for 10-W input, but a grounded-grid amplifier needs 50 W to 100 W of drive to produce 1 kW of RF output. A grounded-grid amplifier has a significant advantage, however: it is less likely to break into unwanted oscillation than a grounded-cathode circuit.

Plate Voltage

The plate voltages (+600 V dc) in the circuits of Figs. 29-4 and 29-5 are given as examples. The amplifiers shown could produce 75- to 150-W signal output provided they receive sufficient drive and are properly biased. An amplifier rated at 1-kW output would require a plate voltage of +2 kV dc to +5 kV dc. In high-power radio and TV broadcast transmitters producing in excess of 50-kW RF output, even higher dc plate voltages are used.

Cathode-Ray Tubes

Many TV receivers, and some desktop computer monitors, use *cathode-ray tubes* (CRTs). So do older oscilloscopes, spectrum analyzers, and radar sets.

Electron Beam

In a CRT, a specialized cathode called an *electron gun* emits an electron beam that is focused and accelerated as it passes through positively charged *anodes*. The beam then strikes a glass screen whose inner surface is coated with *phosphor*. The phosphor glows visibly, as seen from the face of the CRT, because of the effect of the high-speed electrons striking it.

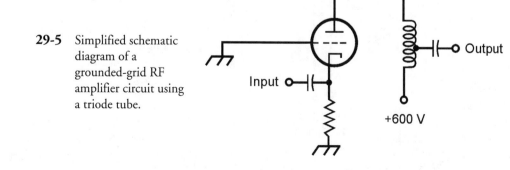

29-5 Simplified schematic diagram of a grounded-grid RF amplifier circuit using a triode tube.

The beam *scanning* pattern is controlled by magnetic or electrostatic fields. One field causes the beam to scan rapidly across the screen in a horizontal direction. Another field moves the beam vertically. When complex waveforms are applied to the electrodes that produce the deflection of the electron beam, a display pattern results. This pattern can be the graph of a signal wave, a fixed image, an animated image, a computer text display, or any other type of visible image.

Electromagnetic CRT

A simplified cross-sectional drawing of an *electromagnetic CRT* is shown in Fig. 29-6. There are two sets of *deflecting coils*, one for the horizontal plane and the other for the vertical plane. (To keep the illustration reasonably clear, only one set of deflecting coils is shown.) The greater the current in the coils, the greater the intensity of the magnetic field, and the more the electron beam is deflected. The electron beam is bent at right angles to the magnetic lines of flux.

In an *oscilloscope*, the horizontal deflecting coils receive a sawtooth waveform. This causes the beam to scan, or *sweep*, at a precise, adjustable speed across the screen from left to right as viewed from in front. After each timed left-to-right sweep, the beam returns, almost instantly, to the left side of the screen for the next sweep. The vertical deflecting coils receive the waveform to be analyzed. This waveform makes the electron beam move up and down. The combination of vertical and horizontal beam motion produces a display of the input waveform as a function of time.

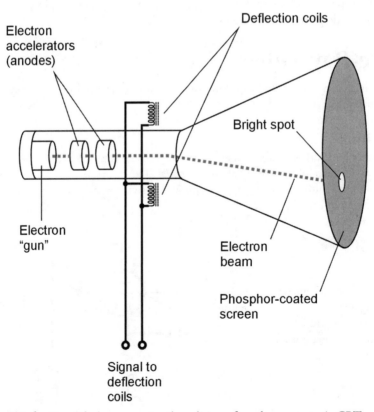

29-6   Simplified cross-sectional rendition of an electromagnetic CRT.

Electrostatic CRT

In an *electrostatic CRT*, charged metal plates, rather than current-carrying coils, are used to deflect the electron beam. When voltages appear on these *deflecting plates*, the beam is bent in the direction of the electric lines of flux. The greater the voltage applied to a deflecting plate, the stronger the electric field, and the greater the extent to which the beam is deflected.

The principal advantage of an electrostatic CRT is the fact that it generates a far less intense magnetic field than an electromagnetic CRT. This so-called extremely low frequency (ELF) energy is a cause for concern, because it might have adverse effects on people who use CRT-equipped devices, such as desktop computers, for extended periods of time. In recent years, with the evolution of *liquid crystal displays* (LCDs) and *plasma displays* as alternatives to the CRT type of display, ELF has become a much less significant concern.

Camera Tubes

Some video cameras use a form of electron tube that converts visible light into varying electric currents. The two most common types of *camera tube* are the *vidicon* and the *image orthicon.*

Vidicon

In the vidicon, a lens focuses the incoming image onto a photoconductive screen. An electron gun generates a beam that sweeps across the screen as a result of the effects of deflecting coils, in a manner similar to the operation of an electromagnetic CRT. The sweep in the vidicon is synchronized with any CRT that displays the image.

As the electron beam scans the photoconductive surface, the screen becomes charged. The rate of discharge in a certain region on the screen depends on the intensity of the visible light striking that region. A simplified cutaway view of a vidicon tube is shown in Fig. 29-7.

The main advantage of the vidicon is its small physical size and mass. A vidicon is sensitive, but its response can be sluggish when the level of illumination is low. This causes images to persist for a short while, resulting in poor portrayal of fast-motion scenes.

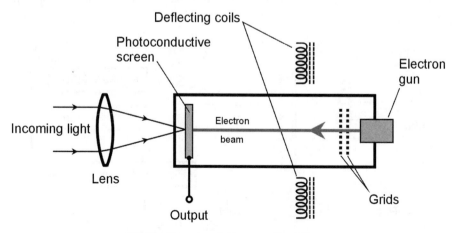

29-7   Functional diagram of a vidicon.

Image Orthicon

Another type of camera tube, also quite sensitive but having a quicker response to image changes, is the image orthicon. It is constructed much like the vidicon, except that there is a target electrode behind the *photocathode* (Fig. 29-8). When a single electron from the photocathode strikes the target electrode, multiple secondary electrons are emitted as a result. The image orthicon thus acts as a video signal amplifier, as well as a camera.

A fine beam of electrons, emitted from the electron gun, scans the target electrode. The secondary electrons cause some of this beam to be reflected back toward the electron gun. Areas of the target electrode with the most secondary electron emission produce the greatest return beam intensity, and regions with the least emission produce the lowest return beam intensity. The greatest return beam intensity corresponds to the brightest parts of the video image. The return beam is modulated as it scans the target electrode and is picked up by a receptor electrode.

One significant disadvantage of the image orthicon is that it produces considerable noise in addition to the signal output. But when a fast response is needed and the illumination ranges from dim to very bright, the image orthicon is the camera tube of choice.

Photomultiplier

A *photomultiplier* is a vacuum tube that generates a variable current depending on the intensity of the light that strikes it. It multiplies its own output, thereby obtaining high sensitivity. Photomultipliers can be used to measure light intensity at low levels.

The photomultiplier consists of a photocathode, which emits electrons in proportion to the intensity of the light striking it. These electrons are focused into a beam, and this beam strikes an electrode called a *dynode*. The dynode emits several secondary electrons for each electron that strikes it. The resulting beam is collected by the anode.

A photomultiplier can have several dynodes, resulting in high gain. The extent to which the sensitivity can be increased by cascading dynodes is limited by the amount of background electron emission or *dark noise* from the photocathode.

Dissector

A *dissector*, also known as an *image dissector*, is a form of photomultiplier in which the light is focused by a lens onto a translucent photocathode. This surface emits electrons in proportion to the

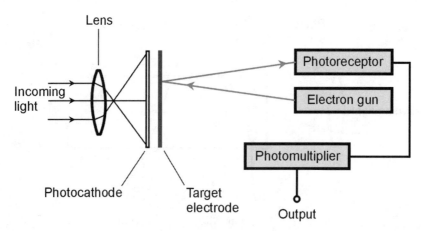

29-8   Functional diagram of an image orthicon.

light intensity. The electrons from the photocathode are directed to a barrier containing a small aperture. The vertical and horizontal deflection plates, supplied with synchronized scanning voltages, move the beam from the photocathode across the aperture. The electron stream passing through the aperture is modulated depending on the light and dark nature of the image.

The *image resolution* of the dissector tube depends on the size of the aperture. The smaller the aperture, the sharper the image, down to a certain limiting point. However, there is a limit to how small the aperture can be, while still allowing enough electrons to pass, and avoiding the generation of *interference patterns*. The image dissector tube produces very little dark noise, and this allows for excellent sensitivity.

Tubes for Use above 300 MHz

Specialized vacuum tubes are required for RF operation at frequencies above 300 MHz. These bands are known as *ultrahigh frequency* (UHF) *band*, which ranges from 300 MHz to 3 GHz, and the *microwave band*, which ranges from 3 GHz up. The *magnetron* and the *Klystron* are examples of tubes that are used to generate and amplify signals at these frequencies.

Magnetron

A magnetron contains a cathode and a surrounding anode. The anode is divided into sections, or *cavities*, by radial barriers. The output is taken from an opening in the anode, and passes into a waveguide that serves as a transmission line for the RF output energy.

The cathode is connected to the negative terminal of a high-voltage source, and the anode is connected to the positive terminal. Therefore, electrons flow radially outward. A magnetic field is applied lengthwise through the cavities. As a result, the electron paths are bent into spirals. The electric field produced by the high voltage, interacting with the longitudinal magnetic field and the effects of the cavities, causes the electrons to bunch up into *clouds*. The swirling movement of the electron clouds causes a fluctuating current in the anode. The frequency depends on the shapes and sizes of the cavities. Small cavities result in the highest oscillation frequencies; larger cavities produce oscillation at relatively lower frequencies.

A magnetron can generate more than 1 kW of RF power at a frequency of 1 GHz. As the frequency increases, the realizable power output decreases. At 10 GHz, a typical magnetron generates about 20 W of RF power output.

Klystron

A Klystron has an electron gun, one or more cavities, and a device that modulates the electron beam. There are several different types. The most common are the *multicavity Klystron* and the *reflex Klystron.*

In a multicavity Klystron, the electron beam is *velocity-modulated* in the first cavity. This causes the density of electrons (the number of particles per unit volume) in the beam to change as the beam moves through subsequent cavities. The electrons tend to bunch up in some regions and spread out in other regions. The intermediate cavities increase the magnitude of the electron beam modulation, resulting in amplification. Output is taken from the last cavity. Peak power levels in some multicavity Klystrons can exceed 1 MW (10^6 W), although the average power is much less.

A reflex Klystron has a single cavity. A *retarding field* causes the electron beam to periodically reverse direction. This produces a phase reversal that allows large amounts of energy to be drawn from the electrons. A typical reflex Klystron can produce signals on the order of a few watts at frequencies of 300 MHz and above.

Quiz

Refer to the text in this chapter if necessary. A good score is at least 18 correct. Answers are in the back of the book.

1. A major difference between a triode tube and an N-channel FET is the fact that
    (a) triodes work with lower signal voltages.
    (b) triodes are more compact.
    (c) triodes need higher power-supply voltages.
    (d) triodes don't need filaments.

2. The control grid of a vacuum tube is the electrical counterpart of the
    (a) source of a MOSFET.
    (b) collector of a bipolar transistor.
    (c) anode of a diode.
    (d) gate of an FET.

3. In a tetrode tube, the charge carriers are
    (a) free electrons, which pass among the electrodes.
    (b) holes, which are conducted within the electrodes.
    (c) holes or electrons, depending on whether the device is P type or N type.
    (d) nuclei of whatever elemental gas happens to exist in the tube.

4. Which factor is most significant in limiting the maximum frequency at which a tube can operate?
    (a) The power-supply voltage
    (b) The capacitance among the electrodes
    (c) The physical size of the tube
    (d) The current passing through the filament

5. In a tube with a directly heated cathode,
    (a) the filament is separate from the cathode.
    (b) the grid is connected to the filament.
    (c) the filament serves as the cathode.
    (d) there is no filament.

6. In a tube with a cold cathode,
    (a) the filament is separate from the cathode.
    (b) the grid is connected to the filament.
    (c) the filament serves as the cathode.
    (d) there is no filament.

7. A screen grid enhances tube operation by
    (a) increasing the gain, helping the circuit to oscillate more easily.
    (b) decreasing the plate voltage required to produce oscillation.
    (c) minimizing the risk that a tube amplifier will break into oscillation.
    (d) pulling excess electrons from the plate.

8. A tube with three grids is called a
    (a) triode.
    (b) tetrode.
    (c) pentode.
    (d) hexode.

9. A tube with four grids is called a
    (a) triode.
    (b) tetrode.
    (c) pentode.
    (d) hexode.

10. An advantage of a grounded-grid RF power amplifier over a grounded-cathode RF power amplifier is the fact that the grounded-grid circuit
    (a) has excellent sensitivity.
    (b) exhibits high input impedance.
    (c) produces little or no noise in the input.
    (d) is more stable.

11. A heptode tube has
    (a) one plate.
    (b) two plates.
    (c) three plates.
    (d) four plates.

12. The electron gun in a CRT is another name for its
    (a) cathode.
    (b) anode.
    (c) control grid.
    (d) screen grid.

13. The electron beam in an electrostatic CRT is bent by
    (a) magnetic fields produced by current-carrying coils.
    (b) electric fields produced by charged electrodes.
    (c) a variable voltage on the screen grid.
    (d) visible light striking the electrodes.

14. A grounded-grid RF power amplifier
    - (a) requires more driving power, for a given RF power output, than a grounded-cathode RF power amplifier.
    - (b) requires less driving power, for a given RF power output, than a grounded-cathode RF power amplifier.
    - (c) oscillates at a more stable frequency than a grounded-cathode RF power amplifier.
    - (d) oscillates at a less stable frequency than a grounded-cathode RF power amplifier.

15. In a Klystron, the electron-beam density varies as a result of
    - (a) amplitude modulation.
    - (b) pulse modulation.
    - (c) velocity modulation.
    - (d) frequency modulation.

16. A vidicon camera tube is noted for its
    - (a) poor signal-to-noise ratio.
    - (b) large size and heavy weight.
    - (c) slow response to image movement in dim light.
    - (d) excellent selectivity and electrical ruggedness.

17. The plate in a tetrode tube is normally connected to
    - (a) a positive dc power-supply voltage.
    - (b) a negative dc power-supply voltage.
    - (c) electrical ground.
    - (d) RF ground.

18. The screen grid in a tetrode tube is normally connected to
    - (a) a positive dc power-supply voltage.
    - (b) a negative dc power-supply voltage.
    - (c) electrical ground.
    - (d) RF ground.

19. Which of the following is most suitable for measuring the intensity of dim light?
    - (a) A triode gas-filled tube
    - (b) A photomultiplier tube
    - (c) An electrostatic CRT
    - (d) An electromagnetic CRT

20. In a dissector tube, the aperture size is directly related to the
    - (a) operating voltage.
    - (b) signal-to-noise ratio.
    - (c) response speed.
    - (d) image resolution.

30
CHAPTER

Transducers, Sensors, Location, and Navigation

IN THIS CHAPTER, YOU'LL LEARN ABOUT ELECTRONIC DEVICES THAT CONVERT ENERGY FROM ONE form to another, devices that can detect phenomena and measure their intensity, systems that can help you find out where you are (or where some other object is), and devices that facilitate navigation for vessels such as ships, aircraft, and robots.

Wave Transducers

In electronics, *wave transducers* convert ac or dc into acoustic or electromagnetic (EM) waves. They can also convert these waves into ac or dc signals.

Dynamic Transducer for Sound

A *dynamic transducer* is a coil-and-magnet device that translates mechanical vibration into varying electrical current, and can also do the reverse. The most common examples are the *dynamic microphone* and the *dynamic speaker*.

Figure 30-1 shows a functional diagram of a dynamic transducer. A *diaphragm* is attached to a coil that is mounted so it can move back and forth rapidly along its axis. A permanent magnet is placed inside the coil. Sound waves cause the diaphragm to move; this moves the coil, which causes fluctuations in the magnetic field within the coil. The result is ac output from the coil, having the same waveform as the sound waves that strike the diaphragm.

If an audio signal is applied to the coil, it generates a magnetic field that produces forces on the coil. This causes the coil to move, pushing the diaphragm back and forth, creating acoustic waves in the surrounding medium.

Electrostatic Transducer for Sound

An *electrostatic transducer* takes advantage of the forces produced by electric fields. Two metal plates, one flexible and the other rigid, are placed parallel to each other and close together (Fig. 30-2).

In an *electrostatic pickup*, incoming sound waves vibrate the flexible plate. This produces small, rapid changes in the spacing, and therefore the capacitance, between the two plates. A dc voltage is

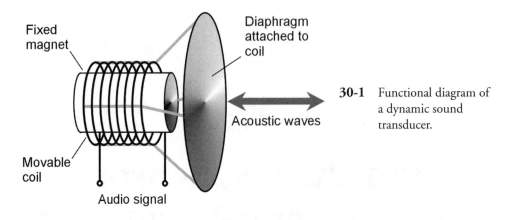

30-1   Functional diagram of a dynamic sound transducer.

applied between the plates. As the interplate capacitance varies, the electric field intensity between them fluctuates. This produces variations in the current through the primary winding of the transformer. Audio signals appear across the secondary.

In an *electrostatic emitter*, fluctuating currents in the transformer produce changes in the voltage between the plates. This results in electrostatic field variations, pulling and pushing the flexible plate in and out. The motion of the flexible plate produces sound waves.

Electrostatic transducers can be used in most applications where dynamic transducers are employed. Advantages of electrostatic transducers include light weight and good sensitivity. The relative absence of magnetic fields can also be an asset in certain situations.

Piezoelectric Transducer for Sound and Ultrasound

Figure 30-3 shows a *piezoelectric transducer*. This device consists of a *crystal* of quartz or ceramic material, sandwiched between two metal plates. When sound waves strike one or both of the plates, the metal vibrates. This vibration is transferred to the crystal. The crystal generates weak electric currents when subjected to this mechanical stress. Therefore, an ac voltage develops between the two metal plates, with a waveform similar to that of the sound.

If an ac signal is applied to the plates, it causes the crystal to vibrate in sync with the current. The metal plates vibrate also, producing an acoustic disturbance.

Piezoelectric transducers can function at higher frequencies than can dynamic or electrostatic transducers. For this reason, they are favored in ultrasonic applications, such as intrusion detectors and alarms.

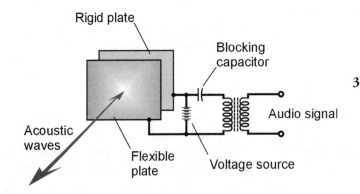

30-2   Functional diagram of an electrostatic sound transducer.

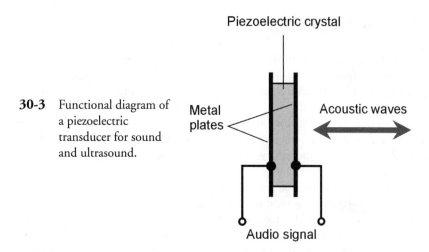

30-3   Functional diagram of a piezoelectric transducer for sound and ultrasound.

Transducers for RF Energy

The term *radio-frequency* (RF) *transducer* is a fancy name for an *antenna*. There are two basic types: the *receiving antenna* and the *transmitting antenna*. You learned about antennas in Chap. 27.

Transducers for IR and Visible Light

Many wireless devices transmit and receive energy at IR wavelengths. Infrared energy has a frequency higher than that of radio waves, but lower than that of visible light. Some wireless devices transmit and receive their signals in the visible range, although these are encountered much less often than IR devices.

The most common IR transmitting transducer is the infrared-emitting diode (IRED). Fluctuating dc is applied to the device, causing it to emit IR rays. The fluctuations in the current constitute modulation, and this produces rapid variations in the intensity of the rays emitted by the semiconductor P-N junction. The modulation contains information, such as which channel your television set should seek, or whether the volume is to be raised or lowered. Infrared energy can be focused by optical lenses and reflected by optical mirrors. This makes it possible to *collimate* IR rays (make them parallel) so they can be transmitted for distances up to several hundred meters.

Infrared receiving transducers resemble photodiodes or photovoltaic cells. The fluctuating IR energy from the transmitter strikes the P-N junction of the receiving diode. If the receiving device is a photodiode, a current is applied to it, and this current varies rapidly in accordance with the signal waveform on the IR beam from the transmitter. If the receiving device is a photovoltaic cell, it produces the fluctuating current all by itself, without the need for an external power supply. In either case, the current fluctuations are weak, and must be amplified before they are delivered to whatever equipment (television set, garage door, oven, security system, etc.) is controlled by the wireless system. Infrared wireless devices work best on a line of sight.

Displacement Transducers

A *displacement transducer* measures a distance or angle traversed, or the distance or angle separating two points. Conversely, it can convert a signal into movement over a certain distance or angle. A device that measures or produces movement in a straight line is a *linear displacement transducer*. If it measures or produces movement through an angle, it is an *angular displacement transducer*.

Pointing and Control Devices

A *joystick* is a control device capable of producing movement, or controlling variable quantities, in two dimensions. The device consists of a movable lever and a ball bearing within a control box. The lever can be moved by hand up and down, and to the right and left. Joysticks are used in computer games, for entering coordinates into a computer, and for the remote control of robots. In some joysticks, the lever can be rotated, allowing control in a third dimension.

A *mouse* is a peripheral commonly used with personal computers. By sliding the mouse around on a flat surface, a cursor or arrow is positioned on the display. Pushbutton switches on the top of the unit actuate the computer to perform whatever function the cursor or arrow shows. These actions are called *clicks*.

A *trackball* resembles an inverted mouse, or a two-dimensional joystick without the lever. Instead of the device being pushed around on a surface, the user moves a ball bearing, causing the display cursor to move vertically and horizontally. Pushbutton switches on a computer keyboard, or on the trackball box itself, actuate the functions.

An *eraser-head pointer* is a rubber button approximately 5 mm in diameter, usually placed in the center of a computer keyboard. The user moves the cursor on the display by pushing against the button. Clicking and double clicking are done with button switches on the keyboard.

A *touch pad* is a sensitive plate that is about the size of a business card. The user places an index finger on the plate and moves the finger around. This results in intuitive movement of the display cursor. Clicking and double clicking are done in the same way as with the trackball and eraser-head pointer.

Electric Motor

An *electric motor* converts electrical energy into angular (and in some cases linear) mechanical energy. Motors can operate from ac or dc, and range in size from tiny devices used in microscopic robots to huge machines that pull passenger trains.

The basics of dc motors were discussed in Chap. 8. In a motor designed to work with ac, there is no commutator. The alternations in the current keep the polarity correct at all times, so the shaft does not lock up. The rotational speed of an ac motor depends on the frequency of the applied ac. With 60-Hz ac, for example, the rotational speed is 60 revolutions per second (60 rps) or 3600 revolutions per minute (3600 rpm).

When a motor is connected to a load, the rotational force required to turn the shaft increases. The greater the required force becomes, the more power is drawn from the source.

Stepper Motor

A *stepper motor* turns in small increments, rather than continuously. The *step angle*, or extent of each turn, varies depending on the particular motor. It can range from less than 1° of arc to a quarter of a circle (90°). A stepper motor turns through its step angle and then stops, even if the current is maintained. When a stepper motor is stopped with a current going through its coils, the shaft resists external rotational force.

Conventional motors run at hundreds, or even thousands, of revolutions per minute (rpm). But a stepper motor usually runs at much lower speeds, almost always less than 180 rpm. A stepper motor has the most turning power when it is running at its slowest speeds, and the least turning power when it runs at its highest speeds.

When a pulsed current is supplied to a stepper motor, the shaft rotates in increments, one step for each pulse. In this way, a precise speed can be maintained. Because of the braking effect, this speed is constant for a wide range of mechanical turning resistances. Stepper motors can be controlled using microcomputers. This type of motor is especially well suited for point-to-point motion. Complicated, intricate tasks can done by computer-controlled robots using stepper motors.

Selsyn

A *selsyn* is an indicating device that shows the direction in which an object is oriented. It consists of a transmitting unit and a receiving (or indicator) unit. As the shaft of the transmitting unit rotates, the shaft of the receiving unit, which is a stepper motor, follows along exactly. A common application is as a direction indicator for a wind vane (Fig. 30-4). When the wind vane rotates, the indicator unit shaft moves through the same number of angular degrees as the transmitting unit shaft. A selsyn for azimuth (compass) bearings has a range of 0° to 360°. A selsyn for elevation bearings has range of 0° to 90°.

A *synchro* is a selsyn used for the control of mechanical devices. Synchros are well suited for robotic *teleoperation*, or remote control. Some synchros are programmable. The operator inputs a number into the generator, and the receiver changes position accordingly. Synchros are commonly used as rotators and direction indicators for directional communications antennas such as the Yagi or dish.

Electric Generator

An *electric generator* is constructed in much the same way as an ac motor, although it functions in the opposite sense. Some generators can also be used as motors; devices of this type are called *motor/generators*.

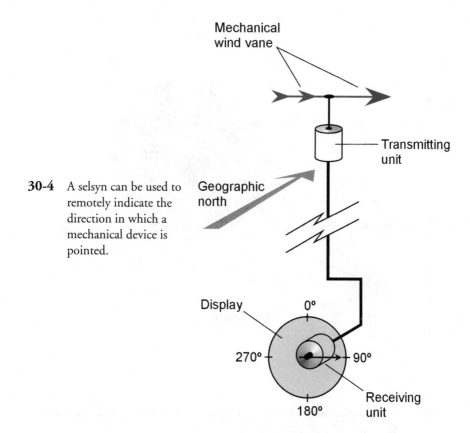

30-4   A selsyn can be used to remotely indicate the direction in which a mechanical device is pointed.

A typical generator produces ac from the mechanical rotation of a coil in a strong magnetic field. Alternatively, a permanent magnet can be rotated within a coil of wire. The rotating shaft can be driven by a gasoline engine, a steam turbine, a water turbine, a windmill, or even by human power. A commutator can be used with a generator to produce pulsating dc output, which can be filtered to obtain pure dc for use with electronic equipment.

Small portable gasoline-powered generators, capable of delivering a few kilowatts, can be purchased in department stores or home-and-garden stores. Larger generators, which usually burn propane or methane ("natural gas"), allow homes or buildings to keep their electrical power in the event of an interruption in the utility. The largest generators are found in power plants, and can produce many kilowatts.

Small generators can be used in synchro systems. These specialized generators allow remote control of robotic devices. A generator can be used to measure the speed at which a vehicle or rolling robot moves. The shaft of the generator is connected to one of the wheels, and the generator output voltage and frequency vary directly with the angular speed of the wheel. This is an example of a *tachometer*, a device familiar to people with automotive experience.

Optical Encoder

In digital radios, frequency adjustment is done in discrete steps. A typical increment is 10 Hz for shortwave radios and 200 kHz for FM broadcast radios. An alternative to mechanical switches or

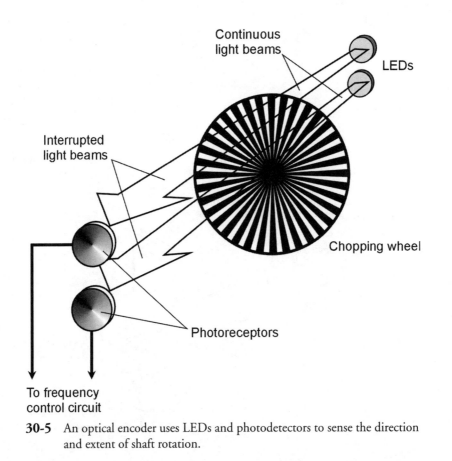

30-5 An optical encoder uses LEDs and photodetectors to sense the direction and extent of shaft rotation.

gear-driven devices, which wear out with time, is the *optical encoder*, also called the *optical shaft encoder*.

An optical encoder consists of two LEDs, two photodetectors, and a device called a *chopping wheel*. The LEDs shine on the photodetectors through the wheel. The wheel has radial bands, alternately transparent and opaque (Fig. 30-5). The wheel is attached to the tuning shaft, which is attached to a large knob. As the tuning knob is rotated, the light beams are interrupted. Each interruption causes the frequency to change by a specified increment. The difference between "frequency up" and "frequency down" (clockwise and counterclockwise shaft rotation, respectively) is determined according to which photodetector senses each sequential beam interruption first.

Detection and Measurement

A *sensor* employs one or more transducers to detect and/or measure phenomena such as temperature, humidity, barometric pressure, texture, proximity, and the presence of certain substances.

Capacitive Pressure Sensor

A *capacitive pressure sensor* is shown in Fig. 30-6. Two metal plates are separated by a layer of dielectric (electrically insulating) foam, forming a capacitor. This component is connected in parallel with an inductor. The resulting inductance/capacitance (*LC*) circuit determines the frequency of an oscillator. If an object strikes the sensor, the plate spacing momentarily decreases. This increases the capacitance, causing a drop in the oscillator frequency. When the object moves away from the transducer, the foam springs back, the plates return to their original spacing, and the oscillator frequency returns to normal.

The output of a capacitive pressure sensor can be converted to digital data using an *analog-to-digital converter* (ADC). This signal can be sent to a microcomputer such as a robot controller. Pressure sensors can be mounted in various places on a mobile robot, such as the front, back, and sides. Then, for example, physical pressure on the sensor in the front of the robot can send a signal to the controller, which tells the machine to move backward.

A capacitive pressure sensor can be fooled by massive conducting or semiconducting objects in its vicinity. If such a mass comes near the transducer, the capacitance may change even if direct contact is not made. This phenomenon is known as *body capacitance*. When the effect must be avoided, an *elastomer* device can be used for pressure sensing.

30-6 A capacitive pressure sensor. When force is applied, the spacing between the plates decreases, causing the capacitance to increase and the oscillator frequency to go down.

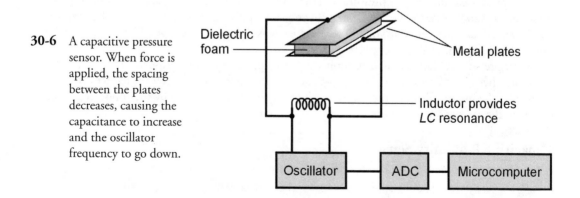

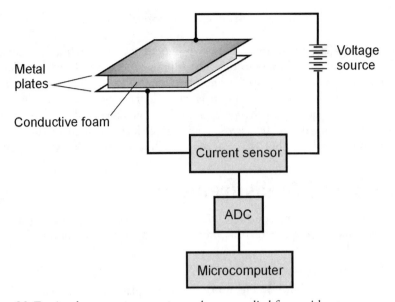

30-7 An elastomer pressure sensor detects applied force without unwanted capacitive effects.

Elastomer

An *elastomer* is a flexible substance resembling rubber or plastic that can be used to detect the presence or absence of mechanical pressure. Figure 30-7 illustrates how an elastomer can be used to detect, and locate, a pressure point. The elastomer conducts electricity fairly well, but not perfectly. It has a foam-like consistency, so that it can be compressed. Conductive plates are attached to the pad.

When pressure appears at some point in the elastomer pad, the material is compressed, and this lowers its electrical resistance. This is detected as an increase in the current between the plates. The greater the pressure becomes, the more the elastomer is compressed, and the greater is the increase in the current. The current-change data can be sent to a microcomputer such as a robot controller.

Back-Pressure Sensor

A motor produces a measurable pressure that depends on the torque being applied. A back-pressure sensor detects and measures the torque that the motor is applying at any given time. The sensor produces a signal, usually a variable voltage, that increases as the torque increases. Figure 30-8 is a functional block diagram of a back-pressure sensor.

Back-pressure sensors are used to limit the force applied by robot grippers, arms, drills, hammers, or other *end effectors*. The *back voltage*, or signal produced by the sensor, reduces the torque applied by the motor. This can prevent damage to objects being handled by the robot. It also helps to ensure the safety of people working around the robot.

Capacitive Proximity Sensor

A *capacitive proximity sensor* uses an RF oscillator, a frequency detector, and a metal plate connected into the oscillator circuit (Fig. 30-9). The oscillator is designed so that a change in the capacitance

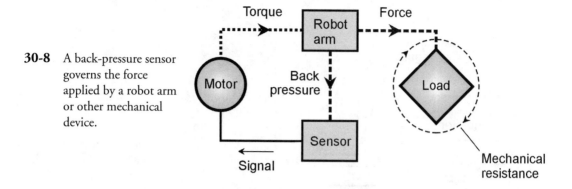

30-8 A back-pressure sensor governs the force applied by a robot arm or other mechanical device.

of the plate, with respect to the environment, causes the oscillator frequency to change. This change is sensed by the frequency detector, which sends a signal to a microcomputer or robot controller.

Substances that conduct electricity to some extent, such as metal, saltwater, and living tissue, are sensed more easily by capacitive transducers than are materials that do not conduct, such as dry wood, plastic, glass, or dry fabric.

Photoelectric Proximity Sensor

Reflected light can provide a way for a mobile robot to tell if it is approaching something. A *photoelectric proximity sensor* uses a light-beam generator, a photodetector, a frequency-sensitive amplifier, and a microcomputer (Fig. 30-10).

The light beam reflects from the object and is picked up by the photodetector. The light beam is modulated at a certain frequency, say 1000 Hz. The amplifier responds only to light modulated at that frequency. This prevents false imaging that can otherwise be caused by lamps or sunlight. (Such light sources are unmodulated, and will not actuate a sensor designed to respond only to modulated light.) If the robot is approaching an object, its controller senses that the reflected, modulated beam is getting stronger. The robot can then steer clear of the object.

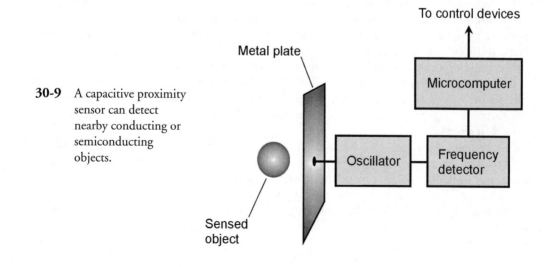

30-9 A capacitive proximity sensor can detect nearby conducting or semiconducting objects.

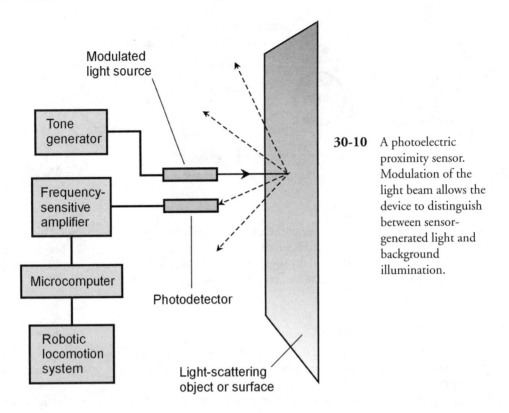

30-10 A photoelectric proximity sensor. Modulation of the light beam allows the device to distinguish between sensor-generated light and background illumination.

This method of proximity sensing does not work for objects that do not reflect light, or for windows or mirrors approached at a sharp angle. In these scenarios, the light beam is not reflected back toward the photodetector, so the object is invisible.

Texture Sensor

Texture sensing is the ability of a machine to determine whether a surface is shiny or rough (matte). Basic texture sensing involves the use of a laser and several light-sensitive sensors.

Figure 30-11 shows how a laser (L) and sensors (S) can be used to tell the difference between a shiny surface (at A) and a rough or matte surface (at B). The shiny surface, such as the polished hood of a car, reflects light at the incidence angle only. But the matte surface, such as a sheet of paper, scatters light in all directions. The shiny surface reflects the beam back entirely to the sensor in the path of the beam whose reflection angle equals its incidence angle. The matte surface reflects the beam back to all the sensors. A microcomputer can be programmed to tell the difference.

Certain types of surfaces can confuse the texture sensor shown in Fig. 30-11. For example, a pile of glass marbles can be defined as shiny on a small scale but irregular on a large scale. Depending on the diameter of the laser beam, the texture sensor might interpret such a surface as either shiny or matte. The determination can also be affected by motion of the sensor relative to the surface. A surface that is interpreted as shiny when standing still relative to the sensor might be interpreted as matte when moving relative to the sensor.

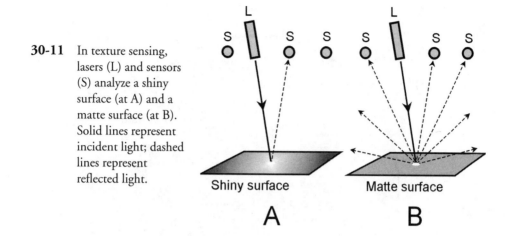

30-11 In texture sensing, lasers (L) and sensors (S) analyze a shiny surface (at A) and a matte surface (at B). Solid lines represent incident light; dashed lines represent reflected light.

Location Systems

Transducers and sensors can operate over long distances. The above-described devices are intended mainly for short-range applications (with the exception of RF antennas). In this section, a few medium-range and long-range applications of transducers and sensors are described. These applications fall into the broad category of *location systems.*

Radar

The term *radar* is derived from the words *radio detection and ranging.* Electromagnetic (EM) waves having certain frequencies reflect from various objects, especially if those objects contain metals or other good electrical conductors. By ascertaining the direction(s) from which radio signals are returned, and by measuring the time it takes for a pulsed beam of EM energy to travel from the transmitter location to a target and back again, it is possible to pinpoint the geographic positions of distant objects. During the Second World War in the 1940s, this property of radio waves was put to use for the purpose of locating aircraft.

In the years following the war, it was discovered that radar can be useful in a variety of applications, such as measurement of automobile speed (by the police), weather forecasting (rain and snow reflect radar signals), and even the mapping of the moon and the planet Venus. Radar is extensively used in aviation, both commercial and military. In recent years, radar has also found uses in robot guidance systems.

A complete radar set consists of a transmitter, a directional antenna with a narrow main lobe and high gain, a receiver, and an indicator or display. The transmitter produces intense pulses of radio microwaves at short intervals. The pulses are propagated outward in a narrow beam from the antenna, and they strike objects at various distances. The reflected signals, or *echoes,* are picked up by the antenna shortly after the pulse is transmitted. The farther away the reflecting object, or target, the longer the time before the echo is received. The transmitting antenna is rotated so that all azimuth bearings (compass directions) can be observed.

A typical circular radar display consists of a CRT or LCD. Figure 30-12 shows the basic display configuration. The observing station is at the center of the display. Azimuth bearings are indicated

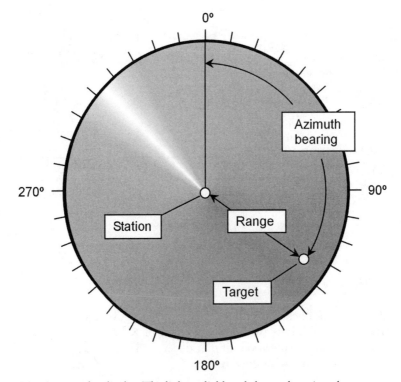

30-12   A radar display. The light radial band shows the azimuth
direction in which the microwave beam is currently transmitted
and received. (Not all radar displays show this band.)

in degrees clockwise from true north, and are marked around the perimeter of the screen. The distance, or *range*, is indicated by the radial displacement of the echo; the farther away the target, the farther from the display center the echo or blip. The radar display is, therefore, a set of polar coordinates. In the drawing, a target is shown at an azimuth of about 125° (east-southeast). Its range is near the maximum for the display.

The maximum range of a radar system depends on the height of the antenna above the ground, the nature of the terrain in the area, the transmitter output power and antenna gain, the receiver sensitivity, and the weather conditions in the vicinity. Airborne long-range radar can detect echoes from several hundred kilometers (km) away under ideal conditions. A low-power radar system, with the antenna at a low height above the ground, might receive echoes from only 50 to 70 km.

The fact that precipitation reflects radar echoes is a nuisance to aviation personnel, but it is invaluable for weather forecasting and observation. Radar has made it possible to detect and track severe thunderstorms and hurricanes. A *mesocyclone*, which is a severe thunderstorm likely to produce tornadoes, causes a hook-shaped echo on radar. The eye of a hurricane, and the eyewall and rainbands surrounding it, all show up clearly on a radar display.

Some radar sets can detect changes in the frequency of the returned pulse, thereby allowing measurement of wind speeds in hurricanes and tornadoes. This is called *Doppler radar*. This type of radar is also employed to measure the speeds of approaching or receding targets.

Sonar

Sonar is a medium-range method of proximity sensing. The acronym derives from the words *sonic navigation and ranging*. The basic principle is simple: Bounce acoustic waves off of objects, and measure the time it takes for the echoes to return.

An elementary sonar system consists of an ac pulse generator, an acoustic emitter, an acoustic pickup, a receiver, a delay timer, and an indicating device such as a numeric display, CRT, LCD, or pen recorder. The transmitter sends out acoustic waves through the medium, usually water or air. These waves are reflected by objects, and the echoes are picked up by the receiver. The distance to an object is determined on the basis of the echo delay, provided the speed of the acoustic waves in the medium is known.

A simple sonar system is diagrammed in Fig. 30-13A. A *computer map* can be generated on the basis of sounds returned from various directions in two or three dimensions. This can help a mobile robot or vessel navigate in its environment. However, the system can be fooled if the echo delay is equal to or longer than the time interval between pulses, as shown at B. To overcome this, a computer can instruct the generator to send pulses of various frequencies in a defined sequence. The computer keeps track of which echo corresponds to which pulse.

Acoustic waves travel faster in water than in air. The amount of salt in water makes a difference in the propagation speed when sonar is used on boats, for example, in depth finding. The density of water can vary because of temperature differences as well. If the true speed of the acoustic waves is not accurately known, false readings will result. In freshwater, the speed of sound is about 1400 meters per second (m/s), or 4600 feet per second (ft/s). In saltwater, it is about 1500 m/s (4900 ft/s). In air, sound travels at approximately 335 m/s (1100 ft/s).

In the atmosphere, sonar can make use of audible sound waves, but ultrasound is often used instead. Ultrasound has a frequency too high to hear, ranging from about 20 kHz to more than 100 kHz. One advantage of ultrasound is that the signals will not be heard by people working around machines equipped with sonar. Another advantage is the fact that it is less likely to be fooled by people talking, machinery operating, and other noises. At frequencies higher than the range of human hearing, acoustical disturbances do not normally occur as often, or with as much intensity, as they do within the hearing range.

In its most advanced forms, sonar can rival *vision systems* (also called *machine vision*) as a means of mapping the environment for a mobile robot or vessel. Sonar has one significant limitation: all acoustic waves, including sound and ultrasound, require a gaseous or liquid medium in order to propagate. Therefore, sonar is useless in outer space, which is practically a vacuum.

Signal Comparison

A machine or vessel can find its geographical position by comparing the signals from two fixed stations whose positions are known, as shown in Fig. 30-14A. By adding 180° to the bearings of the sources X and Y, the machine or vessel (small square block) obtains its bearings as seen from the sources (round dots). The machine or vessel can determine its direction and speed by taking two readings separated by a certain amount of time.

In the old days, diagrams such as the one in Fig. 30-14A were actually plotted on maps by the captains of oceangoing vessels and aircraft. Nowadays, computers do that work, with more accurate results.

Figure 30-14B is a block diagram of an *acoustic direction finder* such as can be used by a mobile robot. The receiver has a signal-strength indicator and a servo that turns a directional ultrasonic

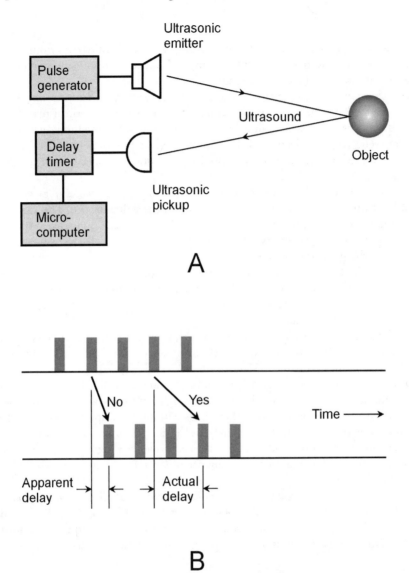

A

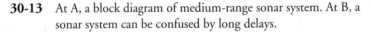

B

30-13   At A, a block diagram of medium-range sonar system. At B, a
sonar system can be confused by long delays.

transducer. There are two signal sources at different frequencies. When the transducer is rotated so
the signal from one source is maximum, a bearing is obtained by comparing the orientation of the
transducer with some known standard such as a magnetic compass. The same is done for the other
source. A computer determines the precise location of the robot, based on this data.

Radio Direction Finding (RDF)

A radio receiver, equipped with a signal-strength indicator and connected to a rotatable, directional
antenna, can be used to determine the direction from which signals are coming. *Radio direction find-*

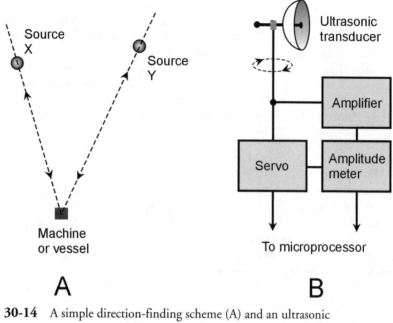

30-14 A simple direction-finding scheme (A) and an ultrasonic direction finder (B).

ing (RDF) equipment aboard a mobile vehicle facilitates determining the location of a transmitter. An RDF receiver can also be used to find one's own position with respect to two or more transmitters operating on different frequencies.

In an RDF receiver for use at frequencies below about 300 MHz, a small loop or loopstick antenna is used. It is shielded against the electric component of radio waves, so it picks up only the magnetic part of the EM field. The loop is rotated until a sharp dip, or null, occurs in the received signal strength, indicating that the axis of the loop lies along a line toward the transmitter. When readings are taken from two or more locations separated by a sufficient distance, the transmitter can be pinpointed by finding the intersection point of the azimuth bearing lines on a map.

At frequencies above approximately 300 MHz, a directional transmitting and receiving antenna, such as a Yagi, quad, dish, or helical type, gives better results than a small loop. When such an antenna is employed for RDF, the azimuth bearing is indicated by a signal peak rather than by a null.

Navigational Methods

Navigation involves the use of location devices over a period of time, thereby deriving a function of position versus time. This technique can be used to determine whether or not a vessel is on course. It can also be used to track the paths of military targets, severe thunderstorms, and hurricanes.

Fluxgate Magnetometer

When conventional position sensors do not function in a particular environment for a mobile robot, a *fluxgate magnetometer* can be used. This system employs sensitive magnetic receptors and a microcomputer to sense the presence of, and detect changes in, an artificially generated magnetic field. Navigation within a room can be done by checking the orientation of magnetic lines of flux generated by electromagnets in the walls, floor, and ceiling of the room. For each point in the room, the magnetic flux lines have a unique direction and intensity. There is a one-to-one correspondence between the magnetic flux intensity/direction and the points within the robot's operating environment. This correspondence can be represented as a two-variable mathematical function of every location in the room. The robot controller is programmed to know this function. This makes it possible for the machine to pinpoint its position with extreme accuracy, in some cases to within a few millimeters.

Epipolar Navigation

Epipolar navigation works by evaluating the way an image changes as viewed from a moving perspective. Suppose that you are piloting an aircraft over the ocean. The only land in sight is a small island. The on-board computer sees an image of the island that constantly changes shape. Figure 30-15 shows three sample sighting positions (A, B, C) and the size/shape of the island as seen by a machine vision system in each case. The computer has the map data, so it knows the true size, shape, and location of the island. The computer compares the shape and size of the image it sees at each point in time, from the vantage point of the aircraft, with the actual shape and size of the island from the map data. From this, the computer can ascertain the altitude of the aircraft, its speed and direction of movement relative to the surface, its latitude, and its longitude.

Log Polar Navigation

In *log polar navigation*, a computer converts an image in *polar coordinates* to an image in *rectangular coordinates*. The polar radius is mapped onto the vertical rectangular axis, and the polar angle is mapped onto the horizontal rectangular axis.

Radial coordinates are unevenly spaced in the polar map, but are uniform in the rectangular map. During the transformation, the logarithm of the radius is taken to maximize the geographical area that the system can observe. As a result of this logarithmic transformation (that's where the "log" comes from in the term), long-distance resolution is sacrificed, but the close-in resolution is enhanced.

The so-called log polar transform distorts the way a scene appears to human observers, but translates images and motions into data that can be efficiently dealt with by a computerized scanning system.

Loran

The term *loran* is an acronym derived from the words *long range navigation*. Loran is one of the oldest electronic navigation schemes, and is still used by some ships and aircraft. The system employs RF pulse transmission at low frequencies from multiple transmitters at specific geographic locations.

In loran, a computer on board a vessel can determine the location of the ship by comparing the time difference in the arrival of the signals from two different transmitters at known locations. Based on the fact that radio waves propagate at the speed of light in free space (approximately 299,792 km/s or 186,282 mi/s), it is possible to determine the distance to each transmitter, and from this, the location of the ship relative to the transmitters.

In recent years, loran has been largely supplanted by the *Global Positioning System* (GPS).

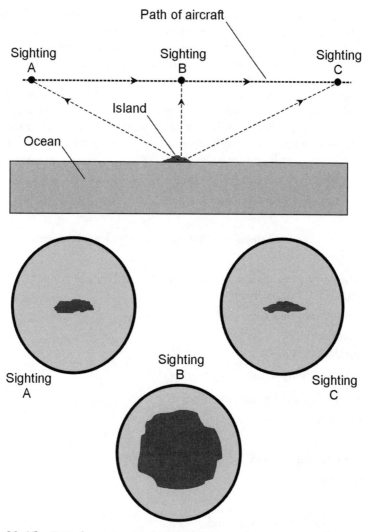

30-15 Epipolar navigation is the optoelectronic counterpart of human spatial perception.

Global Positioning System (GPS)

The Global Positioning System (GPS) is a network of radiolocation/radionavigation apparatus that operates on a worldwide basis. The system employs several satellites, and allows determination of latitude, longitude, and altitude.

All GPS satellites transmit signals in the microwave part of the radio spectrum. The signals are modulated with codes that contain timing information used by the receiving apparatus to make measurements. A GPS receiver determines its location by measuring the distances to several different satellites. This is done by precisely timing the signals as they travel between the satellites and the receiver. The process is similar to triangulation, except that it is done in three dimensions (in space) rather than in two dimensions (on the surface of the earth).

There is some reduction in the propagation speed of EM microwaves in the ionosphere, as compared with the propagation speed in free space. The extent of this reduction depends on the signal frequency. The GPS employs dual-frequency transmission to compensate for this effect. A GPS receiver uses a computer to process the information received from the satellites. From this information, it can give the user an indication of position down to a few meters.

Increasingly, automobiles, trucks, and pleasure boats come with GPS receivers installed as standard equipment. If you are driving in a remote area and you get lost, you can use GPS to locate your position. Using a cell phone, citizens band (CB) radio transceiver, or amateur (ham) radio transceiver, you can call for help and inform authorities of your exact position, which is displayed on a detailed map of the area in which you are located.

Quiz

Refer to the text in this chapter if necessary. A good score is at least 18 correct. Answers are in the back of the book.

1. Which of the following devices would most likely be used for adjusting the frequency setting of a digital radio transmitter or receiver?
    (a) A piezoelectric transducer
    (b) A dynamic transducer
    (c) An optical shaft encoder
    (d) A capacitive proximity sensor

2. Collimation of IR rays can be done by means of
    (a) an ordinary lens.
    (b) a selsyn.
    (c) a log polar transform system.
    (d) any of the above.

3. A computer map
    (a) can help a robot find its way around.
    (b) requires the use of IR beacons.
    (c) is generated using optical encoders.
    (d) requires the use of a low-frequency RDF loop antenna.

4. The distance of a target from a radar station is called
    (a) the resolution.
    (b) the azimuth.
    (c) the range.
    (d) the transform.

5. Which of the following devices is best for use as an ultrasonic pickup?
    (a) A back-pressure transducer
    (b) A piezoelectric transducer

    (c) An elastomer pressure transducer

    (d) A capacitive pressure transducer

6. What do loran and the GPS have in common?

    (a) They both operate using IR energy.

    (b) They both employ loop antennas at the transmitting and receiving stations.

    (c) They both involve measurement of azimuth angles to determine the location of a transmitter.

    (d) They both involve measurement of distances to determine the location of a receiver.

7. A bright spot close to, and directly to the left of, the center of a conventional radar display indicates the presence of a target

    (a) right over the observing station.

    (b) at close range, west of the observing station.

    (c) at close range, south of the observing station.

    (d) far from the observing station, azimuth unknown.

8. Which of the following systems enables a vessel to navigate entirely from its own frame of reference, without the need for any electronic devices external to itself?

    (a) An epipolar navigation system

    (b) The GPS

    (c) Loran

    (d) Forget it! There is no such thing.

9. Ultrasonic waves travel through a vacuum

    (a) at the same speed as they travel in air.

    (b) at a slightly higher speed than they travel in air.

    (c) at a slightly lower speed than they travel in air.

    (d) Forget it! No form of acoustic wave can travel through a vacuum.

10. Which of the following devices or techniques makes use of artificially generated magnetic fields to facilitate location within a work environment?

    (a) A dynamic transducer

    (b) Epipolar navigation

    (c) Capacitive proximity sensing

    (d) None of the above

11. Which of the following is not an example of a transducer?

    (a) A microphone

    (b) A radio antenna

    (c) A headset

    (d) All of the above are examples of transducers.

12. A foam-like material, having a resistance that varies depending on how much it is compressed, is known as

    (a) an elastomer.

    (b) a fluxgate magnetometer.

    (c) a piezoelectric substance.

    (d) none of the above.

13. A stepper motor

    (a) has torque that increases as its speed increases.

    (b) resists applied torque when stopped with current going through its coils.

    (c) has turning power that does not depend on the speed.

    (d) is easily fooled by body capacitance.

14. A permanent magnet is a key component of

    (a) an elastomer pressure sensor.

    (b) a capacitive pressure sensor.

    (c) a dynamic microphone.

    (d) a piezoelectric microphone.

15. An electric motor is an example of

    (a) an electromechanical transducer.

    (b) a pressure sensor.

    (c) a proximity sensor.

    (d) a texture sensor.

16. A capacitive proximity sensor should not be expected to detect the presence of

    (a) a wooden chair.

    (b) a metal desk.

    (c) a large dog.

    (d) a steel post.

17. A well-designed photoelectric proximity sensor

    (a) is easily fooled by loud noises.

    (b) cannot detect the presence of objects that reflect light.

    (c) is sensitive to magnetic fields.

    (d) is not easily fooled by stray sources of light.

18. An acoustic transducer can translate an ac electrical signal into

    (a) visible light.

    (b) sound or ultrasound.

    (c) a dc electrical signal.

    (d) mechanical torque.

19. Body capacitance can be a problem with

    (a) electrostatic transducers.

    (b) elastomer pressure sensors.

    (c) capacitive pressure sensors.

    (d) all of the above.

20. Which of the following devices can be used to limit the torque or force applied by an electromechanical device such as a robot?

    (a) A back-pressure sensor

    (b) A capacitive proximity sensor

    (c) A dynamic transducer

    (d) An optical shaft encoder

31
CHAPTER

Acoustics, Audio, and High Fidelity

IN SOUND RECORDING AND REPRODUCTION, ESPECIALLY WITH MUSIC, FIDELITY SUPERSEDES ALL other considerations. In these applications, low distortion and "sound esthetics" are more important than amplifier efficiency or gain. In this chapter, we'll look at basic acoustics principles, and outline some *high-fidelity* (hi-fi) components and design concepts.

Acoustics

Acoustics is the science of sound waves. Sound consists of molecular vibrations at audio frequency (AF), ranging from about 20 Hz to 20 kHz. Young people can hear the full range of AF; older people lose hearing sensitivity at the upper and lower extremes.

Audio Frequencies

In music, the AF range is divided into three broad, vaguely defined parts, called *bass* (pronounced "base"), *midrange*, and *treble*. The bass frequencies start at 20 Hz and extend to 150 or 200 Hz. Midrange begins at this point, and extends up to 2 or 3 kHz. Treble consists of the audio frequencies higher than midrange. As the frequency increases, the wavelength becomes shorter.

In air, sound travels at about 1100 feet per second (ft/s), or 335 meters per second (m/s). The relationship between the frequency f of a sound wave in hertz, and the wavelength λ_{ft} in feet, is as follows:

$$\lambda_{ft} = 1100/f$$

The relationship between f in hertz and λ_m in meters is given by:

$$\lambda_m = 335/f$$

This formula is also valid for frequencies in kilohertz and wavelengths in millimeters.

A sound disturbance in air at 20 Hz has a wavelength of 55 ft (17 m). A sound of 1.0 kHz produces a wave measuring 1.1 ft (34 cm). At 20 kHz, a sound wave in air is only 0.055 ft (17 mm) long. In substances other than air at sea level, such as air at extreme altitudes, freshwater, saltwater, or metals, the preceding formulas do not apply.

Waveforms

The frequency, or *pitch*, of a sound is only one of several variables that acoustic waves can possess. Another important factor is the shape of the wave itself. The simplest acoustic waveform is a sine wave (or *sinusoid*), in which all of the energy is concentrated at a single frequency. Sinusoidal sound waves are rare in nature. A good artificial example is the beat note, or heterodyne, produced by a steady carrier in a communications receiver.

In music, most of the notes are complex waveforms, consisting of energy at a specific fundamental frequency and its harmonics. Examples are sawtooth, square, and triangular waves. The shape of the waveform depends on the distribution of energy among the fundamental and the harmonics. There are infinitely many different shapes that a wave can have at a single frequency such as 1 kHz. As a result, there is infinite variety in the *timbre* that a single musical note can have.

Path Effects

A flute, a clarinet, a guitar, and a piano can each produce a sound at 1 kHz, but the tone quality is different for each instrument. The waveform affects the way a sound is reflected from objects. Acoustics engineers must consider this when designing sound systems and concert halls. The goal is to make sure that all the instruments sound realistic everywhere in the room.

Suppose you have a sound system set up in your living room, and that, for the particular placement of speakers with respect to your ears, sounds propagate well at 1, 3, and 5 kHz, but poorly at 2, 4, and 6 kHz. This affects the way musical instruments sound. It distorts the sounds from some instruments more than the sounds from others. Unless all sounds, at all frequencies, reach your ears in the same proportions that they come from the speakers, you do not hear the music the way it originally came from the instruments.

Figure 31-1 shows a listener, a speaker, and three sound reflectors, also known as *baffles*. The waves reflected by the baffles (X, Y, and Z), along with the direct-path waves (D), add up to some-

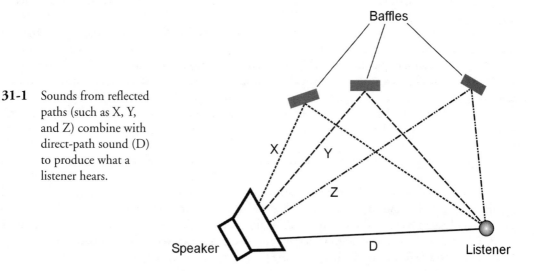

31-1 Sounds from reflected paths (such as X, Y, and Z) combine with direct-path sound (D) to produce what a listener hears.

thing different, at the listener's ears, for each frequency of sound. This phenomenon is impossible to prevent. That is why it is so difficult to design an acoustical room, such as a concert auditorium, to propagate sound well at all frequencies for every listener.

Loudness and Phase

You do not perceive the *loudness* (also called *volume*) of sound in direct proportion to the power contained in the disturbance. Your ears and brain sense sound levels according to the logarithm of the actual intensity. Another variable is the *phase* with which waves arrive at your ears. Phase allows you to perceive the direction from which a sound is coming, and it also affects perceived sound volume.

The Decibel in Acoustics

You have already learned about decibels in terms of signal voltage, current, and power. Decibels are also used in acoustics, and in this application, they are considered in terms of relative power. If you change the volume control on a hi-fi set until you can just barely tell the difference, the increment is one decibel (1 dB). If you use the volume control to halve or double the actual acoustic-wave power coming from a set of speakers, you perceive a change of 3 dB.

For decibels to have meaning in acoustics, there must be a reference level against which everything is measured. Have you read that a vacuum cleaner produces 80 dB of sound? This is determined with respect to the *threshold of hearing*, which is the faintest sound that a person with good hearing can detect in a *quiet room* specially designed to have a minimum of background noise.

Phase in Acoustics

Even if there is only one sound source, acoustic waves reflect from the walls, ceiling, and floor of a room. In Fig. 31-1, imagine the baffles as two walls and the ceiling in a room. As is the case with baffles, the three sound paths X, Y, and Z are likely to have different lengths, so the sound waves reflected from these surfaces will not arrive in the same phase at the listener's ears. The direct path (D), a straight line from the speaker to the listener, is always the shortest path. In this situation, there are at least four different paths by which sound waves can propagate from the speaker to the listener. In some practical scenarios, there are dozens.

Suppose that, at a certain frequency, the acoustic waves for all four paths happen to arrive in exactly the same phase in the listener's ears. Sounds at that frequency will be exaggerated in volume. The same phase coincidence might also occur at harmonics of this frequency. This is undesirable because it causes acoustic peaks, called *antinodes*, distorting the original sound. At certain other frequencies, the waves might mix in phase opposition. This produces acoustic nulls called *nodes* or *dead zones*. If the listener moves a few feet, the volume at any affected frequency will change. As if this isn't bad enough, a new antinode or node might then present itself at another set of frequencies.

One of the biggest challenges in acoustical design is the avoidance of significant antinodes and nodes. In a home hi-fi system, this can be as simple as minimizing the extent to which sound waves reflect from the ceiling, the walls, the floor, and the furniture. Acoustical tile can be used on the ceiling, the walls can be papered or covered with cork tile, the floor can be carpeted, and the furniture can be upholstered with cloth. In large auditoriums and music halls, the problem becomes more complex because of the larger sound propagation distances involved, and also because of the fact that sound waves reflect from the bodies of the people in the audience!

Technical Considerations

Regardless of its size, a good hi-fi sound system must have certain characteristics. Here are two of the most important technical considerations.

Linearity

Linearity is the extent to which the output waveform of an amplifier is a faithful reproduction of the input waveform. In hi-fi equipment, all the amplifiers must be as linear as the state of the art allows.

If you connect a dual-trace oscilloscope (one that lets you observe two waveforms at the same time) to the input and output terminals of a hi-fi audio amplifier with good linearity, the output waveform is a vertically magnified duplicate of the input waveform. When the input signal is applied to the horizontal scope input and the output signal is applied to the vertical scope input, the display is a straight line. In an amplifier with poor linearity, the instantaneous output-versus-input function is not a straight line. The output waveform is not a faithful reproduction of the input, and distortion occurs. In some RF amplifiers this is all right. In hi-fi audio systems, it is not.

Hi-fi amplifiers are designed to work with input signals up to a certain peak (maximum instantaneous) amplitude. If the peak input exceeds this level, the amplifier becomes nonlinear, and distortion is inevitable. In a hi-fi system equipped with VU or distortion meters, excessive input causes the needles to kick up into the red range of the scale during peaks.

Dynamic Range

Dynamic range is a prime consideration in hi-fi recording and reproduction. As the dynamic range increases, the sound quality improves for music or programming having a wide range of volume levels. Dynamic range is expressed in decibels (dB).

At low volume levels, the limiting factor in dynamic range is the *background noise* in the system. In an analog system, most of this noise comes from the audio amplification stages. In tape recording, there is also some *tape hiss*. A scheme called *Dolby* (a trademark of Dolby Laboratories) is used in professional recording studios, and also in high-end consumer tape equipment, to minimize this hiss. Digital recording systems produce less internal noise than analog systems.

At high volume levels, the power-handling capability of an audio amplifier limits the dynamic range. If all other factors are equal, a 100-W audio system can be expected to have greater dynamic range than a 50-W system. The speaker size is also important. As speakers get physically larger, their ability to handle high power improves, resulting in increased dynamic range. This is why serious audio enthusiasts sometimes purchase sound systems with amplifiers and speakers that seem unnecessarily large.

Components

There are myriad ways to set up a hi-fi system. A true *audiophile* ("sound lover") assembles a complex system over a period of time. Here are some basic considerations that can serve as guidelines when choosing system components.

Configurations

The simplest type of home stereo system is contained in a single box, with an AM/FM radio receiver and a *compact disk* (CD) player. The speakers are generally external, but the connecting cables are short. The assets of a so-called compact hi-fi system are small size and low cost.

More sophisticated hi-fi systems have separate boxes containing components such as the following:

- An AM tuner
- An FM tuner
- An amplifier or pair of amplifiers
- A CD player
- A computer and its peripherals

The computer is optional, but it facilitates downloading music files or *streaming audio* from the Internet, creating ("burning") CDs, and composing and editing electronic music. A *satellite radio* receiver, a *tape player*, a *turntable*, or other nonstandard peripheral may also be included. The individual hardware units in this type of system, known as a *component hi-fi system*, are interconnected with shielded cables. A component system costs more than a compact system, but the sound quality is better, you get more audio power, you can do more tasks, and you can tailor the system to your preferences.

Some hi-fi manufacturers build all their equipment cabinets to a standard width so they can be mounted one above the other in a *rack*. A so-called rack-mounted hi-fi system saves floor space and makes the system look professional. The rack can be mounted on wheels so the whole system, except for the external speakers, can be rolled from place to place.

Figure 31-2 is a block diagram of a typical home stereo hi-fi system. The amplifier chassis is grounded to minimize hum and noise, and to minimize susceptibility to interference from external

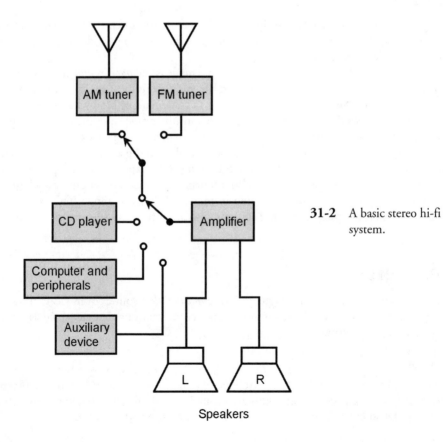

31-2 A basic stereo hi-fi system.

sources. The AM antenna is usually a loopstick built into the cabinet or mounted on the rear panel. The FM antenna can be an indoor type, such as television rabbit ears, or a directional outdoor antenna equipped with lightning protection hardware.

The Tuner

A *tuner* is a radio receiver capable of receiving signals in the standard AM broadcast band (535 to 1605 kHz) and/or the standard FM broadcast band (88 to 108 MHz). Tuners don't have built-in amplifiers. A tuner can provide enough power to drive a headset, but an amplifier is necessary to provide sufficient power to a pair of speakers.

Modern hi-fi tuners employ frequency synthesizers and have digital readouts. Most tuners have several *memory channels.* These are programmable, and allow you to select your favorite stations with a push of a single button, no matter where the stations happen to be in the frequency band. Most tuners also have *seek* and/or *scan* modes that allow the radio to automatically search the band for any station strong enough to be received clearly.

The Amplifier

In hi-fi, an amplifier delivers medium or high audio power to a set of speakers. There is at least one input, but more often there are three or more: one for a CD player, another for a tuner, and still others for auxiliary devices such as a tape player, turntable, or computer. Input requirements are a few milliwatts; the output can range up to hundreds of watts.

Amplifier prices increase with increasing power output. A simplified hi-fi amplifier forms the basis for a public-address system. Massive amplifiers are used by popular music bands. Some such systems employ vacuum tubes, because tubes offer electrical ruggedness and excellent linearity.

Speakers and Headsets

No amplifier can deliver sound that is better than the speakers will allow. Speakers are rated according to the audio power they can handle. It's a good idea to purchase speakers that can tolerate at least twice the audio output power that the amplifier can deliver. This will ensure that speaker distortion will not occur during loud, low-frequency sound bursts. It will also prevent physical damage to the speakers that might otherwise result from accidentally overdriving them.

Good speakers contain two or three individual units within a single cabinet. The *woofer* reproduces bass. The *midrange speaker* handles medium and, sometimes, treble (high) audio frequencies. A *tweeter* is designed especially for enhanced treble reproduction.

Headsets are rated according to how well they reproduce sound. This is a subjective consideration. Equally expensive headsets can, and often do, exhibit huge differences in the quality of the sound that they put out. Not only that, but people disagree about what constitutes good sound.

Balance Control

In hi-fi stereo sound equipment, the *balance control* allows adjustment of the relative volumes of the left and right channels.

In a basic hi-fi system, the balance control consists of a single rotatable knob connected to a pair of potentiometers. When the knob is rotated counterclockwise, the left-channel volume increases and the right-channel volume decreases. When the knob is rotated clockwise, the right-channel volume increases and the left-channel volume decreases. In more sophisticated sound systems, the balance is adjusted by means of two independent volume controls, one for the left channel and the other for the right channel.

Proper balance is important in stereo hi-fi. A balance control can compensate for such factors as variations in speaker placement, relative loudness in the channels, and the acoustical characteristics of the room in which the equipment is installed.

Tone Control

The amplitude versus frequency characteristics of a hi-fi sound system are adjusted by means of a tone control or controls. In its simplest form, a tone control consists of a single knob or slide device. The counterclockwise, lower, or left-hand settings of this control result in strong bass and weak treble audio output. The clockwise, upper, or right-hand settings result in weak bass and strong treble. When the control is set to midposition, the audio response of the amplifier is more or less flat; that is, the bass, midrange, and treble are in roughly the same proportions as in the recorded or received signal.

Figure 31-3A shows how a single-knob tone control can be incorporated into an audio amplifier. The amplifier is designed so that the treble response is exaggerated. The potentiometer attenuates the treble to a variable extent.

A more versatile tone control has two capacitors and two potentiometers, as shown in Fig. 31-3B. One combination is in series, and the other is in parallel. The series-connected resistance-capacitance (*RC*) circuit is connected in parallel with the audio output, and it attenuates the treble to a variable extent. The parallel *RC* circuit is in series with the audio path, and it attenuates the bass to a variable extent. The two potentiometers can be adjusted separately, although there is some interaction.

Audio Mixer

If you simply connect two or more audio sources to the same input of an amplifier, you can't expect good results. Different signal sources (such as a computer, a tuner, and a CD player) are likely to

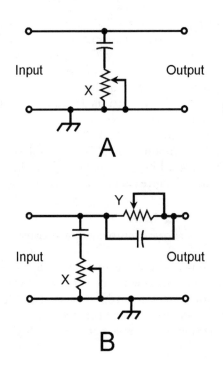

31-3   Methods of tone control. At A, a single potentiometer/capacitor combination (X) provides treble attenuation only. At B, one potentiometer/capacitor combination (X) attenuates the treble, and the other (Y) attenuates the bass.

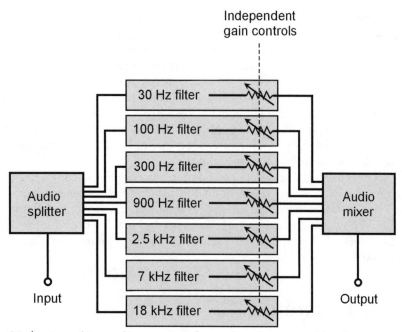

31-4 A graphic equalizer consists of an audio splitter, several bandstop filters, several gain controls, and an audio mixer.

have different impedances. When connected together, the impedances appear in parallel. This can cause impedance mismatches for most or all of the sources, as well as at the amplifier input. The result will be degradation of system efficiency and poor overall performance.

Another problem arises from the fact that the signal amplitudes from various sources almost always differ. A microphone produces minuscule audio-frequency currents, whereas some tuners produce enough power to drive a pair of small loudspeakers. Connecting both of these together will cause the microphone signal to be obliterated by the signal from the tuner. In addition, the tuner output audio might damage the microphone.

An *audio mixer* eliminates all of the problems involved with the connection of multiple devices to a single channel. First, it isolates the inputs from each other, so there is no impedance mismatch or competition among the sources. Second, the gain at each input can be varied independently. This allows adjustment of amplitudes so the signals blend in the desired ratio.

Graphic Equalizer

A *graphic equalizer* is a device for adjusting the relative loudness of audio signals at various frequencies. It allows tailoring of the amplitude versus frequency output of hi-fi sound equipment. Equalizers are used in recording studios and by serious hi-fi stereo enthusiasts. There are several independent gain controls, each one affecting a different part of the audible spectrum. The controls are slide potentiometers with calibrated scales. The slides move up and down, or, in some cases, left to right. When the potentiometers are set so that the slides are all at the same level, the audio output or response is flat, meaning that no particular range is amplified or attenuated with respect to the whole AF spectrum. By moving any one of the controls, the user can adjust the gain within a certain frequency range without affecting the gain outside that range. The positions of the controls on the front panel provide an intuitive graph of the output or response curve.

Figure 31-4 is a block diagram of a hypothetical graphic equalizer with seven gain controls. The input is fed to an *audio splitter* that breaks the signal into seven paths of equal impedance, and prevents interaction among the circuits. The seven signals are fed to *audio attenuators*, also called *band-stop filters*, each filter having its own gain control. (The center frequencies of the attenuators in this example are at 30, 100, 300, and 900 Hz, and 2.5, 7, and 18 kHz. These are not standard frequencies, and are given here only for illustrative purposes.) The slide potentiometers affect the extent to which each filter affects the gain within its frequency range. Finally, the signals pass through an audio mixer, and the composite is sent to the output.

There are several challenges in the design and proper use of graphic equalizers. The filter gain controls must not interact. Judicious choice of filter frequencies and responses is important. The filters must not introduce distortion. The active devices must not generate significant audio noise. Graphic equalizers are not built to handle high power, so they must be placed at low-level points in an audio amplifier chain. In a multichannel circuit such as a stereo sound system, a separate graphic equalizer can be used for each channel.

Specialized Systems

Mobile and portable hi-fi systems operate at low dc voltages. Typical audio power levels are much lower than in home hi-fi systems. Speakers are much smaller also. In portable systems, headsets are often used in place of speakers.

Mobile Systems

Mobile hi-fi systems, designed for cars and trucks, usually have four speakers. The left and right channels each supply a pair of speakers. The left stereo channel drives the left front and left rear speakers; the right stereo channel drives the right front and right rear speakers. The balance control adjusts the ratio of sound volume between the left and right channels for both the front and rear speaker sets. Another control adjusts the ratio of sound volume between the front and rear sets.

A mobile hi-fi system typically has an AM/FM receiver and a CD player. Some older vehicles have systems with cassette tape players. Newer, high-end cars and trucks have satellite radio receivers as well as conventional AM/FM receivers and CD players. One note of caution: compact disks and cassettes are heat-sensitive, so they should not be stored in a car or truck that will be left out in the sun on a warm day.

Portable Systems

Portable hi-fi systems operate from sets of dry cells or rechargeable cells. The most well-known portable hi-fi set is the so-called headphone radio. There are literally hundreds of different designs. Some include only an FM radio; some have AM/FM reception capability. Some have a small box with a cord that runs to the headset; others are entirely contained within the headset. There are portable CD players and even portable satellite radio receivers. The sound quality from these systems can be excellent. The defining factor is the quality of the headset.

Another form of portable hi-fi set, sometimes called a *boom box*, can produce several watts of audio output, and delivers the sound to a pair of speakers built into the box. A typical boom box is about 8 in high by 18 in wide by 6 in deep. It includes an AM/FM radio and a CD player. The system gets its name from the loud bass acoustic energy peaks its speakers can deliver.

Quadraphonic Sound

Quadraphonic sound refers to four-channel audio recording and reproduction. It is also called *quad stereo* or *four-channel stereo*. Each of the four channels is independent of the other three. In a well-designed quadraphonic sound system, the speakers should be level with the listener, equidistant from the listener, and separated by angles of 90° from the listener's point of view. If the listener is facing north, for example, the left front speaker is to the northwest, the right front speaker is to the northeast, the left rear speaker is to the southwest, and the right rear speaker is to the southeast. This provides optimum balance, and also facilitates the greatest possible left-to-right and front-to-rear contrast in the perceived sounds.

Recorded Media

Methods of recording sound, particularly music, have evolved dramatically since the ascent of digital technology. Several types of media are available, but the compact disk (CD) is the most common. Other media include *analog audio tape*, *digital audio tape*, and (of historical interest) *vinyl disk*.

Compact Disk

A compact disk (often spelled "disc" in hi-fi applications), also called a CD, is a plastic disk with a diameter of 4.72 in (12.0 cm), on which data can be recorded in digital form. Any kind of data can be digitized: sound, images, and computer programs and files. This data can also be stored on other digital media of sufficient capacity, such as a computer hard drive or a backup tape drive.

Digital sound, recorded on the surface of a CD, is practically devoid of the hiss and crackle that bedevil recordings on analog media. This is because the information on the disk is binary: a bit (binary digit) is either 1 (high) or 0 (low). The distinction between these two states is more clear-cut than the subtle fluctuations of an analog signal.

When a CD is prepared, the sound is first subjected to *analog-to-digital* (A/D) *conversion*. This changes the continuously variable audio waves into logic bits. These bits are then burned (literally) into the surface of the disk in the form of microscopic pits. The pits are arranged in a spiral track that would, if unwound, measure several kilometers in length. *Digital signal processing* (DSP) minimizes the noise introduced by environmental factors such as microscopic particles on the disk or random electronic impulses in circuit hardware. A *scrambling* process "smears" recordings throughout the disk, rather than burning the pits in a direct linear sequence. This further improves the signal-to-noise (S/N) ratio.

Compact-disk players recover the sound from a disk without any hardware physically touching the surface. A laser beam scans the disk. The beam is scattered by the pits and is reflected from the unpitted plastic. The result is a digitally modulated beam that is picked up by a sensor and converted into electrical currents. These currents proceed to a *descrambling* circuit, a *digital-to-analog* (D/A) *converter*, a DSP circuit, and audio amplifiers. Speakers or headphones convert the audio currents into sound waves.

With a CD player, the track location processes are entirely electronic, and they can all be done quickly. Tracks are assigned numbers that you select by pressing buttons. It is impossible to damage the CD, no matter how much you skip around among the songs. You can move instantly to any desired point within an individual track. You can program the system to play only those tracks you want, ignoring the others.

Analog Audio Tape

Analog audio tape recorders can be classified as either cassette or reel-to-reel. A typical audio cassette plays for 30 minutes on each side; longer-playing cassettes allow recording for as much as 60 minutes per side. The longer tapes are thinner and more subject to stretching than the shorter tapes.

A reel-to-reel tape feed system resembles that of an old-fashioned movie projector. The tape is wound on two flat spools called the *supply reel* and the *take-up reel*. The reels rotate counterclockwise as the tape passes through the recording/playback mechanism. When the take-up reel is full and the supply reel is empty, both reels can be flipped over and interchanged. This allows recording or playback on the other side of the tape. (Actually, the process takes place on the same side of the tape, but on different tracks.) The speed is usually 1.875, 3.75, or 7.5 inches per second (ips).

Figure 31-5 is a simplified functional diagram of an analog audio tape recording/playback mechanism.

In the *record mode*, the tape moves past the *erase head* before anything is recorded on it. If the tape is not blank (that is, if magnetic impulses already exist on it), the erase head removes these before anything else is recorded. This prevents *doubling*, or the simultaneous presence of two programs on the tape. The erase head can be disabled in some tape recorders when doubling is desired. The *recording head* is a small electromagnet. It generates a fluctuating magnetic field whose instantaneous flux density is proportional to the instantaneous level of the audio input signal. This magnetizes the tape in a pattern that duplicates the waveform of the signal. The *playback head* is usually not activated in the record mode. However, the playback head and circuits can be used while recording to create an echo effect.

In the *playback mode*, the erase head and recording head are not activated. The playback head acts as a sensitive magnetic-field detector. As the tape moves past it, the playback head is exposed to a fluctuating magnetic field whose waveform is identical to that produced by the recording head when the audio was originally recorded on the tape. This magnetic field induces weak alternating

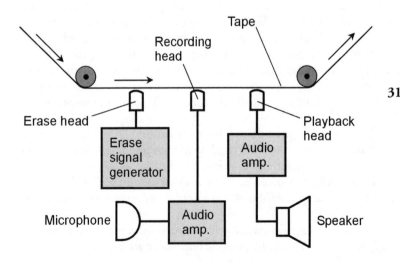

31-5 Simplified functional diagram of the recording and playback hardware in an analog audio tape recorder/player.

currents in the playback head. These currents are amplified and delivered to a speaker, headset, or other output device.

Digital Audio Tape

Digital audio tape (DAT) is magnetic recording tape on which binary digital data can be recorded. In digital audio recording, tape noise is practically eliminated because such noise is analog in nature. Some electronic noise is generated in the analog amplification stages following D/A conversion, but this is minimal compared with the noise generated in older, fully analog systems. The reduced noise in DAT equipment provides more true-to-life reproduction than is possible with analog methods.

With DAT, multigeneration copies can be made with practically no degradation in audio fidelity. The reason for this is the same as the reason a computer can repeatedly read and overwrite data on a magnetic disk. On DAT, the bits are represented by distinct magnetized regions on the tape. While analog signals are fuzzy in the sense that they vary continuously, digital signals are crisp. Imperfections in the recording apparatus, the tape itself, and the pickup head affect digital signals less than they affect analog signals. Digital signal processing can eliminate the minute flaws that creep into a digital signal each time it is recorded and played back.

Vinyl Disk

Vinyl disks were superseded years ago by CDs and Internet downloads, but some audiophiles are still intrigued by vinyl. Some vinyl disks, and the turntables that can play them, have attained considerable value as collectors' items. The main trouble with vinyl is that it can be physically damaged by even the slightest mishandling. Electrostatic effects can produce noise when the humidity is low, as in alpine or far northern regions in the winter.

Vinyl disks require a turntable that spins at speeds of 33 and 45 revolutions per minute (rpm). There are various drive systems. These are called *rim drive, belt drive,* and *direct drive.* The best type is a matter of individual taste because there are various factors to consider, such as cost, audio quality, ruggedness, and durability.

Electromagnetic Interference

As hi-fi equipment becomes more sophisticated and complex, the circuits become more susceptible to interference from outside sources, particularly electromagnetic (EM) fields. This problem is known as *electromagnetic interference* (EMI). Sometimes it is called *radio-frequency interference* (RFI).

If a radio transmitter is operated near a hi-fi stereo system, the radio signals can be intercepted by the hi-fi wiring and peripherals, and delivered to the amplifier. Unshielded interconnecting cables act as radio receiving antennas. This problem is exacerbated if any of the connecting cables happen to resonate at the operating frequency of the radio transmitter. In the hi-fi amplifier, the RF currents are rectified, causing changes in the audio gain. Sometimes the signal data can be heard in the speakers or headset.

In most cases when EMI takes place in a hi-fi setup, the fault exists in the stereo system design, not in the radio transmitter. The transmitter system is merely doing its job: generating and radiating electromagnetic signals!

Several steps can be taken when installing a stereo hi-fi system to minimize the likelihood that EMI will occur:

- Connect the stereo amplifier chassis to a good electrical ground.
- Use shielded interconnecting cables as much as possible.
- Use shielded (coaxial) speaker cables.
- Keep all cables as short as possible.

If you have an amateur or citizens band (CB) radio station in your house and it causes EMI to your hi-fi system, perform these additional steps:

- Locate the radio transmitting antenna as far from the hi-fi equipment as possible.
- Use the lowest possible transmitter output power that will ensure reliable communications.

Unfortunately, EMI problems can sometimes prove nigh impossible to eliminate. This can be especially troublesome for amateur and CB radio operators when it damages relations with neighbors. In these cases, old-fashioned diplomacy may work better than engineering-based attempts at resolution.

Quiz

Refer to the text in this chapter if necessary. A good score is 18 correct. Answers are in the back of the book.

1. A young person can hear sounds at frequencies as high as approximately which of the following?

(a) 20 Hz

(b) 20 kHz

(c) 20 MHz

(d) 20 GHz

2. Electromagnetic interference to a hi-fi amplifier can occur in the presence of

(a) a nearby radio broadcast station.

(b) improperly designed receiving antennas.

(c) excessive utility voltage.

(d) improper balance between the left and right channels.

3. The midrange audio frequencies

(a) are exactly halfway between the lowest and highest audible frequencies.

(b) represent sounds whose volume levels are not too loud or too soft.

(c) are above the treble range but below the bass range.

(d) None of the above are true.

4. In the acoustical design of a room intended for a home audio system,

(a) the use of small speakers can minimize distortion.

(b) reflection of sound waves from walls should be minimized.

(c) the walls should all intersect at perfect 90° angles.

(d) wooden furniture, without upholstery, should be used.

5. A change of +10 dB in an audio signal represents

    (a) a doubling of acoustic power.

    (b) a threefold increase in acoustic power.

    (c) a tenfold increase in acoustic power.

    (d) no change in acoustic power, but a change in frequency.

6. What is the frequency of an acoustic disturbance whose wavelength is 120 mm in air?

    (a) 279 Hz

    (b) 2.79 kHz

    (c) 35.8 Hz

    (d) 358 Hz

7. What is the frequency of a sound wave that propagates at a speed of 1100 ft/s?

    (a) 33.5 Hz

    (b) 335 Hz

    (c) 3.35 kHz

    (d) Forget it! The frequency of a sound wave is independent of the propagation speed.

8. The relative phase of two acoustic waves from the same source at the same time, one wave direct and one wave reflected from a wall, can affect

    (a) the positions of antinodes and nodes.

    (b) the perceived frequency.

    (c) the positions of antinodes and nodes, and the perceived frequency.

    (d) neither the positions of the antinodes and nodes, nor the perceived frequency.

9. In an acoustic sine wave,

    (a) the frequency and phase are identical.

    (b) the sound power is inversely proportional to the frequency.

    (c) the sound power is directly proportional to the frequency.

    (d) all of the sound power is concentrated at a single frequency.

10. Vinyl disks are

    (a) susceptible to physical damage.

    (b) useful primarily in high-power sound systems.

    (c) digital media.

    (d) preferred for off-the-air sound recording.

11. If an amplifier introduces severe distortion in the waveforms of input signals, then that amplifier is

    (a) not delivering enough power.

    (b) operating at the wrong frequency.

    (c) operating in a nonlinear fashion.

    (d) being underdriven.

12. Suppose a 10-W amplifier is used with speakers designed for a 100-W amplifier. Which of the following statements is true?

    (a) The speakers are capable of handling the amplifier output.

    (b) The amplifier might be damaged by the speakers.

    (c) Electromagnetic interference is more likely to occur than would be the case if the speakers were designed for a 10-W amplifier.

    (d) The speakers are likely to produce excessive distortion.

13. Which of the following frequencies cannot be received by an AM/FM tuner?

    (a) 830 kHz

    (b) 95.7 kHz

    (c) 90.1 MHz

    (d) 107.3 MHz

14. Which of the following statements about woofers is true?

    (a) They are especially useful for reproducing the sounds of barking dogs.

    (b) They are designed to handle short, intense bursts of sound.

    (c) They should not be used with graphic equalizers.

    (d) They are specifically designed to reproduce low-frequency sounds.

15. Suppose you have an amateur radio station and its transmitter causes EMI to your hi-fi system. Which of the following would almost certainly *not* help?

    (a) Build a new transmitter that works on the same frequencies with the same power output as your existing transmitter.

    (b) Reduce the transmitter output power.

    (c) Install shielded speaker wires in the hi-fi system, and be sure the system is well grounded.

    (d) Move the amateur radio transmitting antenna to a location farther away from the hi-fi system.

16. In an analog audio tape recorder/player, the recording head

    (a) converts sound waves to radio signals.

    (b) converts sound waves to fluctuating electric current.

    (c) converts AF current to a fluctuating magnetic field.

    (d) converts dc to AF current.

17. A rack-mounted hi-fi system

    (a) can save floor space.

    (b) is more susceptible to EMI than a compact system.

    (c) is cheaper than a compact system.

    (d) is designed especially for use with vinyl disks and turntables.

18. An audio mixer

    (a) cannot match impedances among interconnected components.

    (b) cannot increase the audio output of an amplifier.

(c) eliminates EMI problems in all but the worst cases.

(d) allows a microphone to be used as a speaker.

19. Which of the following devices or circuits ordinarily employs D/A conversion?

(a) A portable CD player

(b) A turntable for use with vinyl disks

(c) An analog audio tape recording/playback system

(d) An *RC* tone control

20. Which of the following is an advantage of digital audio over analog audio?

(a) Digital media can be used to record and play back audio at higher frequencies than can analog media.

(b) Multigeneration copies of digital audio programs can be made without degradation in fidelity, but this is not true of analog audio programs.

(c) Digital audio is compatible with vinyl disks and old-fashioned reel-to-reel and cassette tape, but analog audio is not.

(d) There are no genuine advantages of digital audio over analog audio. In fact, analog audio is superior to digital audio in every respect.

32
CHAPTER

Personal and Hobby Wireless

THE TERM *WIRELESS* AROSE IN THE EARLY 1900S WHEN INVENTORS AND EXPERIMENTERS BEGAN sending and receiving messages using electromagnetic fields. Gradually, the terminology changed, and wireless became known as *radio*, *television*, and *electromagnetic communications*. In the 1980s and 1990s, the term *wireless* emerged again, this time in a consumer context.

Cellular Communications

Wireless telephone sets are used in a specialized communications system called *cellular*. Originally, the cellular communications network was used mainly by traveling business people. Nowadays, many people regard *cell phones* as necessities, and some of them come equipped with extra features such as text messaging, Web browsing, and digital cameras.

How Cellular Systems Work

A cell phone looks like a cross between a cordless telephone receiver and a walkie-talkie, but smaller. The unit contains a radio transmitter and receiver combination called a *transceiver*. Transmission and reception take place on different frequencies, so you can talk and listen at the same time, and easily interrupt the other party if necessary. This capability is known as *full duplex*.

In an ideal cellular network, all the transceivers are always within range of at least one *repeater*. The repeaters pick up the transmissions from the portable units and retransmit the signals to the telephone network and to other portable units. The region of coverage for any repeater (also known as a *base station*) is called a *cell*.

When a cell phone is in motion, say in a car or on a boat, the set goes from cell to cell in the network. This situation is shown in Fig. 32-1. The curved, dashed line is the path of the vehicle. Base stations (dots) transfer access to the cell phone. This is called *handoff*. The hexagons show the limits of the transmission/reception range for each base station. All the base stations are connected to the regional telephone system. This makes it possible for the user of the portable unit to place

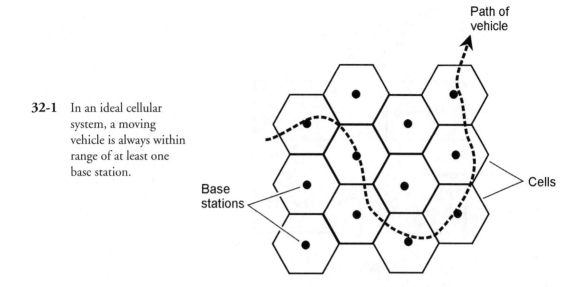

32-1 In an ideal cellular system, a moving vehicle is always within range of at least one base station.

calls to, or receive calls from, anyone else in the system, whether those other subscribers have cell phones or regular phones.

Older cellular systems occasionally suffer from call loss or breakup when signals are handed off from one repeater to another. This problem has been largely overcome by a technology called *code-division multiple access* (CDMA). In CDMA, the repeater coverage zones overlap significantly, but signals do not interfere with each other because every phone set is assigned a unique signal code. Rather than abruptly switching from one base-station zone to the next, the signal goes through a region in which it is sent through more than one base station at a time. This *make-before-break* scheme gets rid of one of the most annoying problems inherent in cellular communications.

In order to use a cellular network, you must purchase or rent a transceiver and pay a monthly fee. The fees vary, depending on the location and the amount of time per month you use the service. When using such a system, it is important to keep in mind that your conversations are not necessarily private. It is easier for unauthorized people to eavesdrop on wireless communications than to intercept wire or cable communications.

Cell Phones and Computers

You can connect a *personal computer* (PC) to a cell phone with a portable *modem* that converts incoming computer data from analog to digital form, and also converts outgoing data from digital to analog form. In this way, you can access the Internet from anywhere within range of a cellular base station. Figure 32-2 is a block diagram of this scheme.

Most commercial aircraft have telephones at each row of seats, complete with jacks into which you can plug a modem. If you plan to access the Internet from an aircraft, you must use the phones provided by the airline, not your own cell phone, because radio transceivers can cause interference to flight instruments. You must also observe the airline's restrictions concerning the operation of electronic equipment while in flight.

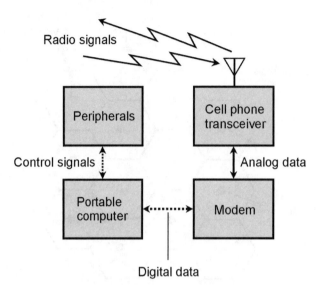

32-2 A cell phone can be equipped with a modem, allowing portable or mobile access to online computer networks.

Satellites and Networks

A satellite system is like a huge cellular network with the base stations (repeaters) located in space rather than on the earth's surface. The zones of coverage are large, and they change in size and shape if the satellite moves relative to the earth's surface.

Geostationary-Orbit Satellites

You learned about geostationary orbits in Chap. 25. Geostationary satellites are used in television (TV) broadcasting, in telephone and data communication, for gathering weather and environmental data, and for radiolocation.

In geostationary satellite networks, earth-based stations can communicate through a single "bird" only when the stations are both on a line of sight with the satellite. If two stations are nearly on opposite sides of the planet, say in Australia and Wisconsin, they must operate through two satellites to obtain a link (Fig. 32-3). In this situation, signals are relayed between the two satellites, as well as between either satellite and its respective earth-based station.

A potential problem with geostationary satellite links is the fact that the signal path is long enough so that perceptible propagation delays occur. This delay, and its observed effect, is known as *latency*. This doesn't cause problems with casual communications or Web browsing, but it slows things down when computers are linked with the intention of combining their processing power.

Low-Earth-Orbit Satellites

The earliest communications satellites orbited only a few hundred miles above the earth. They were *low-earth-orbit* (LEO) *satellites*. Because of their low orbits, LEO satellites took only about 90 minutes to complete one revolution. Communication was spotty, because a satellite was in range of any given ground station for only a few minutes at a time. This is the main reason why geostationary satellites became predominant once rocket technology progressed to the point where the necessary altitude and orbital precision could be obtained.

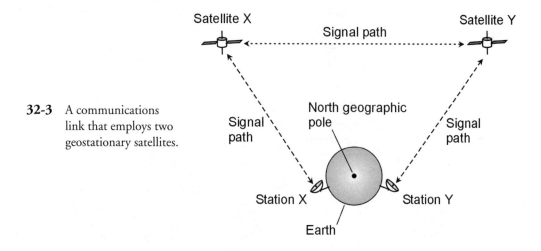

32-3 A communications link that employs two geostationary satellites.

However, geostationary satellites have certain limitations. A geostationary orbit requires constant adjustment, because a tiny change in altitude will cause the satellite to get out of sync with the earth's rotation. Geostationary satellites are expensive to launch and maintain. When communicating through them, there is always a delay because of the path length. It takes high transmitter power, and a sophisticated, precisely aimed antenna, to communicate reliably. These problems with geostationary satellites have brought about a revival of the LEO scheme. Instead of one single satellite, the new concept is to have a large fleet of them.

A good LEO satellite system is launched and maintained in such a way that, for any point on the earth, there is always at least one satellite in direct line-of-sight range. The satellites can relay messages throughout the fleet. Thus, any two points on the surface can always make, and maintain, contact through the satellites. The satellites are placed in *polar orbits* (routes that pass over or near the earth's geographic poles) to optimize the geographical coverage. Even if you're at or near the north geographic pole or the south geographic pole, you can use a LEO satellite system. This is not true of geostationary satellite networks, where the regions immediately around the geographic poles are not seen by the satellites.

A LEO satellite wireless communications link is easier to access and use than a geostationary satellite link. A small, simple antenna will suffice, and it doesn't have to be aimed in any particular direction. The transmitter can reach the network using only a few watts of power. The latency is less than 100 milliseconds (ms), compared with as much as 400 ms for geostationary satellite links.

Medium-Earth-Orbit Satellites

Some satellites revolve in orbits higher than those normally considered low-earth, but at altitudes lower than the geostationary level of 22,300 mi (36,000 km). These intermediate birds are called *medium-earth-orbit* (MEO) *satellites*. A MEO satellite takes several hours to complete each orbit. MEO satellites operate in fleets, in a manner similar to the way LEO satellites are deployed. Because the average MEO altitude is higher than the average LEO altitude, each bird can cover a larger region on the surface at any given time. A fleet of MEO satellites can be smaller than a comparable fleet of LEO satellites, and still provide continuous, worldwide communications.

The orbits of geostationary satellites are essentially perfect circles, and most LEO satellites orbit in near-perfect circles. But MEO satellites often have elongated, or elliptical, orbits. The point of

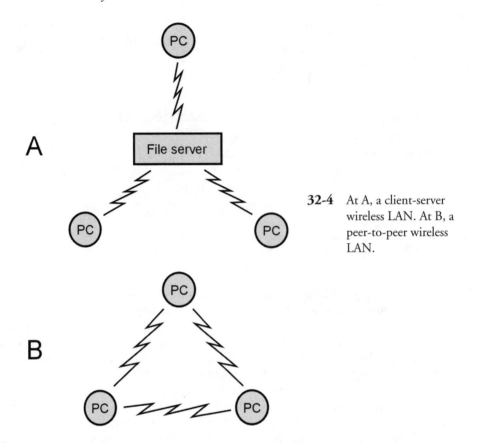

32-4   At A, a client-server wireless LAN. At B, a peer-to-peer wireless LAN.

lowest altitude is called *perigee*; the point of greatest altitude is called *apogee*. The apogee can be, and often is, much greater than the perigee. Such a satellite orbits at a speed that depends on its altitude. The lower the altitude, the faster the satellite moves. A satellite with an elliptical orbit crosses the sky rapidly when it is near perigee, and slowly when it is near apogee; it is easiest to use when its apogee is high above the horizon, because then it stays in the visible sky for a long time.

Every time a MEO satellite completes one orbit, the earth rotates beneath it. The rotation of the earth rarely coincides with the orbital period of the satellite. Therefore, successive apogees for a MEO satellite occur over different points on the earth's surface. This makes the tracking of individual satellites a complicated business, requiring computers programmed with accurate orbital data. For a MEO system to be effective in providing worldwide coverage without localized periodic blackouts, the orbits must be diverse, yet coordinated in a precise and predictable way. In addition, there must be enough satellites so that each point on the earth is always on a line of sight with one or more of the satellites, and preferably, there should be at least one bird in sight near apogee at all times.

Wireless Networks

A *local area network* (LAN) is a group of computers linked together within a building, campus, or other small region. The interconnections in early LANs were made with wire cables, but wireless links are increasingly common. A *wireless LAN* offers flexibility, because the computer users can move around without having to bother with plugging and unplugging cables. This arrangement is ideal when notebook computers (also known as laptops) are used. The way in which a LAN is arranged is called the *LAN topology*. There are two major wireless LAN topologies: the *client-server wireless LAN* and the *peer-to-peer wireless LAN*.

In the client-server topology (Fig. 32-4A), there is one large, powerful, central computer called a *file server*, to which all the smaller personal computers (labeled PC) are linked. The file server has enormous computing power, high speed, and large storage capacity, and can contain all the data for every user. End users do not communicate directly. All the data must pass through the file server.

In a peer-to-peer LAN (Fig. 32-4B), all of the computers in the network are PCs with more or less equal computing power, speed, and storage capacity. Each user maintains his or her own data. Subscribers can, and almost always do, communicate directly without the data having to pass through any intermediary. This offers greater privacy and individuality than the client-server topology, but it is slower when a large number of users need to share data.

Client-server LANs are favored by large institutions. Small businesses and schools, or departments within a larger corporation or university, prefer to use peer-to-peer LANs, mainly because they are cheaper and easier to maintain. In these illustrations, only three PCs are shown in the networks. But any LAN can have as few as two, or as many as several dozen, computers.

Home Internet users sometimes employ a modified version of the arrangement shown in Fig. 32-4A. In place of the file server, a device called a *wireless router* provides a hub through which the computers can communicate. The router is connected to the Internet by a high-speed interface such as a *cable modem*, allowing several computers in a household to have Internet access at the same time.

Amateur and Shortwave Radio

In most countries of the world, people can obtain government-issued licenses to send and receive messages by radio for nonprofessional purposes. In America, this hobby is called *amateur radio* or *ham radio*. If you want only to listen to communications and broadcasting, and not to transmit signals, you do not need a license in the United States (although you do need one in some countries).

Who Uses Amateur Radio?

Anyone can use ham radio, provided they can pass the tests necessary to obtain a license. Amateur radio operators (or hams) can communicate in numerous modes, including voice, Morse code, television, and various forms of text messaging. Text messaging can be done in real time, or by posting messages similar to *electronic mail* (e-mail). Radio amateurs have set up their own radio networks. Some of these networks have Internet gateways. This mode is known as *packet* radio.

Some radio hams chat about anything they can think of (except business matters, which are illegal to discuss via ham radio). Others like to practice emergency communications skills, so they can be of public service during crises such as hurricanes, earthquakes, or floods. Still others like to go out into the wilderness and talk to people far away while sitting out under the stars and using battery power.

Amateur radio operators communicate from cars, boats, aircraft, and bicycles; this is called *mobile operation*. When transceivers are used while walking or hiking, it is known as *portable* or *handheld operation*.

Amateur Equipment and Licensing

A simple ham radio station has a transceiver (transmitter/receiver), a microphone, and an antenna. A small station can fit on a desktop, and is about the size of a home computer or hi-fi stereo system.

Accessories can be added until a ham "rig" is a large installation, comparable to a small commercial broadcast station.

Figure 32-5 shows an example of a fixed, computer-controlled amateur radio station. The computer can be used to control the functions of the transceivers, and also to communicate using digital modes with other hams who own computers. The station can be equipped to interface with the telephone services, also called *landline*. The computer can control the antennas for the station, and can keep a log of all stations that have been contacted.

The best way to learn about ham radio in the United States is to contact the headquarters of the American Radio Relay League (ARRL), 225 Main Street, Newington, CT 06111. They maintain a Web site at www.arrl.org. If you live outside the United States, the ARRL can direct you to an organization in your home country that can help you obtain a license and get on the air.

Shortwave Listening

The high-frequency (HF) portion of the radio spectrum, at frequencies between 3 and 30 MHz, is sometimes called the *shortwave band*. This is a misnomer by contemporary standards; the waves are long compared with ultrahigh frequencies (UHF), microwaves, and IR, which are commonly used in wireless devices. In free space, a frequency of 3 MHz corresponds to a wavelength of 100 m; 30 MHz corresponds to 10 m. In the early years of radio when the shortwave band got its name, the wavelengths between 3 and 30 MHz were short compared with the wavelengths of most broadcast and communications signals, which had wavelengths in the kilometer range.

Anyone can build or obtain a shortwave or general-coverage radio receiver, install a modest outdoor antenna, and listen to signals from all around the world. This hobby is called *shortwave listening* or *SWLing*. In the United States, the proliferation of computers and online communications has, to some extent, overshadowed SWLing, and many young people grow up today ignorant of a

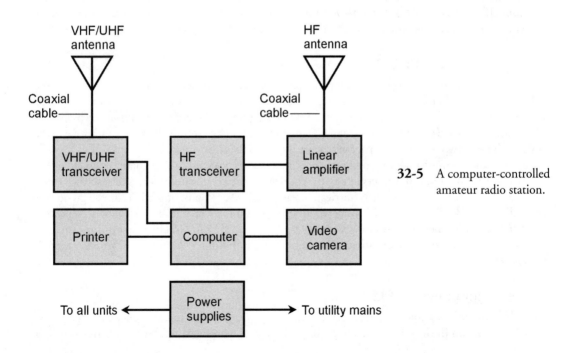

32-5   A computer-controlled amateur radio station.

realm of broadcasting and communications that still predominates in much of the world. But some people are still fascinated by the fact that people can contact each other using wireless devices alone, without the need for any human-made infrastructure other than an antenna at the source and another antenna at the destination. The ionosphere returns shortwave signals to the earth's surface, and allows reliable global broadcasting and communication to take place today, just as it has since the early 1900s.

Various commercially manufactured shortwave receivers are available. Most electronics stores carry one or more models, along with antenna equipment, for a complete installation. Amateur radio conventions, called "hamfests," can be a good source of shortwave receiving equipment at bargain prices. For information about events of this sort in your area, you can contact the American Radio Relay League at www.arrl.org, or pay a visit to your local amateur radio club.

Security and Privacy

People are becoming more and more concerned about the security and privacy of electronic communications, particularly wireless. When a wireless system is compromised, the intrusion is not detected until harm has been done to the system or to its subscribers.

Wireless versus Wired

Wireless eavesdropping differs from conventional *wiretapping* in two fundamental ways. First, eavesdropping is easier to do in wireless systems than in hard-wired systems. Those old-fashioned hard-wired phone sets might not always be convenient, but your privacy is more likely to be maintained than in the case with a system that uses any form of wireless. Second, eavesdropping of a wireless link is nigh impossible to physically detect, but a tap can usually be found in a hard-wired system.

If any portion of a communications link is done by wireless, then an eavesdropping receiver can be positioned within range of the RF transmitting antenna (Fig. 32-6) and the signals intercepted. The existence of a *wireless tap* has no effect on the electronic characteristics of any equipment in the system.

Levels of Security

There are four levels of telecommunications security, ranging from zero (no security) to the most secure connections technology allows.

No Security (Level 0): In a communications system with *level 0 security*, anyone can eavesdrop on a connection at any time, provided they are willing to spend the money and time to obtain the necessary equipment. Two examples of level 0 links are amateur radio and citizens band (CB) voice communications.

Wire Equivalent Security (Level 1): An end-to-end hard-wired connection requires considerable effort to tap, and sensitive detection apparatus can usually reveal the existence of any wiretap. A communications system with *level 1 security* must have certain characteristics in order to be effective and practical:

- The cost must be affordable.
- The system must be reasonably safe for transactions such as credit-card purchases.
- When network usage is heavy, the degree of privacy afforded to each subscriber should not decrease, relative to the case when network usage is light.

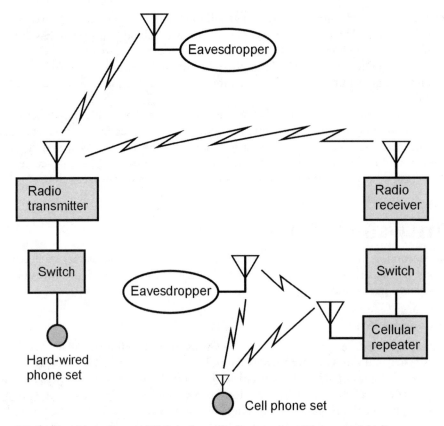

32-6   Eavesdropping on RF links in a telephone system. Heavy, straight lines represent wires or cables; zigzag lines represent RF signals.

- Ciphers, if used, should be unbreakable for at least 12 months, and preferably for 24 months or more.
- Encryption technology, if used, should be updated at least every 12 months, and preferably every six months.

Security for Commercial Transactions (*Level 2*): Some financial and business data demands protection even beyond the wire equivalent level. Many companies and individuals refuse to transfer money by electronic means because they fear criminals will gain access to an account. In a communications system with *level 2 security*, the encryption used in commercial transactions should be such that it would take a potential intruder (also called a *hacker*) at least 10 years, and preferably 20 years or more, to break the cipher. The technology should be updated at least every 10 years, but preferably every 3 to 5 years.

Military Level Security (*Level 3*): Security to *military specifications* (also called *mil spec*) involves the most sophisticated encryption available. Technologically advanced countries, and entities with economic power, have an advantage here. However, as technology gains ever more (and arguably too much) power over human activities, aggressor nations and terrorists might injure powerful nations by seeking out, and striking at, the weak points in communications infrastructures. In a communications system with *level 3 security*, the encryption scheme should be such that engineers believe it would take a hacker at least 20 years, and preferably 40 years or more, to break the cipher. The technology should be updated as often as economics allow.

Extent of Encryption

Security and privacy in wireless networks and communications systems can be achieved by means of *digital encryption*. The idea is to render signals readable only to receivers with the necessary *decryption key*. This makes it difficult for unauthorized people to gain access to the system.

For level 1 security, encryption is required only for the wireless portion(s) of the circuit. The cipher should be changed at regular intervals to keep it fresh. The block diagram of Fig. 32-7A shows *wireless-only encryption* for a hypothetical cellular telephone connection.

For security at levels 2 and 3, *end-to-end encryption* is necessary. The signal is encrypted at all intermediate points, even those for which signals are transmitted by wire or cable. Figure 32-7B shows this scheme in place for the same hypothetical cellular connection as depicted at A.

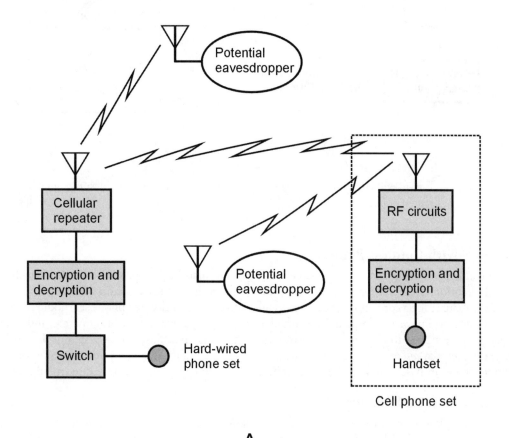

A

32-7A Wireless-only encryption. Heavy, straight lines represent wires or cables; zigzag lines represent RF signals.

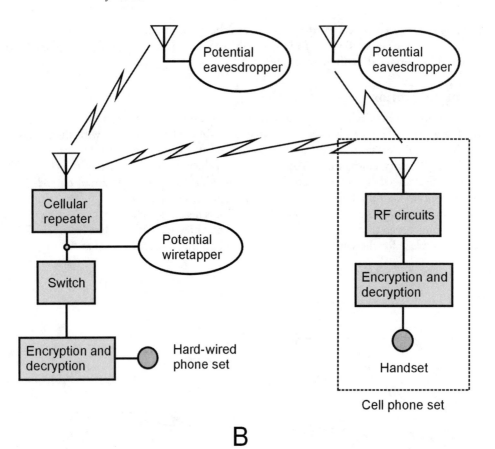

B

32-7B End-to-end encryption. Heavy, straight lines represent wires or cables; zigzag lines represent RF signals.

Security with Cordless Phones

Most cordless phones are designed to make it difficult for unauthorized people to pirate a telephone line. Prevention of eavesdropping is a lower priority, except in expensive cordless systems. If there is concern about using a cordless phone in a particular situation, a hard-wired phone set should be used.

If someone knows the frequencies at which a cordless handset and base unit operate, and if that person is determined to eavesdrop on conversations that take place using that system, it is possible to place a *wireless tap* on the line. The conversation can be intercepted at a point near the cordless phone set and its base unit, and then transmitted to a remote site (Fig. 32-8) and recorded there.

Security with Cell Phones

Cellular telephones are, in effect, long-range cordless phones. The wider coverage of cellular repeaters, as compared with cordless base units, increases the risk of eavesdropping and unauthorized use. Some cell phone vendors advertise their systems as "snoop proof." Some of these claims have more merit than others. The word "proof" (meaning "immune") should be regarded with skepticism. Digital encryption is the most effective way to maintain privacy and security of cellular communications. Nothing short of this is really of any use against a determined hacker.

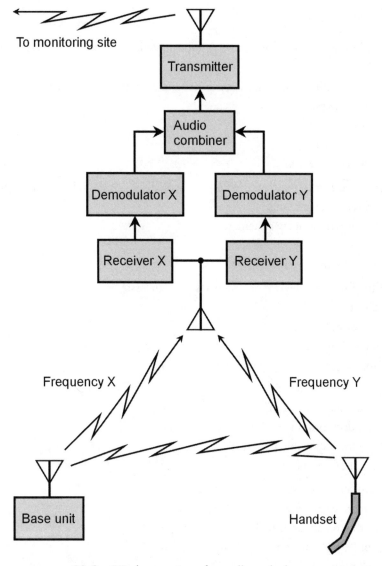

To monitoring site

32-8 Wireless tapping of a cordless telephone.

Access and privacy codes, as well as data, must be encrypted if a cell phone system is to be maximally secure. If an unauthorized person knows the code with which a cell phone set accesses the system (the "name" of the set), rogue cell phone sets can be programmed to fool the system into thinking they belong to the user of the authorized set. This is known as *cell phone cloning*. The

In addition to digital encryption of data, *user identification* (user ID) must be employed. The simplest is a *personal identification number* (PIN). More sophisticated systems can employ *voice-pattern recognition*, in which the phone set functions only when the designated user's voice speaks into it. *Hand-print recognition*, *electronic fingerprinting*, or *iris-print recognition* can also be employed. These are examples of *biometric security* measures.

Quiz

Refer to the text in this chapter if necessary. A good score is 18 correct. Answers are in the back of the book.

1. A network that employs one central computer and multiple personal computers is called
   (a) a wireless network.
   (b) a local area network.
   (c) a client-server network.
   (d) a peer-to-peer network.

2. In order for a receiver to render a digitally encrypted signal intelligible, that receiver must
   (a) have a directional antenna.
   (b) have a biometric security system.
   (c) have a decryption key.
   (d) be licensed by the government.

3. Which of the following devices or systems is *not* wireless?
   (a) An SWL station
   (b) An amateur radio station
   (c) A cordless phone set
   (d) A computer with a telephone modem

4. Eavesdropping of a wireless link
   (a) requires digital encryption.
   (b) requires a decryption key.
   (c) is difficult or impossible to detect.
   (d) includes all of the above.

5. In the United States, a license is required for
   (a) using a receiver on amateur radio frequencies.
   (b) transmitting on amateur radio frequencies.
   (c) using cordless phone sets.
   (d) using cell phone sets.

6. In a LEO satellite system, the repeaters are located
   (a) in space.
   (b) at strategic locations in most countries of the world.
   (c) at the north and south geographic poles.
   (d) over the equator.

7. In a geostationary satellite link, the repeater is located
   (a) on a buoy in the ocean.
   (b) at strategic locations in most countries of the world.
   (c) at the north and south geographic poles.
   (d) over the equator.

8. The term *shortwave*, in reference to radio, refers to signals having wavelengths of approximately

    (a) 10 mm to 100 mm.

    (b) 100 mm to 1 m.

    (c) 1 m to 10 m.

    (d) 10 m to 100 m.

9. Which of the following is illegal to do in the United States using a ham radio station?

    (a) Advertise used cars for sale

    (b) Transmit Morse code

    (c) Receive outside the amateur radio bands

    (d) Receive signals without a license

10. A file server is

    (a) a sensitive radio receiver equipped with D/A conversion.

    (b) a radio transmitter equipped with digital signal processing (DSP).

    (c) a central computer in a local area network.

    (d) a repeater in a LEO satellite system.

11. A device consisting of a receiver and transmitter in the same case or enclosure, and commonly used for wireless communications, is called

    (a) a file server.

    (b) a transceiver.

    (c) a transverter.

    (d) an encoder/decoder.

12. End-to-end encryption provides

    (a) maximum range in a wireless LAN.

    (b) a higher degree of security than wireless-only encryption.

    (c) simultaneous transmission and reception in an amateur radio station.

    (d) an easy means for hackers to clone cell phone sets.

13. A LAN in which each computer has more or less equal status, and in which each computer stores its own data, is known as a

    (a) wireless router.

    (b) wide-area LAN.

    (c) LAN topology.

    (d) peer-to-peer LAN.

14. With respect to security, the term *mil spec* refers to

    (a) wireless-only encryption.

    (b) a peer-to-peer network.

    (c) the use of a wireless router.

    (d) the highest obtainable level of security.

15. A radio signal is considered to be in the shortwave band if its frequency is which of the following:

    (a) 5 MHz
    (b) 50 MHz
    (c) 500 MHz
    (d) Any of the above

16. In a cellular network, the region covered by a single base station is called a

    (a) zone.
    (b) cell.
    (c) radius.
    (d) link.

17. An advantage of a conventional hard-wired telephone over a cell phone is the fact that a hard-wired phone set

    (a) affords better privacy.
    (b) offers greater portability.
    (c) is more convenient for use around the house.
    (d) employs digital encryption.

18. An advantage of a cell phone over a hard-wired phone is the fact that a cell phone set

    (a) affords better privacy.
    (b) costs less.
    (c) offers better mobility.
    (d) provides higher data speed.

19. In a two-way communications system with full duplex,

    (a) either party can hear the other at all times.
    (b) a cell phone transmits on two frequencies at the same time.
    (c) end-to-end encryption is necessary.
    (d) a file server or router is necessary.

20. An advantage of LEO satellite systems over geostationary satellite systems is the fact that

    (a) the latency is much greater with LEO satellite systems.
    (b) the equatorial regions are covered by LEO satellite systems, but not by geostationary satellite systems.
    (c) the regions near the geographic poles are covered by LEO satellite systems, but not by geostationary satellite systems.
    (d) All of the above are advantages of LEO satellite systems over geostationary satellite systems.

A Computer and Internet Primer

COMPUTERS ARE USED FOR CALCULATIONS, COMMUNICATIONS, WORD PROCESSING, DATA PROCESSING, drawing, photo processing, music composition and editing, location, navigation, information searches, robot control, and many other purposes. Figure 33-1 is a block diagram showing the major parts of a typical computer system used in a home or small business.

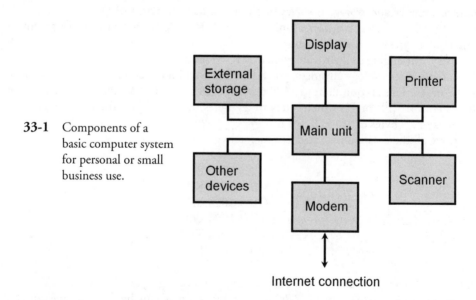

33-1 Components of a basic computer system for personal or small business use.

The Central Processing Unit

The *microprocessor* is the IC, or chip, that forms the core of your computer's "brain." It coordinates all the action and does all the calculations. It is located on the *motherboard*, or main circuit board. This board is sometimes called the *logic board*. The microprocessor, together with various other circuits, comprises the *central processing unit* (CPU). Auxiliary circuits can be fabricated onto the same chip as the microprocessor, but they are often separate. The external chips contain *memory, storage,* and *programming* instructions.

You might think of the microprocessor as the computer's conscious mind, which directs the behavior of the machine by deliberate control. The CPU, controlled by the microprocessor, represents the PC's entire mind, conscious and subconscious. All the ancillary circuits, in conjunction with the CPU, make up the nervous system. Peripherals such as printers, disk drives, mice, speech recognition/synthesis apparatus, modems, and displays are the hands, ears, eyes, and mouth of the system. In advanced computer systems there can also be robots, vision systems, various home appliances, surveillance apparatus, medical devices, and other exotic equipment under the control of the CPU.

Units of Digital Data

You learned about bits and bytes in Chap. 26. Let's examine them again in more detail, with emphasis on their relevance to computers. Recall that one bit (1 b) is a single binary digit, and one byte (1 B) is a unit of digital data consisting of a string of eight contiguous bits (8 b) in most systems. One byte constitutes roughly the same amount of data as one character, such as a letter, numeral, punctuation mark, or space.

Memory and Storage Capacity

Computer memory and storage involves files that are huge in terms of bytes. Therefore, *kilobytes* (units of $2^{10} = 1024$ bytes), *megabytes* (units of $2^{20} = 1,048,576$ bytes), and *gigabytes* (units of $2^{30} = 1,073,741,824$ bytes) are used. The abbreviations for these units are KB, MB, and GB, respectively. Alternatively you might see them abbreviated as K, M, and G.

As computer technology advances, you'll hear more and more about a unit of data called a *terabyte* (TB or T). This is equivalent to 2^{40} bytes, or 1,048,576 MB. Someday we will commonly use the terms *petabyte* (PB or P), which refers to 2^{50} bytes or 1,048,576 GB, and *exabyte* (EB or E), which refers to 2^{60} bytes or 1,048,576 TB.

Here are all these data units listed as a hierarchy:

$$1\ KB = 1024\ B$$
$$1\ MB = 1024\ KB$$
$$1\ GB = 1024\ MB$$
$$1\ TB = 1024\ GB$$
$$1\ PB = 1024\ TB$$
$$1\ EB = 1024\ PB$$

Computer memory is usually specified in megabytes or gigabytes. The same holds true for removable data storage media. The *hard drive* in a computer generally has capacity measured in gigabytes, although a few get into the terabyte range. Some external storage media, used for data *archiving* (saving it for long-term reference), have capacity measured in terabytes and petabytes.

Data Speed

The speed at which computers send digital data to and from each other is almost always expressed in *bits per second* (bps) and large multiples thereof. Multiples of bits per second involve the same prefixes as multiples for bytes, but when talking or writing about bits per second, these prefixes refer to powers of 10 rather than powers of 2.

A *kilobit per second* (1 kbps) is equal to $10^3 = 1000$ bps, a *megabit per second* (1 Mbps) is equal to $10^6 = 1,000,000$ bps, and a *gigabit per second* (1 Gbps) is equal to $10^9 = 1,000,000,000$ bps. (You won't often hear of data speed units larger than the gigabit per second.) Hierarchically:

$$1 \text{ kbps} = 1000 \text{ bps}$$
$$1 \text{ Mbps} = 1000 \text{ kbps}$$
$$1 \text{ Gbps} = 1000 \text{ Mbps}$$

Note the lowercase "k" in reference to "kilo-" meaning 10^3 or 1000, and the uppercase "K" in reference to "kilo-" meaning 2^{10} or 1024. Note also that for the other prefix multipliers, uppercase letters are always used. These are not typos! The "k versus K" distinction is a notational peculiarity, and is often ignored or overlooked. You will sometimes see "Kbps" rather than "kbps," or "kB" rather than "KB," in technical papers and other documents. As long as you know whether the author is writing about data speed (bits per second) or memory/storage (bytes), it should not be a problem. But if you want to be rigorous, this peculiarity is worth remembering.

The Hard Drive

A hard drive, also known as a *hard disk*, is a common form of mass storage for computer data. The drive consists of several disks, called *platters*, arranged in a stack. They are made of rigid, durable material that is coated with a ferromagnetic substance similar to that used in audio or video tape. The platters are spaced a fraction of a centimeter apart. Each has two sides (top and bottom) and two *read/write heads* (one for the top and one for the bottom). The assembly is enclosed in a sealed cabinet. Figure 33-2A is an edgewise, cutaway view of the platters and heads in a typical hard drive.

Drive Action

When the computer is switched off, the hard-drive mechanism locks the heads in a position away from the platters. This prevents damage to the heads and platters if the computer is moved. When the computer is powered up, the platters spin at several thousand revolutions per minute (rpm). The heads hover a few millionths of a centimeter above and below the platter surfaces.

When you type a command or click on an icon telling the computer to read or write data, the hard-drive mechanism goes through a series of rapid, complex, and precise movements. The head positions itself over the particular spot on the platter where the data is located or is to be written; then the head detects the magnetic fields and translates them into tiny electric currents. All this takes place in a small fraction of a second.

Data Arrangement and Capacity

The data on a hard drive is arranged in concentric, circular *tracks*. There are hundreds or even thousands of tracks per radial centimeter of the platter surface. Each circular track is broken into a number of arcs called *sectors*. A *cylinder* is the set of equal-radius tracks on all the platters in the drive. Tracks and sectors are set up on the hard drive during the initial *formatting* process. There are also data units called *clusters*. These are units consisting of one to several sectors, depending on the arrangement of data on the platters. Figure 33-2B is a face-on view of a single hard-disk platter, showing a track and one of its constituent sectors.

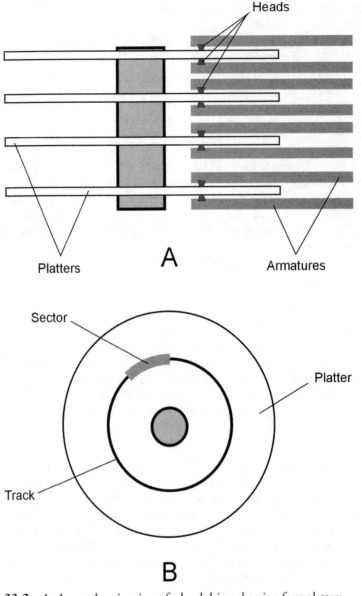

33-2 At A, an edgewise view of a hard drive, showing four platters. At B, a broadside view of a platter, showing one track and one sector.

When you buy a computer, whether it is a desktop, notebook (also called laptop), or portable (also called handheld) unit, it will have a hard drive built in. The drive comes installed and formatted. Most new computers are sold with several commonly used programs installed on the hard drive. Some computer users prefer to buy new computers with only the operating system, by means of which the programs run, installed; this frees up hard-drive space and gives the user control over which programs to install (or not to install).

External Storage

There are several types of *external storage* (besides the hard drive) in which data can be kept in large quantities. Computer experts categorize external storage in two ways: *access time* and *cost per megabyte* (or gigabyte, or terabyte).

Disk Media

Disk media offer the advantage of speed, convenience, and reliability. For personal computers, there are many forms of external disks. Here are three of the most well-known.

An *external hard drive* is exactly what its name suggests. This type of device is exceptionally fast, and has storage capacity similar to the hard drives inside computers. They can be easily connected to any personal computer with a short cable. Most of these devices require a power supply, often found in the form of a "brick" that contains a transformer, rectifier, and filter that converts 120 V ac into the necessary dc for the device.

Compact disk recordable (CD-R) and *compact disk rewritable* (CD-RW) are popular for backing up and archiving computer data. You can buy these disks for various other applications, too, such as storage for digital photos and home videos. They are the same size, physically, as conventional CD-ROMs, which are used for commercial software, databases, and digital publications. The main asset of CD-R and CD-RW is moderately large capacity and long shelf life. Most new computers have built-in drives for these disks.

Diskettes, also called (imprecisely) "floppies," are about 9 cm (3.5 in) in diameter and enclosed in a rigid, square case about 4 mm (0.15 in) thick. Their capacity, individually, is limited. They are all but obsolete. Increasingly, new computers are sold without drives for floppies.

Tape Media

The earliest computers used magnetic tape to store data. This is still done in some systems. You can get a *tape drive* for making an emergency backup of the data on your hard drive, or for archiving data you rarely need to use. Magnetic tape has high storage capacity. There are microcassettes that can hold more than 1 GB of data; standard cassettes can hold many gigabytes. But tapes are extremely slow because, unlike their disk-shaped counterparts, they are a *serial-access* storage medium. This means that the data bits are written in a string, one after another, along the entire length of the tape. The drive might have to mechanically rewind or fast-forward through a football field's length of tape to get to a particular data bit, whereas on a disk medium, the read/write head never has to travel farther than the diameter of the disk to reach a given data bit.

Flash Memory

Flash memory is an all-electronic form of storage that is useful especially in high-level graphics, big-business applications, and scientific work. The capacity is comparable to that of a small hard drive, but there are no moving parts. Because there are no mechanical components, flash memory is faster than any other mass-storage scheme, provided it does not cause a *software conflict* with other programs in the computer. (Software conflicts can cause a computer to slow down or "freeze up.")

Flash memory is available in small modules roughly the size of your index finger, and can be plugged directly into one of the *Universal Serial Bus* (USB) ports provided in all new computers. Some flash memory modules come in the form of *PC cards* (also called *PCMCIA cards*), which are credit-card-sized, removable components.

Memory

In a computer, the term *random access memory* (RAM), often called simply *memory*, refers to ICs that store working data. The amount, and speed, of memory are crucial factors in determining what a computer can and cannot do.

Data Flow

Figure 33-3 shows how data moves in a computer among an internal hard drive, an external hard drive, and a CD-RW storage medium. This process is controlled by the CPU. When you open a file on any medium, the data goes immediately into the memory. The CPU, under direction of the microprocessor, manipulates the data in the memory as you work on the file. Thus, the data in memory changes from moment to moment.

When you hit a key to add a character, or drag the mouse to draw a line that shows up on your display, that character or line goes into memory at the same time. If you hit the backspace key to delete a character, or drag the mouse to erase a line on the screen, it disappears from the memory.

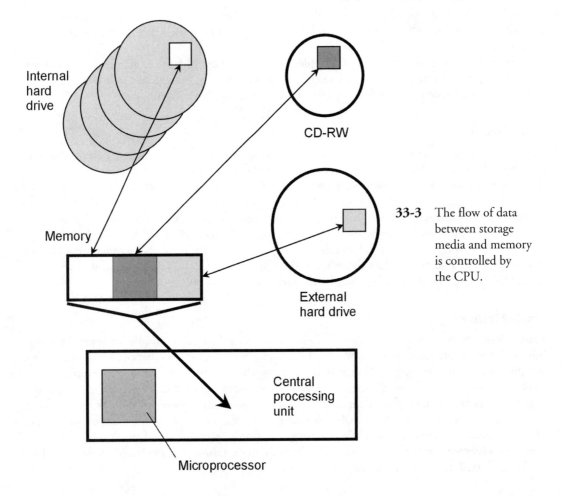

33-3 The flow of data between storage media and memory is controlled by the CPU.

During this time the original file on the storage medium stays as it was before you accessed it. No change is made to that data until you specifically instruct the computer to overwrite it. When you're done working on a file, you tell the microprocessor to close it. Then the data leaves memory and goes back to the medium from which it came, or to some other place, as you direct. If you tell the computer to overwrite the file on the medium from which it came, many programs send the new data (containing the changes you have made) to unused space there; the old data (as it was before you opened the file) stays in its old location. This is a safeguard, in case you decide to undo the changes you made.

All the data passing between the storage media and memory, and between the memory and the CPU, is in *machine language.* This consists of binary digits (bits) 0 and 1. But the data passing between you and the CPU is in plain English (or whatever other language you prefer), or in some high-level programming language, having been translated by the machine into a form you can understand.

Memory Capacity

The maximum number of megabytes or gigabytes of data that can be stored in a computer's memory is known as the *memory capacity.* The main factor that determines memory capacity is the number of transistors that can be fabricated onto a single memory chip. Other factors, such as microprocessor speed, have a practical effect on the usable memory capacity.

A gigantic memory is of little practical value if the microprocessor is slow. Nor is a fast microprocessor worth much if the memory capacity is too small for the applications (programs) you want to run. Most software packages tell you how much memory you need to run the programs they contain. They'll often quote two specifications: a minimum memory requirement and a figure for optimum performance (approximately twice the minimum requirement). If possible, you should equip your computer with enough memory for optimum performance.

When buying a new computer, keep in mind that this year's high-end machine may become next year's ordinary one when it comes to popular software, and in a few years it will no longer be adequate to run many of the programs that will be available. If your computer lacks the memory to run a given application, you can usually add more. But this can only be done up to a certain point. Eventually, your microprocessor will no longer be able to run contemporary software at reasonable speed, no matter how much memory you have.

How much memory do you think your machine will require to run two or three of your favorite applications at the same time? Double or triple it, and you will come close to the amount you are likely to need for the next two or three years, or until you are overcome by the urge to buy a new computer again.

The Display

The visual interface between you and your computer is known as the *display.* In desktop computers, an external display is often called a *monitor.* There are two popular types in use for personal computing today.

A *cathode-ray tube* (CRT) *monitor* resembles a television set without the tuning or volume controls. This type of display is large and heavy, and although some people consider CRTs obsolete, other computer users prefer them, especially the larger ones, which can have a diagonal screen measure of 53 cm (21 in) or more.

A *liquid crystal display* (LCD) is lightweight and thin. This type of display is used in notebook and portable computers. It has become increasingly popular for desktop computers because the technology has improved and has become somewhat less expensive than it used to be. Light weight and a shallow desk profile are decided advantages of this type of display. A good one can provide image crispness and color at least as good as the best CRT display. An LCD also consumes less power.

Other types of displays exist. Of special note is the *plasma display*, which obtains its images from the properties of electric charges passing through rarefied gases. These displays are often found in large department stores, where you can see them hanging from the ceilings, displaying advertisements.

Resolution

The *image resolution*, often called simply the *resolution*, of a computer display is one of its most important specifications. This is the extent to which it can show detail. The better the resolution, the sharper the image.

Resolution can be specified in terms of *dot size* or *dot pitch*. This is the diameter, in millimeters (mm), of the individual elements in the display—the "smallest unbreakable pieces." A good display has a dot pitch that is a small fraction of a millimeter. A typical CRT has a dot pitch of 0.25 or 0.26 mm. The smaller the number, the higher the resolution and, all other factors being equal, the crisper the image in an absolute sense.

Image resolution can be specified in a general sense as a pair of numbers, representing the number of *pixels* (picture elements) the screen shows horizontally and vertically. For a particular screen size, the greater the number of pixels the unit can display, the crisper the image. In personal computers, typical displays have 800×600 resolution (800 pixels wide by 600 pixels high) or 1024×768 resolution. A few can work up to 1600×1200 or even higher.

Screen Size

Screen sizes are given in terms of diagonal measure; a popular size is 43 cm (17 in). This will work quite well at 800×600 resolution. For higher resolution, a larger screen is preferable, such as 48 cm (19 in) or 53 cm (21 in). A high-end display is crucial for doing graphics work, when doing serious research on the Internet, in remote-control robotics, and in computer *gaming*. Besides these practical advantages, a sharp display is more pleasant to work with than a marginal one.

Along with memory and hard-drive capacity, the display is one of the most important parts of a computer from a user-friendliness standpoint. On the job, long hours at a computer can get tedious even if the machine is perfect. An inadequate display can give rise to eye strain and headaches, and can also degrade the quality and accuracy of work done by people using the computer.

Interlacing and Refresh Rate

Another important consideration in the choice of a display is whether or not it uses *interlacing*. Interlacing increases the obtainable resolution, but it also results in a lower *refresh rate* (number of times the entire image is renewed). A low refresh rate can cause noticeable flickering in the image, and can be especially disruptive in applications where rapid motion must be displayed, such as high-end computer gaming.

A good refresh rate specification is 70 Hz or more. For applications not involving much motion, 60 or 66 Hz is adequate for some people, but others complain of eye fatigue because they can vaguely sense the flicker. Most people find 56-Hz refresh rates too slow for comfort in any application.

ELF Fields

Extremely low frequency (ELF) fields are electromagnetic (EM) fields that fluctuate or alternate more slowly than conventional radio waves. Such fields are produced by various consumer electronic devices and appliances. An ELF field is nothing like X rays or gamma rays, which cause radiation sickness. Nor is ELF radiation like ultraviolet (UV), which can cause skin cancer over long periods. An ELF field cannot make anything radioactive. In computer systems, the ELF that you've heard about is emitted mainly by electromagnetic CRT monitors. Other parts of a computer system are not responsible for much ELF energy. The LCDs used in many systems today produce essentially none.

As you learned in Chap. 29, the characters and images in a CRT are created as electron beams strike a phosphor coating on the inside of the glass. The electrons constantly change direction as they sweep from left to right, and from top to bottom, on your screen. The sweeping is caused by deflecting coils that steer the beam across the screen. The coils generate magnetic fields that interact with the negatively charged electrons, forcing them to change direction. Because of the positions of the coils, and the shapes of the fields surrounding them, there is more magnetic energy radiated from the sides of an electromagnetic CRT than from the front. If there's any health hazard with ELF, therefore, it is greater for someone sitting off to the side of an electromagnetic CRT monitor, and less for someone watching the screen from directly in front at the same distance.

If you're concerned about the ELF fields produced by CRT monitors, you might consider buying an electrostatic CRT unit that has been designed to minimize ELF fields, or buying a stand-alone LCD display panel. It's a good idea to arrange your workstation so your eyes are at least 0.5 m (about 18 in) away from the screen of a CRT monitor, and if you are working near other computer users, workstations should be at least 1 m (3 ft) apart.

The Printer

There are several types of *printer* in common use in personal and business computing applications today. The two most common are the *inkjet printer* and the *laser printer.* Less often used are the *thermal printer* and the *dot matrix printer.*

Inkjet Printers

In an inkjet printer, tiny nozzles spray ink onto the paper. Most of these printers are comparatively slow to produce an image, and the ink needs time to dry even after the image comes out, but the quality can be excellent. Inkjet printers are available in single-color and multicolor designs. The best color machines produce images of photo quality. In fact, some are designed especially to print digital photos.

Inkjet printers require periodic replacement of the ink cartridges. These can be quite expensive, and for this reason, inkjet printers are not well suited for high-volume printing. Inkjet printers require paper with low fiber content. This keeps the ink from being carried along by capillary action before it dries, muddling or blurring the printout. Look specifically for "inkjet paper" in computer supply stores and department stores.

Laser Printers

A laser printer works like a photocopy machine. The main difference is that, while a photocopier creates a copy of a real image (the paper original), a laser printer makes a copy of a digital computer image.

When data arrives at the printer from the computer, the encoded image is stored in the printer memory. The memory then sends it along to the laser and other devices. The laser blinks rapidly while it scans a cylindrical *drum*. The drum has special properties that cause it to attract the printing chemical, called *toner*, in some places but not others, creating an image pattern that will ultimately appear on the paper. A sheet of paper is pulled past the drum and also past an *electrostatic charger*. Toner from the drum is attracted to the paper. The image thus goes onto the paper, although it has not yet been permanently fused, or bonded, to the paper. The *fuser*, a hot pair of roller/squeezers, does this job, completing the printing process.

The main asset of laser printers is their ability to produce a large number of copies at high speed. They generally have excellent print and graphics quality for grayscale. Color laser printers are available as well, but they can be quite expensive.

The image resolution of a laser printer ranges from about 300 dots per inch (dpi) for older units to 1200 dpi, 2400 dpi, or even higher resolutions in state-of-the-art machines. As far as the untrained eye can tell, 600 dpi is as good as a photograph. Laser printers can handle graphics and text equally well. If an image can be rendered on a photocopy machine, it can be rendered just as well on a laser printer.

Thermal Printers

A thermal printer uses temperature-sensitive dye and/or paper to create hard-copy text and images. Some thermal printers produce only grayscale images, while others can render full color. Thermal printers are often preferred by traveling executives who use portable computers, because these printers are physically small and light.

A simple grayscale thermal printer employs special paper that darkens when it gets hot. A color thermal printer uses thick, heat-sensitive dyes of the primary pigments: magenta (pinkish red), yellow, and cyan (bluish green). Sometimes black dye is also used, although it can be obtained by combining large, equal amounts of the primary pigments. The print head uses heat to liquefy the dye, so it bleeds onto the paper. This is done for each pigment separately.

Some, if not most, thermal printouts fade after awhile. Have you ever pulled out an old store receipt and found that it was washed-out or blank? Thermal printers can be convenient in a pinch, but you should be aware that some of them have this problem. If you're keeping a receipt for tax purposes or for proof-of-purchase and it has been printed on thermal paper, make a photocopy or a digital scan of the receipt right away. You can recognize the output of a thermal printer because the paper curls up when it's fresh out of the machine.

Dot Matrix Printers

The dot matrix printer is the horse and buggy of the printing family. This type of printer is the least expensive, in terms of both the purchase price and the long-term operating cost. Dot matrix printers produce fair text quality for most manuscripts, reports, term papers, and theses. The mechanical parts are rugged, and maintenance requirements are minimal. Dot matrix printers can render some simple graphic images, but the quality is fair at best, and it can take a long time to print a single image. Dot matrix printers cannot reproduce detailed artwork or photographs with acceptable quality.

The Scanner

A *scanner* is an electromechanical device that converts hard-copy text and graphics into digital form for processing and storage in a computer.

Basic Features

Scanners can save untold hours of tedious manual retyping. Suppose you wrote a book a long time ago, and have lost the digital files (or used an old-fashioned typewriter, and never had any digital files!). The hard copy of that book sits in your basement, awaiting the massive editing that can turn it into a great novel. It needs the power of your computer's word processor. But you can't deal with the prospect of retyping its 1000 pages. A scanner, equipped with *optical character recognition* (OCR), can do away with most of the hard labor involved in getting a hard-copy manuscript onto disk.

A good scanner can be had for a couple of hundred dollars. But the value of optical scanning is hard to measure. For many entrepreneurs, it can make the difference between staying afloat and going bankrupt. It can do the work of one or two full-time typists for a tiny fraction of the long-term cost. Big companies can save money, too. Lawyers and doctors find scanners invaluable for backing up files of all kinds. Aside from storing text, scanners can make digital copies of vital papers, records, and receipts, which can be easily backed up on CD-R or CD-RW media.

A typical scanner can render color images, text, photographs, and everything else needed to make a complete, accurate digital record of any document. Color scanners use three different light beams (red, blue, and green) to get three different images, which are processed and combined in much the same way as a color television camera works. The image resolution of a scanner is measured in dots per inch (dpi), just as is done with printers. The higher the dpi specification, the more detail the scanner can see.

For reliable scanning of text and most images, a resolution of at least 300 dpi is recommended. Virtually all scanners meet this requirement. For images, greater detail translates into more memory consumed. Color increases the amount of memory or storage that an image takes up, if the image resolution remains constant.

Configurations

Scanners come in three basic configurations. The scanner that's best for you will depend on what you want to do with it, and on how much money you're willing to spend for it.

The cheapest type of scanner is a *handheld scanner*. It looks something like a miniature vacuum-cleaner head, or one of the bar-code readers in retail stores. You roll the unit over the paper containing the text and/or graphics you want to scan. Because the unit is not as wide as most pages, you'll have to make two or three passes over the page. Handheld scanners are preferred by people who scan small images, such as snapshots. They are light in weight, and need almost no desk space. One potential problem is that you might try to scan too fast. Some handheld scanners have speed indicators that tell you if you're going too fast. Another potential difficulty is not getting a straight-line scan. Most handheld units have built-in guides (like miniature rolling pins) that minimize this problem.

If you want to scan a book or magazine, a *flatbed scanner* is much easier to use than a handheld scanner. The unit looks something like a photocopier. Using a flatbed scanner is similar to working a small photocopy machine. You lay the page, photo, or sheet down on a clear glass, and the scanning head moves past it, picking up the image. Flatbed scanners consume desk space, which, if you have a couple of printers and a fax machine, might already be at a premium.

A *sheet scanner*, also called a *feedthrough scanner*, resembles a fax machine (and in fact, many of these units can do double duty as fax machines). As its name implies, this type of scanner pulls sheets of paper through, one by one. You can stack several pages, one on top of the other, and the machine will automatically feed and scan them. However, you can't scan bound books or magazines as you can with a flatbed scanner—unless you're willing to rip out individual pages.

Precautions

Even the best scanners make some mistakes when used with OCR. This is especially true if text contains nonstandard symbols. Highly technical material presents the worst problems. Some mathematical symbols are so esoteric that the average person (let alone a machine) is befuddled by them. Ink spots, stray markings, and smudges on a page can cause scanning errors, in much the same way as background noise confuses a speech recognition system.

A human reader can often tell what a printed letter should be, even when it is severely mutilated. But computers lack human intuition. To some extent this can be corrected by a built-in *spell checker.* Some OCR programs have spell checking, but this introduces its own set of problems because it, too, is imperfect.

If a scanner doesn't recognize a character, it will usually print a "tag"—a blank space, underline, or default symbol such as @ or #. Scanned text must always be carefully proofread, and corrections made with word-processing software, after the data has been stored on the hard drive.

The Modem

The term *modem* is a contraction of *modulator/demodulator.* A modem interfaces a computer to a telephone line, *digital subscriber line* (DSL), cable system, fiber-optic network, wireless network, or radio transceiver, allowing you to communicate with other computer users and to "surf the Internet."

Data Speed

Modems work at various speeds, usually measured in kilobits per second (kbps) or megabits per second (Mbps). Sometimes you'll hear about speed units called the *baud* and *kilobaud.* (A kilobaud is 1000 baud.) Baud and bits per second are almost the same units, but they are not identical. People sometimes use the term *baud* when they really mean *bps.*

Modem speeds, particularly in cable, fiber-optic, and wireless networks, keep increasing as computer communications technology advances. Modems are rated according to the highest data speed they can handle. A typical *telephone modem* works at about 56 kbps. Modems for more advanced connections operate much faster.

Slow Modems

A computer works with binary *digital* signals, which are rapidly fluctuating direct currents. For digital data to be conveyed over a telephone or radio circuit, the data must be converted to *analog* form. In a telephone modem or radio-transceiver modem, this is done by changing the digit 1 into an audio tone, and the digit 0 into another tone with a different pitch. The result is an extremely fast back-and-forth alternation between the two tones.

In *modulation,* digital data from the computer is changed into analog data for transmission over the telephone line or radio medium. The modulator is therefore a digital-to-analog converter (D/A converter or DAC). *Demodulation* changes the analog signals from the telephone line or radio medium back to digital signals that a computer can understand. The demodulator is thus an analog-to-digital converter (A/D converter or ADC). The highest practical speed for this type of modem is approximately 56,000 bits per second (bps), or 56 kilobits per second (kbps).

Amateur radio operators use a variety of digital modes at considerably slower speeds than 56 kbps for Internet-like communications. The principal advantage of amateur radio lies in the fact that it can work when all else fails. Such communications are not fast, but no infrastructure is needed. This makes it possible for amateur radio operators to provide emergency communications into and out of areas stricken by natural disasters that destroy the conventional communications infrastructure.

Internal versus External Modems

An *internal modem* is a printed-circuit board, also called a *card*. Virtually all computers, both desktop and notebook, are sold with internal modems that can interface with the telephone line at speeds of up to 56 kbps.

An *external modem* is a self-contained unit. It has a cord that runs to either the computer's *serial data port* (also called the *communications port*) or one of the USB ports, and another cord that runs to the telephone line, cable system, or satellite dish. If you want to use a DSL, cable, or satellite Internet connection, you will usually have to buy or rent an external modem from the service provider.

The Internet

The *Internet* is a worldwide system, or *network*, of computers. It got started in the late 1960s, originally conceived as a network that could survive nuclear war. Back then it was called *ARPAnet*, named after the Advanced Research Project Agency (ARPA) of the United States federal government.

Protocol and Packets

When people began to connect their computers into ARPAnet, the need became clear for a universal set of standards, called a *protocol*, to ensure that all the machines "speak the same language." The modern Internet is such that you can use any computer to take advantage of all the network's resources.

Internet activity consists of computers "talking" to one another. This occurs in machine language. However, the situation is vastly more complicated than when data goes from one place to another within a single computer. In the Internet (often called simply *the Net*), data must often go through several different computers to get from the transmitting or *source* computer to the receiving or *destination* computer. These intermediate computers are called *nodes*, *servers*, *hosts*, or *Internet service providers* (ISPs).

Millions of people are simultaneously using the Net. The most efficient route between a given source and destination can change from moment to moment. The Net is set up in such a way that signals always try to follow the most efficient route. If you are connected to a distant computer, say a machine at the National Hurricane Center, the requests you make of it, and the data it sends back to you, are broken into small units called *packets*. Each packet coming to you has, in effect, your computer's name written on it. But not all packets necessarily travel the same route through the network. Ultimately, all the packets are reassembled into the data you want (such as the infrared satellite image of a hurricane), even though they do not arrive in the same order they were sent.

Figure 33-4 is a simplified drawing of Internet data transfer for a hypothetical file containing five packets transferred during a period of *heavy Net traffic* (meaning that a lot of people are using the Internet at that particular time, so it is operating at near capacity). Nodes are shown as black dots surrounded by circles. In this example, some packets pass through more nodes, and/or over a much greater physical distance, than others. If Net traffic were light, all the packets might follow the same route, or pass through fewer nodes. This is why it takes longer to acquire data on the Net during peak hours of use, as compared with times when there are comparatively few people connected

Destination data

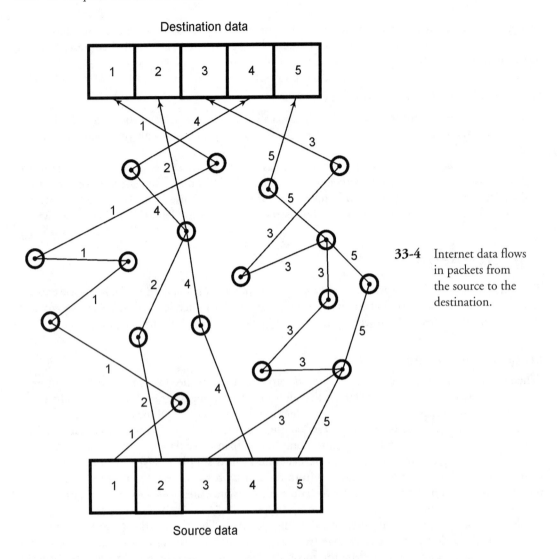

33-4   Internet data flows in packets from the source to the destination.

Source data

into it, no matter how fast your connection happens to be. A file cannot be completely reconstructed until all the packets have arrived and the destination computer has ensured that there are no errors.

E-mail and Newsgroups

For many computer users, communication by means of *electronic mail* (*e-mail*) and/or *newsgroups* has practically replaced the postal service. You can leave messages for, and receive them from, friends and relatives scattered throughout the world.

To effectively use e-mail or newsgroups, everyone must have an *Internet address*. These tend to be arcane. An example is sciencewriter@tahoe.com. The first part of the address, before the @ symbol, is the *username*. The word after the @ sign and before the period (or dot) represents the *domain name.* The three-letter abbreviation after the dot is the *domain type.* In this case, "com" stands for "commercial." Tahoe.com is thus a commercial provider. Other common domain types include "net" (network), "org" (organization), "edu" (educational institution), and "gov" (government). In recent years, country abbreviations have been increasingly used at the ends of Internet addresses, such as "us" for United States, "de" for Germany, "uk" for United Kingdom, and "jp" for Japan. Other abbreviations are constantly coming into common usage, such as "info" for "informational site" and "biz" for "business."

Internet Conversations

You can carry on a "teletype-style" (real-time text) conversation with other computer users using the Internet. When done among users within a single service provider, this is called *chat.* When done among people connected to different service providers, it is called *Internet relay chat* (IRC). Typing messages to, and reading them from, other people in real time is more personal than letter writing, because your addresses get their messages immediately. But it's less personal than talking on the telephone, because you cannot hear, or make, vocal inflections.

It is possible to digitize voice signals and transfer them via the Internet. This has given rise to hardware and software schemes that claim to provide virtually toll-free long-distance telephone communications. As of this writing, this is similar to cellular telephone communications in terms of reliability and quality of connection. When Net traffic is light, such connections can be good. But when Net traffic is heavy, the quality is marginal. Audio signals, like any other form of Internet data, are broken into packets. All, or nearly all, the packets must be received and reassembled before a good signal can be heard. This takes variable time, depending on the route each packet takes through the Net. If many of the packets arrive disproportionately late, the destination computer can only do its best to reassemble the signal. In the worst case, the signal does not get through at all.

Getting Information

One of the most important features of the Internet is the fact that it can connect you with millions upon millions of sources of information (and misinformation). Data is transferred among computers by means of a protocol that allows the files on the hard drives of distant computers to become available exactly as if the data were stored on your own computer's hard drive, except the access time is slower. You can also store files on distant computers' hard drives. When using the Internet for obtaining information, you should be aware of the time at the remote location, and avoid, if possible, accessing files during the peak hours at the remote computer. Peak hours usually correspond to working hours, or approximately 8:00 a.m. to 5:00 p.m. local time, Monday through Friday. You must take time differences into account if you're not in the same time zone as the remote computer.

The *World Wide Web* (also called *WWW* or *the Web*) is one of the most powerful information servers you will find on the Internet, and in common usage, the terms *Internet* and *Web* have become almost synonymous. The outstanding feature of the Web is *hypertext*, a user-friendly scheme for cross-referencing of documents. In fact, the names of Web sites generally begin with the four letters "http," which stands for *hypertext transfer protocol.* Certain words, phrases, and images make up *links.* When you select a link in a *Web page* or *Web site* (a document containing text, graphics, and often other types of files), your computer is transferred to another site dealing with the same or a related subject. This site will usually also contain numerous links. Before long, you might find yourself surfing the Web for hours, going from site to site. The word *surfing* derives from the similarity of this activity to television "channel surfing."

Quiz

Refer to the text in this chapter if necessary. A good score is 18 correct. Answers are in the back of the book.

1. One megabyte is the same amount of data as
   (a) 1024 bytes.
   (b) 1024 KB.
   (c) 1024 GB.
   (d) 1/1024 KB.

2. The Web should be expected to work fastest for a user in New York City at
   (a) 2:00 a.m. local time on a Tuesday.
   (b) 4:00 p.m. local time on a Wednesday.
   (c) 12:30 p.m. local time on a Thursday.
   (d) any time; it doesn't matter.

3. The sharpness of the image on a computer display can be specified in terms of
   (a) refresh rate.
   (b) interlace rate.
   (c) wavelength.
   (d) dot pitch.

4. The term *cylinder* refers to
   (a) a set of tracks in a hard drive.
   (b) a particular type of memory chip.
   (c) a drum-shaped data storage medium.
   (d) the spindle that turns a CD-R or CD-RW.

5. An example of a mass-storage device is a
   (a) hard drive.
   (b) microprocessor.
   (c) modem.
   (d) read-write head.

6. The character string *stangibilisco@rushmore.com* would most likely represent
   (a) a Web site.
   (b) the location of data in a computer's memory.
   (c) an e-mail address.
   (d) a computer's serial number.

7. The megabit per second (Mbps) is a common unit that expresses
   (a) the memory capacity of a chip.
   (b) the storage capacity of an external hard drive.

    (c) the image resolution of a printer.

    (d) the speed of an Internet connection.

8. A platter is

    (a) part of a hard drive.

    (b) a unit of memory.

    (c) an element of a digital image.

    (d) a semiconductor chip.

9. Protocol ensures that

    (a) a hard drive runs smoothly.

    (b) a display or printer reproduces color accurately.

    (c) a display or printer generates a clear image.

    (d) computers can exchange data.

10. A packet is

    (a) a computer memory module.

    (b) a unit of 2^{10} bytes.

    (c) a piece of data sent over the Net.

    (d) a picture element in a display.

11. The main microprocessor in a computer is located

    (a) on the motherboard.

    (b) in the external hard drive.

    (c) in the memory chip.

    (d) in the power supply.

12. Cross-referencing among Web pages is done with

    (a) digital signal processing.

    (b) an analog-to-digital converter.

    (c) Internet relay chat.

    (d) hypertext links.

13. Web page addresses usually begin with the letters "http," which signifies a form of

    (a) digital signal processing.

    (b) protocol.

    (c) modem configuration.

    (d) color rendition.

14. A telephone modem contains

    (a) an internal CR-R or CD-RW drive.

    (b) a microprocessor.

    (c) an A/D converter.

    (d) an image resolver.

15. An advantage of a flatbed scanner over a feedthrough scanner is the fact that
    (a) a flatbed scanner can be combined with a fax machine, but a feedthrough scanner cannot.
    (b) a flatbed scanner can be used to scan magazines or books intact, but a feedthrough scanner cannot.
    (c) a flatbed scanner can reproduce color drawings and photographs, but a feedthrough scanner cannot.
    (d) a flatbed scanner takes up less desktop space than a feedthrough scanner.

16. Which of the following types of external storage provide the fastest access time, provided no software conflicts occur?
    (a) A magnetic tape drive
    (b) A CD-R or CD-RW drive
    (c) A flash memory module
    (d) An external hard drive

17. Which of the following is a serial-access medium?
    (a) Magnetic tape
    (b) A diskette
    (c) A hard drive
    (d) A CD-R or CD-RW

18. Which of the following character strings represents the proper format for an e-mail address?
    (a) http://www.sciencewriter.net
    (b) www.mcgraw-hill.com
    (c) blackhills.com
    (d) None of the above

19. Which of the following devices is best suited for animated graphics work involving fast motion, such as high-end gaming?
    (a) A laser printer
    (b) An external hard drive or flash memory module
    (c) Hypertext transfer protocol
    (d) A display that does not use interlacing

20. A thermal printer might be an ideal choice for
    (a) a salesperson who is on the road.
    (b) someone working with animated graphics.
    (c) an author who needs to print a huge text document.
    (d) a photographer who needs top-quality color printouts.

34

Monitoring, Robotics, and Artificial Intelligence

ELECTRONIC AND COMPUTER SYSTEMS FIND APPLICATIONS IN MANY SCIENTIFIC AND TECHNOLOGICAL fields. Increasingly, devices and systems formerly known only to geeks and techies are becoming commonplace in the consumer and hobby market.

Keeping Watch

Most *monitoring systems* consist of radio transmitters and receivers. Some systems employ lasers at IR or visible wavelengths. A few employ wire or cable links. Some have mechanical hardware, such as robots, that are controlled by the signals reaching the receiver. Here are a few specific examples of monitoring systems.

Baby Monitor

A short-range AM or FM radio transmitter and receiver can be used to listen at a distance to the sounds in an infant's room. The transmitter contains a sensitive microphone, a whip antenna, and a power supply. The receiver is battery-powered and portable. It has a short antenna, similar to the antennas on cordless telephone sets. The receiver can pick up signals from the transmitter at distances of up to about 50 m (165 ft). The signals pass easily through the walls, ceilings, and floors in frame houses.

A so-called baby monitor is subject to interference from other units that might be operating nearby on the same channel. Some baby monitors have multiple, selectable channels to help combat this problem. If interference occurs, the channel can be changed. Communications privacy and security are not a concern.

Smoke Detector

Smoke and fire change the characteristics of the atmosphere. Smoke consists of solid particles, and fire burns away oxygen and produces other gases such as carbon dioxide, carbon monoxide, and sulfur dioxide. These changes can be sensed, and alarms set off if the changes exceed certain limits.

A *smoke detector* senses changes in two characteristics of the air: the *dielectric constant* (an expression of the extent to which pollutants increase the capacitance of the air) and the *ionization potential* (an expression of the extent to which pollutants change the voltage necessary to produce a spark that jumps through the air). Smoke and fire almost always affect one or both of these parameters.

Figure 34-1 is a functional diagram of a smoke detector that operates by sensing changes in the dielectric constant of the air. Two electrically charged plates are positioned a small distance apart. The plates, and the air between them, form a capacitor. A source of dc is connected to the plates. Normally, the plates retain a constant charge, and the current in the circuit is zero. If the dielectric constant of the air increases, the capacitance changes, causing a small, momentary electric current to flow. This current can be detected, and the resulting signal sets off an alarm. The signal can also actuate a robotic system, such as a group of water sprinklers.

Quality Control

Lasers are useful in industrial monitoring and control applications. An example is the *quality control* (QC) checking of bottles for height as they move along an assembly line. A laser/robot combination can find and remove bottles that are not of the correct height. The principle is shown in Fig. 34-2. If a bottle is too short, both laser beams reach the photodetectors. If a bottle is too tall, neither laser beam reaches the photodetectors. In either of these situations, a *robot arm*, equipped with a *gripper*, picks the faulty bottle off the line and discards it. Only when a bottle is within a narrow range of heights (the acceptable range) does the top laser reach its photodetector while the bottom laser is blocked. Then the bottle is allowed to pass.

Of course, proper operation of the QC machine shown in Fig. 34-2 depends on the reliability of the lasers and photodetectors. If, for example, the lower laser burns out and there is no way for the system operator to know about it, the machine will pass all bottles, whether or not they are of the correct height.

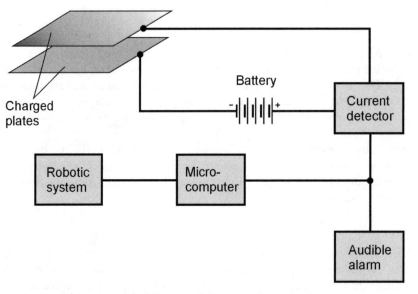

34-1   Simplified functional diagram of a smoke detector.

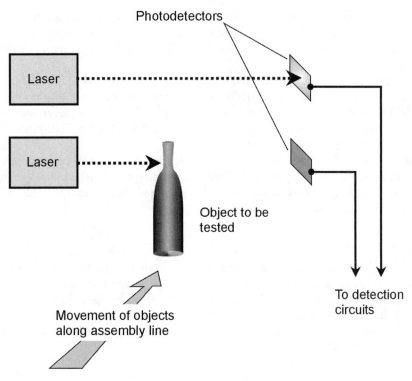

34-2 A method of quality control (QC) for manufactured items.

Tracking People

Suppose that a person is sentenced to house arrest. Compliance can be monitored by having the person carry a conventional beeper (pager). A police or probation officer can page the person at random times; the person must then call the officer within a couple of minutes. The call can be traced, and the location of the telephone verified. This is a simple method of electronically tracking the whereabouts of a person.

A more secure method of ascertaining that a person is at a certain place, at a certain time, is by means of a short-range radio transmitter and receiver. The person wears the transmitting unit. Tamper-proof receiving units are placed at the convict's home, in the car, and at the place of work. The transmitter range is similar to that of a baby monitor. Receiver signals are sent to a central monitoring point. The signals are encoded so the monitoring personnel (or computers) know whether the person is at home, in the car, or at work. Any deviation from normal patterns can be detected.

Radiolocation provides another way to keep track of people. A *transponder* can be carried or worn by the person to be tracked; continuous signals can be sent to the unit asking for a position fix, and the unit can respond through a wireless network such as the cellular telephone system.

Electronic Bug

An *electronic bug* consists of a tiny radio transmitter that can be hidden in a room, placed in a shirt pocket, or planted in a car. The antenna is a length of thin, almost invisible wire. A receiver can be located nearby. The device operates at a low RF power level (on the order of a few milliwatts) to conserve battery energy. If the transmitter is near a wireless repeater that connects into a larger system, eavesdropping can be done anywhere within the coverage of the wireless system. With the advent of low-earth-orbit (LEO) satellite systems, it is theoretically possible to bug a room on the other side of the world.

Wireless electronic bugs can be detected by means of a device called an *RF field-strength meter.* This instrument consists of a microammeter connected to a short whip antenna through a semiconductor diode. The diode rectifies the RF signal, producing dc that shows up as an indication on the meter. When the meter is close to the bug, the current rises. When the bug is within a few centimeters of the meter, the needle may go to full-scale.

The presence of RF fields in your house does not necessarily mean you are being bugged. Many appliances and electronic devices produce RF fields, including computers, radio-operated remote-control units, cell phones, cordless phone sets, and certain medical devices. Even radio receivers and TV sets emit some RF energy.

Electric Eye

The simplest device for detecting an unwanted visitor is an *electric eye.* Narrow beams of IR or visible light are shone across all reasonable points of entry, such as doorways and window openings. A photodetector receives energy from each beam. If, for any reason, the photodetector stops receiving its assigned beam, an alarm is actuated. A person breaking into a property cannot avoid interrupting at least one beam if every possible point of entry has a large enough number of electric eyes spaced at suitable intervals.

IR Motion Detector

A common intrusion alarm device employs an *IR motion detector.* Two or three wide-angle IR pulses are transmitted at regular intervals; these pulses cover most of the room in which the device is installed. A receiving transducer picks up the returned IR energy, normally reflected from the walls, floor, ceiling, and furniture. The intensity of the received pulses is noted by a microprocessor. If anything in the room changes position, there is a change in the intensity of the received energy. The microprocessor detects this change and triggers an alarm (Fig. 34-3). This type of device consumes very little power in regular operation, so batteries can serve as the power source.

Radiant Heat Detector

Certain devices can detect changes in the indoor environment by directly sensing the IR energy (often called *radiant heat*) emanating from objects. Humans, and all warm-blooded animals, emit IR. So does fire. A simple IR sensor, in conjunction with a microprocessor, can detect rapid or large increases in the amount of radiant heat in a room. The time threshold can be set so that gradual or small changes will not trigger the alarm, while significant changes, such as are caused by a person entering the room, will trigger it. The temperature-change threshold can be set so that a small animal will not actuate the alarm, while a full-grown person will. This type of device, like the IR motion detector, can operate from batteries.

The main limitation of radiant-heat detectors is the fact that they can be fooled. False alarms are a risk. The sun, coming out on an overcast day, might suddenly shine directly on the sensor and trigger the alarm. It is also possible that a person clad in a winter jacket, thermal pants, insulated boots, hood, and face mask, entering from a cold outdoor environment, will fail to set off the alarm. For this reason, radiant-heat sensors are used more often as fire-alarm actuators than as intrusion detectors.

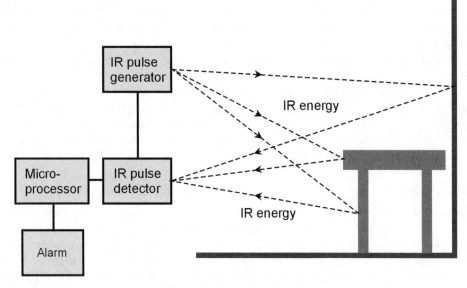

34-3 An IR motion detector.

Ultrasonic Motion Detector

Motion in a room can be detected by sensing the changes in the relative phase of acoustic waves. An *ultrasonic motion detector* employs a set of transducers that emit acoustic waves at frequencies above the range of human hearing (higher than 20 kHz). Another set of transducers picks up the reflected acoustic waves, whose wavelength is on the order of a few millimeters. If anything in the room changes position, the relative phase of the waves, as received by the various acoustic pickups, will change. This data is sent to a microprocessor, which can trigger an alarm and/or notify the police.

Robot Generations and Laws

Some researchers have analyzed the evolution of robots, marking progress according to so-called robot generations. One of the first engineers to make formal mention of robot generations was the Japanese engineer *Eiji Nakano*.

First Generation

According to Nakano, a *first-generation robot* is a simple mechanical arm. Such machines have the ability to make precise motions at high speed, many times, for a long time. They have found widespread industrial application and have been in existence since the middle of the twentieth century. These are the fast-moving systems that install rivets and screws in assembly lines, that solder connections on printed circuits, and that, in general, have taken over tedious, mind-numbing chores that would otherwise have to be done by humans.

First-generation robots can work in groups if their actions are synchronized. The operation of these machines must be constantly watched, because if they get out of alignment and are allowed to keep operating anyway, the result can be a series of bad production units.

Second Generation

A *second-generation robot* has some level of *artificial intelligence* (AI), also called *machine intelligence.* Such a machine is equipped with various sensors that keep it informed about goings-on in the *work environment.* A computer called a *robot controller* processes the data from the sensors and adjusts the operation of the robot accordingly. The earliest second-generation robots came into common use around 1980.

Second-generation robots can stay synchronized with each other, without having to be overseen constantly by a human operator. Periodic checking is needed, however, because things can always go wrong. In fact, as a system becomes more complex, the number of ways in which it can malfunction increases. This is why human beings will never find themselves out of work because of robots. Someone has to make sure the robots keep working properly!

Third Generation

Nakano gave mention to *third-generation robots*, but in the years since the publication of his original paper, some things have changed. Two major avenues are developing for advanced robot technology. These are the *autonomous robot* and the *insect robot.* An autonomous robot is a single machine that works on its own. It contains a controller and can do things largely without supervision, either by an outside computer or by a human being. A good example of this type of third-generation robot is the *personal robot* about which technophiles dream. An insect robot is one of a set of several (or many) identical units that act together to perform a specific task.

Fourth Generation and Beyond

Nakano did not write about anything past the third generation of robots. But we might mention a *fourth-generation robot:* a machine of a sort yet to be deployed. An example is a fleet or population of robots that can reproduce themselves, and perhaps even a system that can evolve to meet changing conditions in its work environment. Past that, we might say that a *fifth-generation robot* is something humans haven't even imagined yet.

Asimov's Three Laws

In one of his early science-fiction stories, the famous author Isaac Asimov first mentioned the word *robotics*, along with three fundamental rules that, in his opinion, all robots ought to obey. These rules were first coined in the 1940s, but *Asimov's three laws of robotics* are still considered valid today:

- First law: A robot must not injure, or allow the injury of, any human being.
- Second law: A robot must obey all orders from humans, except orders that would contradict the first law.
- Third law: A robot must protect itself, except when to do so would contradict the first law or the second law.

Robot Arms

A robot arm, in conjunction with an *end effector* (hand, gripper, or tool), is called a *manipulator.* Some robots, especially industrial robots, are nothing more than sophisticated manipulators. A robot arm can be categorized according to its geometry. Some manipulators resemble human arms. The joints in these machines can be given names like "shoulder," "elbow," and "wrist." Other manipulators are so much different from human arms that these names don't make sense.

Degrees of Freedom

The term *degrees of freedom* refers to the number of different ways in which a robot manipulator can move. Most manipulators move in three dimensions, but often they have more than three degrees of freedom.

You can use your own arm to get an idea of the degrees of freedom that a robot arm might have. Extend your right arm straight out toward the horizon. Extend your index finger so it is pointing. Keep your arm straight, and move it from the shoulder. You can move your shoulder joint in three ways. Up-and-down movement is called *pitch*. Movement to the right and left is called *yaw*. You can rotate your whole arm from the shoulder (albeit to a limited extent); this motion is called *roll*. Your shoulder therefore has three degrees of freedom: pitch, yaw, and roll.

Now move your arm from the elbow only. Holding your shoulder in the same position constantly, you will see that your elbow joint has the equivalent of pitch in your shoulder joint. But that is all. Your elbow, therefore, has one degree of freedom.

Extend your arm toward the horizon again. Now move only your arm below the elbow. Your forearm and wrist can bend up and down, side to side, and it can also twist. Your lower arm has three degrees of freedom.

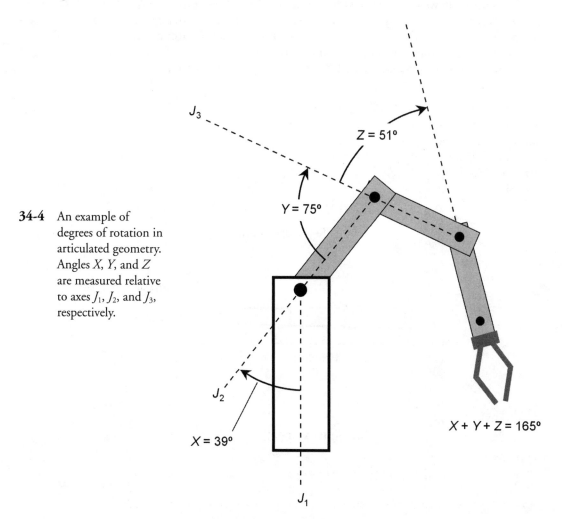

34-4 An example of degrees of rotation in articulated geometry. Angles *X*, *Y*, and *Z* are measured relative to axes J_1, J_2, and J_3, respectively.

J_3

$Z = 51°$

$Y = 75°$

J_2

$X = 39°$

$X + Y + Z = 165°$

J_1

In total, your arm has seven degrees of freedom: three in the shoulder, one in the elbow, and three in the arm below the elbow.

It is tempting to suppose that a robot should never need more than three degrees of freedom, because, after all, space has only three dimensions. But the extra possible motions, provided by multiple joints, give a robot arm (and a human arm) versatility that it could not have with only three degrees of freedom.

Degrees of Rotation

The term *degrees of rotation* refers to the extent to which a robot joint, or a set of robot joints, can turn clockwise or counterclockwise with respect to a prescribed linear axis. Some reference point is always used, and the angles are given in degrees with respect to that joint. Rotation in one direction (usually clockwise) is represented by positive angles; rotation in the opposite direction is specified by negative angles. Thus, for example, if angle $X = 58°$, it refers to a rotation of $58°$ clockwise with respect to the reference axis. If angle $Y = -74°$, it refers to a rotation of $74°$ counterclockwise.

Figure 34-4 shows a robot arm with three joints. The reference axes are J_1, J_2, and J_3 for rotation angles X, Y, and Z, respectively. The individual angles add together. To move this robot arm to a certain position within its *work envelope* (the region in space that the arm can reach and actually manipulate things), the operator enters data into a computer. This data includes the measures of angles X, Y, and Z. The operator has specified $X = 39°$, $Y = 75°$, and $Z = 51°$.

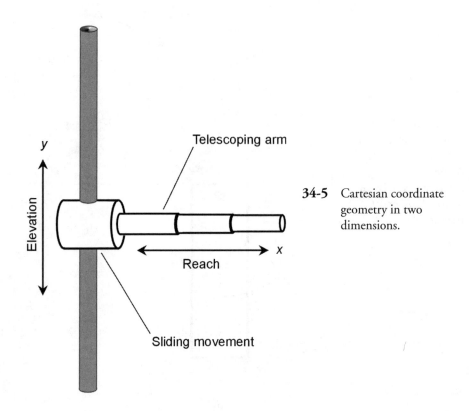

34-5 Cartesian coordinate geometry in two dimensions.

Articulated Geometry

The word *articulated* means "broken into sections by joints." A robot arm with *articulated geometry* bears some resemblance to the arm of a human. The versatility is defined in terms of the number of degrees of freedom. For example, an arm might have three degrees of freedom: base rotation (the equivalent of azimuth), elevation angle, and reach (the equivalent of radius). If you're a mathematician, you might recognize this as a system of *spherical coordinates*. There are several different articulated geometries for any given number of degrees of freedom. Figure 34-4 is a simplified drawing of a robot arm that uses articulated geometry.

Cartesian Coordinate Geometry

Another mode of robot arm movement is known as *Cartesian coordinate geometry* or *rectangular coordinate geometry*. This term comes from the *Cartesian coordinate system* often used for graphing mathematical functions. The axes are always perpendicular to each other. Variables are assigned the letters *x* and *y* in a two-dimensional *Cartesian plane*, or *x*, *y*, and *z* in *Cartesian three-space*. The dimensions are called *reach* for the *x* variable, *elevation* for the *y* variable, and *depth* for the *z* variable. Figure 34-5 is a simplified rendition of a robot arm capable of moving in two dimensions using Cartesian coordinate geometry.

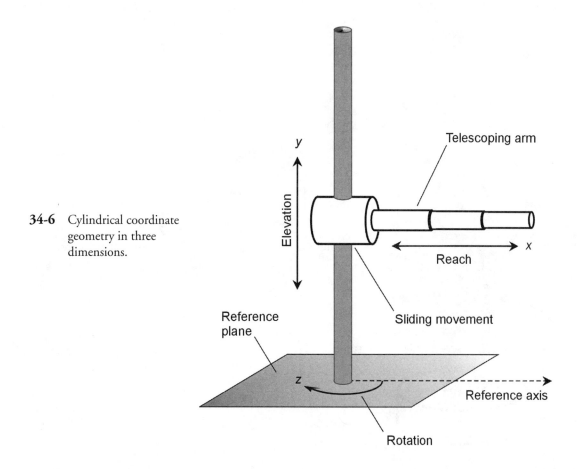

34-6 Cylindrical coordinate geometry in three dimensions.

Cylindrical Coordinate Geometry

A robot arm can be guided by means of a two-dimensional *navigator's polar coordinate system* with an elevation dimension added (Fig. 34-6). This is known as *cylindrical coordinate geometry*. In this system, a *reference plane* is used. An origin point is chosen in this plane. A *reference axis* is defined, running away from the origin in the reference plane. In the reference plane, the position of any point can be specified in terms of reach x, elevation y, and rotation z. The rotation is defined as the angle that the reach arm subtends relative to the reference axis. In this example, it is in the clockwise sense. Note that this is just like the situation for two-dimensional Cartesian coordinate geometry shown in Fig. 34-5, except that the sliding movement is also capable of rotation.

The rotation angle z can range from 0° to 360° clockwise from the reference axis. In some systems, the range is specified as 0° to +180° (up to a half circle clockwise from the reference axis), and 0° to −180° (up to a half circle counterclockwise from the reference axis).

Revolute Geometry

A robot arm capable of moving in three dimensions using *revolute geometry* is shown in Fig. 34-7. The whole arm can rotate through a full circle (360°) at the base point, or *shoulder*. There is also an elevation joint at the base that can move the arm through 90°, from horizontal to vertical. A joint in the middle of the robot arm, at the *elbow*, moves through 180°, from a straight position to dou-

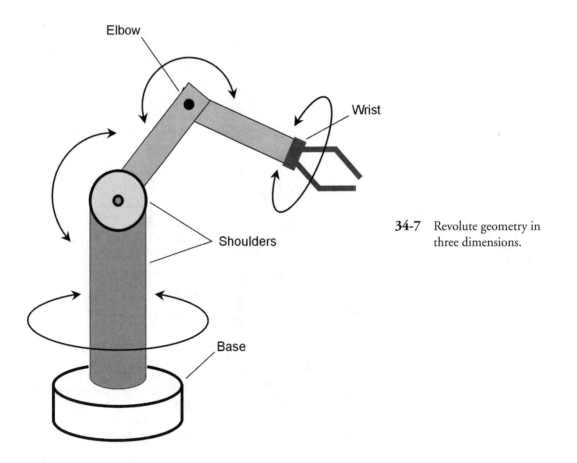

34-7 Revolute geometry in three dimensions.

bled back on itself. There might be, but is not always, a *wrist* joint that can flex like the elbow and/or twist around and around.

A 90° elevation revolute robot arm can reach any point within a half sphere. The radius of the half sphere is the length of the arm when its elbow and wrist (if any) are straightened out. A 180° elevation revolute arm can be designed that will reach any point within a fully defined sphere, with the exception of the small obstructed region around the base.

Robot Hearing and Vision

Machine hearing involves detection of acoustic waves, along with amplification and analysis of the resulting audio signals. Machine vision involves the interception of visible, infrared (IR), or ultraviolet (UV) radiation, and translating this energy into electronic images. Machine hearing and vision can allow robots to locate, and in some cases classify or identify, objects in the environment.

Binaural Hearing

Even with your eyes closed, you can usually tell from which direction a sound is coming. This is because you have *binaural hearing*. Sound arrives at your left ear with a different intensity, and in a different phase, than it arrives at your right ear. Your brain processes this information, allowing you to locate the source of the sound, with certain limitations. If you are confused, you can turn your head until the direction becomes apparent.

Robots can be equipped with binaural hearing. Two acoustic transducers are positioned, one on either side of the robot's head. A microprocessor compares the relative phase and intensity of signals from the two transducers. This lets the robot determine, within certain limitations, the direction from which sound is coming. If the robot is confused, it can turn until the confusion is eliminated and a meaningful bearing is obtained. If the robot can move around and take bearings from more than one position, a more accurate determination of the source location is possible if the source is not too far away.

Visible-Light Vision

A visible-light robotic *vision system* must have a device for receiving incoming images. This is usually a *charge coupled device* (CCD) video camera, similar to the type used in home video cameras. The camera receives an analog video signal. This is processed into digital form by an ADC. The digital signal is clarified by means of DSP. The resulting data goes to the robot controller.

The moving image, received from the camera and processed by the circuitry, contains an enormous amount of information. It's easy to present a robot controller with a detailed and meaningful moving image. But getting the robot controller to know what's happening, and to determine whether or not these events are significant, is another problem altogether.

Optical Sensitivity and Resolution

Optical sensitivity is the ability of a machine vision system to see in dim light or to detect weak impulses at invisible wavelengths. In some environments, high optical sensitivity is necessary. In others, it is not needed and might not be wanted. A robot that works in bright sunlight doesn't need to be able to see well in a dark cave. A robot designed for working in mines, pipes, or caverns must be able to see in dim light, using a system that might be blinded by ordinary daylight.

Optical resolution is the extent to which a machine vision system can differentiate between objects that are close together in the field of vision. The better the optical resolution, the keener the vision. Human eyes have excellent optical resolution, but machines can be designed to have superior resolution.

In general, the better the optical resolution, the more confined the field of vision must be. To understand why this is true, think of a telescope. The higher the magnification, the better its optical resolution will be, up to a certain maximum useful magnification. Increasing the magnification reduces the angle, or field, of vision. Zeroing in on one object or zone is done at the expense of other objects or zones.

Optical sensitivity and resolution are interdependent. If all other factors remain constant, improved sensitivity causes a sacrifice in resolution. Also, the better the optical resolution, the more incident light it requires to function well. In this case, a good analogy is camera film (the old-fashioned kind). The fastest films require more light than slow ones. The corollary to this is the fact that if you want excellent detail in a photograph, you will have to expose the film for a comparatively long period of time.

Invisible and Passive Vision

Robots have an advantage over people when it comes to vision. Machines can see at wavelengths to which humans are blind.

Human eyes are sensitive to EM waves whose length ranges from 390 to 750 nanometers (nm). The nanometer is a thousand-millionth (10^{-9}) of a meter. The longest visible wavelengths look red. As the wavelength gets shorter, the color changes through orange, yellow, green, blue, and indigo. The shortest waves look violet. Infrared (IR) energy is at wavelengths somewhat longer than 750 nm. Ultraviolet (UV) energy is at wavelengths somewhat shorter than 390 nm.

Machines, and most nonhuman living species, can see energy in a range of wavelengths that differs somewhat from the range of wavelengths to which human eyes respond. For example, insects can see UV that humans cannot, but insects are blind to red and orange light that humans can see. (Have you used orange bug lights when camping to keep the flying pests from coming around at night, or those UV devices that attract bugs and then zap them?)

A robot can be designed to see IR and/or UV, as well as (or instead of) visible light, because video cameras can be sensitive to a range of wavelengths much wider than the range humans can see. Robots can be made to see in an environment that is dark and cold, and that radiates too little energy to be detected at any electromagnetic wavelength. In these cases the robot provides its own illumination. This can be a simple lamp, a laser, an IR device, or a UV device. Radar and sonar can also be used.

Binocular Vision

Binocular machine vision is the analog of binocular human vision. It is sometimes called *stereo vision* or *stereoscopic vision*. In humans, binocular vision allows perception of depth. With only one eye— that is, with monocular vision—you can infer depth only to a limited extent, and that perception is entirely dependent on your knowledge of the environment or scene you are observing. Almost everyone has had the experience of being fooled when looking at a scene with one eye covered or blocked. A nearby pole and a distant tower might seem to be adjacent, when in fact they are a city block apart.

Figure 34-8 shows the basic concept of binocular robot vision. High-resolution video cameras, and a sufficiently powerful robot controller, are essential components of such a system.

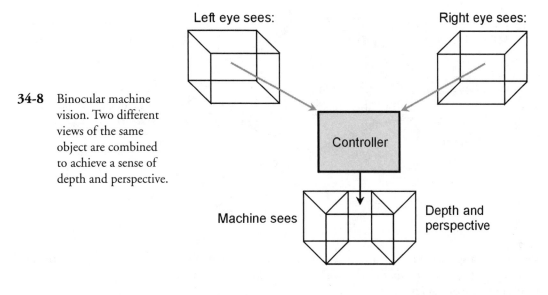

34-8 Binocular machine vision. Two different views of the same object are combined to achieve a sense of depth and perspective.

Color Sensing

Robot vision systems often function only in grayscale, like old-fashioned 1950s television. But color sensing can be added, in a manner similar to the way it is added to television systems. Color sensing can help a robot with AI figure out what an object is. Is that horizontal surface a parking lot, or is it a grassy yard? Sometimes, objects have regions of different colors that have identical brightness as seen by a grayscale system. Such objects, obviously, can be evaluated in more detail with a color-sensing system than with a vision system that sees only shades of gray.

In a typical color-sensing vision system, three grayscale cameras are used. Each camera has a color filter in its lens. One filter passes red light, another passes green light, and another passes blue light. These are the three *primary colors*. All possible hues, levels of brightness, and levels of saturation are made up of these three colors in various ratios. The signals from the three cameras are processed by a microcomputer, and the result is fed to the robot controller.

The Eye-in-Hand System

In order to assist a robot gripper (hand) in finding its way, a camera can be placed in the mechanism. The camera must be equipped for work at close range, from about 1 m down to 1 mm or less. The positioning error must be kept as small as possible. To be sure that the camera gets a good image, lamps are included in the gripper along with the camera (Fig. 34-9). This so-called eye-in-hand system can be used to precisely measure the distance between the gripper and the object it is seeking. It can also make positive identification of the object.

The eye-in-hand system takes advantage of the properties of a *servo*. The robot is equipped with, or has access to, a computer that processes the data from the camera and sends instructions back to the gripper. Most eye-in-hand systems use visible light for guidance and manipulation. Infrared (IR) can be used when it is necessary for the robot gripper to sense differences in temperature.

The Flying Eyeball

In environments hostile to humans, robots find many uses, from manufacturing to exploration. One such device, especially useful underwater, has been called a *flying eyeball*. A cable, containing

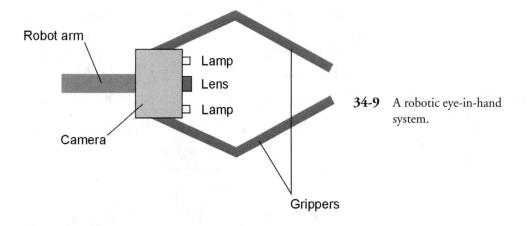

34-9 A robotic eye-in-hand system.

the robot in a special launcher housing, is dropped from a boat. When the launcher gets to the desired depth, it lets out the robot, which is connected to the launcher by a tether. The tether and the drop cable convey data back to the boat.

In some cases, the tether for a flying eyeball can be eliminated, and a wireless link can be used to convey data from the robot to the launcher. The link is usually in the IR or visible red portion of the spectrum. The robot contains a video camera and one or more lamps to illuminate the underwater environment. It also has a set of thrusters (jets or propellers) that let it move around according to control commands sent from the boat. Human operators on board the boat watch the images and guide the robot.

Robot Navigation

Mobile robots must get around in their environment without wasting motion, without running into things, and without tipping over or falling down a flight of stairs. The nature of a *robot navigation system* depends on the size of the work area, the type of robot used, and the sorts of tasks the robot is required to perform. In this section, we'll look at four common methods of robot navigation.

Clinometer

A *clinometer* is a device for measuring the steepness of a sloping surface. Mobile robots use clinometers to avoid inclines that might cause them to tip over or that are too steep for them to ascend while carrying a load.

The floor in a building is almost always horizontal. Thus, its incline is zero. But sometimes there are inclines such as ramps. A good example is the kind of ramp used for wheelchairs, in which a very small elevation change occurs. A rolling robot cannot climb stairs, but it can use a wheelchair ramp, provided the ramp is not so steep that it would upset the robot's balance or cause it to lose its payload.

In a clinometer, a transducer produces an electrical signal whenever the device is tipped from the horizontal. The greater the angle of incline, the greater the electrical output, as shown in the graph of Fig. 34-10A. A clinometer might also show whether an incline goes down or up. A downward slope might cause a negative voltage at the transducer output, and an upward slope a positive voltage, as shown in the graph at Fig. 34-10B.

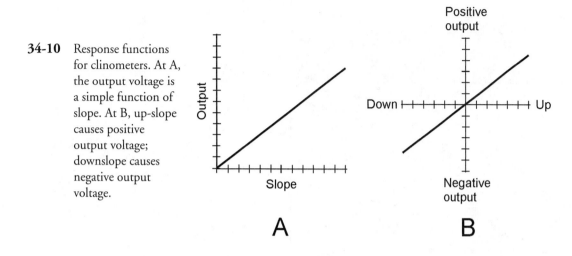

34-10 Response functions for clinometers. At A, the output voltage is a simple function of slope. At B, up-slope causes positive output voltage; downslope causes negative output voltage.

Edge Detection

The term *edge detection* refers to the ability of a robot vision system to locate boundaries. It also refers to the robot's knowledge of what to do with respect to those boundaries. A robot car, bus, or truck can use edge detection to see the edges of a road and use the data to keep itself on the road. But it must stay a certain distance from the right-hand edge of the pavement to avoid crossing into the lane of oncoming traffic (Fig. 34-11). It also must stay off the road shoulder. It must be able to tell the difference between pavement and other surfaces, such as gravel, grass, sand, and snow.

The interior of a home or business contains straight-line edge boundaries of all kinds, and each boundary represents a potential point of reference for a mobile robot's edge detection system. The controller in a personal home robot must be programmed to know the difference between, say, the line where carpet ends and tile begins, and the line where a flight of stairs begins. The vertical line produced by the intersection of two walls would present a different situation than the vertical line produced by the edge of a doorway, even though they might appear identical. Thus, edge detection cannot function very well without a certain amount of AI in the robot controller.

Embedded Path

An *embedded path* is a means of guiding a robot along a specific route. This scheme is commonly used by a mobile robot called an *automated guided vehicle* (AGV). A common embedded path consists of a buried, current-carrying wire. The current in the wire produces a magnetic field that the robot can follow. This method of guidance has been suggested as a way to keep a car on a highway, even if the driver isn't paying attention. The wire needs a constant supply of electricity for this guidance method to work. If this current is interrupted for any reason, the robot will lose its way unless some backup navigation method (or human control) is substituted.

Alternatives to wires, such as colored or reflective paints or tapes, do not need a supply of power, and this gives them an advantage. Tape is easy to remove and put somewhere else; this is difficult to do with paint and practically impossible with wires embedded in concrete. However, tape is obscured by snowfall; and at night, glare from oncoming headlights might be confused for reflections from the tape.

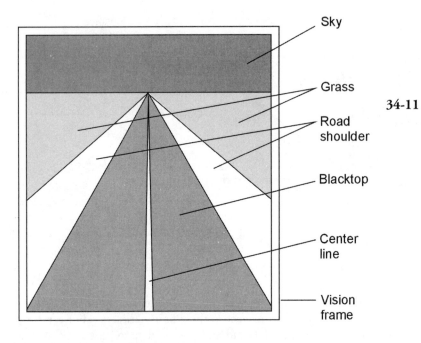

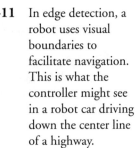

34-11 In edge detection, a robot uses visual boundaries to facilitate navigation. This is what the controller might see in a robot car driving down the center line of a highway.

Range Sensing and Plotting

Range sensing is the measurement of distances to objects in a robot's environment in a single dimension. *Range plotting* is the creation of a graph of the distance (range) to objects, as a function of the direction in two or three dimensions.

In linear or one-dimensional (1D) range sensing, a signal is sent out, and the robot measures the time it takes for the echo to come back. This signal can be sound, in which case the device is sonar. Or it can be a radio wave; this constitutes radar. Laser beams can also be used. Close-range, one-dimensional range sensing is known as *proximity sensing*.

Two-dimensional (2D) range plotting involves mapping the distance to various objects, as a function of direction in a geometric plane. The echo return time for a sonar signal, for example, might be measured every few degrees around a complete circle in the horizontal plane, resulting in a set of range points. A better plot would be obtained if the range were plotted every degree, every tenth of a degree, or even every hundredth of a degree. But no matter how detailed the direction resolution, a 2D range plot is done in only one plane, such as the floor level in a room, or some horizontal plane above the floor. The greater the number of echo samples in a complete circle (that is, the smaller the angle between samples), the more detail can be resolved at a given distance from the robot, and the greater the distance at which a given amount of detail can be resolved.

Three-dimensional (3D) range plotting is done in spherical coordinates: azimuth (compass bearing), elevation (degrees above the horizontal), and range (distance). The distance must be measured for a large number of diverse orientations. In a furnished room, a 3D sonar range plot would show ceiling fixtures, things on the floor, objects on top of a desk, and other details not visible with a 2D plot. The greater the number of echo samples in a complete sphere surrounding the robot, the more detail can be resolved at a given distance, and the greater the range at which a given amount of detail can be resolved.

Telepresence

Telepresence is a refined, advanced form of robot remote control. The robot operator gets a sense of being "on location," even if the remotely controlled machine, or *telechir*, and the operator are miles apart. Control and feedback are done by means of *telemetry* sent over wires, optical fibers, or radio.

What It's Like

Here is an example of a telepresence scenario. The robot is autonomous and has a humanoid form. The control station consists of a suit that you wear or a chair in which you sit with various manipulators and displays. Sensors can give you feelings of pressure, sight, and sound. You wear a helmet with a viewing screen that shows whatever the robot camera sees. When your head turns, the robot head, with its vision system, follows, so you see an image that changes as you turn your head, as if you were in a space suit or diving suit at the location of the robot. Binocular machine vision provides a sense of depth. Binaural machine hearing allows you to perceive sounds. Special vision modes let you see UV or IR; special hearing modes let you hear ultrasound or infrasound.

A telechir can be propelled by means of a *track drive* (similar to those used by military tanks), a *wheel drive*, or *robot legs*. If the propulsion uses legs, you propel the telechir by literally walking around a room in your telepresence suit! In the case of track drive or wheel drive, you sit in a chair and drive the robot like a tank or a car. The telechir, which is a form of android, has two arms, each with grippers resembling human hands. When you want to pick something up, you go through the normal motions with your own hands. Back-pressure sensors and position sensors let you feel (to a limited extent) what's going on. If an object masses 1 kg, it feels as if it masses 1 kg because of resistance provided by the back-pressure sensors. But it will be as if you're wearing thick gloves; you won't be able to feel texture. You might throw a switch, and something that masses 10 kg feels as if it masses only 1 kg. This might be called "strength × 10" mode. If you switch to "strength × 100" mode, a 100-kg object seems to mass only 1 kg.

Functional Description

Figure 34-12 is a simplified diagram of a robot telepresence system using an android and a wireless control link.

At the control end of the system (where the human operator is), electromechanical transducers convert the human operator's movements into electrical signals. The modem converts these signals into analog form. The analog signals modulate the RF produced by the transceiver, which in turn transmits commands by wireless to the telechir end of the system. Radio signals from the telechir arrive at the operator-end transceiver. These signals are converted into analog impulses, and are translated by the modem into electrical signals that power transducers, giving the operator a sense of what is going on.

At the telechir end of the system, the transceiver receives control signals from the human operator. These signals are translated by the modem into impulses that drive electromechanical transducers, propelling or manipulating the telechir. Data obtained by the telechir, such as its visual sense of position, environmental sounds, or the apparent mass of a lifted object, are converted by the modem into electrical signals. These signals modulate the RF energy produced by the transceiver. Signals are thus sent back to the operator end by wireless.

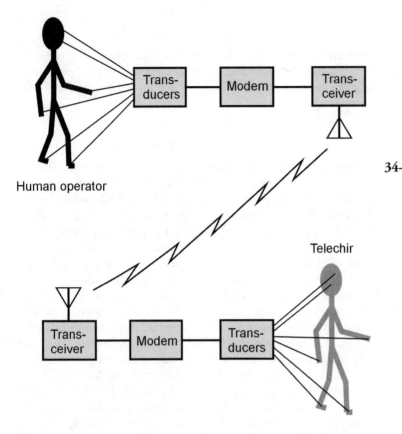

Human operator

Telechir

34-12 Telepresence combines remote monitoring and control, giving the operator the feeling of being onsite at a distant location.

Applications

Here are a few potential applications for a telepresence system using a human operator and an android:

- Working in extreme heat or cold
- Working under high pressure, such as on the sea floor
- Working in a vacuum, such as in space
- Working where there is dangerous radiation
- Disarming bombs
- Handling toxic substances
- Serving in high-risk police or military situations

Of course, the robot must be able to survive conditions at its location. Also, it must have some way to recover if it falls or gets knocked over.

Limitations

The technology for telepresence has existed for some time. But certain logistical problems have always bedeviled engineers committed to developing such systems.

The most serious limitation is the fact that telemetry cannot, and never will, travel faster than the speed of light in free space. This is not a problem over short-range links (a few hundred kilometers or less), but it is slow in outer space, and nigh impractical in interplanetary operations. The moon is approximately 1.3 light-seconds away from the earth; that means that any command sent to a telechir from the earth to the moon takes 1.3 s to get there, and any data from the telechir takes another 1.3 s to get back. For planets in the solar system, the delay is on the order of several minutes to several hours. On an interstellar scale, telepresence is out of the question. The nearest stars are at distances of several light-years.

Another problem is the resolution of the robot's vision. A human being with good eyesight can see things in considerable detail. To send images in real-life detail, at reasonable speed, requires a signal with broad bandwidth. There are engineering problems (and cost problems) that go along with this. However, if one is willing to deal with the cost and accommodate the signal bandwidth, robot vision systems can be designed that offer optical resolution superior to human eyesight.

Still another limitation is best put as a question: How will a robot be able to feel something and transmit these impulses to the human brain? For example, an apple feels smooth, a peach feels fuzzy, and an orange feels shiny yet bumpy. How can this sense of texture be realistically transmitted to the human brain?

The Mind of the Machine

A simple electronic calculator doesn't have AI. But a machine that can learn from its mistakes, or that can show reasoning power, does. Between these extremes, there is no precise dividing line. As computers become more powerful, people tend to set higher standards for what they call AI. Things that were once thought of as AI are now ordinary. Things that seem fantastic now will someday be humdrum. There is a tongue-in-cheek axiom: We can call "computer intelligence" true AI only as long as it remains a little bit mysterious.

Robotics and AI

Robotics and AI complement each other. Scientists have dreamed for more than a century about building *smart androids:* robots that look like people, act like people, and can even reason like people. Androids exist, but they aren't very smart. Powerful computers exist, but they lack mobility.

If a machine has the ability to move around under its own power, to lift things, and to move things, it seems reasonable that it should do so with some degree of intelligence if it is to accomplish anything worthwhile. Conversely, if a computer is to manipulate anything, it must be able to cause a machine to do physical work according to a precise program.

Expert Systems

The term *expert systems* refers to a method of reasoning in AI. Sometimes this scheme is called the *rule-based system.* Expert systems are used in the control of smart robots.

The heart of an expert system is a set of facts and rules. In the case of a robotic system, the facts consist of data about the robot's work environment, such as a factory, an office, or a kitchen. The rules are statements of the logical form "If X, then Y," similar to many of the statements in high-level programming languages. An *inference engine* decides which logical rules should be applied in various situations and instructs the robot to carry out certain tasks. But the operation of the system can only be as sophisticated as the data supplied by human programmers.

Expert systems can be used in computers to help people do research, make decisions, and make forecasts. A good example is a program that assists a physician in making a diagnosis. The computer asks questions and arrives at a conclusion based on the answers given by the patient and doctor.

One of the biggest advantages of expert systems is the fact that reprogramming is easy. As the environment changes, the robot can be taught new rules, and supplied with new facts.

How Smart a Machine?

Experts in the field of AI have been disappointed in the past few decades. Computers have been designed that can handle tasks no human could ever contend with, such as navigating a space probe. Machines have been built that can play board games well enough to compete with human masters. Modern machines can understand, as well as synthesize, much of any spoken language. But these abilities, by themselves, don't count for much in the dreams of scientists who hope to create *artificial life.*

The human mind is incredibly complicated. A circuit that would have occupied, and used all the electricity in, a whole city in 1940 can now be housed in a box the size of a vitamin pill and run by a battery. Imagine this degree of miniaturization happening again, then again, and then again. Would that begin to approach the level of sophistication in your body's nervous system?

Is the human brain nothing more than an amazingly complicated digital switching network? Or is there something more to the human mind? No electronic device yet constructed has come anywhere near human intelligence in every respect. Some experts think that a machine will someday be built that is smarter than its creators. Others insist that the very idea is ridiculous.

It is tempting to extrapolate: If technological trends of the past few decades continue indefinitely, will the only limit on machine intelligence be defined by human imagination?

Quiz

Refer to the text in this chapter if necessary. A good score is 18 correct. Answers are in the back of the book.

1. An android takes the form of
   - (a) an insect.
   - (b) the human body.
   - (c) a simple robot arm.
   - (d) a stereo vision system.

2. According to Asimov's three laws, under what circumstances is it all right for a robot to injure a human being?
   - (a) Under no circumstances
   - (b) When the human being specifically requests it
   - (c) In case of an accident
   - (d) In case the robot controller is infected with a computer virus

3. An RF field strength meter can be used to
   - (a) test the performance of a binaural hearing system.
   - (b) detect the presence of ionized air.

(c) measure the dielectric constant of the air.

(d) detect the presence of a wireless bugging system.

4. The extent to which a machine vision system can differentiate between two objects that are close together is called the

(a) optical magnification.

(b) optical sensitivity.

(c) optical selectivity.

(d) optical resolution.

5. A robot car or truck can best keep itself traveling down a specific lane of traffic by means of

(a) stereoscopic machine hearing.

(b) epipolar navigation.

(c) edge detection.

(d) proximity sensing.

6. A rule-based system is also known as

(a) a logic gate.

(b) an expert system.

(c) a back-pressure sensor.

(d) a telechir.

7. Suppose you are using a battery-powered, multichannel baby monitor, and you hear one end of a two-way radio conversation on the receiver. You check the baby's room, and it is quiet. How might this problem be resolved?

(a) Put the receiver in a different location.

(b) Switch the monitor to a different channel.

(c) Interchange the transmitting and receiving units.

(d) Use ac power instead of battery power.

8. In robotics, the term *manipulator* refers to

(a) a robot propulsion system.

(b) a robot arm, and the device at its end (such as a gripper).

(c) the system used to remotely control a telechir.

(d) a computer that guides a fleet of mobile robots.

9. A device with an IR sensor can be used to detect the presence of

(a) slow-moving objects.

(b) RF signals.

(c) ionized air.

(d) warm or hot objects.

10. Proximity sensing is most closely akin to
    (a) direction measurement.
    (b) edge detection.
    (c) range plotting.
    (d) binaural machine hearing.

11. A telechir is always used in conjunction with a specialized system of
    (a) track drive.
    (b) wheel drive.
    (c) remote control.
    (d) ionization potential measurement.

12. A limit to the distance over which telepresence is practical is imposed by
    (a) the speed of EM wave propagation.
    (b) the image resolution of the vision system.
    (c) the ability of a robot to determine texture.
    (d) all of the above.

13. The ionization potential of the air can be determined in order to
    (a) detect smoke.
    (b) plot distances and directions.
    (c) measure slope.
    (d) detect boundaries.

14. Two-dimensional range plotting
    (a) takes place along a single geometric line.
    (b) takes place in a single geometric plane.
    (c) is done using spherical coordinates.
    (d) requires an ultrasonic sonar system.

15. Spherical coordinates can uniquely define the position of a point in up to
    (a) one dimension.
    (b) two dimensions.
    (c) three dimensions.
    (d) four dimensions.

16. The total number of ways in which a robot arm can move is known as
    (a) functional orientation.
    (b) degrees of freedom.
    (c) dimensional versatility.
    (d) coordinate geometry.

17. The region in space throughout which a robot arm can accomplish tasks is called its
    (a) coordinate geometry.
    (b) reference axis.
    (c) reference frame.
    (d) work envelope.

18. A robot arm that moves along three independent axes, each of which is straight and perpendicular to the other two, employs
    (a) revolute geometry.
    (b) spherical coordinate geometry.
    (c) Cartesian coordinate geometry.
    (d) cylindrical coordinate geometry.

19. Fill in the blank to make the following sentence true: "A color vision system can use three grayscale cameras, equipped with filters that pass _____ light."
    (a) red, yellow, and blue
    (b) blue, red, and green
    (c) white, black, and gray
    (d) orange, green, and violet

20. A robot typically determines the steepness of a slope by means of
    (a) an epipolar navigation system.
    (b) a clinometer.
    (c) an end effector.
    (d) a proximity sensor.

Test: Part 4

DO NOT REFER TO THE TEXT WHEN TAKING THIS TEST. A GOOD SCORE IS AT LEAST 37 CORRECT. Answers are in the back of the book. It's best to have a friend check your score the first time, so you won't memorize the answers if you want to take the test again.

1. Which of the following tasks can a displacement transducer perform?
   (a) It can detect the peak amplitude of an ac signal and display it on a screen.
   (b) It can convert acoustic energy into radio signals.
   (c) It can detect radiated IR energy and display IR images on a screen.
   (d) It can convert an electrical signal into mechanical rotation through a defined angle.
   (e) It can be used to measure the frequency of a complex wave.

2. Which of the following is *not* an advantage of an IC, compared to a circuit built with discrete components (individual resistors, capacitors, inductors, diodes, and transistors)?
   (a) Compactness
   (b) Reliability
   (c) Ease of maintenance
   (d) Low power consumption
   (e) Unlimited power handling capacity

3. Computer memory is typically measured in
   (a) kilobits, megabits, and gigabits.
   (b) kilobits per second, megabits per second, and gigabits per second.
   (c) kilobytes, megabytes, and gigabytes.
   (d) kilobytes per second, megabytes per second, and gigabytes per second.
   (e) any of the above.

4. A mode of communications in which either party can hear while talking, thus allowing one party to instantly interrupt the other at any time, is called

(a) half simplex.

(b) full simplex.

(c) half duplex.

(d) full duplex.

(e) nothing, because there is no such mode.

5. An FM stereo tuner generally covers a frequency range of

(a) 535 kHz to 1605 kHz.

(b) 88 MHz to 108 MHz.

(c) 3 MHz to 30 MHz.

(d) 9 kHz to 300 MHz.

(e) 144 MHz to 148 MHz.

6. Suppose an antenna has a radiation resistance of 35 Ω and a loss resistance of 15 Ω. What is the efficiency of the antenna?

(a) 20 percent

(b) 30 percent

(c) 43 percent

(d) 70 percent

(e) Impossible to calculate from this information

7. The peak power output from a multicavity Klystron

(a) can be much greater than the average power output.

(b) can be slightly greater than the average power output.

(c) is about the same as the average power output.

(d) is slightly less than the average power output.

(e) is much less than the average power output.

8. To which of the following does the term *platter* apply?

(a) An individual disk in a hard drive

(b) An individual disk in CD-R media

(c) An individual disk in CD-RW media

(d) An individual disk in flash memory

(e) An individual disk in RAM

9. Why can signals in the so-called shortwave band sometimes propagate for thousands of kilometers without the need for satellites, repeaters, or any other infrastructure?

(a) Because EM waves at these frequencies are often returned to the earth by the ionosphere

(b) Because waves at these frequencies can propagate through the ground or the ocean as electric currents

(c) Because EM waves at these frequencies are reflected back to the earth by the moon

(d) Because EM waves at these frequencies propagate along the boundary between the troposphere and the stratosphere

(e) Forget it! Signals in the shortwave band cannot propagate beyond a line of sight without satellites, repeaters, or other human-made systems.

10. What is the purpose of the filament in a vacuum tube?

(a) It prevents secondary electrons from causing excessive screen grid current.

(b) It causes holes to flow more easily in the collector.

(c) It heats the cathode, thereby increasing the electron emission.

(d) It glows, making the tube easy to locate when servicing is necessary.

(e) Forget it! Vacuum tubes do not have filaments.

11. Figure Test4-1 is a schematic diagram of

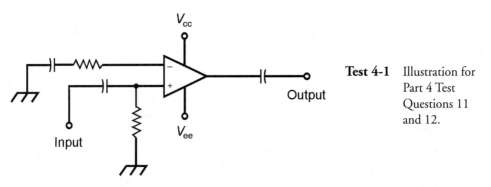

Test 4-1   Illustration for Part 4 Test Questions 11 and 12.

(a) a closed-loop op amp circuit with positive feedback.

(b) a closed-loop op amp circuit with negative feedback.

(c) an open-loop op amp circuit.

(d) an op amp oscillator circuit.

(e) an op amp modulator circuit.

12. Suppose that, in the circuit of Fig. Test4-1, a resistor is connected between the output and the inverting input. What effect will this have?

(a) It will increase the gain.

(b) It will decrease the gain.

(c) It will increase the frequency.

(d) It will decrease the frequency.

(e) It will have no effect.

13. Which of the following is an example of a serial-access storage medium for computer data?

(a) An external hard drive

(b) An internal hard drive

(c) A CD-R

(d) A CD-RW

(e) A magnetic tape

14. Which of the following is an undesirable characteristic of ELF radiation?

(a) It can cause radiation burns like those produced by overexposure to X rays.

(b) It can increase the susceptibility of a CRT to jitter and poor resolution.

(c) It can reduce the efficiency of an antenna system.

(d) It can result in undesirable modulation in a radio transmitter.

(e) None of the above apply.

15. A zepp antenna measuring $\lambda/2$ can be oriented vertically, and the feed line placed so it lies in the same line as the radiating element. This antenna is known as

(a) a vertical dipole.

(b) a Yagi.

(c) a collinear array.

(d) an end-fire array.

(e) a J pole.

16. The GPS can be used to determine the position of a point in

(a) one dimension.

(b) two dimensions.

(c) three dimensions.

(d) four dimensions.

(e) five dimensions.

17. In a component-type hi-fi system, the cables connecting the various devices should

(a) be resonant at the intended operating frequency.

(b) consist of single conductors only.

(c) be connected in parallel.

(d) consist of two parallel wires whenever possible.

(e) be shielded whenever practicable.

18. In a radar display, the azimuth of a target is equal or proportional to

(a) the distance of the blip representing the target from the center of the display.

(b) the clockwise angle between a radial line from the center toward geographic north, and a radial line from the center through the blip representing the target.

(c) the diameter of a circle whose center coincides with the center of the display, and that passes through the blip representing the target.

(d) the distance of the blip representing the target from the outer edge of the display, or from a circle whose diameter is equal to the diameter of the display.

(e) Forget it! A radar set cannot determine the azimuth of a target.

19. Which of the following is an example of wireless technology?
    (a) The Global Positioning System
    (b) A fiber-optic communications system
    (c) Digital-to-analog conversion
    (d) Serial-to-parallel conversion
    (e) All of the above

20. A smoke detector can function by sensing a change in the
    (a) temperature of the air.
    (b) relative humidity of the air.
    (c) barometric pressure of the air.
    (d) ionization potential in the air.
    (e) amount of nitrogen in the air.

21. Computer disk storage is typically measured in
    (a) kilobits, megabits, and gigabits.
    (b) kilobits per second, megabits per second, and gigabits per second.
    (c) kilobytes, megabytes, and gigabytes.
    (d) kilobytes per second, megabytes per second, and gigabytes per second.
    (e) any of the above.

22. Which of the following antennas would most likely be used in space communications?
    (a) A half-wave dipole antenna
    (b) A ground-plane antenna
    (c) A helical antenna
    (d) A small loop or loopstick antenna
    (e) A coaxial antenna

23. The use of a cell phone may be prohibited in a commercial aircraft in flight because
    (a) it can interfere with other people's computers.
    (b) it can interfere with other people's cell phones.
    (c) the interior of the aircraft produces a dangerous RF resonant cavity.
    (d) the area of coverage will not be wide enough to be of any use.
    (e) it can interfere with flight instruments and communications.

24. The circuit shown in Fig. Test4-2 shows two NPN transistors connected in
    (a) reverse series.
    (b) cascade.

+12 V

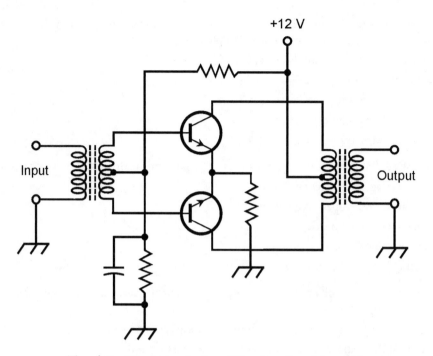

Test 4-2 Illustration for Part 4 Test Questions 24 and 25.

Input

Output

    (c) inverse parallel.

    (d) push-push.

    (e) push-pull.

25. The circuit in Fig. Test4-2 would be a bad choice for use as a medium-power audio amplifier because

    (a) bipolar transistors do not work well in audio applications.

    (b) transformers should not be used in audio amplifiers.

    (c) this type of circuit invariably produces a lot of distortion.

    (d) this type of circuit cannot produce appreciable output power.

    (e) Hold it! The circuit in Fig. Test4-2 would be a good choice for use as a medium-power audio amplifier.

26. In an electrostatic CRT, the electron beam can be diverted up and down by

    (a) synchronization pulses applied to the deflecting coils.

    (b) ELF radiation from the accelerating anodes.

    (c) charged plates that attract or repel moving electrons.

    (d) modulation of the signal on the control grid.

    (e) variable input to the electron gun.

27. A device that translates 500-Hz sound waves into ac by means of the interaction between a permanent magnet and a coil of wire is an example of
    (a) an electromechanical loudspeaker.
    (b) a passive pressure sensor.
    (c) a dynamic transducer.
    (d) an inductive proximity sensor.
    (e) a piezoelectric transducer.

28. Which of the following considerations is most important in a hi-fi audio power amplifier?
    (a) Optimizing the efficiency
    (b) Minimizing the operating voltage
    (c) Maximizing the linearity
    (d) Maximizing the driving power
    (e) Biasing beyond the cutoff point

29. A loopstick antenna, oriented so its ends point toward the eastern and western horizons, exhibits
    (a) nulls in the response to signals coming from the east and west.
    (b) peaks in the response to signals coming from the east and west.
    (c) nulls in the response to signals coming from the north and south.
    (d) an omnidirectional response in three-dimensional space.
    (e) a peak in the response to signals coming from any horizontal direction, and a null in the response to signals coming from the zenith (directly overhead).

30. Fill in the blank to make the following sentence true: "An _____ functions by sensing variations in the relative phase of acoustic waves reflected from objects in the environment."
    (a) IR transducer
    (b) acoustic transponder
    (c) electrostatic loudspeaker
    (d) ultrasonic motion detector
    (e) acoustic pressure sensor

31. Which of the following is an advantage of a large LCD over a large CRT display?
    (a) Lower cost
    (b) Lighter weight
    (c) Greater ELF radiation
    (d) The use of electron beams and deflecting coils
    (e) All of the above

32. An op amp can produce or facilitate
    (a) a low signal-to-noise ratio in an oscillator.
    (b) improved efficiency in an antenna system.

(c) high signal gain over a wide range of frequencies.

(d) a match between a feed line and an antenna.

(e) none of the above.

33. Which of the following is an application of a gas-filled tube?

(a) A high-frequency RF power amplifier

(b) A microwave oscillator

(c) An audio oscillator or amplifier

(d) Decorative lighting

(e) Impedance matching in an antenna system

34. For a communications satellite with an elliptical orbit around the earth, the point at which the altitude is lowest is the

(a) geominimum.

(b) proxima.

(c) perigee.

(d) approach.

(e) optimum.

35. Suppose an antenna has a radiation resistance of 35 Ω and its feed line has a characteristic impedance of 50 Ω. What is the efficiency of the antenna?

(a) 15 percent

(b) 41 percent

(c) 59 percent

(d) 70 percent

(e) Impossible to calculate from this information

36. Which of the following is a characteristic of a stepper motor?

(a) It rotates in defined increments, not continuously.

(b) Its turning power increases as its speed increases.

(c) It cannot rotate faster than approximately 1 rpm.

(d) It is not suitable for use in mechanical devices, but only in sensors.

(e) It operates without an external source of power.

37. With respect to ICs, terms such as MSI, LSI, and VLSI define the

(a) maximum number of transistors on the chip.

(b) maximum frequency at which the chip can operate.

(c) maximum power output the chip can produce.

(d) maximum gain the chip can produce.

(e) maximum number of switching operations the chip can perform per second.

38. In a tetrode vacuum tube amplifier, the screen grid
    (a) reduces the capacitance between the control grid and the plate.
    (b) serves as an auxiliary output in case the plate circuit is overloaded.
    (c) conducts holes away from the control grid.
    (d) prevents undesirable reverse bias between the cathode and the control grid.
    (e) Forget it! Tetrode tubes do not have screen grids.

39. Fill in the blank in the following sentence to make it true: "On the Web, data is transmitted by means of a ____ that allows the files on distant computers to appear as if they are on your own computer."
    (a) protocol
    (b) memory chip
    (c) microprocessor
    (d) CD-R
    (e) tape drive

40. A robot arm with articulated geometry
    (a) has only one degree of freedom.
    (b) has joints, similar to those in a human arm.
    (c) can function in only two dimensions.
    (d) can rotate and move up and down, but cannot bend.
    (e) can perform only coarse movements.

41. The instantaneous output voltage of a differentiator is proportional to
    (a) the peak instantaneous input voltage.
    (b) the rate at which the instantaneous input voltage changes.
    (c) the accumulated input voltage as a function of time.
    (d) the difference between the instantaneous voltages of the two input signals.
    (e) the sum of the instantaneous voltages of the two input signals.

42. What types of devices require particular care in handling, so they are not destroyed by electrostatic discharges that can build up on a technician's body?
    (a) Vacuum tubes
    (b) Ferrite resistors
    (c) Ceramic capacitors
    (d) MOS components
    (e) Rectifier diodes

43. According to Isaac Asimov's rules for the behavior of robots, a robot must not injure, or allow the injury of, any human,
    (a) unless the robot is forced to prevent its own destruction.
    (b) unless the human is committing a crime.

(c) unless the human is trying to turn off the robot's power supply.

(d) unless it is the result of an accident.

(e) under any circumstances.

44. Computer data speed is typically measured in

(a) kilobits, megabits, and gigabits.

(b) kilobits per second, megabits per second, and gigabits per second.

(c) kilobytes, megabytes, and gigabytes.

(d) kilobytes per second, megabytes per second, and gigabytes per second.

(e) any of the above.

45. Doppler radar is useful for measuring or estimating

(a) the frequency of a sine wave that is modulated with a complex signal.

(b) the wind speed in the funnel cloud of a tornado.

(c) the distance between a robot gripper and a tool or object to be manipulated.

(d) the depth of the ocean at a particular location.

(e) the direction from which a thunderstorm is coming.

46. A device that translates a 30-kHz ac voltage into ultrasonic waves by producing stress on a crystal, thereby causing the crystal to vibrate, is an example of

(a) an electromechanical loudspeaker.

(b) a passive pressure sensor.

(c) a dynamic transducer.

(d) an inductive proximity sensor.

(e) a piezoelectric transducer.

47. Fill in the blank in the following sentence to make it true: "A _____ antenna is a vertical radiator, usually measuring $\lambda/4$, elevated above the surface of the earth, and operated against a system of $\lambda/4$ horizontal or slightly drooping radials."

(a) dipole

(b) coaxial

(c) collinear

(d) ground-plane

(e) vertizontal

48. Figure Test4-3 shows a robotic system that employs

(a) Cartesian coordinate geometry.

(b) spherical coordinate geometry.

(c) polar coordinate geometry.

(d) cylindrical coordinate geometry.

(e) Riemannian coordinate geometry.

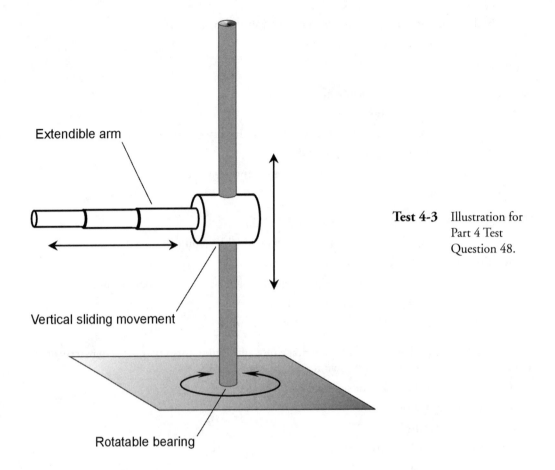

Extendible arm

Vertical sliding movement

Rotatable bearing

Test 4-3 Illustration for Part 4 Test Question 48.

49. A satellite in a LEO system is usually placed in
    (a) an equatorial orbit.
    (b) a geosynchronous orbit.
    (c) a retrograde orbit.
    (d) a polar orbit.
    (e) an orbit midway between the earth and the orbit of the moon.

50. Sound waves in the atmosphere consist of
    (a) rotating magnetic fields.
    (b) alternating electric fields.
    (c) variable electron beams.
    (d) moving molecules.
    (e) rotating electron orbits within atoms.

Final Exam

1. A wire can be coiled around a magnetic compass to make
    (a) an ac wattmeter.
    (b) an ac voltmeter.
    (c) a dc wattmeter.
    (d) a dc ohmmeter.
    (e) a dc galvanometer.

2. The inductive reactance of a fixed air-core coil of wire, assuming the wire has zero dc resistance (it conducts perfectly),
    (a) does not change as the frequency changes.
    (b) increases as the frequency increases.
    (c) decreases as the frequency increases.
    (d) alternately increases and decreases as the frequency increases.
    (e) is zero at all frequencies.

3. *Power* is defined as
    (a) the rate at which current flows in a circuit.
    (b) the product of voltage and resistance in a circuit.
    (c) the rate at which energy is radiated or dissipated.
    (d) the accumulation of energy over time.
    (e) the amount of heat generated in a circuit.

4. In an amplifier that employs a P-channel JFET, the device can usually be replaced with an N-channel JFET having similar specifications, provided that

    (a) all the resistors are reversed in polarity for the circuit in question.

    (b) the power supply polarity is reversed for the circuit in question.

    (c) the drain, rather than the source, is placed at signal ground.

    (d) the input is supplied to the drain, rather than to the gate.

    (e) the output is taken from the source, rather than from the drain.

5. At the exact moment a 60-Hz ac sine wave is at its positive peak voltage, the instantaneous rate of change in the voltage is

    (a) large and positive.

    (b) small and positive.

    (c) large and negative.

    (d) small and negative.

    (e) zero.

6. In order for a bipolar transistor to conduct under conditions of no signal input, the bias must be

    (a) in the forward direction at the emitter-base (E-B) junction, sufficient to cause forward breakover.

    (b) in the reverse direction at the E-B junction, but not sufficient to cause avalanche effect.

    (c) such that the application of a signal would cause the transistor to go into a state of cutoff.

    (d) such that the application of a signal would cause the transistor to go into a state of saturation.

    (e) such that the application of a signal would cause the transistor to become nonlinear.

7. The risk of electrocution to personnel who service "live" electrical or electronic equipment increases in proportion to the

    (a) maximum amount of power, in watts, that the equipment demands.

    (b) maximum amount of energy, in watt-hours, that the equipment consumes.

    (c) maximum amount of current, in amperes, that the equipment draws.

    (d) maximum potential difference, in volts, that exists in the equipment.

    (e) maximum resistance, in ohms, that exists in the equipment.

8. In a logical AND gate with four inputs, the output is high if and only if

    (a) none of the inputs are high, and all four are low.

    (b) one of the inputs is high, and the other three are low.

    (c) two of the inputs are high, and the other two are low.

    (d) three of the inputs are high, and the other one is low.

    (e) all four of the inputs are high, and none are low.

9. Fill in the blank in the following sentence to make it true: "A _____ receiver derives its audio output by mixing incoming signals with the output of a tunable local oscillator."

    (a) crystal set

    (b) direct conversion

(c) phase-locked

(d) very low frequency

(e) frequency modulation

10. In a power supply filter, the term *L section* refers to a combination of

(a) a capacitor in parallel with the rectifier output, and a choke in series.

(b) a capacitor in parallel with the rectifier output, and another capacitor in series.

(c) a choke in parallel with the rectifier output, and a capacitor in series.

(d) a diode in parallel with the transformer output, and another diode in series.

(e) any of the above.

11. What is the frequency of a pure sine wave that has a period of 50 nanoseconds? (A nanosecond is equal to one thousand-millionth, or 10^{-9}, of a second.)

(a) 2.0 MHz

(b) 20 MHz

(c) 0.20 GHz

(d) 2.0 GHz

(e) 20 GHz

12. Fill in the blank in the following sentence to make it correct: "The peak inverse voltage rating of a rectifier diode is the maximum instantaneous reverse-bias voltage that it can withstand without _____ taking place."

(a) forward breakover

(b) avalanche effect

(c) harmonic generation

(d) oscillation

(e) thermal destruction

13. Suppose a coil and capacitor are connected in series, with $jX_L = j50$ and $jX_C = -j90$, and there is no resistance. What is the complex impedance of this combination?

(a) $0 - j40$

(b) $0 + j40$

(c) $40 + j50$

(d) $50 - j90$

(e) It cannot be determined without more information.

14. In an IC, the term *linear* arises from the fact that

(a) all of the etchings on the semiconductor material are straight lines.

(b) the amplification factor generally decreases in direct proportion to the instantaneous input amplitude.

(c) the amplification factor generally increases in direct proportion to the instantaneous input amplitude.

(d) the amplification factor is generally constant as the instantaneous input amplitude varies.

(e) the device never exhibits distortion under any circumstances.

15. Suppose a dc circuit has 10 kΩ of resistance, and a 12-V battery is connected to it. How much current is drawn from the battery?

   (a) 1.2 μA

   (b) 12 μA

   (c) 1.2 mA

   (d) 12 mA

   (e) None of the above

16. The high input impedance of a MOSFET makes this type of device ideal for use in

   (a) weak-signal amplifiers.

   (b) high-power oscillators.

   (c) high-current rectifiers.

   (d) antenna tuning networks.

   (e) graphic equalizers.

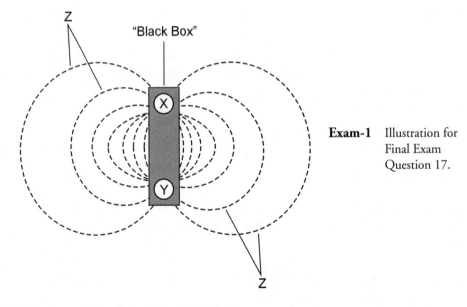

Exam-1   Illustration for Final Exam Question 17.

17. In Fig. Exam-1, the dashed lines labeled (Z) can represent

   (a) the electric lines of flux in the vicinity of two electrically charged objects (X and Y inside the black box) having opposite polarity.

   (b) the magnetic lines of flux in the vicinity of two electrically charged objects (X and Y inside the black box) having opposite polarity.

   (c) the electric lines of flux in the vicinity of two electrically charged objects (X and Y inside the black box) having the same polarity.

   (d) the magnetic lines of flux in the vicinity of two electrically charged objects (X and Y inside the black box) having the same polarity.

   (e) any of the above.

18. The drain of a JFET is the analog of the

    (a) plate of a vacuum tube.

    (b) emitter of a bipolar transistor.

    (c) cathode of a diode.

    (d) positive electrode in a solar cell.

    (e) substrate of a MOSFET.

19. Imagine a circuit with 21 Ω of inductive reactance and 28 Ω of resistance. What is the absolute-value impedance of this circuit?

    (a) 7 Ω

    (b) 24 Ω

    (c) 35 Ω

    (d) 49 Ω

    (e) None of the above

20. Suppose you want to obtain a 12-V battery by connecting 1.5-V size D flashlight cells together. This can be done by

    (a) connecting 8 cells together in parallel, making sure to always connect the terminals so they are plus-to-plus and minus-to-minus.

    (b) connecting 8 cells together in series, making sure to always connect the terminals so they are plus-to-minus.

    (c) either connecting 8 cells together in parallel as described in (a), or connecting 8 cells together in series as described in (b).

    (d) connecting two sets of 4 cells together in series, making sure to always connect the terminals so they are plus-to-minus, and then connecting these sets together in parallel, making sure to connect the terminals so they are plus-to-plus and minus-to-minus.

    (e) connecting two sets of 4 cells together in parallel, making sure to always connect the terminals so they are plus-to-plus and minus-to-minus, and then connecting these sets together in series, making sure to connect the terminals so they are plus-to-minus.

21. One of the technical limitations of capacitive proximity sensors is the fact that they

    (a) are not very sensitive to objects that are poor electrical conductors.

    (b) are insensitive to objects that reflect light.

    (c) are insensitive to metallic objects.

    (d) cannot be used with oscillators.

    (e) require extreme voltages in order to function properly.

22. A flute sounds different than a violin, even if the two instruments are played at the same pitch, because of a difference in the

    (a) phase.

    (b) chamber length.

    (c) frequency.

    (d) waveform.

    (e) bias.

23. In a personal computer, the internal mass-storage medium is nearly always

    (a) a CD (compact disc) in some form.

    (b) a diskette drive.

    (c) a tape drive.

    (d) read-only memory (ROM).

    (e) none of the above.

24. A device to which signals can be sent asking for data, and that responds with the requested information by means of radio or a wireless network, is known as a

    (a) transceiver.

    (b) transponder.

    (c) repeater.

    (d) duplexer.

    (e) signal reflector.

25. Suppose you want to use a transformer to obtain exactly 30 V rms ac from an electrical outlet that supplies exactly 120 V rms ac. The primary-to-secondary turns ratio of the transformer should be

    (a) exactly 1:16.

    (b) exactly 1:4.

    (c) variable.

    (d) exactly 4:1.

    (e) exactly 16:1.

26. Fill in the blank in the following statement to make it true: "Two problems can occur with the ____ type of intrusion alarm. First, the sun, suddenly emerging from clouds or from shadow, might shine on the sensor and trigger the device when there is no threat. Second, someone with malicious intent, wearing winter clothing and entering on a cold day, might fail to set off the alarm in the presence of a real threat."

    (a) IR radiation detector

    (b) electromagnetic detector

    (c) ultrasonic detector

    (d) ultraviolet detector

    (e) fluxgate detector

27. Suppose that five resistors, each one having the same ohmic value as all the others, are connected in series, and the entire combination is connected to a 12-V battery. How does the voltage across any one of the resistors compare with the battery voltage, assuming no other loads are connected to the circuit?

    (a) The voltage across any one of the resistors is 5 times the battery voltage.

    (b) The voltage across any one of the resistors is the same as the battery voltage.

    (c) The voltage across any one of the resistors is $1/5$ of the battery voltage.

    (d) The voltage across any one of the resistors is $1/25$ of the battery voltage.

    (e) It is impossible to answer this without more information.

28. When the carrier and one of the sidebands are removed from an amplitude-modulated signal, the result is

    (a) a frequency-modulated signal.

    (b) a phase-modulated signal.

    (c) a frequency-shift-keyed signal.

    (d) a pulse-modulated signal.

    (e) a single-sideband signal.

29. In log polar navigation, a computer converts a polar coordinate image to

    (a) a Riemannian coordinate image.

    (b) a spherical coordinate image.

    (c) a cylindrical coordinate image.

    (d) a curvilinear coordinate image.

    (e) a rectangular coordinate image.

30. The charged particles in the nucleus of an atom are

    (a) electrons.

    (b) protons.

    (c) positrons.

    (d) neutrons.

    (e) negatrons.

31. Which of the following functions (a), (b), (c), or (d), if any, is a semiconductor diode never used to perform?

    (a) Rectification

    (b) Detection

    (c) Frequency multiplication

    (d) Signal mixing

    (e) A semiconductor diode can perform any of the above functions.

32. The reactance of a fixed capacitor, assuming it has no leakage conductance (that is, it has a theoretically infinite dc resistance),

    (a) does not change as the frequency changes.

    (b) increases negatively as the frequency increases.

    (c) decreases negatively (approaches zero) as the frequency increases.

    (d) alternately becomes negative and positive as the frequency increases.

    (e) is theoretically infinite at all frequencies.

33. What is the dc resistance of a component with a dc conductance of 10^{-4} S?

    (a) $0.0001 \ \Omega$

    (b) $0.01 \ \Omega$

    (c) $100 \ \Omega$

    (d) $10 \ k\Omega$

    (e) None of the above

34. A common-collector transistor circuit is often used

    (a) to provide high gain and sensitivity over a wide range of frequencies.

    (b) to match a high impedance to a low impedance.

    (c) as a high-fidelity audio power amplifier.

    (d) as an oscillator at microwave frequencies.

    (e) as the rectifier in a dc power supply.

35. The internal conductance (expressed in siemens) of an ammeter is generally

    (a) low.

    (b) directly proportional to the current.

    (c) inversely proportional to the current.

    (d) high.

    (e) any value; it doesn't matter.

36. A half-wave length of wire, fed at one end with a quarter-wave section of parallel-wire transmission line, is known as

    (a) a zeppelin antenna.

    (b) a dipole antenna.

    (c) a Yagi antenna.

    (d) a random wire antenna.

    (e) an end-fire antenna.

37. A twin T oscillator is commonly used for generating

    (a) AF signals.

    (b) high-frequency RF signals.

    (c) microwave RF signals.

    (d) powerful bursts of RF energy.

    (e) FM signals.

38. Suppose that five resistors, each one having the same ohmic value as all the others, are connected in series, and the entire combination is connected to a 12-V battery. How does the current through any one of the resistors compare with the current drawn from the battery?

    (a) The current through any one of the resistors is 5 times the current drawn from the battery.

    (b) The current through any one of the resistors is the same as the current drawn from the battery.

    (c) The current through any one of the resistors is $1/5$ as great as the current drawn from the battery.

    (d) The current through any one of the resistors is $1/25$ as great as the current drawn from the battery.

    (e) It is impossible to answer this without more information.

39. When a nonzero net reactance exists in an ac circuit, the volt-ampere (VA) power, also known as the *apparent power*, is

(a) zero.

(b) equal to the true power.

(c) less than the true power.

(d) greater than the true power.

(e) infinite.

40. When an impurity, also called a *dopant*, is added to a semiconductor material, it can contain either an excess of electrons or a deficiency of electrons. A dopant with an excess of electrons, also known as a *donor impurity*, results in the production of

(a) an NPN transistor.

(b) a PNP transistor.

(c) a P-N junction.

(d) a P type semiconductor.

(e) an N type semiconductor.

41. Suppose an audio amplifier produces 100 W rms output when the input is 5.00 W rms. This represents a power gain of

(a) 1.30 dB.

(b) 2.60 dB.

(c) 13.0 dB.

(d) 20.0 dB.

(e) 26.0 dB.

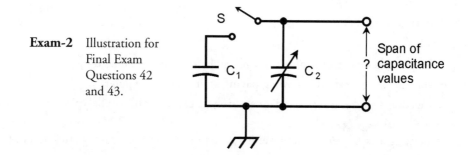

Exam-2   Illustration for Final Exam Questions 42 and 43.

42. In Fig. Exam-2, suppose the value of the fixed capacitor, C_1, is 330 pF, and the range of the variable capacitor, C_2, is 10 to 365 pF. What is the span of capacitance values (minimum to maximum) that can be obtained with this parallel combination, with switch S either open or closed, as desired?

(a) 10 to 365 pF, over a continuous range

(b) 10 to 181 pF, but with a gap in the range

(c) 10 to 181 pF, over a continuous range

(d) 10 to 695 pF, but with a gap in the range

(e) 10 to 695 pF, over a continuous range

43. In Fig. Exam-2, suppose the value of the fixed capacitor, C_1, is 220 pF, and the range of the variable capacitor, C_2, is 5 to 100 pF. What is the span of capacitance values (minimum to maximum) that can be obtained with this parallel combination, with switch S either open or closed, as desired?

(a) 5 to 100 pF, over a continuous range

(b) 5 to 320 pF, but with a gap in the range

(c) 5 to 320 pF, over a continuous range

(d) 5 to 69 pF, but with a gap in the range

(e) 5 to 69 pF, over a continuous range

44. Which of the following units is most often used for specifying data speed in digital communications systems?

(a) Megabytes per second

(b) Megabauds per second

(c) Megabits per second

(d) Megahertz per second

(e) Any of the above units are used equally often when specifying data speed in digital communications systems.

45. Imagine four 100-µH inductors connected in a 2 × 2 series-parallel combination. Suppose there is no mutual inductance among them. What is the net inductance of this matrix?

(a) 25 µH

(b) 50 µH

(c) 100 µH

(d) 200 µH

(e) 400 µH

46. In the output of a dc power supply, voltage regulation can be obtained by

(a) connecting a Zener diode in series, paying attention to the polarity.

(b) connecting a choke in series and an electrolytic capacitor in parallel, paying attention to the polarity.

(c) connecting a capacitor in parallel and a Zener diode in series, paying attention to the polarity.

(d) connecting a Zener diode in parallel, paying attention to the polarity.

(e) connecting a resistor in parallel and another resistor in series.

47. Imagine an inductor and resistor connected in series, such that the inductive reactance and the resistance are both equal to 400 Ω at a frequency of 100 MHz. How are the instantaneous current and the instantaneous voltage related?

(a) They are in phase.

(b) They are 180° out of phase.

(c) The current leads the voltage by 90°.

(d) The current lags the voltage by 90°.

(e) None of the above are true.

48. Suppose you want to use a transformer to match the output of an audio amplifier to a speaker. The amplifier has a purely resistive output impedance of exactly 128 Ω. The speaker has a purely resistive impedance of exactly 8 Ω. The primary-to-secondary turns ratio of the transformer should be

(a) exactly 1:16.

(b) exactly 1:4.

(c) variable.

(d) exactly 4:1.

(e) exactly 16:1.

49. Imagine a string of 20 holiday lights, all connected in parallel and plugged into a 120-V rms ac wall outlet. If one of the bulbs is short-circuited, what will happen?

(a) All of the remaining 19 bulbs will shine a little more brightly than before.

(b) All of the remaining 19 bulbs will shine a little less brightly than before.

(c) The fuse or circuit breaker will blow, and all of the remaining 19 bulbs will go out.

(d) Most of all of the remaining 19 bulbs will burn out, and some may explode.

(e) The remaining 19 bulbs will continue to shine exactly as before—no more or less brightly.

50. Suppose a coil and capacitor are connected in series, with $jX_L = j40$ and $jX_C = -j70$, and the coil has an internal resistance of 10 Ω. Suppose the frequency of operation is 12.5 MHz. This circuit will exhibit resonance at

(a) some frequency below 12.5 MHz.

(b) some frequency above 12.5 MHz.

(c) 12.5 MHz; it is resonant under the conditions stated.

(d) any and all frequencies.

(e) no frequency, because of the internal resistance.

51. Imagine a transmission line consisting of two wires, each having a diameter of exactly 1 mm, and uniformly spaced exactly 10 mm apart with nothing but air as the dielectric between them. If the two wires are moved so they are uniformly spaced exactly 50 mm apart, and the dielectric between them is still nothing but air, what happens to the characteristic impedance?

(a) It does not change.

(b) It increases.

(c) It decreases.

(d) It is impossible to say without knowing what is connected to the line.

(e) It is impossible to say without knowing the frequency.

52. In an FM signal, deviation is

(a) the difference between the highest instantaneous carrier frequency and the lowest instantaneous carrier frequency.

(b) the maximum extent to which the instantaneous carrier frequency differs from the unmodulated-carrier frequency.

(c) the bandwidth of the signal, expressed in hertz, kilohertz, or megahertz.

(d) the rate at which the phase of the signal varies, expressed in hertz per second, kilohertz per millisecond, or megahertz per microsecond.

(e) None of the above is true.

53. A typical angular speed for an old-fashioned vinyl disk in a hi-fi sound system is

(a) 12 rpm.

(b) 24 rpm.

(c) 33 rpm.

(d) 60 rpm.

(e) 90 rpm.

54. An autotransformer can be recognized by the fact that it has

(a) a continuously adjustable turns ratio.

(b) a solenoidal core that can be moved in and out of a coil.

(c) a single, tapped winding.

(d) a special set of terminals for use with an automotive battery.

(e) a toroidal core.

55. One of the purposes of the suppressor grid in a pentode vacuum tube amplifier is to

(a) keep the cathode from emitting too many electrons, which could drive the control grid current to excessive levels.

(b) ensure that the plate is maintained at a constant negative voltage, so electrons will not be attracted to it.

(c) prevent the screen grid from shorting out to the plate, which could cause the power supply to burn out.

(d) repel secondary electrons emitted by the plate, keeping the screen grid current from getting too high.

(e) prevent distortion caused by overdrive in the control grid, screen grid, or cathode circuits.

56. Consider two sine waves having identical frequency. Suppose that one wave leads the other by exactly $1/20$ of a cycle. What is the phase difference between the two waves?

(a) 9°

(b) 12°

(c) 18°

(d) 20°

(e) 24°

57. In a cellular communications system, a technology in which cells overlap, but signals do not interfere with each other because every phone set is assigned a unique signal code, is known as

(a) code-division multiple access.

(b) diversity reception.

(c) a phase-locked loop.

(d) analog signal mixing.

(e) digital diversity encryption.

58. A steady magnetic field can be produced by

(a) a straight wire carrying a constant direct current.

(b) a loop of wire carrying a constant direct current.

(c) a coil of wire carrying a constant direct current.

(d) a constant-intensity stream of protons in free space, moving in a straight line.

(e) any of the above.

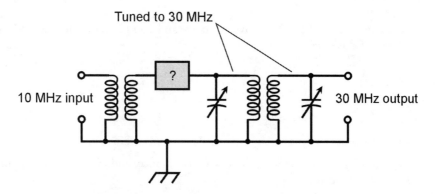

Exam-3 Illustration for Final Exam Question 59.

59. In Fig. Exam-3, what type of component can the box with the query symbol (?) contain in order for the circuit to function as indicated?

(a) A capacitor

(b) A resistor

(c) A battery

(d) An inductor

(e) A diode

60. The binary number 1001101 represents which decimal number?

(a) 44

(b) 55

(c) 66

(d) 77

(e) 88

61. Suppose a dc circuit has 10 kΩ of resistance, and a 12-V battery is connected to it. How much power is drawn from the battery?

(a) 1.2 μW

(b) 12 μW

(c) 1.2 mW

(d) 12 mW

(e) None of the above

62. In which of the following applications would you never find an op amp used as the main active circuit component?

(a) A differentiator

(b) An integrator

(c) A highpass audio filter

(d) A lowpass audio filter

(e) A microwave oscillator

63. Imagine a circuit with 21 Ω of inductive reactance and 28 Ω of resistance. What is the complex impedance of this circuit?

(a) 7 Ω

(b) 24 Ω

(c) 35 Ω

(d) 49 Ω

(e) None of the above

64. In a circuit containing conductance, inductive susceptance, and capacitive susceptance, a condition of resonance exists if and only if

(a) the conductance is zero.

(b) the conductance is infinite.

(c) the inductive susceptance is infinite and the capacitive susceptance is zero.

(d) the inductive and capacitive susceptances are both zero.

(e) the inductive and capacitive susceptances cancel out.

65. Imagine a perfect square wave with no dc component. That means the positive and negative portions have equal amplitudes and equal durations. Suppose you are told that the voltage is 10.0 V pk-pk. What is the average voltage?

(a) 0.00 V

(b) 5.00 V

(c) 7.07 V

(d) 10.0 V

(e) It is impossible to calculate this without more information.

66. Fill in the blank in the following statement to make it true: "It takes a certain minimum ____ voltage, known as the *forward breakover voltage*, for conduction to occur through a P-N junction."

(a) RF

(b) AF

(c) forward bias

(d) negative

(e) positive

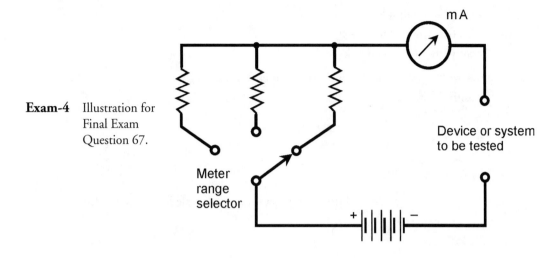

Exam-4 Illustration for
Final Exam
Question 67.

67. Figure Exam-4 is a generic diagram of a device that can be used to measure
    (a) dc voltage.
    (b) dc resistance.
    (c) dc power.
    (d) ac voltage.
    (e) ac power.

68. Suppose an audio amplifier produces 8.0 V rms output with 80 mV rms input. Suppose the amplifier input impedance is the same as the output load impedance. What is the gain of this amplifier in this situation?
    (a) 20 dB
    (b) 40 dB
    (c) 100 dB
    (d) 200 dB
    (e) We must know the actual input and output impedance values, in ohms, in order to determine the gain in this case.

69. The maximum gain obtainable with an amplifier that employs a given bipolar transistor
    (a) gradually increases as the frequency increases.
    (b) stays the same as the frequency increases.
    (c) gradually decreases as the frequency increases.
    (d) alternately increases and decreases as the frequency increases.
    (e) remains constant up to a certain frequency, and then abruptly drops to zero above that frequency.

70. A phase comparator would most likely be found in
    (a) an AF oscillator.
    (b) an AM detector.

    (c) an RF power amplifier.

    (d) a PLL frequency synthesizer.

    (e) An antenna tuning network.

71. An atom with 7 protons and 5 electrons is an example of

    (a) a positive isotope.

    (b) a negative isotope.

    (c) a positive ion.

    (d) a negative ion.

    (e) a neutral ion.

72. Electrical energy can be derived from hydrogen on a small scale in a device called

    (a) an electrolytic cell.

    (b) an alkaline cell.

    (c) a fission cell.

    (d) a fusion cell.

    (e) a fuel cell.

73. The output wave of a common-gate amplifier circuit with a pure sine-wave input

    (a) is in phase with the input wave.

    (b) lags the input wave by 90° of phase.

    (c) leads the input wave by 90° of phase.

    (d) is 180° out of phase with the input wave.

    (e) is inverted with respect to the input wave.

74. Which of the following capacitor types is polarized?

    (a) Electrolytic

    (b) Paper

    (c) Ceramic

    (d) Mica

    (e) Air variable

75. The maximum radiation and response from a full-wave loop antenna, assuming there are no nearby conductors or obstructions to distort the pattern, occurs

    (a) in the plane of the loop.

    (b) perpendicular to the plane of the loop.

    (c) at 45° angles to the plane of the loop.

    (d) parallel to the horizon, in all directions of the compass.

    (e) in all directions in three dimensions; it is a true omnidirectional antenna.

76. Suppose that five resistors, each having a different ohmic value from any of the others, are connected in series, and the entire combination is connected to a 12-V battery. How does the current through any one of the resistors compare with the current drawn from the battery?

(a) The current through any one of the resistors is more than the current drawn from the battery.

(b) The current through any one of the resistors is the same as the current drawn from the battery.

(c) The current through any one of the resistors is less than the current drawn from the battery.

(d) It depends on the relative ohmic values of the resistors.

(e) None of the above statements are true.

77. Suppose a battery-powered circuit has 1.00 Ω of net resistance, and 15.0 A of current flows through it. How much power is demanded from the battery?

(a) 225 W

(b) 150 W

(c) 22.5 W

(d) 66.7 mW

(e) It is impossible to calculate this, based on the information given.

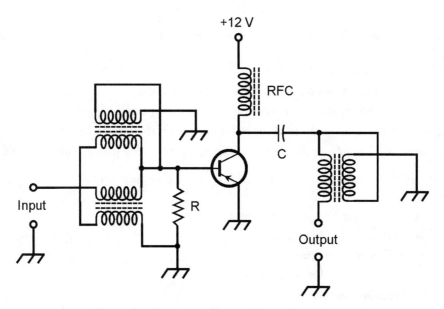

Exam-5 Illustration for Final Exam Question 78.

78. Figure Exam-5 shows a generic circuit that can be used for broadband RF power amplification. What is wrong with the circuit as shown?

(a) The resistor, R, should be replaced with a capacitor.

(b) The capacitor, C, should be replaced with a resistor.

(c) The +12-V power source should be replaced with a −12-V power source.

(d) The choke, RFC, should be replaced with a capacitor.

(e) Nothing is wrong with this circuit.

79. Imagine a string of 20 holiday lights, all connected in parallel and plugged into a 120-V rms ac wall outlet. If one of the bulbs is unscrewed and removed from the circuit, what will happen?

(a) All of the remaining 19 bulbs will shine a little more brightly than before.

(b) All of the remaining 19 bulbs will shine a little less brightly than before.

(c) The fuse or circuit breaker will blow, and all of the remaining 19 bulbs will go out.

(d) Most of all of the remaining 19 bulbs will burn out, and some may explode.

(e) The remaining 19 bulbs will continue to shine exactly as before—no more or less brightly.

80. When a volatile RAM chip is used in a computer,

(a) the contents of the memory can be easily overwritten.

(b) the contents of the memory remain intact even if power is removed.

(c) data in memory cannot be transferred to or from external media.

(d) the contents of memory never last more than a few minutes.

(e) excessive ELF radiation is produced.

81. A nickel-cadmium (NICAD) battery should never be discharged until it is totally dead because

(a) the battery may explode.

(b) the battery will have excessive voltage after it is recharged.

(c) the battery will develop unwanted inductance after it is recharged.

(d) the polarity of one or more of the cells may reverse, ruining the battery.

(e) Forget it! There is no problem with discharging a NICAD battery until it is totally dead.

82. Which of the following is an advantage of a CRT display over an LCD for a computing workstation?

(a) You can get a bigger screen for a lower price.

(b) The CRT emits less ELF energy.

(c) The CRT is less massive.

(d) The CRT takes up less space on the desk.

(e) All of the above are true.

83. The gauss is a unit of

(a) charge carrier flow speed.

(b) magnetic flux density.

(c) electrostatic field strength.

(d) electromagnetic field intensity.

(e) electrical charge quantity.

84. Imagine an inductor and resistor connected in series, such that the inductive reactance is 300 Ω and the resistance is 500 Ω at a particular frequency. If the frequency is decreased, what happens to the relative phase of the instantaneous current and the instantaneous voltage?

  (a) Their relative phase does not change.

  (b) They become more nearly in phase.

  (c) They become more out of phase.

  (d) They go alternately into and out of phase.

  (e) It is impossible to say anything about this without more information.

85. Imagine two sine waves, one with a frequency of 50.00 Hz and the other with a frequency of 60.00 Hz. The phase difference between these two waves is

  (a) not definable.

  (b) 10.00°.

  (c) 10.00%.

  (d) 16.67%.

  (e) $\frac{5}{6}$ of a radian.

86. Consider a variable capacitor connected in parallel with a fixed toroidal inductor. Suppose a signal of constant frequency is applied to the combination. If the variable capacitor is adjusted so its value changes from 200 pF to 100 pF, what happens to the reactance of the inductor?

  (a) It doubles.

  (b) It becomes half as great.

  (c) It quadruples.

  (d) It becomes $\frac{1}{4}$ as great.

  (e) It does not change.

87. When two inductors having the same value are connected in series with a small opposing mutual inductance (say the coefficient of coupling is 5 percent or so), the total reactance of the combination at a specific, constant frequency is

  (a) equal to the reactance of either inductor.

  (b) slightly greater than the reactance of either inductor.

  (c) slightly less than the reactance of either inductor.

  (d) slightly greater than twice the reactance of either inductor.

  (e) slightly less than twice the reactance of either inductor.

88. The power factor in an ac circuit is defined as

  (a) the actual power divided by the maximum power the circuit can handle.

  (b) the ratio of the real power to the imaginary power.

  (c) the ratio of the apparent power to the true power.

  (d) the ratio of the true power to the apparent power.

  (e) the ratio of the imaginary power to the apparent power.

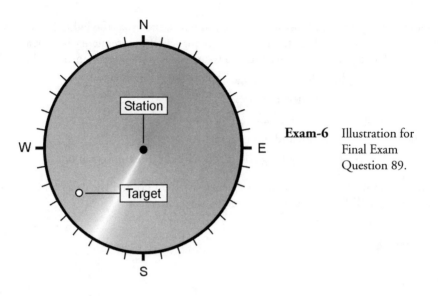

Exam-6 Illustration for Final Exam Question 89.

89. Figure Exam-6 shows a radar display and a single blip, indicating the position of a target. What is the approximate azimuth of this target?

(a) 122°

(b) 212°

(c) 238°

(d) 148°

(e) More information is needed to answer this.

90. Which of the following is an e-mail address in the proper format?

(a) www.sciencewriter.net

(b) members.authorsguild.net/stangib

(c) sciencewriter@tahoe.com

(d) http://google.com

(e) All of the above

91. A robot can be guided along a specific route by means of a current-carrying wire beneath the surface. The robot detects, and follows, the magnetic field produced by the current, thus following the wire. This is called

(a) an epipolar system.

(b) a fluxgate magnetometer.

(c) a servo system.

(d) an electromagnetic router.

(e) an embedded path.

92. Which of the following substances is sometimes used as the semiconductor material in junction field-effect transistors (JFETs)?

    (a) Gallium arsenide

    (b) Mica

    (c) Glass

    (d) Polystyrene

    (e) Porcelain

93. Suppose that five resistors, each one having the same ohmic value as all the others, are connected in parallel, and the entire combination is connected to a 12-V battery. How does the current through any one of the resistors compare with the current drawn from the battery?

    (a) The current through any one of the resistors is 5 times the current drawn from the battery.

    (b) The current through any one of the resistors is the same as the current drawn from the battery.

    (c) The current through any one of the resistors is $^1/_5$ as great as the current drawn from the battery.

    (d) The current through any one of the resistors is $^1/_{25}$ as great as the current drawn from the battery.

    (e) It is impossible to answer this without more information.

94. Figure Exam-7 illustrates a method of

    (a) wireless-only encryption.

    (b) end-to-end encryption.

    (c) wireless tapping of a cell phone set.

    (d) wireless tapping of a cordless phone set.

    (e) dual-diversity shortwave reception.

95. In Fig. Exam-7, the box at the top contains a query symbol (?). What type of device should this be, in order for the system to perform its intended task?

    (a) A low-frequency receiver

    (b) A microwave transponder

    (c) A video and data receiver

    (d) A wireless transmitter

    (e) A GPS receiver

96. Some substances cause magnetic lines of flux to bunch closer together than they would be if the magnetic field existed in a vacuum. This property is known as

    (a) electromagnetism.

    (b) diamagnetism.

    (c) flux magnification.

    (d) flux constriction.

    (e) ferromagnetism.

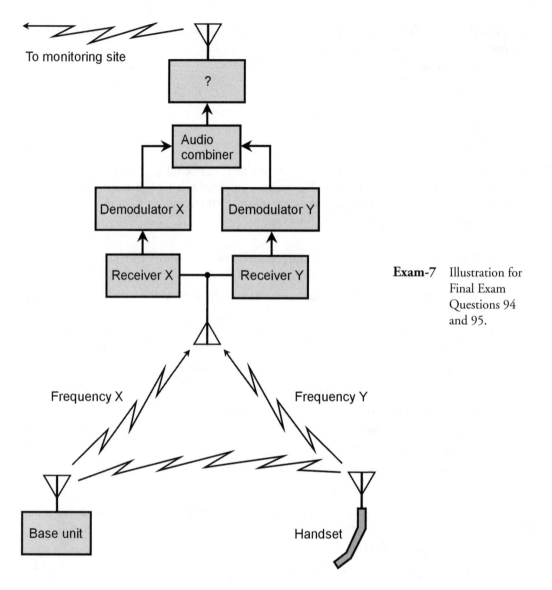

Exam-7 Illustration for Final Exam Questions 94 and 95.

97. In digital audio tape recording and reproduction, the noise can be practically eliminated because
    (a) noise is analog in nature, and can be nulled or filtered out without degrading the digital signals.
    (b) noise is digital in nature, and can be eliminated by means of specialized computer programs.
    (c) music and voices are digital in nature, and can be reproduced multiple times without distortion.
    (d) music and voices are analog in nature, while noise is digital and can therefore be canceled out.
    (e) Forget it! In a digital audio tape recording system, noise is a more serious problem than it is in an analog system.

98. Imagine a capacitor and resistor connected in series, such that the capacitive reactance is equal to $-200\ \Omega$ and the resistance is equal to $200\ \Omega$ at a frequency of 70 MHz. How are the instantaneous current and the instantaneous voltage related?

    (a) They are in phase.

    (b) They are 90° out of phase.

    (c) The current leads the voltage by 45°.

    (d) The current lags the voltage by 45°.

    (e) None of the above are true.

99. The amount of current that a silicon photodiode can deliver in direct sunlight depends on

    (a) the forward breakover voltage.

    (b) the thickness of the substrate.

    (c) the surface area of the P-N junction.

    (d) the applied voltage.

    (e) the reverse bias.

100. Fill in the blank to make the following sentence true: "Electrical current can be expressed in terms of the number of _____ passing a given point per unit time."

    (a) neutrons

    (b) webers or gauss

    (c) charge carriers

    (d) wave cycles

    (e) isotopes

A
APPENDIX

Answers to Quiz, Test, and Exam Questions

Chapter 1

1. b	2. c	3. d	4. d	5. d
6. a	7. b	8. c	9. a	10. b
11. c	12. a	13. d	14. c	15. c
16. a	17. c	18. b	19. c	20. d

Chapter 2

1. b	2. c	3. d	4. b	5. c
6. a	7. c	8. c	9. a	10. d
11. d	12. a	13. b	14. b	15. c
16. d	17. b	18. c	19. a	20. c

Chapter 3

1. b	2. a	3. c	4. c	5. a
6. d	7. c	8. a	9. d	10. c
11. a	12. b	13. a	14. c	15. d
16. b	17. d	18. a	19. c	20. b

Chapter 4

1. a	2. c	3. d	4. a	5. b
6. d	7. c	8. a	9. c	10. a
11. d	12. b	13. d	14. b	15. a
16. c	17. c	18. a	19. d	20. b

Chapter 5

1. b	2. d	3. c	4. c	5. a
6. b	7. d	8. a	9. b	10. d
11. b	12. c	13. b	14. c	15. a
16. c	17. b	18. c	19. b	20. c

Chapter 6

1. c	2. a	3. a	4. c	5. d
6. b	7. b	8. c	9. b	10. c
11. d	12. c	13. a	14. c	15. a
16. a	17. b	18. c	19. b	20. a

Chapter 7

1. c	2. c	3. b	4. c	5. d
6. b	7. a	8. b	9. a	10. d
11. b	12. a	13. b	14. c	15. d
16. c	17. c	18. c	19. a	20. d

Chapter 8

1. c	2. a	3. d	4. b	5. c
6. b	7. b	8. c	9. a	10. c
11. c	12. d	13. b	14. a	15. d
16. d	17. c	18. b	19. b	20. a

Test: Part 1

1. a	2. a	3. c	4. b	5. a
6. b	7. a	8. e	9. e	10. c
11. c	12. c	13. b	14. c	15. d
16. b	17. a	18. a	19. e	20. e
21. d	22. d	23. d	24. c	25. a
26. c	27. d	28. a	29. c	30. c
31. c	32. d	33. e	34. d	35. e
36. b	37. d	38. d	39. c	40. a
41. a	42. d	43. c	44. e	45. c
46. c	47. d	48. a	49. e	50. c

Chapter 9

1. c	2. c	3. a	4. c	5. d
6. d	7. a	8. b	9. c	10. b
11. d	12. b	13. b	14. d	15. d
16. a	17. c	18. c	19. d	20. a

Chapter 10

1. b	2. a	3. d	4. b	5. b
6. c	7. d	8. c	9. a	10. d
11. b	12. d	13. a	14. a	15. d
16. b	17. a	18. c	19. a	20. d

Chapter 11

1. d	2. a	3. a	4. a	5. c
6. d	7. b	8. a	9. d	10. d
11. b	12. c	13. b	14. a	15. b
16. c	17. d	18. a	19. b	20. b

Chapter 12

1. c	2. a	3. b	4. b	5. d
6. b	7. d	8. c	9. a	10. b
11. b	12. d	13. c	14. a	15. c
16. b	17. d	18. c	19. d	20. c

Chapter 13

1. c	2. a	3. b	4. d	5. c
6. a	7. b	8. c	9. d	10. c
11. d	12. d	13. a	14. b	15. d
16. b	17. d	18. c	19. b	20. a

Chapter 14

1. b	2. d	3. a	4. c	5. b
6. a	7. b	8. a	9. a	10. d
11. c	12. b	13. b	14. b	15. c
16. b	17. c	18. d	19. a	20. c

Chapter 15

1. b	2. b	3. a	4. d	5. b
6. c	7. a	8. c	9. d	10. a
11. a	12. a	13. c	14. a	15. d
16. b	17. d	18. c	19. c	20. b

Chapter 16

1. b	2. a	3. d	4. d	5. d
6. b	7. c	8. c	9. a	10. c
11. b	12. b	13. a	14. b	15. b
16. d	17. c	18. a	19. b	20. a

Chapter 17

1. c	2. c	3. a	4. d	5. a
6. b	7. c	8. d	9. b	10. c
11. c	12. a	13. d	14. a	15. c
16. b	17. d	18. d	19. c	20. a

Chapter 18

1. c	2. a	3. d	4. b	5. b
6. a	7. c	8. d	9. a	10. d
11. b	12. b	13. c	14. d	15. c
16. a	17. c	18. d	19. b	20. b

Test: Part 2

1. e	2. e	3. a	4. a	5. e
6. d	7. c	8. d	9. a	10. e
11. c	12. a	13. d	14. c	15. c
16. e	17. e	18. d	19. c	20. b
21. b	22. d	23. c	24. c	25. e
26. c	27. b	28. a	29. e	30. c
31. a	32. a	33. a	34. a	35. b
36. e	37. c	38. d	39. c	40. b
41. b	42. c	43. c	44. c	45. d
46. d	47. e	48. e	49. a	50. a

Chapter 19

1. b	2. d	3. c	4. a	5. c
6. a	7. d	8. b	9. c	10. b
11. d	12. b	13. b	14. a	15. b
16. c	17. a	18. c	19. d	20. c

Chapter 20

1. b	2. d	3. c	4. b	5. a
6. a	7. d	8. c	9. b	10. d
11. b	12. a	13. c	14. a	15. c
16. d	17. b	18. a	19. c	20. a

Chapter 21

1. c	2. d	3. b	4. a	5. b
6. c	7. a	8. b	9. d	10. c
11. a	12. d	13. d	14. b	15. c
16. a	17. c	18. b	19. a	20. c

Chapter 22

1. d	2. a	3. b	4. d	5. d
6. c	7. b	8. c	9. b	10. d
11. c	12. a	13. c	14. d	15. a
16. d	17. b	18. c	19. b	20. b

Chapter 23

1. b	2. d	3. a	4. b	5. c
6. a	7. b	8. c	9. d	10. b
11. c	12. a	13. d	14. a	15. c
16. d	17. b	18. c	19. a	20. a

Chapter 24

1. a	2. b	3. b	4. c	5. d
6. c	7. d	8. c	9. d	10. b
11. c	12. b	13. d	14. c	15. c
16. d	17. b	18. a	19. b	20. c

Chapter 25

1. c	2. c	3. c	4. c	5. a
6. d	7. b	8. a	9. c	10. d
11. b	12. a	13. a	14. b	15. d
16. a	17. a	18. c	19. a	20. d

Chapter 26

1. b	2. a	3. b	4. c	5. c
6. a	7. d	8. d	9. a	10. d
11. c	12. a	13. a	14. d	15. a
16. d	17. c	18. d	19. b	20. c

Test: Part 3

1. e	2. b	3. a	4. c	5. a
6. c	7. a	8. e	9. e	10. d
11. d	12. e	13. c	14. c	15. d
16. c	17. c	18. e	19. a	20. d
21. c	22. e	23. a	24. c	25. d
26. a	27. b	28. e	29. a	30. b
31. c	32. c	33. b	34. b	35. a
36. d	37. a	38. b	39. b	40. e
41. c	42. a	43. d	44. b	45. a
46. b	47. b	48. c	49. e	50. b

Chapter 27

1. b	2. a	3. b	4. d	5. c
6. a	7. b	8. d	9. c	10. b
11. c	12. b	13. d	14. a	15. b
16. a	17. d	18. c	19. c	20. d

Chapter 28

1. c	2. c	3. a	4. d	5. b
6. d	7. c	8. a	9. d	10. c
11. a	12. c	13. b	14. a	15. c
16. a	17. c	18. d	19. d	20. b

Chapter 29

1. c	2. d	3. a	4. b	5. c
6. d	7. c	8. c	9. d	10. d
11. a	12. a	13. b	14. a	15. c
16. c	17. a	18. a	19. b	20. d

Chapter 30

1. c	2. a	3. a	4. c	5. b
6. d	7. b	8. a	9. d	10. d
11. d	12. a	13. b	14. c	15. a
16. a	17. d	18. b	19. c	20. a

Chapter 31

1. b	2. a	3. d	4. b	5. c
6. b	7. d	8. a	9. d	10. a
11. c	12. a	13. b	14. d	15. a
16. c	17. a	18. b	19. a	20. b

Chapter 32

1. c	2. c	3. d	4. c	5. b
6. a	7. d	8. d	9. a	10. c
11. b	12. b	13. d	14. d	15. a
16. b	17. a	18. c	19. a	20. c

Chapter 33

1. b	2. a	3. d	4. a	5. a
6. c	7. d	8. a	9. d	10. c
11. a	12. d	13. b	14. c	15. b
16. c	17. a	18. d	19. d	20. a

Chapter 34

1. b	2. a	3. d	4. d	5. c
6. b	7. b	8. b	9. d	10. c
11. c	12. a	13. a	14. b	15. c
16. b	17. d	18. c	19. b	20. b

Test: Part 4

1. d	2. e	3. c	4. d	5. b
6. d	7. a	8. a	9. a	10. c
11. c	12. b	13. e	14. e	15. e
16. c	17. e	18. b	19. a	20. d
21. c	22. c	23. e	24. e	25. e
26. c	27. c	28. c	29. a	30. d
31. b	32. c	33. d	34. c	35. e
36. a	37. a	38. a	39. a	40. b
41. b	42. d	43. e	44. b	45. b
46. e	47. d	48. d	49. d	50. d

Final Exam

1. e	2. b	3. c	4. b	5. e
6. a	7. d	8. e	9. b	10. a
11. b	12. b	13. a	14. d	15. c
16. a	17. a	18. a	19. c	20. b
21. a	22. d	23. e	24. b	25. d
26. a	27. c	28. e	29. e	30. b
31. e	32. c	33. d	34. b	35. d
36. a	37. a	38. b	39. d	40. e
41. c	42. e	43. b	44. c	45. c
46. d	47. e	48. d	49. c	50. b
51. b	52. b	53. c	54. c	55. d
56. c	57. a	58. e	59. e	60. d
61. e	62. e	63. e	64. e	65. a
66. c	67. b	68. b	69. c	70. d
71. c	72. e	73. a	74. a	75. b
76. b	77. a	78. c	79. e	80. a
81. d	82. a	83. b	84. b	85. a
86. e	87. e	88. d	89. c	90. c
91. e	92. a	93. c	94. d	95. d
96. e	97. a	98. c	99. c	100. c

B
APPENDIX

Schematic Symbols

ammeter	
amplifier, general	
amplifier, inverting	

amplifier, operational

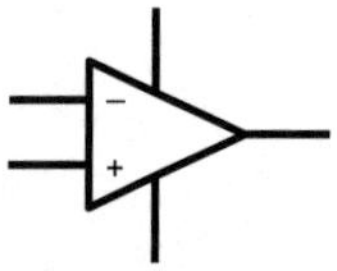

AND gate

antenna, balanced

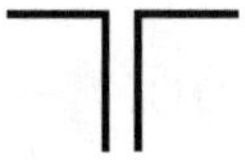

antenna, general

antenna, loop

antenna, loop, multiturn

battery, electrochemical

capacitor, feedthrough

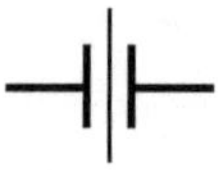

capacitor, fixed

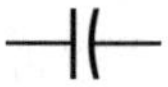

capacitor, variable

capacitor, variable, split-rotor

capacitor, variable, split-stator

cathode, electron-tube, cold

cathode, electron-tube, directly heated

cathode, electron-tube, indirectly heated

cavity resonator

cell, electrochemical

circuit breaker

coaxial cable

crystal, piezoelectric

delay line

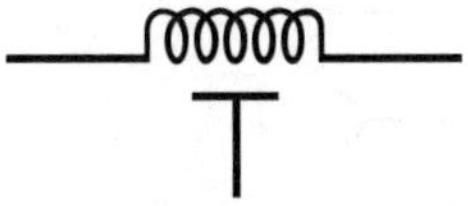

diac

diode, field-effect

diode, general

diode, Gunn

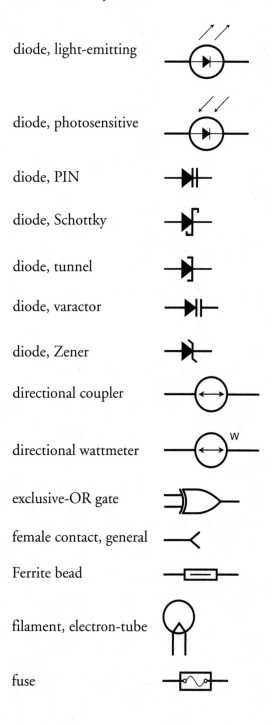

diode, light-emitting

diode, photosensitive

diode, PIN

diode, Schottky

diode, tunnel

diode, varactor

diode, Zener

directional coupler

directional wattmeter

exclusive-OR gate

female contact, general

Ferrite bead

filament, electron-tube

fuse

galvanometer

grid, electron-tube

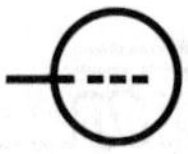

ground, chassis

ground, earth

handset

headset, double

headset, single

headset, stereo

inductor, air core

inductor, air core, bifilar

inductor, air core, tapped

inductor, air core, variable

inductor, iron core

inductor, iron core, bifilar

inductor, iron core, tapped

inductor, iron core, variable

inductor, powdered-iron core

inductor, powdered-iron core, bifilar

inductor, powdered-iron core, tapped

inductor, powdered-iron core, variable

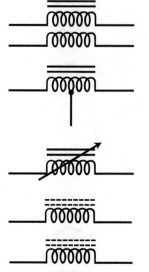

integrated circuit, general

jack, coaxial or phono

jack, phone, 2-conductor

jack, phone, 3-conductor

key, telegraph

lamp, incandescent

lamp, neon

male contact, general

meter, general

microammeter

microphone

microphone, directional

milliammeter

NAND gate

negative voltage connection

NOR gate

NOT gate

optoisolator

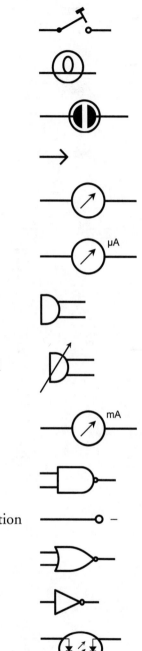

OR gate

outlet, 2-wire, nonpolarized

outlet, 2-wire, polarized

outlet, 3-wire

outlet, 234-volt

plate, electron-tube

plug, 2-wire, nonpolarized

plug, 2-wire, polarized

plug, 3-wire

plug, 234-volt

plug, coaxial or phono

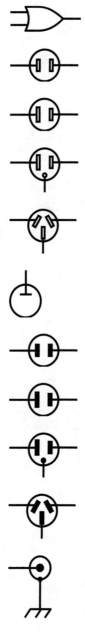

plug, phone, 2-conductor

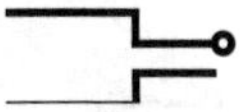

plug, phone, 3-conductor

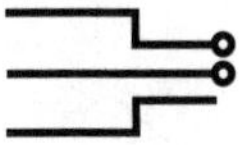

positive voltage connection   +

potentiometer

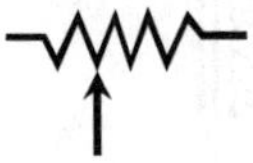

probe, radio-frequency

or

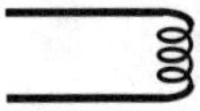

rectifier, gas-filled

rectifier, high-vacuum

rectifier, semiconductor

rectifier, silicon-controlled

relay, double-pole, double-throw

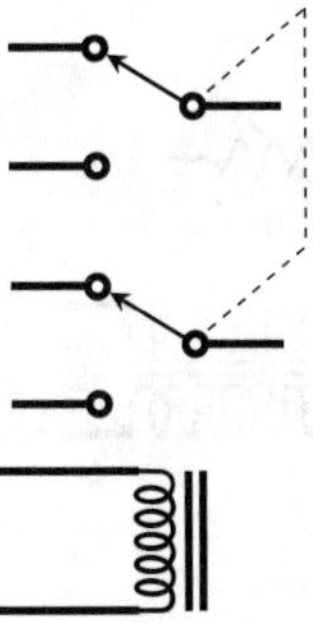

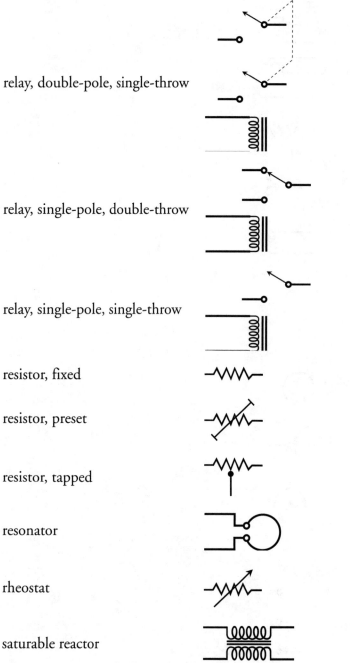

relay, double-pole, single-throw

relay, single-pole, double-throw

relay, single-pole, single-throw

resistor, fixed

resistor, preset

resistor, tapped

resonator

rheostat

saturable reactor

signal generator

solar battery

solar cell

source, constant-current

source, constant-voltage

speaker

switch, double-pole, double-throw

switch, double-pole, rotary

switch, double-pole, single-throw

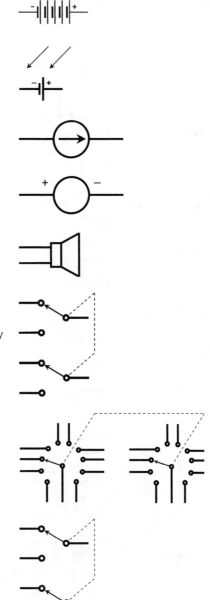

switch, momentary-contact

switch, silicon-controlled

switch, single-pole, double-throw

switch, single-pole, rotary

switch, single-pole, single-throw

terminals, general, balanced

terminals, general, unbalanced

test point

thermocouple

transformer, air core

transformer, air core, step-down

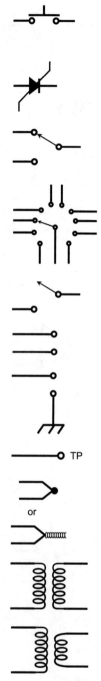

transformer, air core, step-up

transformer, air core, tapped primary

transformer, air core, tapped secondary

transformer, iron core

transformer, iron core, step-down

transformer, iron core, step-up

transformer, iron core, tapped primary

transformer, iron core, tapped secondary

transformer, powdered-iron core

transformer, powdered-iron core, step-down

transformer, powdered-iron core, step-up

transformer, powdered-iron core, tapped primary

transformer, powdered-iron core, tapped secondary

transistor, bipolar, NPN

transistor, bipolar, PNP

transistor, field-effect, N-channel

transistor, field-effect, P-channel

transistor, MOS field-effect, N-channel

transistor, MOS field-effect, P-channel

transistor, photosensitive, NPN

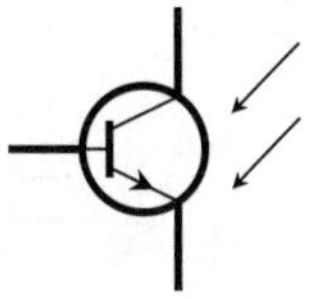

transistor, photosensitive, PNP

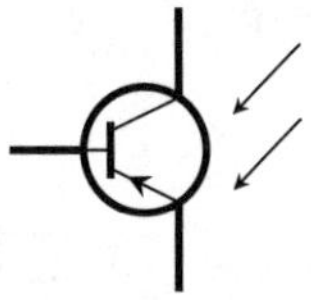

transistor, photosensitive, field-effect, N-channel

transistor, photosensitive, field-effect, P-channel

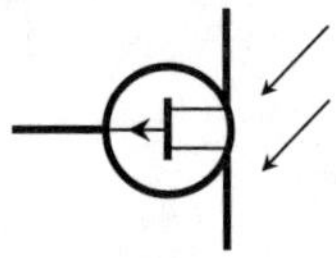

transistor, unijunction

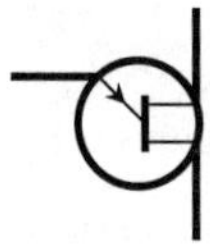

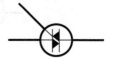

triac

tube, diode

tube, heptode

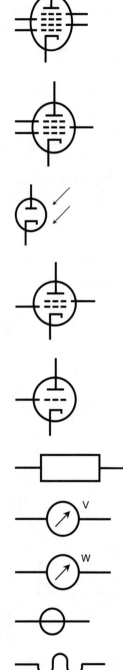

tube, hexode

tube, pentode

tube, photosensitive

tube, tetrode

tube, triode

unspecified unit or component

voltmeter

wattmeter

waveguide, circular

waveguide, flexible

waveguide, rectangular

waveguide, twisted

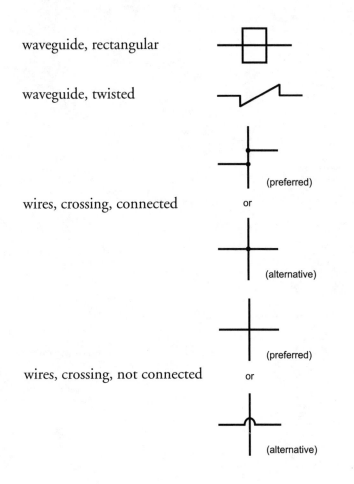

wires, crossing, connected

(preferred)

or

(alternative)

wires, crossing, not connected

(preferred)

or

(alternative)

Suggested Additional Reading

- Brindley, K. *Starting Electronics,* 3d ed. Oxford, England: Newnes, 2004.
- Cutcher, D. *Electronic Circuits for the Evil Genius.* New York: McGraw-Hill, 2005.
- Gibilisco, S. *Electricity Demystified.* New York: McGraw-Hill, 2005.
- Gibilisco, S. *Electronics Demystified.* New York: McGraw-Hill, 2005.
- Horn, D. T. *Basic Electronics Theory with Projects and Experiments,* 4th ed. New York: McGraw-Hill, 1994.
- Long, L. *Personal Computing Demystified.* New York: McGraw-Hill, 2004.
- McComb, G. *Electronics for Dummies.* Hoboken, NJ: John Wiley & Sons, 2005.
- Scherz, P. *Practical Electronics for Inventors.* New York: McGraw-Hill, 2000.
- Slone, G. R. *TAB Electronics Guide to Understanding Electricity and Electronics,* 2d ed. New York: McGraw-Hill, 2000.
- Slone, G. R. *The Audiophile's Project Sourcebook.* New York: McGraw-Hill, 2002.
- Wise, E. *Robotics Demystified.* New York: McGraw-Hill, 2004.

Index

About the Author

Stan Gibilisco is one of McGraw-Hill's most prolific and popular authors. His clear, reader-friendly writing style makes his books accessible to a wide audience, and his experience as an electronics engineer, researcher, and mathematician makes him an ideal editor for reference books and tutorials. Stan has authored several titles for the McGraw-Hill Demystified library of home-schooling and self-teaching volumes, along with more than 20 other books and dozens of magazine articles. His work has been published in several languages. Booklist named his *McGraw-Hill Encyclopedia of Personal Computing* one of the "Best References of 1996," and named his *Encyclopedia of Electronics* one of the "Best References of the 1980s."